NONLINEAR OPTICAL PROPERTIES OF ORGANIC MOLECULES AND CRYSTALS

Volume 1

This is a volume in
QUANTUM ELECTRONICS—PRINCIPLES AND APPLICATIONS

A Series of Monographs

Editors: Paul F. Liao and Paul Kelley
Founding Editor: Yoh-Han Pao (1972–1979)

The complete listing of books in this series is available from the Publisher upon request.

NONLINEAR OPTICAL PROPERTIES OF ORGANIC MOLECULES AND CRYSTALS

Volume 1

Edited by

D. S. Chemla

AT&T Bell Laboratories
Crawford Hill Laboratory
Holmdel, New Jersey

J. Zyss

Centre National d'Etudes des Télécommunications
Laboratoire de Bagneux
Bagneux, France

1987

AT&T

Published by arrangement with AT&T

ACADEMIC PRESS, INC.
Harcourt Brace Jovanovich, Publishers
Orlando San Diego New York
Austin Boston London
Sydney Tokyo Toronto

ACADEMIC PRESS, INC.
Orlando, Florida 32887

United Kingdom Edition published by
ACADEMIC PRESS INC. (LONDON) LTD.
24–28 Oval Road, London NW1 7DX

Library of Congress Cataloging in Publication Data

Nonlinear optical properties of organic molecules
and crystals.

(Quantum electronics—principles and applications)
Includes indexes.
1. Molecular crystals—Optical properties.
2. Chemistry, Physical organic. I. Chemla, D. S.
II. Zyss, J. III. Series.
QD941.N66 1986 547.1'3 86-8070
ISBN 0–12–170611–7 (v. 1 : alk. paper)

PRINTED IN THE UNITED STATES OF AMERICA

86 87 88 89 9 8 7 6 5 4 3 2 1

Contents

Preface: New Materials for a New Age

I. Concepts, Purpose, and Scope

The postindustrial societies, sometimes referred to as information-based societies, urgent demand for an always increasing volume of information processing and communication is continuously challenging modern technology. To date most of the materials used for devices are minerals, i.e., metals or semiconductors. Yet in almost all the cases, the best performing and most reliable systems are living organisms, and the drive of science has always been, consciously or not, to imitate them as closely as possible. The latest step of the physical sciences in this direction is the idea of using artificial "neural networks" for massive parallel computation. On the other hand, the latest developments of molecular biology show that the concentration of microscopic responses at each structural level in a living organism results from the specific properties of organic molecules as such. This last remark leads us naturally to wonder why evolution has lead to the selection of this class of compounds. What particular features bestow them with such selective, sensitive, fast, and intense responses to all sort of external stimulations?

In the wake of these interrogations, it is very tempting to envision a new technology that would exploit the remarkable properties of organic molecules. However, it is now clearly established that the development of a modern technology is necessarily rooted in a detailed understanding of the materials involved and thus requires surpassing a *knowledge-base threshold*. In the four centuries of development of modern science, we have only begun to understand even the simplest physical phenomena. Thus it is natural that technologies based on simple metals and, later, on simple semiconductors were implemented first. Considering the tremendous increase in complexity when moving from simple siliconlike structures to organics and further to biological systems, this knowledge-base threshold is certainly not around the corner for the more complex materials. However, it would be a naive and perhaps dangerous view to reject these two latter domains in an oversimplified, short-sighted, and linear vision of the development of science and technology. This is particularly true since a closer look at organic science demonstrates that it is rooted in ancient and vast, although so far poorly exploited, layers of technological know-how.

The approach toward material optimization in the field of organics runs surprisingly parallel with *but in the opposite sense* to that which currently leads semiconductor-based electrophotonics from bulk materials to quantum wells, quantum wires, quantum dots, or grade-gap microstructures. These new sophisticated semiconductor-based architectures are the up-to-date advances in the evolution from well-mastered bulk technologies of limited flexibility toward microscopically scaled technologies where optical and electronic properties can be manipulated and tailored for applications. This latter stage is backed by the considerable amount of know-how generated at earlier levels of development. We are forced to recognize that however interesting was the theoretical evidence that semiconductor band structure could be engineered at the microscopic scale, such predictions would not of themselves have spurred the development *ex nihilo* of the appropriate growth, fabrication, and preparation techniques. Moreover, manipulations of semiconductor properties are based on the control of composition or of impurity concentration. Thus they involve *brutal* "physical" methods for introducing these activating heteroatoms that tend to disturb the host lattices significantly (e.g., through ion implantation, reactive ion etching, or epitaxial growth). In contrast, the chemical methods used for predictive manipulation at *the molecular level* have developed into a variety of *mild*, well-mastered, industrially oriented techniques based on the virtually unlimited possibilities of organic synthesis. They are also more controllable and flexible than the injection of dopants or compositional changes in semiconductors. However, the development of devices requires the organization of the optimized molecules into larger structures meeting symmetry and shaping constraints imposed by macroscopic physical and operational requirements. The control of this coherent arrangement of microscopic building units is much less developed, and the rich possibilities of organic chemistry are just now starting to be mature enough to open promising avenues in this direction.

For historical reasons a number of transistorlike electronic devices using electric charge transport have developed quite independently from photonic devices using absorption, emission, and modulation of light. It is evident that the two approaches, far from being opposed, are complementary and intimately related. In fact the ultimate goal of an information technology based on electromagnetic forces is a control of the electron state (source and detector of the electromagnetic field) and photon state (vector of the electromagnetic field) limited *only* by quantum mechanics. At their common root stands the symmetric current field ($\mathbf{J} \cdot \mathbf{A} - \rho\phi$) interaction. The ultimate electromagnetic device will fuse both concepts at the microscopic level in the conceptual and technological senses. In semiconductors the doping by impurities introduces a micrometer scale asymmetry that is responsible for the operation of all diodelike devices. By analogy an inter- or intramolecular charge transfer between the donor and acceptor ends of a conjugated molecule introduces a nanometer-scaled asymmetry already used in optical parametric amplifiers and soon to be extended to other functions.

The artificial distinction between linear and nonlinear optics is now obsolete. Optics is in essence nonlinear. Both absorption and emission processes were recognized and analyzed as such before the advent of lasers. Also consistent with this assertion is the early observation of such "nonlinear processes" as the Pockels, Kerr, and Faraday effects at the turn of the century or the theoretical evaluation of two-photon optical transitions at the very beginning of the development of quantum mechanics. It was nevertheless the spectacular observation of frequency-mixing processes in the visible and the further demonstration of sizable quantities of multiphoton processes following the invention of the laser that led pioneers to invent the term "nonlinear optics" in order to make room for new concepts against the mainstream of optics at the time. A quarter of a century afterward, the distinction between " nonlinear optics" and "optics" does not make any more sense than the previous one between "optics" and "classical optics" or the similar one between "music" and "classical music." Furthermore, this usage wrongly induces the nonspecialist to consider this domain as a rather highly specialized if not obscure subfield of optics.

Within this context, the need for a unified presentation of the nonlinear optical effects in organic molecules and crystals emerged quite naturally. The development of the subject matter covered in this book obviously results from a cross-fertilization among weakly related disciplines: chemistry, optics, crystallography, electronics, applied mathematics, system architecture, etc. Thus multidisciplinarity appears in the book because it is actually intrinsic to the subject. The book is meant to be self-consistent. For example, the chemist should find a clear treatment of the physics he lacks just as the physicist should fine the introduction to the chemistry he needs. Furthermore, the solid-state physicist with a previous background in, for example, band structure and the related description of opto-electronic processes in semiconductors should find the relevant material to make the link with the formalism used to describe the corresponding processes in organics. We have also purposely meant in our choice of contents to treat on equal footing technology and fundamental science, since it is well recognized that each benefits from the other to a point that such a distinction has become obsolete.

II. Content Description

The editors had to face a difficult choice between equally relevant organizations for this book either by classes of materials (polymers, molecules, ordered or quasi-ordered phases, etc.) or by types of optical phenomena. The final choice was rather arbitrarily set on the classical distinction between quadratic (i.e., three-wave mixing) and cubic (i.e., four-wave mixing) processes. We are aware of all sorts of limitations linked to this choice, such as the necessity for a nonperturbative and nonlocal description of (nonlinear) optical processes at high excitation levels in strongly delocalized systems. In practice, however, most of the subject matter covered in this book fits readily into such a classification. The materials, the symmetry constraints, and the experimental techniques differ considerably from one section to the other, leading to a natural diversification of working groups following this scheme.

Professor Silbey introduces the book with a general overview of the basic properties of organic matter. The two following chapters by J. Zyss and D. S. Chemla and by J. F. Nicoud and R. L. Twieg are meant to serve a double purpose, bridging the gap between more chemically and more physically oriented readers and orienting them in both fields, from basic notions to state-of-the-art quadratic optics in molecules and molecular crystals. The first of these two chapters emphasizes the physical aspects of the subject, whereas the second concentrates on the chemical aspects. Although readers will understandably tend to consider first the chapter that best fits their own backgrounds, it is highly recommended to try to overcome natural inhibitions and discover a different approach from the usual one so as to experiment with the idea that the benefits of stereoscopic vision originate from doubled vision, in science as well as in real life. Professor J. Morley and Dr. D. Pugh demonstrate in Chapter II-3 the benefits derived from quantum-chemistry-based models so as to provide the synthetic chemist with cost- and timesaving guidelines, which help each chemist find the way within the unlimited but otherwise embarrassing jungle of organic molecules. Chapter II-4 by Drs. J. Badan, R. Hierle, A. Perigaud, and D. Vidakovic deals with the technologically crucial bottleneck of growth and characterization of high quality, bulk organic crystals and waveguided structures. By contrast, Chapter II-5 by A. Barraud and M. Vandevyver introduces the growth of thin films by the Langmuir–Blodgett technique. A case study focusing on a specific material, namely urea, which exemplifies a situation in which transparency in the UV region has been purposely traded for nonlinear efficiency, is presented in Chapter II-6 by Professor C. L. Tang and Dr. J. M. Halbout. This chapter illustrates the successful strategy of a team concentrating all of its efforts on a single molecular crystal as the unique target, with efforts ranging from molecular solution measurements to the first demonstration of optical parametric oscillation in an organic material. In Chapter II-7, Dr. D. J. Williams discusses another type of trade off between the unpredictable orientation of molecules in crystalline media and the possibility of acting on the polarity of liquid-crystalline structures. In the final chapter of Part II, Drs. K. D. Singer,

S. L. Lalama, J.E. Sohn, and R. D. Small introduce and link electro-optic to other effects second-order optical processes with special emphasis on organics. They show the dominant electronic contribution to the electro-optic effect in such media as opposed to inorganic compounds.

Section III in Volume II is dedicated to third-order processes. Here the polydiacetylene family is singled out for the moment because of its many appealing physicochemical features. Foremost among these is its unique topochemical polymerization potential. This family is introduced by Professors M. Schott and G. Wegner in Chapter III-1. Experimental studies of third-order processes in these compounds are presented and discussed in Chapters III-2 and III-3 by Drs. F. Kajzar and J. Messier and by Drs. G. M. Carter, Y. J. Chen, M. F. Rubner, M. K. Thakur, S. K. Tripathy, and D. J. Sandman, respectively. Related theoretical aspects are discussed in Chapters III-4 and III-5: Professor C. Flytzanis presents a general discussion of the essential relation between system dimensionality and optical response. In a complementary approach, Professors J. M. André and J. Delhalle present detailed computations on specific oligomeric systems and resulting trends, strongly suggesting that polymeric structures, beyond polydiacetylenes, can also be "engineered" in very much the same way as was previously established for molecular systems and quadratic effects. The important aspects of multiphoton spectroscopy as a tool to disentangle some of the previously nonaccessible features of molecular-matter-excited states is surveyed in Chapter III-6 by Dr. B. Dick and Professors R. Hochstrasser and H. P. Trommsdorff. The concluding comment by Professor J. M. Lehn points out one of the many avenues to be explored in the near future by applying previously reported, however unrelated, modern concepts of chemistry to optics. The up-to-date appendixes at the end of Volume II, expertly compiled by Drs. J. F. Nicoud and R. Twieg, will certainly provide a much needed and previously missing experimental data bank on powder SHG tests and molecular second-order hyperpolarizability of organics. It is the editors' hope, in view of the large amount of work that remains to be done in organics, that the publication of such data banks will avoid unnecessary and time-consuming duplications.

We anticipate that in such a rapidly evolving field an update or extension of this book will be needed within the next few years. It is, however, our belief that most of the material covered herein will be at the root of these further developments, although unpredictable breakthroughs are also to be expected. We apologize to those research teams who rightfully consider that they should have appeared in the book, but time and space were severe constraints to which we had to yield. We hope that any future edition will correct any serious omission.

Part I

Introduction

Chapter I-1

The Structure and Properties of the Organic Solid State

ROBERT SILBEY

Department of Chemistry
Massachusetts Institute of Technology
Cambridge, Massachusetts 02139

I. INTRODUCTION

In most textbooks on the solid state, molecular solids are given at most a few paragraphs, and organic solids are never mentioned at all (the solid noble gases being the "best example of molecular solids" [1]). This is unfortunate since organic solids have an enormous range of interesting properties that are almost continuously "tunable." For the last 15 years there has been increasing awareness of these systems as their more interesting properties have been uncovered: for example, superconductivity [2], spinless conductivity in "doped" conjugated polymers [3], and large nonlinear optical responses [4].

In this review, I discuss the structure and properties of organic solids in a general and, hence, incomplete way with examples taken from the recent literature. It is clearly impossible to cover this field adequately in a few pages; I hope only to give the flavor of its vastness and variations, and, since my own field of research is the theory of optical properties, conductivity, and relaxation, the review is undoubtedly skewed in these directions. However, I hope the reader will come away with the feeling that understanding the properties of organic materials is an interesting and exciting challenge.

II. STRUCTURE

A. Crystal Structures

Most organic solids are molecular solids; that is, the molecules retain their geometric shape and properties to a large degree upon condensation from the vapor. Thus, benzene, for example, is (almost) hexagonal in the crystal with only small changes from the gas-phase molecule. In addition, the infrared and low-energy optical spectra of the solid are very similar to those of the gas. To be sure, there are changes; these are, however, small compared to those occurring in typical ionic crystals or metals. Finally, organic solids have small cohesive energies. Therefore, most organic solids are collections of molecules held together by weak intermolecular interactions, usually described by van der Waals forces. Because the molecules are non-simply shaped and the forces are weak and anisotropic, organic solids (1) often have more complicated structures than solids normally studied in elementary books on solid state physics, (2) exhibit polymorphism and many phase transitions, and (3) often have many structural defects.

Although a first-principles quantum-mechanical calculation of the equilibrium geometric structure of a typical organic crystal is impossible at present, phenomenological atom–atom two-body potentials to compute intermolecular interactions are available and useful. That is, assuming the molecules have a certain geometry (given perhaps by the gas-phase data), we can compute intermolecular forces by adding atom–atom (central force) interactions. This gives a highly anisotropic molecule–molecule interaction. The atom–atom interactions are usually assumed to be [5, 6] either Lennard–Jones or Buckingham potentials:

$$u_{\mathrm{LJ}}(r) = -\frac{A}{r^6} + \frac{B}{r^{12}} \tag{1a}$$

$$u_{\mathrm{B}}(r) = -\frac{A}{r^6} + B'e^{-Cr} \tag{1b}$$

where the parameters for carbon–carbon, carbon–hydrogen, and hyrdogen–hydrogen interactions are fixed by fitting a few model compounds (typically anthracene, naphthalene, etc.). Using these potentials, it is possible to compute crystal structures, thermodynamic functions, phonon frequencies, and others in reasonable agreement with experimental data. Although various authors have chosen different atom–atom parameters [5,6], the final results of their calculations do not differ markedly. Since the mid 1970s, these methods have also been applied to the calculation of structural defects [7].

When Robertson [8] reviewed the organic crystals studied by X rays in the early 1950s, there were 10^2 structures known; when Dunitz [9] reviewed the field in the late 1970s, there were 10^4 structures known; in the late 1980s, the number is probably $>10^5$. The ability to study complex crystals of complex molecules using automated equipment and sophisticated computer software and hardware has led to an explosion in our knowledge of the static structure of these systems. This does not mean, of course, that our understanding of the other properties of these systems has increased in the same proportion.

Since 1960, the polyacenes have been the model organic molecular crystals for studies of static structure, optical spectra, dynamic studies, etc. Anthracene, naphthalene, and tetracene are relatively simple systems and exhibit many of the typical behaviors of organic crystals. Anthracene is a planar conjugated system with D_{2h} symmetry in the gas phase. The normal crystal structure was determined by Robertson [8] and Cruikshank [10], and the molecule in the solid is remarkably similar to the gas-phase structure (found by electron diffraction) [11]. The crystal structure is monoclinic (space group $P2_1/a$) with two molecules in the unit cell, related to each other by a screw axis (parallel to the b crystal axis). Naphthalene has an almost identical structure to anthracene, while tetracene and pentacene are very similar; however, the two molecules in the unit cell in the latter two cases are not related by screw axis symmetry as in naphthalene and anthracene. The crystal structures of pyrene and α-perylene are different from these by the fact that a dimer of pyrene takes the place of a single anthracene molecule [11]. This structure is quite interesting because the dimer then has considerable π-electron overlap, which is undoubtedly the reason for excimer formation [12] in these systems.

As already mentioned, these crystals often have phase transitions to other, more closely packed structures. (In addition, metastable forms of anthracene crystals [13] have been found under high pressure.)

Although the structure of the polyacenes is quite common in aromatic crystals, another structure is also often found. This is a linear stacked structure often found in halogenated aromatics, as in tetrachlorobenzene and 1,4-dibromonaphthalene. In this case, the closest molecules form a stack of aromatic planes tilted (i.e., nonperpendicular) in general to the stack axis. There may be one or more molecules in the unit cell, either crystallographically equivalent or not. In contrast, the benzene crystal is orthorhombic with four molecules in the unit cell related by screw axes parallel to the three crystal axes.

Kitaigorodsky [5] has discussed the reasons for the frequent occurrence of certain space groups and not others. His major thesis is that purely geometric packing leads only to certain possibilities. Since the unit-cell volume in these structures is often only $\frac{4}{3}$ the volume of the molecules, these molecules must pack very tightly, with the irregularities of shape fitting into one another like jigsaw-puzzle pieces. By attempting to fit irregularly shaped objects in optimal packing together with ~ 12 near neighbors, Kitaigorodsky finds that (1) in general only a *site* inversion symmetry can be retained in the crystal and (2) only a very few crystal space groups are allowed. Although this purely geometric approach cannot be completely general, it is a remarkably good guide to the structures of these systems. A rationale for its success can be found in the strength and steepness of the repulsive forces compared to the attractive forces between these molecules. At the densities of the crystal, the free energy is dominated by the repulsions, at least in the absence of specifically chemical forces.

A word of caution is in order here. Even though the site symmetry only contains the inversion (for example, for benzene) and so, in principle, the molecule is not D_{6h} in the crystal, the distortions from D_{6h} are at the limit of X-ray detectability, so for most purposes we can treat benzene as a regular hexagon even in the crystal.

Given these structures, we find that these crystals can exhibit large anisotropies in their properties. For example, the intermolecular interactions that give rise to exciton states in these crystals are largely due to π electrons. When these are analyzed for lower electronic states of benzene, naphthalene, and 1,4-dibromonaphthalene, we find that benzene can be considered three-dimensional, naphthalene two-dimensional and 1,4-dibromonaphthalene (or tetrachlorobenzene) one-dimensional. By this, we mean that the anisotropies in the exciton bands are such that, to a good approximation, these dimensionalities obtain. Obviously, the same is not true for the interactions responsible for the phonon band structures, so that these materials have some properties that are highly anisotropic, while other properties are not. Since the *optical properties*, at least in the visible and near ultraviolet (UV), probe the π-electron states, we expect to see large anisotropies in the optical response.

B. Vibrations and Phonon Bonds

Because of the complexity of the structure, few experiments on the phonon structure and dynamics in molecular crystals have been done [14]. It is clear, however, that the phonon density of states is not a simple function except at the lowest frequencies.

The number of vibrational modes (at any **k** value) in a molecular crystal is large. For example, the benzene molecule has 30 vibrational modes and three rotations; since there are four molecules in a unit cell, there are 144 vibrational modes (per **k** value) in the solid. Near $\mathbf{k}=\mathbf{0}$, three of these are acoustic modes, nine are translational optical modes, 12 are librational (or rotational) modes, and 120 are internal vibrations. Of course, these modes mix together so the designations are only approximate. The sheer number of modes forces us to consider the vibrational band structures in the simplest theoretical models only [15]. However, it is easy to see that the acoustic bands are $\sim$50–100 cm^{-1} wide and mix with librational modes at relatively low energies. (In naphthalene and anthracene, librations can be seen at $\sim$50 cm^{-1}, while in 1,4-dibromonaphthalene there is a libration at $\sim$20 cm^{-1} and in benzene at $\sim$30 cm^{-1}.)

Recently, inelastic neutron scattering studies of anthracene [16], naphthalene [17], and others have been reported, and the observed density of states has been compared to the predictions of the harmonic model. The agreement is good, suggesting that we can be confident in the details of the calculations.

C. Defects

Structural defects in molecular crystals are common [18]. Point defects such as misorientations, interstitials, vacancies, and impurities are well known. Extended defects such as dislocations and disclinations are often found, as well as stacking faults and others. Finally, polymorphic inclusions and impurity aggregations are known. At the present time, very little is known about these except for their existence. These defects will cause the electronic states of the ideal crystal to be perturbed and perhaps be localized [19]; they will cause inhomogeneous broadening in the optical spectrum; and they will cause the formation of local phonon modes that can interact strongly with electronic states.

III. ELECTRONIC STATES

In order to understand the frequency-dependent polarizability or dielectric function of the crystal in the optical region, we must know the energies of all the optical transitions, their oscillator strengths, and their widths. The energies and oscillator strengths can be computed based on a simple model of the molecular crystal, which treats the molecules as units and neglects overlap and exchange. This works quite well in those cases where it has been tested. The *widths* of the spectral lines depend on exciton–phonon coupling and inhomogeneities, which are less well understood.

A. Exciton States in Crystals

Because the molecules in the molecular crystal retain their identity to a great extent, the description of the electronic states of the crystal is quite different from that of metals or covalently bonded solids. The basic ideas were formulated by Frenkel [20], Peierls [21], and Davydov [22], and elaborated by others [23]. The ground electronic state wave function of the crystal must describe a collection of ground-state molecules and, to a good approximation, is then just a product of ground-state wave functions, one for each molecule:

$$\Psi^{(g)} = \prod_{i=1}^{N} \psi_i^{(g)} \tag{2}$$

Here $\psi_i^{(g)}$ is the ground-state wave function of the molecule at the ith site. Note that this wave function is not the same wave function that the molecule has in the gas phase but is changed by the interactions with the other molecules. The low-lying excited electronic states will be made by exciting a single molecule to an excited state; this is, however, N-fold degenerate in the crystal:

$$\Psi_i^{(e)} = \psi_i^{(e)} \prod_{j \neq 1} \psi_j^{(g)}, \qquad i = 1, 2, \ldots, N \tag{3}$$

The intermolecular interactions will split this degeneracy and form N nondegenerate states. For a perfect crystal with one molecule per unit cell, we find

$$\Psi^{(e)}(\mathbf{k}) = \sum_{i=1}^{N} \frac{e^{i\mathbf{k}\cdot\mathbf{R}_i^0}}{\sqrt{N}} \Psi_i^{(e)} \tag{4}$$

where $\mathbf{R}_i^0$ is the position of the lattice point at which the ith molecule sits. The intermolecular interactions will also cause mixing between the electronic states for each $\mathbf{k}$ value.

In the case that there is more than one molecule per unit cell, we form states as in Eq. (4) for each site and then diagonalize the Hamiltonian in the subspace of each $\mathbf{k}$. For two molecules in a unit cell, as in anthracene, there will then be two electronic states (for each $\mathbf{k}$) for every one in the molecule. In particular, at $\mathbf{k} = 0$ there will be two electronic states split by an energy now known as the Davydov splitting. Since the optical spectrum measures the $\mathbf{k} = 0$ region only, there will be two optical transitions for each one in the gas phase.

The energies of these electronic states can be found, to low order, by diagonalizing the electronic Hamiltonian in the basis of these states. To do this, we must compute the various Hamiltonian matrix elements. We take

$$H = \sum_n h_n + \sum_{n>m} V_{nm} \tag{5}$$

where h_n is the Hamiltonian of the electrons on site n and V_{nm} is the interaction between electrons and nuclei on site n and those on site m. (We are implicitly neglecting exchange here and will also neglect overlap; more careful procedures are possible but do not change the qualitative conclusions.) Since we have constructed the crystal states to be products of molecular states, we find

$$\langle \Psi^{(g)} | H | \Psi^{(g)} \rangle = N\varepsilon^{(g)} + \sum_{n>m} \langle \psi_n^{(g)} \psi_m^{(g)} | V_{nm} | \psi_n^{(g)} \psi_m^{(g)} \rangle \tag{6a}$$

$$\langle \Psi^{(e)}(\mathbf{k}) | H | \Psi^{(e')}(\mathbf{k}') \rangle - \langle \Psi^{(g)} | H | \Psi^{(g)} \rangle$$
$$= \delta_{\mathbf{kk}'} \left[(\varepsilon^{(e)} - \varepsilon^{(g)}) \delta_{e,e'} + \sum_{n \neq m} \exp i\mathbf{k} \cdot (R_n^0 - R_m^0) \langle \psi_n^{(e)} \psi_m^{(g)} | V_{nm} | \psi_n^{(g)} \psi_m^{(e')} \rangle \right] \tag{6b}$$

where $\varepsilon^{(i)}$ is the energy of the ith state of the molecule (since they are assumed to be identical). Note that at this level, van der Waals attractions are missing; they can be easily included by considering excited states in which more than one molecule is excited [24]. The last terms in Eq. (6b) are energy-transfer matrix elements, which in the simplest approximation (nonoverlapping charge distributions) become dipole–dipole interactions in which the dipoles are the transition dipoles:

$$\langle \psi_n^{(e)} \psi_m^{(g)} | V_{nm} | \psi_n^{(g)} \psi_m^{(e')} \rangle = -\mu_{ge} \cdot \left\{ \frac{1}{R_{nm}^3} - \frac{3\mathbf{R}_{nm}\mathbf{R}_{nm}}{R_{nm}^5} \right\} \cdot \mu_{ge'}$$
$$\equiv \hat{V}_{ne,me'}$$

Therefore, we find that the electronic band structure [i.e., $E(\mathbf{k})$] is given by dipole sums, with all their interesting pathologies [25].

In principle, we should take into account excited states with more than one excited molecule. This has been done by Hopfield [26] and Agranovitch [27]. In the dipole approximation—that is, confining attention to states that have a transition dipole moment from the ground state—we can do this easily by second quantizing the Hamiltonian and taking $\varepsilon^{(g)} = 0$ for convenience.

$$H = \sum_{e,n} \varepsilon^{(e)} a_{ne}^+ a_{ne} + \sum_{\substack{n>m \\ e,e'}} V_{ne,me'} (a_{ne}^+ + a_{ne})(a_{me'}^+ + a_{me'}) \tag{7}$$

Here, a_{ne}^+ is an excitation operator that converts the ground state of molecule n to the eth excited state of molecule n. By transforming to excitation operators in $\mathbf{k}$ space and assuming periodic boundary conditions, we find

$$H = \sum_{\mathbf{k},e} \varepsilon^{(e)} a_{\mathbf{k}e}^+ a_{\mathbf{k}e} + \frac{1}{2} \sum_{\mathbf{k},e,e'} \hat{V}_{\mathbf{k}e,\mathbf{k}e'} (a_{-\mathbf{k}e}^+ + a_{\mathbf{k}e})(a_{\mathbf{k}e'}^+ + a_{-\mathbf{k}e'}) \tag{8}$$

where $\hat{V}_{\mathbf{k}e,\mathbf{k}e'}$ is the dipole–dipole sum discussed earlier. At this point, the approximation is made that the exciton operators $a^{+}_{\mathbf{k}e}$ and $a_{\mathbf{k}e}$ obey Boson commutation relations; although this is not strictly correct, it is a good approximation for the low-lying states. The excited state energies can then be found, including the effect of multiply excited states.

When this is done, within the dipole approximation we find that this theory agrees precisely with the classical dipole theory. Thus, it should be clearly stated that the excited-state energies computed in this way can also be found from the classical equations for the frequency-dependent dielectric constant for this physical model of a molecular crystal [28]. If we were to evaluate the intermolecular matrix elements in a less approximate manner (than dipole–dipole interactions), we would change the value of the $V_{\mathbf{k}e,\mathbf{k}e'}$ and the excited-state energies, but the form of the equations would remain the same. To show this form, let us assume that only one excited state is important in Eq. (8), so that the sum over e can be suppressed. Then

$$H = \sum_{\mathbf{k}} \left[\varepsilon a^{+}_{\mathbf{k}} a_{\mathbf{k}} + \frac{1}{2} V_{\mathbf{kk}}(a_{\mathbf{k}} + a^{+}_{-\mathbf{k}})(a_{-\mathbf{k}} + a^{+}_{\mathbf{k}}) \right] \tag{9}$$

We find the eigenenergies by solving the Heisenberg equation of motion for a_k and $a^{+}_{\mathbf{k}}$ (the superdot represents time derivative),

$$\dot{a}_{\mathbf{k}} = i[H, a_{\mathbf{k}}] = -i\varepsilon a_{\mathbf{k}} - iV_{\mathbf{kk}}(a_{\mathbf{k}} + a^{+}_{-\mathbf{k}}) \tag{10a}$$

$$\dot{a}^{+}_{\mathbf{k}} = +i\varepsilon a^{+}_{\mathbf{k}} + iV_{\mathbf{kk}}(a_{-\mathbf{k}} + a^{+}_{\mathbf{k}}) \tag{10b}$$

where we have assumed that $a_{\mathbf{k}}$ and $a^{+}_{\mathbf{k}}$ obey Boson commutation relations. Therefore,

$$(\dot{a}_{\mathbf{k}} + \dot{a}^{+}_{-\mathbf{k}}) = -i\varepsilon(a_{\mathbf{k}} - a^{+}_{-\mathbf{k}}) \tag{11a}$$

$$(\dot{a}_{\mathbf{k}} - \dot{a}^{+}_{-\mathbf{k}}) = -i\varepsilon(a_{\mathbf{k}} + a^{+}_{-\mathbf{k}}) - i2V_{\mathbf{kk}}(a_{\mathbf{k}} + a^{+}_{-\mathbf{k}}) \tag{11b}$$

or

$$\ddot{a}_{\mathbf{k}} + \ddot{a}^{+}_{-\mathbf{k}} = -\{[i\varepsilon]^2 + 2V_{\mathbf{kk}}\}(a_{\mathbf{k}} + a^{+}_{-\mathbf{k}}) \tag{11c}$$

so that the eigenenergy $E(\mathbf{k})$ is given by

$$E(\mathbf{k}) = \{[\varepsilon]^2 + 2V_{\mathbf{kk}}\}^{1/2} \tag{12}$$

In the limit that $|V_{\mathbf{kk}}| \ll \varepsilon$, then

$$E(k) \approx \varepsilon + V_{\mathbf{kk}} \tag{13}$$

which is the usual (i.e., Frenkel or Davydov) result that only includes single excitations (i.e., one molecule excited). In the present formulation, the effect of multiple excitations is included and we can see that effect clearly on the excitation frequency.

Another amusing aspect of this formulation is the idea that (since the excitations are Bosons) the cohesive energy Δ of the molecular crystal can be computed as the change in zero-point energy of the collection of Boson (harmonic) oscillators.

$$\Delta = \frac{1}{2}\sum_{\mathbf{k},e}[E^e(\mathbf{k}) - \varepsilon^{(e)}] \approx \sum_{\mathbf{k},e}\left[\frac{V_{\mathbf{k}e,\mathbf{k}e}}{2} - \frac{V^2_{\mathbf{k}e,\mathbf{k}e}}{4\varepsilon^{(e)}} + \cdots\right]$$

$$\approx -\sum_e \frac{1}{4\varepsilon^{(e)}}\sum_{\mathbf{k}} V^2_{\mathbf{k}e,\mathbf{k}e} = -N\sum_e\sum_n \frac{|V_{ne,me}|^2}{4\varepsilon^{(e)}} \tag{14}$$

Since $V_{ne,me} \approx R_{nm}^{-3}$, this term corresponds to the sum of van der Waals interactions, which is the first term in the cohesive energy [24]. When interactions between different electronic states (e') are included, the computation is more difficult but the principles are the same.

We have written the electronic wave functions without showing the dependence on nuclear coordinates. In fact, these functions are indeed dependent, and using this dependence within the adiabatic approximation allows us to define vibronic functions [29]. In the case of weak electronic coupling (where the spread of electronic energies in **k** space is small compared to the vibrational frequency), each *molecular* wave function is a molecular vibronic function; in the strong electronic coupling case, the crystal electronic state is multiplied by a vibrational function. Thus for the low-lying, weakly allowed states, the weak coupling limit applies, and this modifies the Hamiltonian matrix elements by multiplying them by the overlaps of the vibrational wave functions (Franck–Condon factors). This approximation is equivalent to assigning a label e to each *vibronic* state in the molecule. This enlarges the number of equations to solve but is not a stumbling block to their solution.

An elegant and definitive discussion (and calculation) based on these ideas has been given by Philpott [30], who shows that very good agreement between theory and experiment in anthracene and tetracene can be found. It can thus be said that the theory of the energies of the low-lying dipole-allowed states of molecular crystals is in good shape. The calculation of the energies of nonallowed or weakly allowed transitions is not, on the other hand. Although we know in principle how to compute these energies, in practice we cannot.

B. Exciton–Phonon Coupling

Thus far, we have assumed that the lattice vibrations and librations do not play a role in the optical response of the crystal, so our theory is restricted to $T = 0$ K. Above this temperature, we must include these effects, the most obvious of which is the broadening of spectral lines.

The origin of the exciton–phonon coupling lies in the dependence of the electronic energies in the crystal on the positions of the molecules. It is easy to see that the dipole–dipole interaction depends on intermolecular separation; the excitation energy of a molecule at a given site will also depend on its position because of the van der Waals interactions with other molecules. Thus, if we make a Taylor series expansion of $\varepsilon_n^{(e)}$ and the matrix elements $V_{ne,me'}$ around the equilibrium position of the lattice, we find (in a schematic notation)

$$\varepsilon_n^{(e)} = \varepsilon^{(e)}(0) + \sum_m \left(\frac{\partial \varepsilon_n^{(e)}}{\partial R_m}\right)_0 (R_m - R_m^0) + \cdots \tag{15a}$$

$$V_{ne,me'} = V_{ne,me'}(0) + \sum_p \left(\frac{\partial V_{ne,me'}}{\partial R_p}\right)(R_p - R_p^0) + \cdots \tag{15b}$$

The first terms in these expansions are used in the theory we outlined earlier. The second and higher terms are the exciton–phonon coupling terms. Those couplings occurring in the expansion of ε_n are called (site) diagonal and those in the expansion of $V_{ne,me'}$ nondiagonal couplings. This description is of course based in the site representation; in the **k** representation, there will be terms diagonal in **k** and terms off-diagonal in **k**,

$$H_{\text{ex-phonon}} = \sum_{\mathbf{k}} W_{\mathbf{kk}} a_{\mathbf{k}}^{+} a_{\mathbf{k}} + \sum_{\substack{\mathbf{k},\mathbf{k}' \\ \mathbf{k} \neq \mathbf{k}'}} W_{\mathbf{kk}'} a_{\mathbf{k}}^{+} a_{\mathbf{k}'} \tag{16}$$

where $W_{\mathbf{kk}'}$ is a sum of terms linear in displacements, quadratic in displacements, etc. Consider a linear exciton–phonon interaction term

$$W_{\mathbf{kk}'}^{(\alpha)} a_{\mathbf{k}}^{+} a_{\mathbf{k}'} q_{\mathbf{k}-\mathbf{k}'}^{(\alpha)} \tag{17}$$

where $q_{\mathbf{K}}^{(\alpha)}$ is the coordinate of the phonon of branch α (i.e., labeling the type of phonon) and wave vector **K**. This perturbation allows scattering of the $\mathbf{k}'$ exciton state with the creation or annihilation of a phonon of wave vector $\mathbf{k} - \mathbf{k}'$ (since wave vector is a conserved quantity). If energy is conserved in this process, this will be an allowed event in the lowest order of perturbation theory. Therefore, exciton state $\mathbf{k}'$ can decay by emitting (or creating) a phonon of the correct energy and wave vector, and an exciton state **k** will broaden in a temperature-dependent manner (since the number of phonons of the correct energy and wave vector is temperature-dependent). Multiphonon and dephasing processes can also be important in broadening transitions; this is an area of current interest [31].

Another effect of exciton–phonon interactions is to scatter excitations as they move through crystals, so that the wavelike motion becomes diffusive. This phenomenon has been studied experimentally and theoretically for a number of years and was recently reviewed by Kenkre and Reineker [32].

Two important questions are often discussed when dealing with exciton–phonon coupling. The first has to do with the description of the motion of the excitation: is it bandlike or hopping? Band motion could be described as infrequent scattering of a state (**k**) by phonons such that the mean free path is much larger than a lattice spacing, while hopping is the motion of quasi-localized states with a mean free path on the order of the lattice spacing. The second question has to do with localization of the states. Toyozawa [33] has reviewed this for inorganic crystals. The idea is that strong local (i.e., diagonal) phonon interactions can cause the excitation to be quasi-localized on a site by lowering the energy sufficiently (the analogy to charge localization due to strong polarization is often made). Obviously, these two questions are intertwined; this is also an area of current interest. The possibility of localization is also complicated by the effect of defects and other kinds of disorder.

C. Disorder in Organic Solids

The work of Anderson [34] on the effect of static disorder on the electronic states of solids generated an enormous amount of theoretical and experimental work [35] that still continues. Anderson pointed out that if the site energies [$\varepsilon_n^{(e)}(0)$ in our notation] are not all equal, the states change in character. In particular, if the distribution of site energies is broad enough (greater than a multiple of the exciton bandwidth, in our case), *all* the states are localized. The effect of this on the dynamical properties of the system is dramatic; the effect on the spectral properties may be considerably less. Clearly, the effect of this site energy distribution will be to spread optical *intensity* throughout the band. This effect can be calculated, at least in certain cases, using the coherent potential approximation (CPA) [36]. This method has been applied to the exciton states in molecular crystals [37] where the site energy distribution can be caused by a number of effects. For example, the natural abundance of ^{13}C causes small shifts ($\sim 1-2\ \text{cm}^{-1}$) in the excitation energies of organic molecules; also, since intermolecular interactions are anisotropic, the rotation of a molecule by a few degrees can also change its excitation energy, etc. Even though these effects may be small, the exciton bandwidths are often small, too; therefore, these shifts can have observable and interesting consequences. For example, the optical absorption line shape of the first triplet state of 1,4-dibromonaphthalene at 1.2 K is quite anisotropic (non-Lorentzian); this can be explained by naturally occurring ^{13}C isotopic substitution in the molecule [38].

An important (and unsolved) problem is to understand the optical response of a molecular crystal allowing for exciton interactions, exciton–phonon interactions, and static disorder. Although we understand most of the effects of each (at least in principle) individually, we do not know how all of these interact with each other.

D. Recent Experimental Work on Organic Crystals

It is impossible to cover this topic adequately in less than a book-length manuscript (the journal *Molecular Crystals and Liquid Crystals* publishes an annual list of papers in the field of the spectroscopy of organic solids that normally has over 1000 entries) [39]. However, recent reviews and books [40] cover much of interest. A few topics of current interest are mentioned here, in particular those connected to nonlinear optical properties.

Coherent optical effects, such as photon echos [41] and four-wave mixing [42], have become possible in recent years and have been used to probe the dynamics of excited states in solids. Transient grating methods [43] have been used to attempt to measure exciton coherence and diffusion constants. Recently, Warren and Zewail [44] have proposed that, with the possibility of phase-shifting optical lasers, many of the coherent effects of nuclear magnetic resonance (NMR) can be transferred to the optical domain.

Optical hole burning in inhomogeneously broadened spectral lines has become an exciting field. The possibility of using this technique for information storage has been a motivating force in some of this work; however, the scientific questions as to the efficiency and mechanisms of this process remain unresolved and extremely interesting. An introduction to this is given in Small's paper [44a].

The spectroscopy of charge transfer crystals (usually linear stacks) have been reviewed by Haarer and Philpott [44b]. These are examples of crystals that radically change their electronic structure upon excitation (and charge transfer) but can still be studied within the framework of Frenkel exciton theory because of the localization of the charges.

The mobilities of electrons and holes in organic matter, whether solids, liquids, or crystals, have been measured since the mid 1970s, and interesting temperature and field dependencies have been seen. The detailed explanation of these remains a challenge to theorists, although many attempts have been made. In particular, Duke and Schein [45] suggest that the electrical and optical properties of organic molecular solids remain unexplained quantitatively by conventional theories.

IV. ELECTRONIC STRUCTURE OF CONJUGATED POLYMERS

Conjugated polymers, such as the polydiacetylenes, polyacetylene, and polypyrrole, have been intensely studied over the past few years; their optical and conducting properties are remarkable. It is not possible to give an adequate overview of this immense field, so we will restrict out discussion to a few topics only.

A. Highly Conducting Doped Conjugated Polymers

The conductivity of conjugated polymers such as polyacetylene and polypyrrole is extremely low; however, upon doping with electron acceptors or donors, the conductivity rises, in some cases, by up to 10–15 orders of magnitude [46]. At the same time, the optical properties change dramatically also. Because these systems have the possibility of combining the plastic properties of polymers with the conductivity properties of semiconductors or metals, there has been a tremendous amount of interest in them. In addition, fundamental scientific questions regarding the mechanisms of conductivity remain unanswered, although a qualitative picture may now be emerging.

Since these polymers are made up of covalently bound monomers, they are unlike the molecular crystals we discussed above in that there are large collective effects due to the π-electron conjugation. This changes the basic models we must use in order to describe the electronic states, vibrational states, etc. These systems are more like one-dimensional tight-binding models than molecular crystals.

At first, the large conductivities found in doped polyacetylene were quite puzzling. After all, the usual criteria for such conductivity were absent. The band gap is large (1.4 eV), the structure is rather soft, and the fluctuations and defects are numerous and large. All of these would rule out metallic conductivity and, when it was shown that the spin susceptibility was completely uncharacteristic of electrons moving in bands (i.e., spinless conductivity), physicists and chemists became very excited about understanding this phenomenon. Su *et al.* [47] (using a model based on earlier work by Pople and Walmsley [48] and reviewed by Salem [49]) rationalized much of the experimental data by invoking the idea of charged solitonlike defects being responsible for the high conductivity. This suggestion used the softness of the lattice and the large band gap (which was due in their model to a Peierls dimerization) to understand the physics of the process. Although this model needed some refinements [50], it is the Rosetta stone for understanding the conductivity of these systems.

In the ensuing years, chemists and physicists have collaborated to synthesize and study many other systems [46], all of which differ from polyacetylene in that free solitons (in the SSH [47] sense) do not exist, but bound pairs of such structures do [50]; these are called radical ions, dications (by the chemists), and polarons and bipolarons (by the physicists). Polypyrrole has been the best studied so far, and the experimental results for conductivity, optical spectra, and magnetic resonance are all understandable within this model [51].

In spite of the success of this model in describing the experiments to date, there is no *detailed* theory of conduction in these systems. Such a theory must include interchain transport, the presence of a broad distribution of inhomogeneities, and the existence of traps, in addition to understanding the role of the dopant. At the present time, the theoretical description of the optical properties of these systems is improving, but important questions remain; for example, what is the role of the low-lying g states (two photons allowed) in the conducting and are they related to the soliton states?

B. Polydiacetylenes

The solid-state reaction of a number of diacetylene monomers yields the polymer in a single crystal of macroscopic dimensions [52]. The polymer is fully conjugated and highly anisotropic in its optical properties. The π-electron conjugation in the backbone provides a quasi-one-dimensional electron band structure that should, according to Ducuing [53], yield high nonlinear optical susceptibilities; and, in fact, they do [54].

The initial step in the polymerization of the diacetylene crystals is the formation of a biradical dimer [55]; at a later stage, carbenes and dicarbenes can be formed. By a series of beautiful electron spin resonance (ESR) and optical experiments [55], these intermediates have been isolated and studied.

Polymerization is accompanied by a dramatic change in visible optical properties. The monomer is colorless, while the polymer absorbs strongly, beginning at ~6000 Å. The third-order optical susceptibility of the polymer crystal is comparable to the best inorganic systems [54]; however, recent work on soluble polydiacetylenes in solution [56] has indicated that large two-photon absorption plays a role in this.

These soluble polydiacetylenes, first discovered by Patel *et al.* [57], have been the focus of interesting light-scattering experiments and speculations on chain dynamics and gelation [58]. These solutions change color when nonsolvent (hexane) is added or as a function of temperature, indicating a conformational change in the polymer. This conformational change influences the π-electron conjugation length, which in turn governs the optical spectrum. These results indicate that good theoretical calculations of linear and nonlinear optical susceptibilities are important goals for the near future.

C. Electronic Structure

In order to understand theoretically the electrical and optical properties of these polymers, a variety of quantum chemical methods have been developed. These range from the most primitive Hückel theory [49], through semiempirical methods such as Pariser Parr Pople (PPP) [49] and CNDO/S [59], to pseudo-potential methods like VEH [60], and finally *ab initio* methods [61].

All of these have been applied to these polymers with varying degrees of success. However, it is clear that with the advent of supercomputers, the low-lying electronic states of polymers can be described, at least at the Hartree–Fock level and perhaps beyond. Calculations of this sort are already underway. Semiempirical methods still have their usefulness, since the amount of work (and computer time) for the *ab initio* calculations is still great. For example, in 1984 Ramesesha and Soos [62] developed a method for exactly solving the PPP equations that promises to be extremely useful when applied to polymers. At the present time, only short-chain polyenes have been studied; however, the method will soon be applied to polymers. Since this method gives both the low-lying one-photon and two-photon optically allowed excited states, it may provide a way to compute nonlinear susceptibilities in a reliable way. At the present time, such calculations use crude methods of quantum chemistry, although the results have been extremely encouraging [63].

V. CONCLUSION

In this chapter, I have tried to give the reader a taste of the physics and chemistry of the organic solid state. Organic solids have electronic and optical properties that can be tuned continuously, within certain broad ranges, making them extremely interesting from a practical view. On the other hand, because they are (largely) molecular solids, the standard models of solid-state physics (so successfully used in describing inorganic metals and semiconductors) must be applied with care. For example, because of the weak intermolecular interactions, the electronic band widths are narrow, and therefore inhomogeneities and molecular vibrations must play an important (and as yet unclarified) role in these properties [45]. Another example is the effect of large anisotropies in electronic band structures on optical and conduction properties. This will lead to extremely interesting possibilities, which are just beginning to be explored.

I hope I have convinced the reader that organic materials provide interesting challenges for experiment and theory and will continue to do so for the near future.

REFERENCES

1. N. Ashcroft and D. Mermin, "Solid State Physics," p. 389. Holt, New York, 1976.
2. D. Jerome, A. Mazaud, M. Ribault, and K. Bechgaard, *J. Phys. Lett.* **41,** L95 (1980).
3. T. Skotheim, ed., "The Handbook of Conducting Polymers." Dekker, New York, 1985.

4. D. Williams, ed., "Nonlinear Optical Properties of Organic and Polymeric Materials." Am. Chem. Soc., Washington, D.C., 1983.
5. A. I. Kitaigorodsky, "Molecular Crystals and Molecules." Academic Press, New York, 1973; *J. Chim. Phys.* **63,** 9 (1966).
6. D. E. Williams, *J. Chem. Phys.* **45,** 3770 (1966); **47,** 4680 (1967).
7. See, for example, S. Ramdas, J. M. Thomas, and M. Goringe, *J. Chem. Soc., Faraday Trans. 2* **73,** 551 (1977).
8. J. M. Robertson, "Organic Crystals and Molecules." Cornell Univ. Press, Ithaca, New York, 1979.
9. J. D. Dunitz, "X-Ray Analysis and the Structure of Organic Materials." Cornell Univ. Press, Ithaca, New York, 1979.
10. D. Cruikshank, *Acta Crystallogr.* **9,** 915 (1956); **10,** 470 (1957); *Tetrahedron* **17,** 155 (1962).
11. J. Robertson and J. White, *J. Chem. Soc.,* p. 358 (1947); A. Camerman and J. Trotter, *Acta Crystallogr.* **18,** 636 (1965).
12. See, for example, J. Birks, "Photophysics of Aromatic Molecules." Wiley (Interscience), New York, 1970.
13. G. Parkinson, M. Goringe, S. Ramdas, J. O. Williams, and J. M. Thomas, *J. Chem. Soc. D,* p. 134 (1978); see also D. P. Craig, J. Ogilvie, and P. Reynolds, *J. Chem. Soc., Faraday Trans. 2* **72,** 1603 (1972).
14. M. Ito, M. Suzuki, and T. Yokoyama, "Excitons, Magnons, and Phonons." Cambridge Univ. Press, London and New York, 1968, and references therein.
15. Recent work on the harmonic level is T. A. Krivenko, V. A. Dementjev, E. L. Bokhenkov, A. I. Kolesnikov, and E. F. Sheka, *Mol. Cryst. Liq. Cryst.* **104,** 207 (1984); A. I. Kolesnikov, (V. A. Dement'ev, E. L. Bokhenkov, T. A. Krivenko, and E. F. Sheka, *Sov. Phys.—Solid State, (Engl. Transl.)* **25,** 1663 (1983).
16. E. Bokhenkov, A. Kolesnikov, J. Mayer, I. Natkaniec, V. Fedotov, and E. Sheka, *Sov. Phys.—Solid State (Engl. Transl.)* **25,** 1301 (1983).
17. E. Bokhenkov, A. Kolesnikov, T. Krivenko, E. Sheka, V. Dementjev, and I. Natkaniec, *J. Phys. (Orsay, Fr.)* **42,** C6-605 (1981).
18. E. A. Silinsh, "Organic Molecular Crystals." Springer, New York, 1980.
19. J. M. Ziman, "Models of Disorder." Cambridge Univ. Press, London and New York, 1979.
20. J. J. Frenkel, *Phys. Rev.* **37,** 17, 1276 (1931).
21. R. Peierls, *Ann. Phys. (Leipzig)* [5] **13,** 905 (1932).
22. A. S. Davydov, "Theory of Molecular Excitons." Plenum, New York, 1971.
23. D. A. Craig, *in* "The Physics and Chemistry of the Organic Solid State" (D. Fox and M. Labes, eds.), Vol. 1, (Interscience), New York, 1963; R. S. Knox, "Theory of Excitons." McGraw-Hill, New York, 1963; S. A. Rice and J. Jortner, *in* "The Physics and Chemistry of the Organic Solid State" (D. Fox and M. Labes, eds.), Vol. 3. Wiley (Interscience), New York, 1967.
24. M. Renne and B. Nijboer, *Chem. Phys. Lett.* **1,** 317 (1967); **2,** 35 (1968); A. Lucas, *Physica (Amsterdam)* **35,** 353 (1967); D. Langbein, "Van der Waals Attractions." Springer, New York, 1974.
25. M. Born and M. Bradburn, *Proc. Cambridge Philos. Soc.* **39,** 104 (1942); B. Nijboer and F. de Wette, *Physica (Amsterdam)* **23,** 309 (1957); **24,** 442 (1958).
26. J. J. Hopfield, *Phys. Rev.* **112,** 1555 (1958).
27. V. Agranovitch, *JETP Lett. (Engl. Transl.) 10,* 307 (1960).
28. G. D. Mahan, *J. Chem. Phys.* **43,** 1569 (1965).
29. R. M. Hochstrasser, "Molecular Aspects of Symmetry." Benjamin, New York; E. Lim, ed., "Excited States," Vol. 3. Academic Press, New York, 1978.
30. M. Philpott, *Adv. Chem. Phys.* **23,** 227 (1973).

31. R. Silbey, *Annu. Rev. Phys. Chem.* **27,** 203 (1976); D. Burland and A. Zewail, *Adv. Chem. Phys.* **50,** 369 (1980); B. Jackson and R. Silbey, *J. Chem. Phys.* **77,** 2763 (1982); H. Benk and R. Silbey, *J. Chem. Phys.* **79,** 3487 (1983); D. Hsu and J. Skinner, *J. Chem. Phys.* **81,** 1604 (1984).
32. V. M. Kenkre and P. Reineker, "Exciton Dynamics in Molecular Crystals." Springer, New York, 1982.
33. Y. Toyozawa, *in* "Relaxation of Elementary Excitations" (R. Kubo and E. Hanamura, eds.), p. 3. Springer, New York, 1980.
34. P. W. Anderson, *Phys. Rev.* **109,** 1492 (1958); P. A. Lee and T. Ramakrishnan, *Rev. Mod. Phys.* **57,** 287 (1985).
35. D. Thouless, ed., "Ill Condensed Matter." North-Holland Publ., Amsterdam, 1979.
36. P. Soven, *Phys. Rev.* **156,** 809 (1967); R. Elliott, J. Krumhansl, and P. Leath, *Rev. Mod. Phys.* **46,** 465 (1974).
37. J. Hoshen and J. Jortner, *J. Chem. Phys.* **56,** 5550 (1972); H. Hong and G. W. Robinson, *J. Chem. Phys.* **52,** 825 (1970); R. Kopelman, *in* "Excited States" (E. Lim, ed.), Vol. 2, p. 34. Academic Press, New York, 1975.
38. D. Burland, U. Konzelmann, and R. MacFarlane, *J. Chem. Phys.* **67,** 1926 (1977); H. Port and Silbey, unpublished.
39. E. Sheka, V. Makarova, and T. Krievenko, *Mol. Cryst. Liq. Cryst.* **114,** 306 (1984).
40. V. Agranovitch and R. Hochstrasser, eds., "Spectroscopy and Excitation Dynamics of Condensed Molecular Systems." North-Holland Publ., Amsterdam, 1983.
41. W. Hesselink and D. Wiersma, *in* "Spectroscopy and Excitation Dynamics of Condensed Molecular Systems" (V. Agranovitch and R. Hochstrasser, eds.), p. 249. North-Holland Publ., Amsterdam, 1983.
42. R. M. Hochstrasser and H. P. Trommsdorff, *Acc. Chem. Res.* **17,** 47 (1984).
43. M. Fayer, *in* "Spectroscopy and Excitation Dynamics of Condensed Molecular Systems" (V. Agranovitch and R. Hochstrasser, eds.), p. 185. North-Holland Publ., Amsterdam, 1983.
44. W. Warren and A. Zewail, *J. Chem. Phys.* **78,** 2298 (1983).
44a. G. J. Small, *in* "Spectroscopy and Excitation Dynamics of Condensed Molecular Systems" (V. Agranovitch and R. Hochstrasser, eds.), p. 515. North-Holland Publ., Amsterdam, 1983.
44b. D. Haarer and M. Philpott, *in* "Spectroscopy and Excitation Dynamics of Condensed Molecular Systems" (V. Agranovitch and R. Hochstrasser, eds.), p. 27. North-Holland Publ., Amsterdam, 1983.
45. C. B. Duke and L. Schein, *Phys. Today* **33,** 42 (1980); M. Pope and C. Swenberg, "Electronic Processes in Organic Crystals." Oxford Univ. Press, London and New York, 1982.
46. Ref. 3; Proceedings of Workshop in Synthetic Metals, *Synth. Met.* **9,** 129–346 (1984), and references therein; *Mol. Cryst. Liq. Cryst.* **117–121** (1985).
47. Y. Su, R. Schrieffer, and A. Heeger, *Phys. Rev. Lett.* **42,** 1698 (1979); *Phys. Rev. B* **22,** 2099 (1980); **28,** 1138 (1983).
48. J. Pople and S. Walmsley, *Mol. Phys.* **5,** 15 (1962).
49. L. Salem, "Molecular Orbital Theory of Conjugated Molecules." Benjamin, New York, 1966.
50. A. Bishop, D. Campbell, and K. Fesser, *Mol. Cryst. Liq. Cryst.* **77,** 253 (1981); S. Brazovski and N. Kirova, *JETP Lett. (Engl. Transl.)* **33,** 4 (1981); J. L. Brédas, R. R. Chance, and R. Silbey, *Mol. Cryst. Liq. Cryst.* **77,** 319 (1981); *Phys. Rev. B* **26,** 5843 (1982).
51. J. L. Brédas and G. B. Street, *Acc. Chem. Res.* **18,** 309 (1985).
52. G. Wegner, *in* "Molecular Metals" (W. Hatfield, ed.), p. 209. Plenum, New York, 1979; D. Bloor and R. R. Chance, eds., "NATO Workshop on Polydiacetylenes." Plenum, New York, 1985.

53. J. Ducuing, *in* "Nonlinear Spectroscopy" (N. Bloembergen, ed.), Enrico Fermi Course 44. Academic Press, New York, 1975.
54. C. Sauteret, J.-P. Hermann, R. Frey, F. Pradere, J. Ducuing, R. Baughman, and R. Chance, *Phys. Rev. Lett.* **36,** 956 (1976).
55. H. Sixl, *in* "Nato Workshop on Polydiacetylenes" (D. Bloor and R. R. Chance, eds.), p. 41. Plenum, New York, 1985; *Adv. Polym. Sci* **63,** 49 (1984).
56. M. Shand and R. Chance, *J. Chem. Phys.* **71,** 4482 (1978); R. R. Chance, M. Shand, and R. Silbey, *Chem. Phys. Lett.* **64,** 448 (1979); R. R. Chance, M. Shand, C. Hogg, and R. Silbey, *Phys. Rev.* **22,** 3540 (1980); F. Kajzar, J. Messier, J. Zyss, and I. Ledoux, *Opt. Commun.* **45,** 133 (1983).
57. G. Patel, R. Chance, and J. Witt, *J. Chem. Phys.* **70,** 4387 (1979).
58. K. Lim, C. Fincher, and A. Heeger, *Phys. Rev. Lett.* **50,** 1934 (1983); M. Sinclair, K. Lim, and A. Heeger, *Phys. Rev. Lett.* **51,** 1768 (1983).
59. C. B. Duke, *Mol. Cryst. Liq. Cryst.* **50,** 63 (1979); W. Ford, C. Duke, and W. Salaneck, *J. Chem. Phys.* **77,** 5020 (1982).
60. See, for example, J. L. Brédas, *in* "The Handbook of Conducting Polymers" (T. Skotheim, ed.), p. 859. Dekker, New York, 1985.
61. A. Karpfen, *Phys. Scr.* **T1,** 79 (1982); also see S. Suhai, *Phys. Rev. B* **27,** 3506 (1983).
62. S. Ramesesha and Z. Soos, *J. Chem. Phys.* **80,** 3278 (1984); for earlier work, see A. Lasaga, R. Aeoni, and M. Karplus, *J. Chem. Phys.* **73,** 5230 (1980).
63. G. P. Agrawal, C. Cojan, and C. Flytzanis, *Phys. Rev. B* **17,** 776 (1978); C. Flytzanis, "Nonlinear Optical Properties of Organic Molecules and Crystals," Vol. 2, Chapter III-4. Academic Press, Orlando, 1986.

Part II

Quadratic Nonlinear Optical Effects

Chapter II-1

Quadratic Nonlinear Optics and Optimization of the Second-Order Nonlinear Optical Response of Molecular Crystals

J. ZYSS

Centre National d'Etudes des Télécommunications
9220 Bagneux, France

D. S. CHEMLA

AT&T Bell Laboratories
Holmdel, New Jersey, 07733

I. INTRODUCTION

Most of the early pioneering work in nonlinear optics concentrated on the demonstration and understanding of new multiphoton effects, based on whatever medium of appropriate symmetry and transparency was available at the time. Quartz, potassium dihydrogen phosphate (KDP), lithium niobate, and semiconductors, all initially developed for other purposes, were thoroughly exploited, together with gas and solutions whenever condensed oriented phases were not required. No systematic investigation of new materials, specially designed for efficient nonlinear optical interactions, seems to have been undertaken as such before the advent of the 1970s. However, a number of earlier isolated experimental evidences of the second-harmonic generation efficiency of some organic materials have anticipated further more systematic developments and may have been the result of some early intuition of the potential of organics. Most noteworthy are benzopyrene (Rentzepis and Pao, 1964), hexamethylenetetraamine (Heilmeir *et al.*, 1964), hippuric acid (Orlov, 1966), and benzil (Gott, 1971). All these materials, except for benzil, which can easily be grown in large samples, are now surpassed in terms of nonlinear efficiency, but are note worthy if only to help measure, two decades later, the progress therefrom. At the end of the 1960s, a more systematic approach, encompassing both organic and inorganic structures, was facilitated by the definition of a new standard semiquantitative testing procedure for the second-harmonic efficiency of powder samples (Kurtz and Perry, 1968). At the same time, Hauchecorne *et al.* (1971) demonstrated electric-field-induced second-harmonic generation in centrosymmetric media, such as fluids, opening the way, after some later adaptation, to solution measurement of individual molecular nonlinearities (Levine, 1976; Oudar, 1977). Finally, Jerphagnon and Kurtz (1970) fully developed the analysis of nonlinear interference processes in crystals known as the Maker fringe methods (Maker *et al.*, 1962). Well-tried experimental tools operating at molecular, polycrystalline, and single-crystalline levels were thus fully available a little more than 10 years ago, at a time when the need for new nonlinear materials of higher efficiency became apparent. The domain of organic nonlinear optics emerged from mere random scanning of materials into real scientific status when Davydov *et al.* (1970, 1971) unambiguously established the connection between enhanced nonlinear activity and charge transfer in conjugated molecules: in the wake of this work, MNA, known at the time as *m*-nitromethylaniline, had been tested in powder form by Koreneva, *et al.* (1975), while Jerphagnon (1971) and Bergman *et al.* (1972) were the first to propose and thoroughly study single-crystalline nitrouracil, a material of efficiency comparable to that of lithium niobate. Considerable work was undertaken during the 1970s to help set up the scientific basis of a new approach, known as molecular engineering and aiming, at the design of "tailor-made" materials for nonlinear optics, based on the understanding of structure–property relationships. The evidence that organic structures with their inherent flexibility and synthesis potential were ideal candidates for such an approach did not need much arguing and two research teams, one in France at the Centre National d'Etude des Télécommunications (CNET) and the other in the United States at Bell Laboratories, undertook the exploration. A variety of measurements on molecular systems of various geometries, polarizability, conjugation, and charge-transfer character were performed (see, for example, Chemla *et al.*, 1975; Oudar and Chemla, 1977; Oudar, 1977; Levine, 1975; Levine and Bethea, 1976, 1977). Most of these studies dealt with solutions, except for such remarkable exceptions as the systematic investigation of the nonlinear properties of crystalline meta-disubstituted aromatics (Carenco *et al.*, 1977). In parallel, various theoretical attempts were proposed to connect crystalline susceptibilities with their molecular counterparts (Chemla *et al.*, 1975) or to relate molecular structural features with measured nonlinear trends. The advent of more refined quantum-chemistry-based computational models, although needed to increase numerical accuracy (Zyss, 1979 a and b; Lalama and Garito, 1979), have not out-dated, as far as trends are concerned, earlier simple reference models such as the equivalent internal field model for singly substituted aromatics (Oudar and Chemla, 1975) and the two-level model for charge-transfer systems (Oudar and Chemla, 1977; Oudar, 1977). This phase of research helped substantiate the choice of two interesting single crystals of higher efficiency than lithium niobate by one order of magnitude, namely, MAP (Oudar and Hierle, 1977) and MNA (3-methyl-4-nitroaniline) (Levine *et al.*, 1979). From the late 1970s up to now, fundamental work on solutions continued (Singer and Garito, 1981; Ledoux and Zyss, 1982) but the mainstream of research shifted from molecular to solid-state problems. By that time, molecular mechanism had somewhat been cleared up and speculations on all-optical components for optical communication or computational systems spurred the demand for new materials of ever-increasing nonlinear efficiency. It was then felt urgent, in view of the technological challenge set by purification, growth, and processing of organic crystals up to the quality level required by nonlinear optics, to concentrate on a limited number of these, carefully selected after a decade of exploration. Milestones evidencing some achievements in that direction are the demonstration of tunable parametric oscillation in urea crystals (Halbout, *et al.*, 1980; Donaldson and Tang, 1984) and of tunable parametric amplification and emission in POM (3-methyl-4-nitropyridine-1-oxide) crystals (Zyss and Berthier, 1982; Zyss *et al.*, 1985) and in NPP [(*N*)-(4-nitrophenyl)-(L)-prolinol] crystals (Zyss *et al.*, 1984). NPP, together with the parent structure PNP [nitro-4-pyridino-2-(L)-prolinol] (Twieg and Dirck, 1986), exemplifies the ultimate "optimal" structure for phase-matched nonlinear interactions (Zyss and Oudar, 1982) where molecules are ideally oriented with respect to the crystalline symmetry axis so as to maximize the effective nonlinear coefficient. The measured noncritically phase-matchable d_{21} coefficient of NPP is 200×10^{-9} esu, which corresponds to an enhancement of two orders of magnitude nonlinear efficiency (d^2n^{-3}) over lithium niobate. A new material, NPAN [(*N*)-cyanomethyl-(*N*)-methyl-4-nitroaniline] Nicoud *et al.*, 1986) has been recently invented and grown: its efficiency scales with that of NPP but its mechanical properties are significantly improved over that of NPP. In the future, technological parameters such as tensile module, thermal conductivity, chemical stability, and crystal growth conditions are going to take the lead over classical molecular-engineering parameters. The theoretical "optimum" may be reached by a number of different structures, displaying essentially equivalent nonlinear efficiencies; however, more technology-oriented criteria of the kind reported here will serve as an additional sieve and help select the adequate one for applications. Developments of single-crystalline bulk structures for nonlinear optics still have a way to go, as parametric oscillation, the ultimate test in surface and bulk quality for nonlinear optics, has only been demonstrated so far in the case of urea. As was successfully demonstrated by the Cornell group, concentrating efforts on a single material and facing its own specific and often puzzling technological demands is likely to pay off as well on the way toward applications in the case of other single-crystalline materials of possibly higher efficiency. In parallel, some progress is being reported in the growth and characterization of organic waveguiding structures for nonlinear optics, in the form of either planar waveguides (Ledoux *et al.*, 1986a) or organic cored capillaries (Nayar *et al.*, 1984; Vidakovic *et al.*, 1984). The aim of such efforts in to combine the high nonlinearities of organics with the additional benefit of power confinement and increased interaction length. Moreover, such new configurations are closer than bulk crystals to actual shape and connectivity requirements imposed by modern fiber-optic and integrated-optics technologies.

The composition of this chapter closely follows the historical streamlines briefly recalled in the introduction. We discuss in Section II some basic tools of constant use in nonlinear optics such as the nonlinear propagation equation and phenomenological nonlinear susceptibilities. These are applied in Section III to various experimental situations of relevance to the study of nonlinear materials in crystals or solutions, be it for measurement or device-oriented purpose. The nonlinear properties of organic molecular structures are introduced in Section III with reference to various experiments and theoretical models. Section V is devoted to the nonlinear properties in the crystalline phase, with examples taken from actual crystalline structures such as those of MAP, POM, and NPP. Simultaneous use of chemical, crystallographic, and optical (or dielectric) concepts will be seen to open new insights on the properties of intramolecular charge-transfer crystals. Finally, Section VI, after summing up a number of device-oriented criteria, compares these to measured performances of MAP, POM, and NPP. The last part of Section VI anticipates the properties of presently developed organic waveguiding structures, with special emphasis put on the influence of their high birefringency on phase-matching and group velocity dispersion characteristics.

II. BASIC CONCEPTS IN QUADRATIC NONLINEAR OPTICS

A. Linear Optics in Dense Matter

1. Introduction

A number of nonlinear processes in condensed matter can be described by natural generalization of the semiclassical theory of electromagnetism. In this section we give a brief summary of some well-known results of linear optics, which are useful in developing a good description of nonlinear optics.

2. Constitutive Relations

The Maxwell equations

$$\nabla \times \mathbf{H} = \frac{1}{c}\frac{\partial}{\partial t}\mathbf{D} + \frac{4\pi}{c}\mathbf{J}_{\text{ext}}, \qquad \nabla \times \mathbf{E} = -\frac{1}{c}\frac{\partial}{\partial t}\mathbf{B}$$

$$\nabla \cdot \mathbf{B} = 0, \qquad \nabla \cdot \mathbf{D} = 4\pi\rho \tag{1}$$

involve macroscopic electromagnetic variables **E**, **H**, **D**, **B**, **J**, and ρ_{ext}, which are usually interpreted as averages of microscopic variables over volumes small on a macroscopic scale but large compared to atomic dimensions.

Equation (1) must be completed by further relations among the vectors **E**, **D**, **B**, and **H**. It is customary to introduce the electric-dipole density **P**, the electric-quadrupole density **Q**, the magnetic dipole density **M**, etc. to account for the dielectric and magnetic response of matter:

$$\begin{aligned} \mathbf{D} &= \mathbf{E} + 4\pi\mathbf{P} - 4\pi\nabla \cdot \mathbf{Q} + \cdots \\ \mathbf{B} &= \mathbf{H} + 4\pi\mathbf{M} + \cdots \end{aligned} \tag{2}$$

Omitting the high-order multipoles in the expansion of **D** and **B** may result in difficulties in the description of certain effects such as optical activity.

In optics, one is usually concerned only with nonmagnetic media for which $\mathbf{B} = \mathbf{H}$ and with induced effects for which the external charge density ρ_{ext} and current density J_{ext} vanish. It is then possible to eliminate the magnetic variables in Eq. (1), which in this case implies that

$$\nabla \times \nabla \times \mathbf{E} = -\frac{1}{c^2}\frac{\partial^2}{\partial t^2}\mathbf{D} \tag{3}$$

This is the basic equation we shall use to discuss propagation in linear as well as nonlinear media.

3. Linear Optical Effects in Crystals

Let us first consider the case of "conventional" optics, for which the fields applied to the material system are so small compared to the microscopic fields acting on the charges that the multipole densities can be approximated by a term that is proportional to the applied fields. For instance, one can write

$$\mathbf{P}(\mathbf{r}, t) = \int_{-\infty}^{t} dt' \int d\mathbf{r}' \kappa(\mathbf{r}, t; \mathbf{r}', t') \cdot \mathbf{E}(\mathbf{r}', t') \tag{4}$$

Here $\kappa(\mathbf{r}, t; \mathbf{r}', t')$ is a second-rank tensor called the linear response function of the system for the electric-dipole density. Because of the principles of time invariance and of causality, the tensor depends only on $\tau' = t - t'$, and it vanishes for $\tau' < 0$. In addition, $\kappa(\mathbf{r}, \mathbf{r}', \tau')$ is real as the response of a material system to a real excitation. Finally, for a spatially homogeneous system, $\kappa(\mathbf{r}, \mathbf{r}', \tau')$ depends only on the difference $\mathbf{r} - \mathbf{r}' = \boldsymbol{\rho}'$, i.e., $\kappa(\mathbf{r}, \mathbf{r}', \tau') = \kappa(\boldsymbol{\rho}', \tau')$. The general properties of linear-response functions have been thoroughly studied in electrical engineering, and their optical counterparts fit with the same framework.

Let us suppose that the electric field consists of a superposition of monochromatic waves $\mathbf{E}(r, t) = Re\, \Sigma\, \mathbf{E}(\omega, \mathbf{k}) \exp[i(\omega t - \mathbf{k} \cdot \mathbf{r}]$. Taking the Fourier transform of Eq. (4) implies for each monochromatic wave

$$\mathbf{P}(\omega, \mathbf{k}) = \chi(\omega, \mathbf{k}) \cdot \mathbf{E}(\omega, \mathbf{k}) \tag{5}$$

where $\chi(\omega, \mathbf{k})$ is the linear susceptibility. The **k**-vector dependence of $\chi(\omega, \mathbf{k})$ is related to the nonlocal character of the optical response. In a material system where charges are strongly coupled, the field at one point **r** can induce large changes of polarization at another point **r**′ and give rise to important effects such as Polariton effects. When this is not the case, the polarization at a given point depends only on the field acting at the same point, and the spatial dispersion of χ can be neglected, as well as the effect of higher multipole densities. This is known as the dipole approximation, for $\mathbf{D}(\omega)$ reduces to a simpler expression,

$$\mathbf{D}(\omega) = \mathbf{E}(\omega) + 4\pi \mathbf{P}(\omega) \tag{6a}$$

At this point it is convenient to introduce the dielectric constant tensor, $\varepsilon(\omega) = 1 + 4\pi\chi(\omega)$, such that

$$\mathbf{D}(\omega) = \varepsilon(\omega) \cdot \mathbf{E}(\omega) \tag{6b}$$

Under these very limiting assumptions one can write Eq. (3) in the following form:

$$\nabla \times \nabla \times \mathbf{E}(\omega, \mathbf{k}) - \frac{\omega^2}{c^2} \varepsilon(\omega) \cdot \mathbf{E}(\omega, \mathbf{k}) = 0 \tag{7}$$

We consider the propagation of plane waves, and we use the notations $\mathbf{k} = (n\omega/c)\hat{s}$ and $\mathbf{E}(\omega) = E(\omega)\hat{e}$ where $\hat{s}$ and $\hat{e}$ are, respectively, the unit vector along the normal of the planes of equal phase and the polarization vector of the electric field, and n is the refractive index. Equation (7) transforms to

$$n^2[(\hat{s} \cdot \hat{e})\hat{s} - \hat{e}] + \varepsilon(\omega) \cdot \hat{e} = 0 \tag{8}$$

For transparent isotropic media, $\varepsilon(\omega)$ reduces to a single diagonal component ε_ω so that the electromagnetic wave are transverse, $\hat{s} \cdot \hat{e} = 0$ and $n = \sqrt{\varepsilon_\omega}$. The energy flux is given by the Poynting vector $\mathbf{S} = (c/4\pi)\mathbf{E} \times \mathbf{H}$ and has the same direction as the wave vector $\mathbf{k}$. With our definition of plane waves, the intensity is $I_\omega = (nc/8\pi)|E(\omega, \mathbf{k})|^2$. It is useful to note that some authors use the definition $\mathbf{E}(\omega) = \{E(\omega)\hat{e}\exp[i(\omega t - \mathbf{k} \cdot \mathbf{r})] + c \cdot c\}$ for the field so that $I_\omega = (nc/2\pi)|E(\omega, \mathbf{k})|^2$. This point has no importance in linear optics but implies a number of factors of 2 in the definition of nonlinear susceptibilities, as we shall see later on.

For anisotropic media, the second-rank tensor ε can always be made diagonal by a proper choice of axes. Then Eq. (8) transforms into the Fresnel equation,

$$\frac{1}{n^2} = \frac{s_x^2}{n^2 - \varepsilon_x} + \frac{s_y^2}{n^2 - \varepsilon_y} + \frac{s_z^2}{n^2 - \varepsilon_a} \tag{9}$$

This equation can be represented by the double-sheet surface. In general, there are three principal indices: $n_x = \sqrt{\varepsilon_x}$; $n_y = \sqrt{\varepsilon_y}$, and $n_z = \sqrt{\varepsilon_z}$. For any direction of propagation there are two possible indices, except for the optic axis, which has the directions defined by the intersection of the inner and outer sheets.

The intersections with the $[xy]$, $[yz]$, and $[zx]$ planes are composed of a circle and an ellipse. For instance, if $\hat{s} = \cos\theta\hat{z} + \sin\theta\hat{y}$, then for the ordinary wave $\hat{e}_0 = \hat{x}$, $n_0 = \sqrt{\varepsilon_x}$ and for the extraordinary wave $\hat{e}_e \in [y, z]$, $n_e^{-2}(\theta) = \varepsilon_y^{-1}\cos^2\theta + \varepsilon_z^{-1}\sin^2\theta$. In this case, it is found that the direction of propagation of the phase and that of energy make an angle ρ:

$$\tan\rho = \frac{\varepsilon_y - \varepsilon_z}{\varepsilon_y \tan\theta + \varepsilon_z \operatorname{cotan}\theta} = \tfrac{1}{2}n_e^2(\theta)\left(\frac{1}{\varepsilon_z} - \frac{1}{\varepsilon_y}\right)\sin 2\theta = \frac{d}{d\theta}\log n_e(\theta) \tag{10}$$

In the case of uniaxial media, $\varepsilon_x = \varepsilon_y = \varepsilon_\perp$, $\varepsilon_x = \varepsilon_\parallel$, the index surface decomposes into a sphere and an ellipsoid of revolution.

In a number of experiments, electromagnetic waves are incident on a sample and generate, within the sample through nonlinear processes, other waves, which are detected after the sample. The correct analysis of such experiments relies on the description of the reflection and refraction at the plane boundaries, which are summarized in the following.

At the boundary between two dielectric media, the tangential components of the electric and magnetic fields are continuous. We consider a dielectric medium in the vacuum and use the subscripts I, T, and R to label the incident, transmitted, and reflected waves, respectively. The space and time dependence of the three waves is of the form $\exp[i(\omega t - \mathbf{k}_j \cdot \mathbf{r})]$; $j = \mathrm{I, T}$, or R. In order for a relation among the E_j written at one point on the boundary to remain valid for all points of the boundary at all times, the three space–time dependences must be equal on that plane. Choosing the coordinate axis so that the normal to the boundary plane is along $\hat{z}$ leads to $\mathbf{k}_\mathrm{I} \cdot \mathbf{r} = \mathbf{k}_\mathrm{T} \cdot \mathbf{r} = \mathbf{k}_\mathrm{R} \cdot \mathbf{r}$ with $\mathbf{r} = x\hat{x} + y\hat{y}$. This relation implies that all the wave vectors must be in the same plane and that the ordinary transmitted (OT) and extraordinary transmitted (ET) wave propagate along different directions. The angle between the normal and wave vectors is given by

$$\theta_\mathrm{I} = \theta_\mathrm{R}, \qquad \sin\theta_\mathrm{I} = n_0 \sin\theta_\mathrm{OT} = n_e \sin\theta_\mathrm{ET} \tag{11}$$

The continuity relations of the **E** and **H** fields are different for the ordinary and extraordinary waves. They yield the Fresnel transmission coefficients,

$$\begin{aligned} \frac{E_\mathrm{OT}}{E_\mathrm{I}} &= \frac{\cos\theta_\mathrm{I} - \sqrt{\varepsilon_0 - \sin^2\theta_\mathrm{I}}}{\cos\theta_\mathrm{I} + \sqrt{\varepsilon_0 - \sin^2\theta_\mathrm{I}}} \\ \frac{E_\mathrm{ET}}{E_\mathrm{I}} &= \frac{-\varepsilon_e\cos\theta_\mathrm{I} + \sqrt{\varepsilon_e - \sin^2\theta_\mathrm{I}}}{\varepsilon_e\cos\theta_\mathrm{I} + \sqrt{\varepsilon_e - \sin^2\theta_\mathrm{I}}}. \end{aligned} \tag{12}$$

The case of an absorbing medium requires some care. It is characterized by a complex dielectric constant $\tilde{\varepsilon} = \varepsilon - i\varepsilon'$. We must therefore consider the complex wave vector $\mathbf{k} = (n\omega/c)\hat{s} - i(\alpha/2)\hat{s}'$ with real and imaginary parts directed respectively along the unit vectors $\hat{s}$ and $\hat{s}'$ such that

$$\hat{s} = \cos\theta_\mathrm{T}\hat{z} + \sin\theta_\mathrm{T}\hat{y} \qquad \sin\theta_\mathrm{I} = n\sin\theta_\mathrm{T} \tag{13a}$$

$$\hat{s}' = \hat{z} \tag{13b}$$

Although the planes of constant phase have a normal $\hat{s}$ with direction still given by the Snell law, the planes of constant amplitude are parallel to the boundary plane:

$$\mathbf{E}_\mathrm{T}(\omega, \mathbf{k}) = \mathrm{Re}\left\{E_\mathrm{T}\hat{e}\exp(-\tfrac{1}{2}\alpha z)\exp\left[i\left(\omega t - n\frac{\omega}{c}\hat{s}\cdot\hat{\mathbf{r}}\right)\right]\right\} \tag{14}$$

The parameter α is the absorption coefficient as usually defined by the Beer law. Writing the propagation equation [Eq. (7)] with a complex dielectric constant implies that $\mathbf{k} \cdot \mathbf{k} = \tilde{\varepsilon}(\omega^2/c^2)$ or

$$n^2\frac{\omega^2}{c^2} - \frac{\alpha^2}{4} = \varepsilon\frac{\omega^2}{c^2}, \qquad n\frac{\omega}{c}\alpha\cos\theta_T = \varepsilon'\frac{\omega}{c} \tag{15}$$

The Fresnel transmission coefficients can be calculated as in the case of a transparent dielectric. It turns out that they have exactly the same form as for transparent media. However, they are complex because $\tilde{\varepsilon}$ is complex.

B. Phenomenological Introduction of Nonlinear Susceptibilities

1. Introduction

The electric charges in matter are held in equilibrium by atomic fields of the order of 10^8–10^9 V/cm, and anharmonicity in their motion can only be observed by using perturbing fields that are not negligible compared to these values. Although modern lasers are able to deliver such large fields, most of the nonlinear effects in condensed matter can be observed under much weaker excitation, typically at intensities in the kilowatt to megawatt per square centimeter range. This corresponds to electric fields of the order of 10^3–10^4 V/cm, which are small compared to atomic fields. However, because of coherence, the efficiency of a nonlinear optical process of order n is proportional to

$$\frac{(\text{interaction length}) \times (\text{laser field})^{n-1}}{(\text{wavelength}) \times (\text{atomic field})^{n-1}}$$

so that in the case of second- and third-order processes, efficiencies in the range 10^{-3}–10^{-1} can be observed for interaction lengths of the order of some millimeters for wavelengths in the visible to near infrared domain with fields of the order of 10^3–10^4 V/cm. To describe such configurations, the linear expansions of the multipole densities are no longer sufficient. It is convenient to expand these quantities as power series in the fields (Bloembergen, 1965). For instance, Eq. (4) must be replaced by

$$\mathbf{P}(\mathbf{r}, t) = \mathbf{P}^{(1)}(\mathbf{r}, t) + \mathbf{P}^{(2)}(\mathbf{r}, t) + \mathbf{P}^{(3)}(\mathbf{r}, t) + \cdots \tag{16}$$

where the superscripts in parentheses indicates the order in the fields. As for the linear case, the form of $\mathbf{P}^{(n)}(r, t)$ can be written on the basis of spatial homogeneity, time invariance, causality, and reality principles (Butcher, 1965); Flytzanis, 1975).

$$\mathbf{P}^{(n)}(\mathbf{r}, t) = \int_{-\infty}^{+\infty} dt_1 \int_{-\infty}^{+\infty} dt_2 \cdots \int_{-\infty}^{+\infty} dt_n \int d\mathbf{r}_1 \int d\mathbf{r}_2 \cdots \int d\mathbf{r}_n \times \kappa^{(n)}(\mathbf{r}, t; \mathbf{r}_1, t_1 \cdots \mathbf{r}_n, t_n) \times E(\mathbf{r}_1, t_1)E(\mathbf{r}_2, t_2) \cdots E(\mathbf{r}_n, t_n) \tag{17}$$

The nth-order response function $\kappa^{(n)}$ depends only on the differences $\tau_j = (t - t_j)$; it vanishes whenever one of these differences is negative, and it must be real. Again, since we restrict ourselves to optical effects in nonmagnetic media, the most relevant quantity to consider is the polarization density $\mathbf{P}(\mathbf{r}, t)$. Let us note that although $\mathbf{P}^{(n)}(\mathbf{r}, t)$ is uniquely defined by Eq. (17), the response function $\kappa^{(n)}$ is not unique, because of the $n!$ ways to specify the order of the fields. It is convenient to define a unique nth-order response function by use of a symmetrization procedure (Flytzanis, 1975):

$$K^{(n)}(\mathbf{r}; \mathbf{r}_1\tau_1, \mathbf{r}_2\tau_2 \cdots \mathbf{r}_n\tau_n) = \frac{1}{n!}\sum_p \kappa^{(n)}(\boldsymbol{\rho}_1\tau_1, \boldsymbol{\rho}_2\tau_2, \ldots, \boldsymbol{\rho}_n\tau_n) \tag{18}$$

where the symbol Σ_p refers to a sum over all permutations of the $\mathbf{r}_j$, τ_j.

As for the linear regime, the susceptibility tensor $\chi^{(n)}$ appears when transforming from space–time variables to wave-vector–frequency variables. The transform of Eq. (17) is written

$$\begin{aligned}\mathbf{P}(\mathbf{k}, \omega) = \int d\mathbf{k}\, d\omega \cdots \int d\mathbf{k}_n\, d\omega_n\, \chi^{(n)}(\mathbf{k}\omega; \mathbf{k}_1\omega_1, \ldots, \mathbf{k}_n\omega_n)\\ \times E(\mathbf{k}_1, \omega_1) \cdots E(\mathbf{k}_n, \omega_n)\end{aligned} \tag{19}$$

where

$$\begin{aligned}\chi^{(n)}(\mathbf{k}\omega; \mathbf{k}_1\omega_1; \ldots; \mathbf{k}_n\omega_n) = \int d\rho_1\, d\tau_1 \cdots \int d\rho_n\, d\tau_n\, \kappa^{(n)}(\boldsymbol{\rho}_1\tau_1, \ldots, \boldsymbol{\rho}_n\tau_n)\\ \times \exp[-i(\omega_1\tau_1 + \cdots + \omega_n\tau_n)\\ - i(\mathbf{k}_1\rho_1 + \cdots + \mathbf{k}_n\rho_n)]\end{aligned} \tag{20}$$

The susceptibilities contain the same information as the response functions, and they generalize to the nonlinear regime the notion of transfer function. In the description of nonlinear optical effects, the susceptibilities are easier to handle, because to be observed these effects require the strong fields emitted by lasers, which are narrow superpositions of monochromatic waves. The Fourier transforms of such waves involve Dirac δ-functions, so that, leaving aside the wave-vector dependence for the moment, Eq. (19) can be expressed as

$$P^{(n)}(\omega) = \chi^{(n)}(-\omega; \omega_1, \omega_2, \ldots, \omega_n) \vdots E(\omega_1)E(\omega_2) \cdots E(\omega_n) \tag{21}$$

where the usual notation has been adopted to express the frequency dispersion of $\chi^{(3)}$. The sign attached to a frequency is negative if the photon is emitted and positive if it is absorbed. The sum of the frequencies appearing between brackets vanishes, meaning that $\omega = \omega_1 + \omega_2 + \cdots + \omega_n$. The dots indicate the contracted tensor product, which will be made explicit later on. The physical interpretation of such an expression is the following: when electric charges in matter experience strong fields at frequencies $\omega_1, \omega_2, \ldots, \omega_n$, they oscillate in an anharmonic manner so that dipoles at linear combinations of $\omega_1, \omega_2, \ldots, \omega_n$ are produced. The mechanisms responsible for the anharmonicity depend on the positions of the ω_J with respect to the spectrum of the natural frequencies of the medium.

The linear susceptibility and therefore the refractive index and the linear absorption depend only on a single frequency. Two regions can be distinguished in the absorption spectra of solids: the high-frequency domain, usually in the ultraviolet, where electronic transitions govern the variations of $\chi^{(1)}(\omega)$, and the low-frequency domain, usually in the infrared, where interactions with nuclear vibrations (and eventually rotations) are dominant. Although these two domains can be analyzed independently within the frame of the Born–Oppenheimer approximation, the unified description of the refractive index and of the absorption coefficient of crystals is a formidable task that has never been fully completed. Nonlinear susceptibilities, which are more complicated, are even more difficult to describe. They depend not only on the frequencies $\omega_1, \omega_2, \ldots, \omega_n$ but also on all the intermediate linear combinations that can appear to form ω. To be specific, let us take the example of the mixing of three frequencies ω'_1, ω'_2, ω'_3 to generate the frequency $\omega = \omega'_1 - \omega'_2 + \omega'_3$. The corresponding third-order susceptibility $\chi^{(3)}(-\omega; \omega'_1, -\omega'_2, \omega'_3)$ depends of course on the position of ω, ω'_1, ω'_2, and ω'_3 with respect to the electronic and vibronic transitions of the crystals, but it also depends on $\omega'_1 - \omega'_2$, $\omega'_3 - \omega'_2$, and $\omega'_1 + \omega'_2$. It might well be possible, although ω, ω'_1, ω'_2, and ω'_3 are lying in the transparency range of the crystal, that $\omega'_1 - \omega'_2$ or $\omega'_3 - \omega'_2$ can be close to some vibronic transition frequency and $\omega'_1 + \omega'_3$ close to an electronic one. A pictorial way to interpret these types of resonances is to consider the two-step processes in which two fields interact with a mode of the crystal, say a vibration for $\omega'_1 - \omega'_2$ and $\omega'_3 - \omega'_2$ or an electronic transition for $\omega'_1 + \omega'_3$, which in turn mixes with the third field to generate ω. Of course, the closer $\omega'_1 - \omega'_2$, $\omega'_3 - \omega'_3$, or $\omega'_1 + \omega'_3$ are to the frequency of the mode, the more efficient the two-step processes are. This type of dispersion is specific for nonlinear optics and is not encountered in linear optics.

In addition to the frequency dispersion, the susceptibilities exhibit a spatial dispersion, i.e., a **k**-vector dependence. For most of the material of this chapter, the spatial dispersion can be neglected and the dipole approximation can be assumed. This corresponds to the situation where the wavelength is large as compared to the dimensions of the polarizable entities within the medium. However, this approximation may break down in the case of delocalized charges—for instance, in long conjugated chain compounds, or when the polarizable entities are strongly coupled over large distances. Then, similar to the linear case, the nonlinear response is nonlocal (Agranovitch and Ginzburg, 1966; Flytzanis, 1975). When the **k**-vector dependence can be neglected, the $\chi^{(n)}$ are called dipole susceptibilities.

Let us consider now what the effects associated with $\chi^{(2)}$ and $\chi^{(3)}$ are. To the second-order polarization $P^{(2)}(\omega) = \chi^{(2)}(-\omega; \omega_1, \omega_2) : E(\omega_1)E(\omega_2)$ correspond the sum–frequency generation (SFG) and the second-harmonic generation (SHG) in the particular case where $\omega_1 = \omega_2$, $\omega = 2\omega_1$. To $P^{(2)}(\omega) = \chi^{(2)}(-\omega; \omega_1, -\omega_2) : E(\omega_1)E(\omega_2)^*$ correspond the difference–frequency generation (DFG) and the optical recitification in the case where $\omega_1 = \omega_2$, $\omega = 0$. When $\omega_1 \gg \omega_2 \approx 0$, the nonlinear polarization has the same frequency as the only optical field $P^{(2)}(\omega) = \chi^{(3)}(-\omega; \omega, 0) : E(\omega)E(0)$. This can be interpreted as a variation of the linear polarization induced by the dc field and is known as the linear electro-optic or Pockels effect.

The third-order polarization

$$P^{(3)}(\omega) = \chi^{(3)}(-\omega; \omega'_1, \omega'_2, \omega'_3) \vdots E(\omega'_1)E(\omega'_2)E(\omega'_3)$$

describes a sum–frequency four-wave mixing (FWM) and third-harmonic generation if $\omega'_1 = \omega'_2 = \omega'_3$, $\omega = 3\omega_1$. There can be two types of difference–frequency four-wave mixing, corresponding to

$$P^{(3)}(\omega) = \chi^{(3)}(-\omega; \omega'_1, -\omega'_2, -\omega'_3) \vdots E(\omega'_1)E(\omega'_2)^*E(\omega'_3)^*$$

and $P^{(2)}(\omega) = \chi^{(3)}(-\omega; \omega'_1, -\omega'_2, \omega'_3) \vdots E(\omega'_1)E(\omega'_2)^*E(\omega'_3)$. The second has received more attention, because if $\omega'_1 \approx \omega'_2 \approx \omega'_3$, then $\omega \approx \omega'_1$. The polarization $P^{(3)}(\omega) = \chi^{(3)}(-\omega; \omega'_1, -\omega'_1, \omega) \vdots E(\omega'_1)E(\omega'_1)^*E(\omega)$ can be interpreted as a variation of the linear polarization at ω induced by the field at ω'_1. The real part of $\chi^{(3)}(-\omega; \omega'_1, -\omega'_1, \omega)$ corresponds to a variation of the refractive index at ω. This is known as the optical Kerr effect (OKE) by analogy with the case where $\omega_1 = 0$ and $P^{(3)}(\omega) = \chi^{(3)}(-\omega; 0, 0, \omega) \vdots E(0)E(0)E(\omega)$, which describes the well-known Kerr effect. The imaginary part of $\chi^{(3)}(-\omega; \omega'_1, \omega'_1, \omega)$ corresponds to an induced absorption if its sign is positive and to an induced gain if its sign is negative. It is encountered in such effects as two-photon absorption (TPA) and stimulated Raman scattering ($\omega_1 - \omega \approx \Omega_{VIB}$). Finally, for $\omega'_1 = \omega'_2$, $\omega'_3 \approx 0$, the polarization has the same frequency as in SHG: $P(2\omega'_1) = \chi^{(3)}(-2\omega'_1, \omega'_1, \omega'_1, 0) \vdots E(\omega'_1)E(\omega'_1)E(0)$. It describes a static-field-induced second-harmonic generation.

2. *Intrinsic Symmetries of the Nonlinear Susceptibilities*

A number of properties of the nonlinear susceptibilities can be deduced from very general symmetry principles. First, it must be noticed that the response functions and susceptibilities are tensors; in the particular case of dipole approximation, $\kappa^{(n)}$ and $\chi^{(n)}$ are polar tensors of rank $(n + 1)$, as required by the fact that their contracted product with the n vectors $E(\omega_j)$ results in a vector $\mathbf{P}^{(n)}$. In cartesian representation, they have $3^{(n+1)}$ cartesian components—i.e., 9 for $\chi^{(1)}$, 27 for $\chi^{(2)}$, 81 for $\chi^{(3)}$, etc. We shall adopt the usual convention of summation over repeated indices, so that the contracted products $\chi^{(2)}\colon E(\omega_1)E(\omega_2)$ or $\chi^{(3)} \vdots E(\omega'_1)E(\omega'_2)E(\omega'_3)$ will be denoted $\chi_{ijk}(E_j(\omega_1)E_k(\omega_2)$ or $\chi^{(3)}_{ijkl}E_j(\omega'_1)E_k(\omega'_2)E_l(\omega'_3)$.

The symmetrization procedure we have imposed on the response function [Eq. (18)] induces by Fourier transform that the susceptibility tensors $\chi^{(n)}_{ijk\cdots n}(-\omega - \mathbf{k}; \omega_1\mathbf{k}_1, \omega_2\mathbf{k}_2, \ldots, \omega_n\mathbf{k}_n)$ are invariant under the permutation of the triplets of indices $(j, \omega_1\mathbf{k}_1)$, $(k, \omega_2\mathbf{k}_2)$, etc. The reality condition implies in addition that

$$\chi^{(n)}(-\omega - \mathbf{k}; \omega_1\mathbf{k}_1, \omega_2\mathbf{k}_2, \ldots, \omega_n\mathbf{k}_n) = [\chi^{(n)}(\omega\mathbf{k}; -\omega_1 - \mathbf{k}_1, -\omega_2 - \mathbf{k}_2, \ldots, -\omega_n - \mathbf{k}_n)]^*$$

For the dipole susceptibilities, other symmetry relations may occur when some of their frequencies are equal. For instance, the tensor describing SHG, $\chi^{(2)}_{ijk}(-2\omega; \omega, \omega)$, is invariant in the permutation of its last two indices; that describing THG, $\chi^{(3)}_{ijkl}(-3\omega; \omega, \omega, \omega)$, is invariant in the permutation of its last three indices. Similar invariances hold for the linear electro-optic tensor $\chi^{(2)}_{ijk}(-\omega; \omega, 0)$, the TPA tensor $\chi^{(3)}_{ijkl}(-\omega; \omega_1, -\omega_1, \omega)$, the Kerr-effect tensor $\chi^{(3)}_{ijkl}(-\omega; \omega, 0, 0)$, etc.

The symmetry of the SHG and THG tensor has resulted for historical reasons in a contracted notation analogous to that used in piezoelectricity and elasticity:

$$\chi^{(2)}_{ijk}(-2\omega; \omega, \omega) = d_{im}(\omega) \tag{22a}$$

$$\chi^{(3)}_{ijkl}(-3\omega; \omega, \omega, \omega) = c_{ip}(\omega) \tag{22b}$$

Another category of very general symmetry relations can be deduced from thermodynamic considerations, in particular in the case of nonabsorbing materials where energy is exchanged among the fields and is not dissipated in the medium (Pershan, 1963; Bloembergen, 1965; Shen, 1968; Flytzanis, 1975).

For instance, in the case of three-wave mixing, $\omega = \omega_1 + \omega_2$, one has

$$\begin{aligned} P_i(\omega) &= \chi^{(2)}_{ijk}(-\omega; \omega_1, \omega_2)E_j(\omega_1)E_k(\omega_2) \\ P_j(\omega_1) &= \chi^{(2)}_{ijk}(-\omega; \omega_1, \omega_2)E_i(\omega)E^*_k(\omega_2) \\ P_k(\omega_2) &= \chi^{(2)}_{ijk}(-\omega; \omega_1, \omega_2)E_i(\omega)E^*_j(\omega_1) \end{aligned} \tag{23}$$

from which, by use of the reality condition and comparison with the three relations, one can deduce from Eq. (21) in the case $n = 2$ that

$$\chi^{(2)}_{ijk}(-\omega; \omega_1, \omega_2) = \chi^{(2)}_{jik}(-\omega_1; \omega, -\omega_2) = \chi^{(2)}_{kij}(-\omega_2; \omega, -\omega_1) \tag{24}$$

This relation indicates that only one susceptibility tensor is sufficient to describe the coupling of three fields in a lossless medium. This conclusion has a simple physical interpretation: because no energy is dissipated in the material, every time a photon of the high-frequency field is created, one photon of each low-frequency field is annihilated, and vice versa. Relations of the same type can be written for higher-order processes.

Finally, in the configuration of technological importance where all the frequencies and combinations of frequencies involved in $\chi^{(n)}$ are far from the natural frequencies of the material, an approximate symmetry is verified. The additional relations are known as the Kleinman symmetry relations (Kleinman, 1962); they state that the dispersion of the $\chi^{(n)}$ is negligible and therefore that the susceptibility tensors are invariant under any permutation of their Cartesian indices. For instance, in the case of $\chi^{(2)}$ they are written as

$$\chi_{ijk}^{(2)} = \chi_{ikj}^{(2)} = \chi_{jik}^{(2)} = \chi_{jki}^{(2)} = \chi_{kij}^{(2)} = \chi_{kji}^{(2)} \tag{25}$$

When the Kleinman symmetry relations apply, the number of independent components of $\chi^{(2)}$ reduces from 27 to 10 and that of $\chi^{(3)}$ from 81 to 15.

3. Geometrical Symmetries of the Nonlinear Susceptibilities

Let us consider a transformation of coordinate system from the orthogonal frame $(oxyz)$ to $(OXYZ)$; it is represented by the orthogonal matrix $R = [\cos\theta_{Jj}]$, where θ_{Jj} is the angle between the "new" axis J and the "former" axis j. The transformation of the dipole susceptibility tensor follows the general law of transformation of $(n+1)$-rank polar tensors, i.e.,

$$\chi_{IJ\cdots LM}^{(n)} = \cos\theta_{Ii}\cos\theta_{Jj}\cdots\cos\theta_{Ll}\cos\theta_{Mm}\,\chi_{ij\cdots lm}^{(n)} \tag{26}$$

As for any other physical properties of a medium, the susceptibilities must reflect the spatial symmetries of the medium. Therefore $\chi^{(n)}$ must transform into itself under all the transformations that leave the medium invariant. Since $\chi^{(n)}$ does not depend on translations, this means that $\chi^{(n)\prime} = \chi^{(n)}$ for all transformations R belonging to the point symmetry group of the medium, and results in a lowering of the number of independent components. The first important consequence of the geometric selection rules is that all even-order susceptibilities $\chi^{(2n)}$, and in particular of course $\chi^{(2)}$, vanish for the 10 centrosymmetric classes. The list of nonzero components of $\chi^{(2)}$ and $\chi^{(3)}$ can be found in a number of textbooks (Butcher, 1965; Singh, 1971; Flytzanis, 1975; Kurtz *et al.* 1979).

Other symmetries of the susceptibility tensors result from their transformation under a general rotation. The irreducible representations of the rotation group are labeled by their weight J, an irreducible tensor belonging to this representation having $(2J + 1)$ independent components. As the susceptibility tensor $\chi^{(n)}$ has 3^{n+1} components, it is generally reducible and can be decomposed in a sum of irreducible tensors of weight $J \leqslant n$. It is therefore possible to write

$$\chi^{(n)} = \oplus_{\tau,J}\chi^{(n)}(\tau, J) \tag{27}$$

where the seniority index τ has been introduced to distinguish the linearly independent irreducible tensors that may appear in the spectrum of $\chi^{(n)}$. The importance of the $\chi^{(n)}(\tau, J)$ is that in a general rotation its $(2J + 1)$ components transform among each other but do not mix with those of other irreducible components tensors having different J or different τ. They may consequently be associated with quantities having intrinsic physical meaning.

Irreducible tensors can be represented in a spherical tensor form closely related to the spherical harmonics, but they can also be represented in a cartesian form (Jerphagnon *et al.*, 1978).

$$\chi^{(n)}_{ij\cdots l}(\tau, J) \rightarrow \chi^{(\tau,J)}_{M} \tag{28}$$

For low weight and rank, the two representations are simple to relate. A scalar $J = 0$ has, of course, the same expression in the two cases, and a vector $J = 1$ is represented by its cartesian components u_1, u_2, u_3 or its spherical ones $u_{\pm} = \mp(u_1 \pm u_2)/\sqrt{2},\ u_0 = u_3$. For higher order, the decomposition is obtained from the composition law of angular momentum; i.e., the multiplication of two irreducible tensors of weights J and J' generates tensors of weight $|J - J'|$ to $|J + J'|$.

For example, a second-rank tensor has three components with weight 0, 1, and 2; in particular, $\chi^{(1)}$, being symmetric has no vector part ($J = 1$). Its scalar part is related to the average refractive index as measured, for instance, in a gas or a liquid, whereas its deviator ($J = 2$) is related to the birefringence of nonisotropic media.

Let us now consider the case of $\chi^{(2)}$ and $\chi^{(3)}$. By application of the law of addition of angular momentum, it appears that the 27 components of $\chi^{(2)}$ can be grouped in one pseudo-scalar ($J = 0$), three vectors ($J = 1$), two pseudo-deviators ($J = 2$), and one septor ($J = 3$): $27 = 1 + 3 \times 3 + 2 \times 5 + 7$. The 81 components of $\chi^{(3)}$ form three scalars, six pseudo-vectors, six deviators, three septors, and one nonor ($J = 9$): $81 = 3 + 6 \times 3 + 6 \times 5 + 3 \times 7 + 9$. In the transparency range where Kleinman symmetry is valid, $\chi^{(2)}$ has only one vector and one septor, and $\chi^{(3)}$ has only one scalar, one deviator, and one nonor.

The meaning of all this is simpler than it may seem at first sight. Let us take some examples: $\chi^{(2)}_{333}$ transforms like z^3, so that the sum $\chi^{(2)}_{333} + \chi^{(2)}_{322} + \chi^{(2)}_{311}$ transforms like $z^3 + zy^2 + zx^2 = z|\hat{u}|^2$, i.e., like the z component of a vector.

Therefore, the three sums $\chi^{(2)}_{333} + \chi^{(2)}_{322} + \chi^{(2)}_{311}$, $\chi^{(2)}_{222} + \chi^{(2)}_{211} + \chi^{(2)}_{233}$, and $\chi^{(2)}_{111} + \chi^{(2)}_{133} + \chi^{(2)}_{122}$ can be considered as the three components of a "vector" that is in fact one of the vector parts of $\chi^{(2)}$. It vanishes in all nonpolar media and describes, as we shall see, second-harmonic generation in polar liquids or gases induced by static field orientation. Along the same lines, third-harmonic generation in liquid or gas is described by the scalar part of $\chi^{(3)}$, i.e., the double trace: $\Sigma_{i,j}\chi^{(3)}_{iijj}$.

An application of this decomposition is to suggest relations among different effects. For instance, the pseudo-scalar of $\chi^{(2)}$ that describes three-wave mixing in noncentrosymmetric liquids is closely related to another pseudo-scalar, describing rotatory power (Giordmaine, 1965; Giordmaine and Rentzepis, 1967). Very early in nonlinear optics it was recognized that ferroelectric crystals exhibit large nonlinear coefficients, but the relation between the spontaneous polarization P_s, the rank-1 tensor describing ferroelectricity, and the rank-3 tensor $\chi^{(2)}$ describing second-harmonic generation was obscure. Jerphagnon, who was one of the first to use rotational invariance in nonlinear optics, pointed out that the relation can only be found between P_s and the vector part of $\chi^{(2)}$. He showed empirically that these two vectors were indeed proportional (Jerphagnon and Kurtz, 1970a,b) for a number of insulating crystals.

Finally, let us consider a last aspect of symmetry that is particularly relevant for organic compounds—that is, the question of dimensionality. The selection rules imposed by reducing the dimensionality are of different nature from those imposed by geometry invariance. In very-long-chain molecules, the charges can only move along one dimension, and a number of properties of the molecule can be described as if the molecule were unidimensional, along the same lines by which planar molecules are often described as bidimensional systems. Let us choose the axis of the unidimensional system along y; then the only components of the susceptibility tensor that do not vanish are those that involve only the index y. Therefore, all the tensors have only one component: $\chi^{(1)}_{22}$, $\chi^{(2)}_{222}$, $\chi^{(3)}_{2222}$, etc. For two-dimensional systems, let us choose the z axis perpendicular to the plane of the system. The only nonvanishing components are those that do not involve the index z. The number of components of a tensor or rank n in a two-dimensional space is 2^n, i.e., four components for $\chi^{(1)}$, eight for $\chi^{(2)}$, 16 for $\chi^{(3)}$, etc. This number is further reduced if permutation symmetrics, such as Kleinman's symmetry, are valid. For instance, a tensor of rank n in a two-dimensional space invariant by permutation of its indices has only $(n + 1)$ components, i.e., $\chi^{(1)}_{11}$, $\chi^{(1)}_{12}$, $\chi^{(1)}_{22}$; $\chi^{(2)}_{111}$, $\chi^{(2)}_{112}$, $\chi^{(2)}_{122}$, $\chi^{(2)}_{222}$, etc. Consequences of these reduction are investigated in Sections IV and V of this chapter for two-dimensional systems. In another chapter, Flytzanis will address the general problem of dimensionality reduction and its influence on the nonlinear susceptibilities of periodic systems.

III. SECOND-ORDER NONLINEAR OPTICAL EFFECTS

A. Introduction

Although nonlinear susceptibilities have very small magnitude, they can produce sizeable effects when microscopic dipoles are made to emit coherently. To specify the order of magnitude of the enhancement of coherent emission over the incoherent one, let us consider an assembly of microscopic units in a volume V that acquires a dipole at frequency ω under excitation by fields $\mathbf{E}(\omega_j)$; $\mathbf{p}(\omega) = \kappa^{(n)} \vdots E(\omega_1) \cdots E(\omega_n)$. Each microscopic unit, in turn, emits a field at ω; $E_l e^{i\phi_l}$ where the subscript l is used to label the microscopic units. The total intensity radiated by the assembly is then $I_{\mathrm{T}} \propto \Sigma_{l,m} E_l E_m^* e^{i(\phi_l - \phi_m)}$. If the phases of the dipoles are unrelated, the terms with $(\phi_l - \phi_m) \neq 0$ average to zero and $I_{\mathrm{T}} \approx \Sigma_l (\mathrm{n}c/8\pi)|E_l|^2 = NI$, where I is the intensity radiated by one unit and N is the number of dipoles in the volume V. This kind of incoherent emission is observed, for example, in spontaneous Raman scattering. Its efficiency is very weak, and it requires special equipment to be detected. Conversely, if the microscopic units are all in phase, $\phi_l = \phi_m = \phi$, the total intensity is $I_{\mathrm{T}} \propto |\Sigma_l E_l e^{i\phi}|^2 \approx N^2 I$, i.e., approximately N times the intensity observed for an incoherent emission. For instance, if V is a small cyclinder 1 mm high and 100 μm in diameter containing molecules with typical dimensions of the order of 5 Å, the enhancement factor is $N \approx 10^{16}$–10^{17}. In nonlinear optics, the condition stating that the fields radiated by the microscopic dipoles are all in phase is customarily referred to as the phase-matching condition. It is most often satisfied only within a very small solid angle around a precise direction.

The efficiency of nonlinear optical processes depends strongly on the propagation of the interacting fields. In this section, we discuss propagation effects in quadratic nonlinear optical media. Experimentally, three configurations are encountered. In the first one, the fields propagate in the nonlinear medium where they interact. The medium is considered as continuous and the propagation is described by a set of coupled Maxwell equations that have to be solved. In a second configuration, an assembly of molecules in a small volume or on a surface is excited by some incident fields, and we are interested in the field they emit at some point. Finally, in another situation, the interacting fields propagate through a succession of linear and nonlinear media separated by interfaces where specific boundary conditions have to be solved to correctly describe the propagation.

In order to dicuss these three configurations, we shall use slightly different formalisms that are more adapted to each case. In passing we will point out their similarities and their specificities. We will analyze the propagation of plane waves only; some aspects of the limitations introduced by the finite transverse extension of laser beams will be only briefly discussed. These aspects are important for device design (Bjorkholm, 1966; Kleinman *et al.*, 1966; Asby, 1969a,b). An exhaustive review of second-order interaction of gaussian beams can be found in the paper by Boyd and Kleinman (1968).

1. Coupled Wave Equations in a Nonlinear Medium

In this section we consider a set of fields $\mathbf{E}(\omega_j)$ propagating in a nonlinear medium. At each frequency, the propagation is described by Eq. (3). In the expression of $\mathbf{D}(\omega_j)$ it is possible to separate the linear and nonlinear parts of the polarization: $\mathbf{D}(\omega_j) = \mathbf{E}(\omega_j) + 4\pi\mathbf{P}_{\rm L}(\omega_j) + 4\pi\mathbf{P}_{\rm NL}(\omega_j) = \tilde{\varepsilon}(\omega_j)\mathbf{E}(\omega_j) + 4\pi\mathbf{P}_{\rm NL}(\omega_j)$. The propagation equation then becomes

$$\nabla \times \nabla \times \mathbf{E}(\omega_j) + \frac{\tilde{\varepsilon}(\omega_j)}{c^2}\frac{\partial^2}{\partial t^2}\mathbf{E}(\omega_j) = -\frac{4\pi}{c^2}\frac{\partial^2}{\partial t^2}\mathbf{P}_{\rm NL}(\omega_j) \tag{29}$$

The left-hand side has exactly the same form as that describing the propagation of a free wave in a linear medium.

It is customary to use the so-called slowly varying envelope approximation to solve the propagation problem (Bloembergen, 1965). Since $\mathbf{P}_{\rm NL}(\omega_j)$ is small, we expect that the variations induced by the second term of Eq. (29) over a distance of one wavelength are small as well and that it is possible to state that

$$\mathbf{E}(\omega_j) = E_j(z)\hat{e}\exp[i(\omega_j t - k_j z)] \tag{30}$$

with

$$\frac{\partial^2 E_j(z)}{\partial z^2} \ll k_j \frac{\partial E_j(z)}{\partial z} \tag{31}$$

Here $\hat{z}$ is the unit vector along the direction of propagation. Introducing this expression for $\mathbf{E}(\omega_j)$ in Eq. (29) with $k_j = n_j\omega_j/c$ and $\alpha_j = 2\omega_j\varepsilon_j'/n_j c$, we obtain

$$\left[\frac{d}{dz}E_j(z) + \frac{1}{2}\alpha_j E_j(z)\right]e^{i(\omega_j t - k_j z)} = -\frac{i2\pi\omega_j}{n_j c}\left(\frac{\hat{e}\cdot\mathbf{P}_{\rm NL}(\omega_j)}{(\hat{z}\times\hat{e})^2}\right) \tag{32}$$

It is worth noticing that one must use consistently the slowly varying envelope approximation in Eq. (7). This is equivalent to neglecting $\frac{1}{2}\alpha_j$ as compared to $n_j\omega_j/c$ in Eq. (15), i.e., $\alpha_j\lambda_j \ll 4\pi n_j$, so that $n_j \approx \sqrt{\varepsilon_j}$ even in the presence of absorption.

Equation (32) must be satisfied by all fields interacting in the nonlinear medium, either applied or generated. However, for a given set of applied fields, the frequencies arising from second- and third-order effects are most often distinct, and the selection rules obeyed by $\chi^{(2)}$ and $\chi^{(3)}$ are different. Therefore, in a number of cases, it is possible to analyze the second- and third-order processes independently. It should be kept in mind that sometimes these processes can occur simultaneously and can interfere.

Let us first consider three-wave mixing $\omega_1 + \omega_2 = \omega$. The three coupled wave equations are

$$\frac{d}{dz}E_\omega + \tfrac{1}{2}\alpha E_\omega = -i\kappa E_{\omega_1}E_{\omega_2}\exp(i\,\Delta kz) \tag{33a}$$

$$\frac{d}{dz}E_{\omega_1} + \tfrac{1}{2}\alpha_1 E_{\omega_1} = -i\kappa_1 E_\omega E^*_{\omega_2}\exp(-i\,\Delta kz) \tag{33b}$$

$$\frac{d}{dz}E_{\omega_2} + \tfrac{1}{2}\alpha_2 E_{\omega_2} = \mp i\kappa_2 E_\omega E^*_{\omega_2}\exp(-i\,\Delta kz) \tag{33c}$$

where the nonlinear coupling coefficients are $\kappa_j = (2\pi\omega_j/n_j c)[\hat{e}_j \cdot \chi^{(2)}(-\omega_j;\ \omega_i, \omega_k){:}\,\hat{e}_i\hat{e}_k/(\hat{z} \times \hat{e}_j)^2]$. The scalar quantity obtained by contraction of the susceptibility tensor with the electric field polarization vectors is called the effective nonlinear coefficient: $d_{\text{eff}} = \hat{e}_j \cdot \chi^{(2)}{:}\,\hat{e}_i\hat{e}_k$. The phase mismatch is

$$\Delta k = k - k_1 \mp k_2 \tag{34}$$

The sign in front of the right-hand term of Eq. (33c) and in front of k_2 in Eq. (34) is negative if all the waves propagate in the same direction (forward mixing), and it is positive if the wave at ω_2 propagates in the direction opposite to that of the two other waves (backward mixing). This set of equations has been discussed at length and solved exactly in a number of configurations (Armstrong *et al.*, 1962).

In the absence of absorption ($\alpha_j = 0$), the energy is redistributed among the frequencies via the nonlinear media, resulting in conservation relation first formulated by Manley and Rowe (1959). This is obtained by multiplying the three equations by $E_j^*/n_j\omega_j$, summing them two by two, and using the intrinsic symmetry of the susceptibilities Eq. (24).

$$-\frac{1}{\omega}\frac{d}{dz}I - +\frac{1}{\omega_1}\frac{d}{dz}I_1 = \pm\frac{1}{\omega_2}\frac{d}{dz}I_2 \tag{35}$$

This equation can be interpreted in terms of photon fluxes, $N_{\omega_j} = I(\omega_j)/\hbar\omega_j$, which vary in such a way that

$$\delta N_\omega = \delta N_{\omega_1} = \pm\delta N_{\omega_2} \tag{36}$$

These relations show that in each microscopic process, each time a high-frequency photon is annihilated, one photon at the two other frequencies is created. The sign in front of the last term accounts for the direction of the photon fluxes at ω_2. Photon fluxes are most convenient parameters to utilize when several frequencies are involved in a nonlinear process.

The simplest nonlinear effect is second-harmonic generation, $2\omega = \omega + \omega$. An exhaustive review of harmonic generation has been given by Akhmanov *et al.* (1975). For small losses, the system of Eq. (33) reduces to

$$\frac{d}{dz}E_{2\omega} = -i\kappa_{2\omega}E_{\omega}^{2}e^{i\Delta kz}, \qquad \frac{d}{dz}E_{\omega} = -i\kappa_{\omega}E_{2\omega}E_{\omega}^{*}e^{-\Delta kz} \tag{37}$$

The Manley–Rowe relations for SHG give $2N_{2\omega}(z) + N_{\omega}(z) = \text{const}$. The phase mismatch $\Delta k = k_{2\omega} - 2k_{\omega}$ is often expressed as a function of the coherence length, $l_c = \lambda/4(n_{2\omega} - n_{\omega})$; $\Delta kz = \pi(z/l_c)$. Because of the dispersion of refractive indices in general, it does not vanish. In such a case, the conversion efficiency is small and the intensity at ω can be assumed to remain constant. This is the so-called parametric approximation, under which the intensity at 2ω is given by

$$I_{2\omega} = \frac{128\pi^3\omega^3}{c^3}\frac{d_{\text{eff}}^2}{n_{\omega}^2 n_{2\omega}}z^2 I_{\omega_1}^2(0)\left(\frac{\sin(\Delta kz/2)}{\Delta kz/2}\right)^2 \tag{38}$$

It is an oscillating function of distance, increasing over the first coherence length, then decreasing to zero while the energy is transferred back to the field at ω, and so on.

In the case where $\Delta k = 0$, the conversion efficiency may become large and the parametric approximation breaks down, requiring an exact solution for Eq. (37). It turns out that it is more convenient to consider the photon fluxes at ω and 2ω, which are given by

$$N_{2\omega} = N_{\omega}(0)\tanh^2(gz), \qquad N_{\omega} = N_{\omega}(0)/\cosh^2(gz) \tag{39}$$

where we have introduced the nonlinear gain, $g^2 = (64\pi^3\omega^2/c^3)\,(d_{\text{eff}}^2/n^3)\,I_{\omega}(0)$, and set $n = n_{\omega} = n_{2\omega}$. The quantity $d_{\text{eff}}^2 n^{-3}$ is often called the nonlinear figure of merit of the material; it measures its efficiency in a second-order nonlinear process. Plane-wave theory predicts that total conversion is possible; in actual configurations, it is limited by the finite size of the laser beam.

In crystals, the most convenient technique to fulfill the phase-matching condition is the use of birefringence to overcome dispersion (Maker *et al.*, 1962; Giordmaine, 1962). In the case of SHG, the phase-matching condition reduces to $n_{2\omega} = n_{\omega}$, which can be satisfied if one wave is an extraordinary wave and the other an ordinary one. To be more specific, let us consider as an example the case of an important organic crystal, urea (Halbout *et al.*, 1979). Urea is a positive uniaxial crystal ($n_e > n_0$) belonging to the point group $\bar{4}2m$. The refractive index for an extraordinary ray making an angle θ with the optic axis is $n_e^{-2}(\theta) = n_0^{-2}\cos^2\theta + n_e^{-2}\sin^2\theta$. Since the birefringence ($n_e - n_0 \approx 0.1$) is larger than the dispersion in most of the transparency range $[n_0(2\omega) - n_0(\omega) \approx 0.02]$, it is possible to find an angle θ_m for which an ordinary wave at

2ω has the same refractive index as an extraordinary wave at ω given by

$$\sin^2 \theta_m = [n_0(2\omega)^{-2} - n_0(\omega)^{-2}]/[n_e(\omega)^{-2} - n_0(\omega)^{-2}] \tag{40}$$

Along this direction the waves are phase-matched and a high efficiency occurs. The sensitivity of the phase matching with respect to the orientation of the wave is measured by the variation $\Delta\theta$ around θ_m that corresponds to cancellation of the SHG. The condition defining this acceptance angle is $\pi = (\omega/c)\ z[n_0(2\omega) - n_e(\omega, \theta_m + \Delta\theta)] \approx -(2\pi/\lambda)\ z(dn_e/d\theta)_m \Delta\theta$. It is convenient to introduce the so-called walk-off angle between the propagation of the phase and that of the energy defined by Eq. (10) to express the acceptance angle as measured outside of the crystal ($\Delta\theta_{\text{ext}} = n\Delta\theta$):

$$\Delta\theta_{\text{ext}} = \lambda/2z\rho \tag{41}$$

Typical values for ρ are in the range 10^{-3}–10^{-2} rad, so that for a sample a few millimeters thick, $\Delta\theta_{\text{ext}} \approx 10^{-2}$ rad, showing that phase matching is very sensitive to orientation and even to beam divergence. When $\rho = 0$, the acceptance angle becomes very large and is said to be "noncritical." For reasons that will appear in Section VI.A, we will call it θ-noncritical phase matching. This corresponds to a efficient configuration for interaction of very monochromatic beams, and it avoids another limitation related to ρ and to the finite beam waist of laser output. This latter limitation arises because when phase matching is achieved by the use of birefringence, at least one wave is extraordinary and the energy flow of that wave diverges from the energy flow of the ordinary wave, leading to the spatial separation of the beams. This walk-off effect becomes important for sample thicknesses $z \gg W_0/\rho$, where W_0 is the beam radius. The limitation originating from the walk-off effect has been analyzed in detailed by Boyd and Kleinman (1968); it is characterized by the effective length $l_a = \sqrt{\pi} W_0/\rho$. For $z \ll l_a$, the SHG varies as z^2; for $z \approx l_a$, it varies as zl_a; and for $l_a \ll z$, it remains approximatively constant, $\sim l_a^2$.

Let us consider now the case of absorbing nonlinear media. We assume that the depletion of the fundamental field due to the SHG is small so that it varies as $\exp(-\frac{1}{2}\alpha_\omega z)\exp(ik_\omega z)$. The propagation equation for the harmonic is

$$\frac{d}{dz}E_{2\omega} + \tfrac{1}{2}\alpha_{2\omega}E_{2\omega} = -i\kappa_{2\omega}E_\omega^2 \exp(-\alpha_\omega z)\exp(i\,\Delta kz)$$

It is convenient to introduce the difference of absorption coefficients, $\Delta\alpha = \alpha_\omega - \frac{1}{2}\alpha_{2\omega}$. The intensity at 2ω is

$$I_{2\omega} = \frac{256\pi^3\omega^2}{c^3}\frac{d_{\text{eff}}}{n_\omega^2 n_{2\omega}} I_\omega^2(0)\exp[-(\alpha_\omega + \tfrac{1}{2}\alpha_{2\omega})z]\left[\frac{\cosh(\Delta\alpha z) - \cos(\Delta kz)}{\Delta\alpha^2 + \Delta k^2}\right] \tag{42}$$

This expression shows that the amplitude of the SHG intensity is decreased by a factor $\exp[-(\alpha_\omega + \frac{1}{2}\alpha_2\omega)z]$ and that the phase-matching term is completely modified. In general it is very difficult to ensure that $\Delta\alpha = 0$ and even more difficult to simultaneously achieve $\Delta k = 0$. A classical experimental configuration corresponds to the fundamental frequency well in the transparency range of the material and the harmonic frequency close to the absorption edge ($\alpha_{2\omega} > \alpha_\omega$). Therefore, even when $\Delta k \to 0$, the denominator of the phase-matching term does not vanish. However, the periodicity of the second-harmonic intensity is still governed by Δk.

Let us consider now the case of three-frequency mixing, $\omega = \omega_1 + \omega_2$. In a non–phase-matched configuration, the main results obtained in the case of SHG remain valid. In general, two beams are applied to the sample and a third field is generated. Its intensity presents spatial oscillations characterized by the coherence length $l_c = \pi/(k - k_1 - k_2)$, with (for loosely nonlinear media) eventually a exponential decay.

The phase-matched configuration is more interesting to analyze. Let us consider the case of negligible absorption. As for SHG, it is natural to introduce the photon flux. The Manley–Rowe relations imply

$$\begin{aligned} N_{\omega_1}(z) + N_\omega(z) &= \text{const} \\ \pm N_{\omega_2}(z) + N_\omega(z) &= \text{const} \\ N_{\omega_1}(z) \mp N_{\omega_2}(z) &= \text{const} \end{aligned} \tag{43}$$

These equations suggest the definition of new amplitude variables u_j, given by $E_j = \sqrt{(\omega_j/n_j)}\,u_j e^{\phi_i}$, and the photon fluxes are proportional to the square of the new normalized amplitudes $N_{\omega_j} = (c/8\pi h)\,u_j^2$. The coupled wave of Eq. (33) transforms to

$$\frac{d}{dz}u = \eta u_1 u_2 \sin\Delta\phi$$

$$\frac{d}{dz}u = \mp\eta u u_1 \sin\Delta\phi$$

$$\frac{d}{dz}u_1 = -\eta u u_2 \sin\Delta\phi$$

$$\frac{d}{dz}\Delta\phi = \left(\frac{u_1 u_2}{u} \pm \frac{u u_1}{u_2} + \frac{u u_2}{u_1}\right)\cos\Delta\phi$$

where $\Delta\phi = \phi - \phi_1 \mp \phi_2$ and $\eta = (2\pi/c)\,(\omega\omega_1\omega_2/nn_1n_2)^{1/2}\,\chi_{\text{eff}}^{(2)}$. The two applied fields have well-defined phases, while among all generated waves the one that corresponds to $\Delta\phi = \pi/2$ locks its phase to this value (if $\Delta\phi = \pi/2$ at $z = 0$, $d\,\Delta\phi/dz = 0$) and experiences the maximum gain. Its amplitude grows, and its coupling with the two applied fields overcomes that between the applied fields and the waves with fluctuating phase. Only this wave will eventually be considered. For $\Delta\phi = \pi/2$, the phase-matched system becomes

$$\frac{d}{dz}u = \eta u_1 u_2, \qquad \frac{d}{dz}u_2 = \pm\eta u u_1, \qquad \frac{d}{dz}u_1 = -\eta u u_2$$
$$u^2 + u_1^2 = a_1^2 \qquad u^2 \pm u_2^2 = a_2^2 \qquad u_1^2 \mp u_2^2 = a_3^2 \tag{44}$$

where the three constants of the motion are related by $a_1^2 = a_2^2 + a_3^2$. The solution of this system is the Jacobian elliptic functions $sn(x, m)$, $cn(x, m)$. They are defined as follows: if m is a real number $0 \leq m \leq 1$ and $x = \int_0^\beta dy/(1 - m\sin^2 y)^{1/2}$, then $sn(x, m) = \sin\beta$ and $cn(x, m) = \cos\beta$.

Let us first notice the symmetry between the forward and backward mixing as they correspond to each other by the exchange $u \leftrightarrow u_1$ and an overall sign reversal. Therefore, it is sufficient to discuss the forward three-wave mixing, which is more important in practice. We consider first the case of difference mixing with input at ω and ω_1 to generate ω_2. The frequencies are such that $\omega > \omega_1 > \omega_2$, and it is customary to call the corresponding beams pump, signal and idler, respectively. The solution depends on three parameters: the ratio of the pump photon flux to the sum of the two high frequency photon fluxes at $z = 0$, $m = N_\omega(0)/[N_\omega(0) + N_{\omega_1}(0)]$; the total gain,

$$g^2 = \frac{32\pi^3\omega\omega_2}{c^3}\frac{(\chi_{\text{eff}}^{(2)})^2}{nn_1n_2}I_\omega(0) + \frac{32\pi^3\omega_1\omega_2}{c^3}\frac{(\chi_{\text{eff}}^{(2)})^2}{nn_1n_2}I_{\omega_1}(0) \tag{45}$$

and the length z_0 over which a complete depletion of the pump beam occurs, defined by $sn(gz_0, m) = 1$. In terms of these parameters, the photon fluxes are

$$N_\omega(z) = N_\omega(0)sn^2[g(z + z_0), m] \tag{46a}$$

$$N_{\omega_1}(z) = N_{\omega_1}(0) + N_\omega(0)cn^2[g(z + z_0), m] \tag{46b}$$

$$N_{\omega_2}(z) = N_\omega(0)cn^2[g(z + z_0), m] \tag{46c}$$

These are periodic functions of the distance z with period z_0. Over the first period the pump photon flux N_ω drops to zero while the difference frequency photon flux grows from zero to $N_\omega(0)$ and the signal photon flux experiences a similar gain. Then the process is reversed and the pump is regenerated. The general solution shows that the fields at ω_1 and ω_2 can build up even if there is no input at ω_1 $[N_{\omega_1}(0) = 0]$. This phenomenon is known as parametric fluorescence, which is a spontaneous process. It can be described as a difference mixing between the pump and a noise photon input.

When the pump input is very large compared to the signal input, it is possible to neglect the pump depletion and the signal and idler experience exponential gains,

$$N_{\omega_1}(z) = N_{\omega_1}(0)\cosh^2(gz), \qquad N_{\omega_2}(z) = N_{\omega_1}(0)\sinh^2(gz) \tag{47}$$

where g is now given by the first term of Eq. (45) only.

Three-wave difference mixing can be used in a number of devices to generate infrared radiation. The occurrence of an exponential gain can be exploited in a parameteric oscillator to obtain a tunable coherent source (Byer, 1975). In a parametric oscillator, a nonlinear crystal is placed in an optical resonator with low losses and highly reflecting mirrors for ω_1 or ω_2. If the parametric gain exceeds the losses, the field generated by parametric fluorescence can build up after a sufficient number of round trips in the resonator and reach a sizeable magnitude. The couple of frequencies generated by this device (ω_1, ω_2) are fixed by the phase-matching condition. Therefore, by acting on the refractive indices of the crystal (orientation angle, temperature, etc.), it is possible to tune the output frequency of the oscillator. Devices of this type using inorganic crystals such as $LiNbO_3$ or $LiIO_3$ (Byer, 1975) or the organic crystal urea (Halbout and Tang, Chapter II-6) have been demonstrated.

The general solution of Eq. (33) also describes the sum generation, where the inputs are at ω_1 and ω_2. These new boundary conditions correspond to a translation of Eqs. (46) by one period z_0 and to $m = N_\omega(0)/[N_\omega(0) + N_{\omega_1}(0)]$. In the parametric approximation where $N_{\omega_1}(0) \gg N_{\omega_2}(0)$, the photon fluxes are

$$N_{\omega_2}(z) = N_{\omega_2}(0)\cos^2(gz), \qquad N_\omega(z) = N_{\omega_2}(0)\sin^2(gz) \tag{48}$$

where only the second term in the expression of g is to be considered. This process can be used to transpose an infrared signal in to the visible or ultraviolet domain, where good photodetectors are available. This is called up conversion and has been reviewed in detail by Warner (1975).

2. *Emission from an Assembly of Oscillating Dipoles*

In this section we consider an assembly of molecules in a volume V. A set of monochromatic fields drive molecular dipoles at the frequency ω. We choose the origin of coordinates within the volume V and call $\hat{u}$ the unit vector along the dipole at ω of the molecule located at the point $\mathbf{r}$:

$$\mathbf{p}(\omega, \mathbf{r}) = p(\omega, \mathbf{r})\hat{u} \tag{49}$$

The field radiated by the molecule at a distant point $\mathbf{R}$ is given by

$$\mathbf{E}(\omega, \mathbf{r}, \mathbf{R}) \approx k^2[(\mathbf{R} - \mathbf{r}) \times \mathbf{p}(\omega, \mathbf{r})] \times (\mathbf{R} - \mathbf{r})\frac{\exp(-ik|\mathbf{R} - \mathbf{r}|)}{|\mathbf{R} - \mathbf{r}|^3}$$

where $k = n_\omega \omega/c$ is the wave number at frequency ω in the medium where the molecules are embedded and n_ω is the corresponding refractive index. Since

$R \gg r$, it is possible to write $|\mathbf{R} - \mathbf{r}| \simeq R - \hat{s} \cdot \mathbf{r}$, where $\hat{s}$ is the unit vector in the direction of $\mathbf{R}$. In the radiation zone it is possible to expand the radiated field to first order in r/R as

$$\mathbf{E}(\omega, \mathbf{r}, \mathbf{R}) \approx k^2 p(\omega, \mathbf{r}) \frac{e^{-ikR}}{R} [(\hat{s} \times \hat{u}) \times \hat{s}] \exp(ik\hat{s} \cdot r)$$

The first two terms describe the magnitude of the field, the third its spherical wave character, the fourth its polarization, and the last one the phase shift due to the position off origin of the emitting molecule.

The total field radiated at the point $\mathbf{R}$ by all molecules is

$$\mathbf{E}(\omega, \mathbf{R}) \simeq \sum_{r \in V} \mathbf{E}(\omega, \mathbf{r}, \mathbf{R}) \tag{50a}$$

It is necessary now to consider the phase of the molecular dipoles. For an excitation by monochromatic plane waves, this can be characterized by the wave vector $\mathbf{K}$. To each term of Eq. (68b) corresponds a value for $\mathbf{K}$. For instance, the first term describes a linear excitation such that $\mathbf{p}(\omega, \mathbf{r}) = \alpha \cdot \hat{e} E(\omega) \exp[i(\omega t - \mathbf{k} \cdot \mathbf{r})]$ and $\mathbf{K} = \mathbf{k}$, whereas the second term describes a nonlinear excitation for which $\mathbf{p}(\omega, \mathbf{r}) = \beta : \hat{e}_1 \hat{e}_2 E(\omega_1)\ E(\omega_2) \exp[i(\omega t - (\mathbf{k}_1 + \mathbf{k}_2) \cdot r)]$ and $\mathbf{K} = \mathbf{k}_1 + \mathbf{k}_2$.

We write $p(\omega, \mathbf{r}) = p(\omega, r) \exp[i(\omega t - \mathbf{K} \cdot \mathbf{r})]$ so that the total field is

$$\mathbf{E}(\omega, \mathbf{R}) \approx k^2 \frac{\exp[i(\omega t - kR)]}{R} \sum_{r \in V}{}' p(\omega, \mathbf{r}) [(\hat{s} \times \hat{u}) \times \hat{s}] \exp[i(k\hat{s} - \mathbf{K}) \cdot \mathbf{r}] \tag{50b}$$

The expression can be used to calculate the field at $\mathbf{R}$ for any distribution of molecules in V. In particular, if the molecular orientation is not uniform, the contracted products $\alpha \cdot \hat{e}$, $\beta : \hat{e}_1 \hat{e}_2$ can vary from point to point. For simplicity we assume that the molecules have all the same orientation and that the amplitudes of the exciting fields are constant in V. In this case, $p(\omega, \mathbf{r})\hat{u}$ is a constant that can be extracted from the summation. If the size of the molecules is small compared to V, the sum can be replaced by an integral and

$$\mathbf{E}(\omega, \mathbf{R}) = k^2 \frac{\exp[i(\omega t - kR)]}{R} p(\omega)(\hat{s} \times \hat{u}) \times \hat{s} N \int_V d\mathbf{r} \exp[i(k\hat{s} - \mathbf{K}) \cdot \mathbf{r}] \tag{50c}$$

where N is the uniform density of molecules. This expression explains the origin of phase matching. The field at $\mathbf{R}$ results from the interference of the fields emitted by all the molecules. The interference depends on two phase-shifts: one $(k\hat{s} \cdot \mathbf{r})$ relates to the variation of distances between the emitting molecule and the observation point, and the other one $(-\mathbf{K} \cdot \mathbf{r})$ describes the phase lag of the molecular excitation. It is only when these two factors cancel out that the emission from all the molecules is coherent. In order to obtain results in a closed form, we must specify the shape and size of the volume V. Considering a cylinder of height l and radius W_0, we may use the cylinderical coordinates defined by $\mathbf{r} = z\hat{z} + \rho\cos\phi\hat{x} + \rho\sin\phi\hat{y}$. The phase mismatch Δk is given by

$$\Delta\mathbf{k} = k\hat{s} - \mathbf{K} = \Delta k_{\parallel}\cos\theta\hat{z} + \Delta k_{\perp}\sin\theta(\cos\Psi\hat{x} + \sin\Psi\hat{y}) \tag{51}$$

Using standard integrals appearing in diffraction theory (Born and Wolf, 1965), it is easy to reexpress Eq. (50c) as

$$\mathbf{E}(\omega, \mathbf{R}) \simeq 4k^2 \frac{e^{i(\omega t - kR)}}{R} p(\omega)[(\hat{s} \times \hat{u}) \times \hat{s}](4N\pi W_0^2 l) \times \frac{\sin(\frac{1}{2}\Delta k_{\parallel} l\cos\theta)}{(\frac{1}{2}\Delta k_{\parallel} l\cos\theta)} \frac{J_1(\Delta k_{\perp} W_0 \sin\theta)}{(\frac{1}{2}\Delta k_{\perp} W_0 \sin\theta)} \tag{52}$$

where $J_1(x)$ is the Bessel function of first order. The last three terms of this expansion deserve some comments. The total number of molecules in the volume V is $(4N\pi W_0^2 l)$. The next term has the same form as that describing phase matching in the previous section. It has the same interpretation because it describes the interference originating form the variation of the position of the molecules along the axis of the cylinder. Finally, the last term has exactly the same form as that describing the Frauenhofer diffraction by a circular aperture. It characterizes the coherence *across* the cylinder. To see more precisely the effects of phase matching along and across the volume V, let us consider the two extreme cases of a very elongated cylinder ($l \gg W_0$) and that of a flat one ($W_0 \gg l$).

The first example represents, for instance, a thick sample excited by collinear well collimated laser beams, such as in the configuration analyzed in the previous section. In this case the diffraction terms introduce little effect since $2J_1(x)/x \to 1$ for $x \to 0$. Because the cross-sectional dimension of the cylinder is small, all molecules in a slab perpendicular ot the axis are in phase. The phase-matching reduces to that discussed previously, with field enhancement at $\mathbf{R}$ when

$$\Delta k_{\parallel}\cos\theta = k\hat{s}\cdot\hat{z} - \mathbf{K}\cdot\hat{z} \approx 0 \tag{53}$$

This condition puts restrictions both on the direction of observation ($\hat{s}$) and on the magnitude of the wave vectors (k and K). If Eq. (53) is satisfied, the radiation is emitted in a very narrow pencil of light with angular width given by $\delta(\Delta k_{\parallel}\cos\theta) \approx 2\pi/l$, and its intensity is proportional to the square of the number of molecule in V as expected for a coherent emission.

The other configuration represents molecules on a surface excited by collinear beams of circular cross section. Because of the reduced sample thickness the phase of the molecules does not depend on z and the phase-matching term $\sin(\Delta k_{\parallel} l \cos\theta)/(\Delta k_{\parallel} l \cos\theta)$ is almost unity. The field distribution has the same pattern as that of the diffraction by a circular aperture. For a large radius it is strongly peaked in the direction such that $\Delta k_{\perp} W_0 \sin\theta = 0$ and has an angular width inversely proportional to the square of the number of molecules. It is worth noting that this result also holds for a number of nonuniform excitations. For instance, an excitation by diffraction-limited gaussian beams induces a gaussian distribution of dipoles, which in turn radiate a diffraction-limited beam.

In general, the emission distribution is the product of the two patterns, a transverse diffraction-like distribution and a longitudinal phase-matching distribution. It is even more peaked and collimated than in the two examples we just discussed.

3. Experimental Techniques

Various experimental techniques have been developed, aimed at the measurement of optical nonlinearities in different phases: powder, solution, and crystals. The second-harmonic generation powder test (Kurtz and Perry, 1968) is currently employed to assess at an early stage, before single crystals of significant size are available, the relevance of a given organic synthesis product toward quadratic nonlinear optics. More generally, and regardless of possible optical applications, this sensitive test is often used by crystallographers to help give evidence for or rule-out the existence of a crystal inversion center. When calibrated grain sizes are available, the evolution of the second-harmonic signal with respect to variations of the grain size will detect the occurrence of phase-matched configurations at the wavelength of operation. When varying the wavelength, it is possible, as was actually demonstrated in the case of urea derivatives (Halbout and Tang, Chapter II-6), to single-out θ-noncritical phase-matched configurations that give rise to significant discontinuities of the harmonic signal at the corresponding wavelength. Relative orders of magnitude for a number of interesting organic compounds, referred to the response of a calibrated quartz powder, are found in Table I.

Although this test provides semiquantitative insight on the nonlinear efficiency of a given material, its result will be decisive as to the critical choice of starting a crystal growth effort subsequent to the initial synthesis work. When this test has been found convincing which means, for p-nitroaniline-like compounds, a second-harmonic signal at least three orders of magnitude above that of quartz (see Table I)—and crystals are grown, two kinds of measurements are subsequently required. The first aims at measuring various linear and nonlinear data for the material of interest, such as absorption spectra, refractive indices, and individual nonlinear coefficients. Various possibilities are available to measure nonlinear coefficients ranging from the

TABLE I

SGH Powder Test Results for a Number of Organic Compounds[a]

Compound	Grain caliber (μm)	Efficiency
Urea	120–150	20
Urea	150–200	40
Urea	200–350	70
MAP	200–350	200
MAP	350–500	200
NPP	150–200	2000
NPP	200–350	3500

[a] Scaled to the response of a calibrated quartz powder with a fundamental wavelength of 1.06 μm.

observation of nonphase-matched nonlinear interference patterns known as Maker fringes (Maker *et al.*, 1962; Jerphagnon and Kurtz, 1970a,b) phase-matched second-harmonic generation, or spontaneous parametric fluorescence (Harris *et al.*, 1967). These methods are reviewed at length by Kurtz (1975). The latter type of measurements is closer to device assessment methods and requires a more mature sample fabrication technology; it may include parametric amplification, emission and oscillation, efficient second-harmonic generation, and sum- and difference-mixing experiments. For these experiments, laser wavelength, pulse duration, and repetition rate may be varied, with results more far-reaching than mere measurement of optical nonlinear coefficients. As an example, a most informative experiment is the utilization of subpicosecond sources now available in the visible and near infrared to increase the parametric gain coefficients, varying as $\tau^{-1/2}$, where τ is the pump duration. However, the limited phase-matching bandwidth available at a given angle–wavelength combination may oppose the expected gain increase, and the new concept of λ-noncritical phase-matching configurations will be developed in Section VI.A, together with other device-related considerations, to help demonstrate possible trade-offs.

When the nonlinearity of the individual molecules is of interest, the relevant experiment, very specific of organic compounds and of the molecular engineering approach developed thereafter, is electric-field induced second-harmonic generation (EFISH). EFISH can be applied to assess the nonlinearity of any centrosymmetric phase, be it a gas (Hauchecorne *et al.*, 1971), a centrosymmetric molecular crystal (Hochstrasser, 1973), or an organic solution (Levine and Bethea, 1975). In the case of liquids or solutions, which will be addressed more specifically in this chapter in view of their relevance to organic studies, a dc field is applied to the solution and synchronized to the laser pulse so as to statistically orient the molecules of interest and allow the coherent emission of a second-harmonic signal. A description of the experimental cell geometry was given by Oudar (1977), and we shall develop here the nonlinear wave-propagation phenomena underlying the interpretation of EFISH experiments. The formalism developed hereafter applies to nonlinear measurements in weakly nonlinear nonphase-matched media such as organic solutions or thin slabs used in Maker fringes observations.

4. Some Considerations on Interfaces in Nonlinear Optics

The formalism developed hereafter will be seen to rely strongly on interface phenomena between nonlinear media and can be generalized to include any succession of solid as well as liquid homogeneous phases separated by infinitely thin planar boundaries perpendicular to the beam propagation. Let us assume for the sake of simplicity that harmonic and fundamental beams have parallel polarization and propagation directions along X and Z, respectively. The nonlinear medium is assumed to be isotropic so that walk-off is not to be taken into account. The nonlinear wave propagation equation [Eq. (29)] at 2ω will then reduce to

$$\frac{d^2 E_x^{2\omega}}{dz^2} + \frac{4\omega^2 n_{2\omega}^2}{c^2} E_x^{2\omega} = -\frac{16\pi\omega^2}{c^2} P_{\mathrm{NL},x}^{2\omega} \tag{54}$$

where no further assumption has been made as to the nature of the various field distributions involved; $n_{2\omega}$ is the index of refraction at 2ω and $P_{\mathrm{NL},x}^{2\omega}$ is the nonlinear polarization along x. In the case of EFISH experiments, x will respond to the dc field orientation. This wave-propagation equation has been discussed previously. Introducing new variables $E(z) = E_x^{2\omega} e^{ik_f z}$ and $P_{\mathrm{NL}}(z) = P_{\mathrm{NL},x}^{2\omega} e^{ik_b z}$ with $k_f = 2\omega n_{2\omega}/c$ and $k_b = 2\omega n_\omega/c$, it becomes

$$\frac{d^2 E(z)}{dz^2} - 2ik_f \frac{dE(z)}{dz} = \frac{-16\pi\omega^2}{c^2} P_{\mathrm{NL}}(x) e^{j\Delta kz}$$
$$\Delta k = k_f - k_b \tag{55}$$

If one assumes the slowly varying amplitude approximation, Eq. (55) is further simplified to

$$\frac{dE(z)}{dz} = -\frac{i8\pi\omega^2}{c^2 k_f} P_{\mathrm{NL}}(z) e^{j\Delta kz} \tag{56}$$

which is equivalent to Eq. (32) at 2ω. It is clear that $E(z)$ will vary significantly over distances of the order of $(\Delta k)^{-1}$, so that consistency with the previous slowly varying amplitude approximation demands $(\Delta k)^{-1} \gg k_f^{-1}$ or $(n_{2\omega} - n_\omega)/n_{2\omega} \ll 1$, which corresponds to quasi-phase matching of fundamental and harmonic waves. For the highly dispersive organic solutions considered here, with 2ω possibly close to a resonant excited state, this inequality is not valid and the proper formalism to account for propagation phenomena in such media must not neglect the d^2E/dz^2 term. Equation (54) or equivalently Eq. (55) must be solved, the former by use of the classical theory of second-order differential equations or the latter by the no-less-classical "variation of the constant" method for first-order linear equations. We assume in the following that the pump amplitude is not depleted, in other words, that we propagate in a weakly nonlinear medium so that $P_{\mathrm{NL}}(z) = P_{\mathrm{NL}}$. Using either of the two methods finally yields the same result:

$$E_x^{2\omega} = E_f^+ e^{ik_f z} + E_f^- e^{-ik_f z} + \frac{16\pi\omega^2 P_{\mathrm{NL}}}{c^2(k_f^2 - k_b^2)} e^{-ik_b z} \tag{57}$$

while integration of Eq. (56), corresponding to the slowly varying amplitude approximation, would yield

$$E_x^{2\omega} = E_f^- e^{-ik_f z} + \frac{16\pi\omega^2 P_{\mathrm{NL}}}{c^2 2k_f(k_f - k_b)} e^{-ik_b z} \tag{58}$$

It may be seen that the slowly varying amplitude approximation amounts to assuming $E_f^+ = 0$ and $k_f = k_b$ (or again $n_{2\omega} = n_\omega$) in the general solution. E_f^- is called the forward propagating free wave, E_f^+ the backward free wave, and the remaining nonlinear polarization wave is referred to as the bound wave.

The proper framework for discussing nonlinear propagation effects in highly dispersive (i.e., $n_{2\omega} - n_\omega \approx 0.1$ or more, thus precluding the slowly varying envelope approximation) in weakly nonlinear media is that of Eq. (57), where the amplitudes E_f^+ and E_f^- of backward and forward propagating waves are to be found from boundary conditions. Let us discuss the general situation depicted in Fig. 1, and N different nonlaterally bounded nonlinear media are separated by planar interfaces with air or any other linear medium at each side. Two boundary conditions will connect fields at both sides of each planar boundary, one corresponding to the in-plane continuity of the harmonic E field, the other to that of the harmonic H field. This will yield $2N + 2$ relations corresponding to $N + 1$ interfaces. The unknown harmonic field amplitudes are the incoming backward free harmonic wave $E_f^{(0)+}$, the outgoing forward free wave $E_f^{(N+1)-}$, and two forward $E_f^{(i)-}$ and backward $E_f^{(i)+}$ waves for each of the N media, making a total of $2N + 2$ unknown quantities, and thus compatible with the same number of boundary conditions. The nonlinear polarization field amplitudes $E_b^{(i)}$ are straightforwardly connected via the nonlinear tensor to the fundamental field obtained for each medium (i) by applying proper transmission and dephasing factors to the squared incident fundamental field:

$$E_b^{(i)}(z) = \frac{4\pi}{\Delta\varepsilon^{(i)}} d^{(i)} e^{-ik_b^i z} (E_\omega^0)^2 \prod_{l-1}^{i-1} (T_l^{l+1} e^{jz_l(k_b^i - k_b^l)}) \tag{59}$$

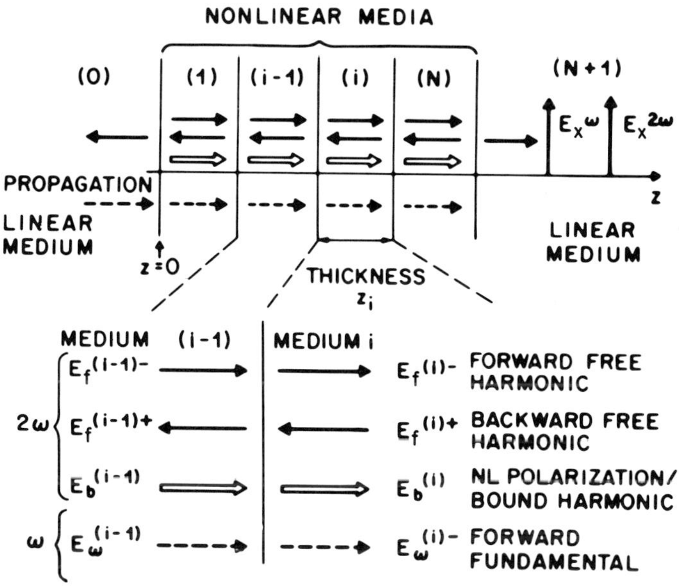

Fig. 1. General stratified nonlinear medium consisting of N nonlinear media separated by planar boundaries. The harmonic and fundamental fields to be considered in corresponding propagation equations are detailed below. Two field amplitudes, namely the forward and backward free harmonic waves, are to be found for each nonlinear medium, while bound polarization waves are known.

where $\Delta\varepsilon^{(i)}$ is the dielectric constant dispersion in medium i, $d^{(i)}$ the effective quadratic nonlinearity in medium i, T_l^{l+1} the squared field transmission factor through the $l/l+1$ interface, z_l the thickness of medium l, and z the current position in medium (i) as measured from the first 0/1 interface. It may be verified that Eq. (59) is coherent with phase continuity of the nonlinear polarization wave through the various interfaces, as imposed by the phase continuity of the transmitted fundamental wave. The general continuity conditions at the $l-1/l$ interface ($z = z_1 + z_2 + \cdots + z_{i-1}$) are

$$E_f^{(i-1)+} e^{ik_f^{i-1}z} + E_f^{i-1-} e^{-ik_f^{i-1}z} + E_b^{i-1}(z) = E_f^{(i)+} e^{ik_f^i z} + E_f^{(i)-} e^{-ik_f^i z} + E_b^{(i)}(z) \tag{60}$$

and

$$n_{2\omega}^{i-1}(E_f^{(i-1)+} e^{ik_f^{i-1}z} - E_f^{(i-1)-} e^{-ik_f^{i-1}z}) + n_\omega^{i-1} E_b^{i-1}(z) = n_{2\omega}^i (E_f^{(i)+} e^{ik_f^i z} - E_f^{(i)-} e^{-ik_f^i z}) + n_\omega^i E_b^i(z) \tag{61}$$

Simpler relations can be straightforwardly established at both extreme interfaces 0/1 and $N/N+1$, remembering that there are no bound waves in linear media and neither forward harmonic free wave in medium (0) nor backward harmonic free wave in medium $(N+1)$. The general relation between free-wave unknown amplitudes and bound-polarization-wave known amplitudes can be cast in the form of a general linear system

$$[T]_{2N+2}^{2N+2}\mathbf{F}_{2N+2} = \mathbf{B}_{2N+2}$$

where T is a $(2N+2) \times (2N+2)$ square matrix essentially composed of complex exponential dephasing factors, $\mathbf{F}$ and $\mathbf{B}$ being $2N+2$-dimensional column vectors refering respectively to free-wave and polarization-wave amplitudes. Solution of the nonlinear propagation problem amounts to inverting the "nonlinear transmission" matrix T. The problem is usually simplified (Oudar, 1977; Meredith, 1983) by considering the backward propagating free harmonic field $E_f^{(i-1)+}$ only at interface $i-1/i$ and disregarding its influence at interface $i-2/i-1$. This assumption allows expression of the forward field $E_f^{(i)-}$ in terms of $E_f^{(i-1)-}$ by elimination of $E_f^{(i-1)+}$ from Eqs. (60) and (61) where E_f^{i+} has been cancelled out. This relation turns out to be the key factor in allowing for a simple solution of the propagation problem in an EFISH cell. The previous assumption corresponds to the neglect of Fabry–Perot reflection for the harmonic wave, while consideration of the sole forward propagating fundamental field implicitly means symmetric neglect of Fabry–Perot effects for the fundamental. Previous partial attempts to deal with these multiple reflections in nonlinear problems (Jerphagnon and Kurtz, 1970a and b) used a ray-optics description of the phenomenon, while the full rigorous Maxwellian treatment requires taking into account at both interfaces of the Fabry–Perot cavity forward and backward propagating free and polarization waves together with an additional uniform polarization field (Kajzar *et al.*, 1985).

IV. QUADRATIC NONLINEAR EFFECTS IN ORGANIC MOLECULES

A. Elementary Introduction to the Electronic and Optical Structure of Organic Molecules

1. Introduction

Organic molecules exhibit very special properties owing to the versatility of the carbon atom, which is capable of engaging in various types of covalent bonds and of articulating in a wide range of combinations. This is the origin of the extraordinarily rich variety of organic compounds and of the very special role they play in biology, chemistry, and physics.

In this section some elementary aspects of the structure of organic molecules are summarized, with emphasis put on charge distribution and optical spectra, which are of importance in nonlinear optics. The goal here is not to give an exhaustive account of quantum organic chemistry, which has been completely covered by a number of review articles and textbooks. The development of modern numerical methods has resulted in more and more accurate descriptions of the electronic structures of molecules. However, the physical insight is often lost in the complexity of the mathematical treatment. The basic ideas of the theory of valence are very simple and have provided, historically, pictorial and beautiful representations of the fundamental mechanisms involved in atomic bonding (Pauling, 1960; Coulson, 1979). In more recent works, advantage has been taken of the capacity of modern computers to handle, as rigorously as possible and in a self-consistent way, all the effects related to electron–electron interactions. Nevertheless, when discussing new properties of materials, such as their nonlinear optical response, the power of the intuitive approach for pointing out important physical parameters remains unrivaled. Once these have been identified and simple semiempirical models have been adapted to a semiquantitative description of experimental data, one may then turn to more sophisticated mathematical treatments to account precisely for the physical processes in their full complexity.

2. *Atomic and Molecular Orbitals*

In molecular orbital (MO) theories, the wave functions of electrons are described as linear combinations of atomic orbitals (LCAO), as for instance in a diatomic molecule A—B, given by

$$\Psi = N(C_A\phi_A + C_B\phi_B) \tag{62}$$

where ϕ_A and ϕ_B are atomic orbitals centered at atoms A and B and the normalization factor $N = (C_A^2 + 2C_AC_BS_{AB} + C_B^2)^{-1/2}$ depends on the overlap integral $S_{AB} = \int \phi_A\phi_B\, d\tau$. The values of the coefficients are chosen to minimize the energy expectation value. It is usual to introduce in addition to S_{AB} the Coulomb integrals $\alpha_A = \int \phi_A H \phi_A\, d\tau$ and α_B and the resonance integral $\beta = \int \phi_A H \phi_B\, d\tau$. Coefficients C_A and C_B must satisfy coupled secular equations, and the energies are solutions of

$$(\alpha_A - E)(\alpha_B - E) - (\beta - ES_{AB})^2 = 0 \tag{63}$$

corresponding to two MOs, one bonding and one antibonding.

Organic molecules are built up primarily from atoms of the second row of the periodic table and may give rise to two types of orbitals with contrasted properties: the σ orbitals and the π orbitals. They are distinguished by their symmetries with respect to the internuclei axis A—B. The σ orbitals exhibit an axial symmetry around it while π orbitals have a nodal plane containing the AB axis. The π orbitals overlap less than σ orbitals and form weaker and more polarizable bonds. As for the inner shells, because of their small extension compared to atomic separation, they are hardly modified by the bonding, and the corresponding MOs retain most of the characteristics of free-atom orbitals.

Because the most stable MOs correspond to maximum overlap, the concept of localized bonds can be generalized to polyatomic molecules. The MOs constructed by LCAO extend over the whole molecule, and the electrons may be considered delocalized. We will see in the next section that this is actually the case for certain π-type MOs; however, for the σ-type MOs, it is possible to form orbitals where the electronic densities are concentrated between atoms and therefore retain the picture of the localized bond initially developed for diatomic molecules. Around a given atom A, the arrangement of the nearest neighbors determines the local geometry; it is then possible to combine the AOs of atom A to form hybrids that point towards the nearest neighbors and that, because of the maximum overlap condition, participate in very stable MOs. The MOs formed from these hybrids maximize the electronic density between pairs of atoms, in agreement with the intuitive picture of a diatomic bond. In addition, the MOs can be orthogonalized so that a number of properties of polyatomic molecules can be accounted for by addition of the corresponding properties at the bond level. This scheme holds for some electronic properties such as the electronic distribution and for some optical properties such as the contribution to optical susceptibilities far from resonances.

Before discussing the properties of conjugated molecules, let us examine the electronic charge distribution of a localized σ MO corresponding to a diatomic bond. The two nuclei are two positive point charges localized at $\pm R/2$: $\rho_{\mathrm{N}} = e[\delta(z + R/2) + \delta(z - R/2)]$. The electronic charge distribution is $\rho_{\mathrm{e}} = -2eN^2|C_{\mathrm{A}}\phi_{\mathrm{A}} + C_{\mathrm{B}}\phi_{\mathrm{B}}|^2$ and has axial symmetry around the inter-nucleus axis z. As with any charge density it can be characterized by its multipole moments: charge, dipole, quadrupole, etc. Let us assume, for simplicity of notation, that the $C_{\mathrm{A,B}}$ and $\phi_{\mathrm{A,B}}$ terms are real. The charge distribution can be expressed as

$$\rho_{\mathrm{e}} = -(eq_{\mathrm{A}}\phi_{\mathrm{A}}^2 + eq_{\mathrm{AB}}\phi_{\mathrm{AB}} + eq_{\mathrm{B}}\phi_{\mathrm{B}}^2) \tag{64}$$

where $q_{\mathrm{A,B}} = 2N^2C_{\mathrm{A,B}}^2$, $q_{\mathrm{AB}} = 4N^2C_{\mathrm{A}}C_{\mathrm{B}}S_{\mathrm{AB}}$; and $\phi_{\mathrm{AB}} = \phi_{\mathrm{A}}\phi_{\mathrm{B}}/S_{\mathrm{AB}}$. This can be viewed as composed of three parts: a charge $-eq_{\mathrm{A}}$ surrounding nucleus A with an "atomic" distribution ϕ_{A}^2, a similar charge $-eq_{\mathrm{B}}$ surrounding B, and a charge $-eq_{\mathrm{AB}}$ distributed over the two nuclei and mostly concentrated in the overlap region. The neutrality of the bond requires that $\int \rho_e \, d\tau = (q_{\mathrm{A}} +$

$q_{AB} + q_B) = 2$. For a homonuclear bond, $q_A = q_B$, whereas for a completely ionic bond A^+B^-, $q_A = 0$ and $q_B = 2$. In a heteronuclear bond there will be a certain flow of charge toward the more electronegative atom. The parameters $(q_B - q_A)/2$ can serve as a measure of the charge dissymmetry. It is often called the polarity parameter. The dipole moment of the bond is uniquely defined because the total charge vanishes; it has for components

$$\mu_z = e(q_A Z_A + q_B Z_B + q_{AB} Z_{AB}), \qquad \mu_x = \mu_y = 0 \tag{65}$$

where $Z_A = \int z\phi_A^2\, d\tau$, $Z_B = \int z\phi_B^2\, d\tau$, and $Z_{AB} = \int z\phi_{AB}\, d\tau$. To define the higher-order moments of the electronic density, some care is required, because the electronic charge does not vanish. It is convenient to introduce the centered coordinates $\hat{r}_j = r_j - \langle r_j \rangle$, where the $\langle r_j \rangle = \int \rho_e r_j\, d\tau / \int \rho_e\, d\tau$ are defined as the mean electronic coordinates averaged over the electronic density. Using centered coordinates, it is possible to define for the electronic distribution semi-invariant moments that are independent of the choice of origin. Because of axial symmetry, the only nonvanishing components of the quadupolar and octupolar moment are

$$\begin{aligned} \langle \hat{z}^2 \rangle &= \langle z^2 \rangle - \langle z \rangle^2, \qquad \langle \hat{x}^2 \rangle = \langle \hat{y}^2 \rangle = \langle x^2 \rangle \\ \langle \hat{z}^3 \rangle &= \langle z^3 \rangle - 3\langle z \rangle \langle z^2 \rangle + 2\langle z \rangle^3 \\ \langle \hat{z}\hat{x}^2 \rangle &= \langle zx^2 \rangle - \langle z \rangle \langle x^2 \rangle \end{aligned} \tag{66}$$

The general expressions are quite complicated, and no conclusion can be drawn without further information about the charge distribution. In order to gain some intuitive understanding of the meaning of the moments, let us consider a very simple distribution peaking close to the nuclei ($Z_A \approx -\frac{1}{2}R$, $Z_B \approx \frac{1}{2}R$) and for which the overlap is small compared to the electron densities around the nuclei ($q_{AB} \ll q_A, q_B$). In such a case it is found that the dipole moment $\mu_z \simeq \frac{1}{2}e(q_B - q_A)R$ is simply proportional to the bond length and to the polarity parameter. The quadrupole moment describes the "extension" of the electronic cloud: $\langle \hat{z}^2 \rangle \simeq (R/2)^2$. The octupole moment is $\langle \hat{z}^3 \rangle = -(q_B - q_A)\{1 - [(q_B - q_A)/2]^2\}R^3/8$, and describes an asymmetry of the distribution more subtle than the dipole. It is proportional to the latter but also to $\{1 - [(q_B - q_A)/2]^2\}$, which vanishes for a completely ionic bond, i.e., $q_B = 2$, $q_A = 0$, or $q_B = 0$, $q_A = 2$. The octupole moment of the electronic distribution vanishes when it has a center of symmetry, either because the two atoms of the bond are identical or because the charges are concentrated around one nucleus only as in the case of a totally ionic distribution.

The trends indicated by this over-simplified model hold, in general. The dipole moment is indeed related to the difference of charges surrounding the nuclei, the quadrupole moment describes the extension of the electronic distribution, and the octupole moment describes the deviation from centrosymmetry. These remarks will be useful in the discussion of the origins of the optical susceptibilities.

3. Conjugated Carbon Compounds

There is a class of molecules for which some electrons are actually completely delocalized. In such molecules, carbon atoms and some heteroatoms are engaged in σ bonds that involve most of their valence electrons. These bonds form a planar framework while the remaining electrons occupy p_z orbitals, where z is the direction perpendicular to the plane of the molecule. The MOs that are constructed from these p_z orbitals are orthogonal to the σ orbitals and have π symmetry. They extend over the whole molecule. The electrons that occupy these orbitals cannot be associated with a particular pair of nuclei. They are very polarizable in the sense that they can "move" over large distances under the influence of weak perturbations.

The principal properties of conjugated molecules are well described by the simple Hückel model (Heilbronner and Bock, 1970). In this model it is assumed that the framework of nuclei, inner shells, and σ bonds provides an effective field in which the π electrons move. It is assumed that the Coulomb integrals are the same for all atoms, $H_{mm} = \alpha$, and that the resonance integrals as well as the overlaps are significant only between adjacent atoms and otherwise vanish: $H_{mm+1} = \beta$ and $S_{mm+1} = S$. The wave functions of the π MOs are then constructed by LCAO from the p_z orbitals of each atom ϕ_m: $\Psi = \Sigma\, C_m \phi_m$. The boundary conditions satisfied by the coefficients C_m are $C_0 = C_{N+1} = 0$ for a chain of N atoms, and $C_0 = C_N$ for a ring of N atoms. It is natural to introduce the reduced energy $\varepsilon = (E-\alpha)/(\beta - ES)$ (for small S it measures the energy in unit β relative to α), in terms of which the secular equation is written

$$C_{m+1} + C_{m-1} - \varepsilon C_m = 0 \tag{67}$$

Let us first consider the case of a chain of atoms. A possible choice for coefficient C_m is $C_m \propto \sin(m\zeta)$, with $\zeta = \arccos(\frac{1}{2}\varepsilon)$ the boundary condition being satisfied if $\sin[(N+1)\zeta] = 0$ or

$$E = \left[\alpha + 2\beta\cos\left(\frac{\pi k}{N+1}\right)\right] \Big/ \left[1 + 2S\cos\left(\frac{\pi k}{N+1}\right)\right]$$

where the integer k can take the values $1, 2, \ldots, N$. For a zero overlap integral, the levels are symmetrically disposed around α, and in this case for N odd there is always one level for which $E = \alpha$. It is customary to call the levels for which $E < \alpha$ bonding levels and those for which $E > \alpha$ antibonding levels. When it exists, the level $E = \alpha$ is called nonbonding. For nonvanishing overlap, this simple symmetry is destroyed. For very long chains of double bonds the levels merge into a quasi-continuum exactly as in the case of a solid. They form a "band" of width 4β. In the ground state the lower orbitals are occupied by spin-paired electrons up to the $N/2$ level for N even, and up to $(N-1)/2$ for N odd with an unpaired electron in the nonbonding level in the latter case. The increase of overlap due to the π-electron shortens the C—C bonds and gives more stability to the planar structure. The model predicts an absorption spectrum starting at the transition from the highest occupied level to the lowest unoccupied one, i.e., approximately at $\Delta E \approx -4\beta \sin[\pi/(2N+1)]$. The energy of the first absorption band decreases very rapidly with the chain length, in quantitative agreement with the experiments for some molecules like cyanine dyes or cumulenes. For other long-chain molecules like polyenes or polyacetylenes, although the Hückel model qualitatively describes the energy decrease of the first absorption band, it overestimates its rate.

In the case of a ring, the boundary condition gives energies of the form

$$E = [\alpha + 2\beta \cos(2k\pi/N)]/[1 + 2S\cos(2k\pi/N)]$$

where the integer k can now take the value $0, \pm 1, \pm 2 \ldots$. The MOs are doubly degenerate except for $k = 0$ and $k = N$ (odd). The coefficients of the AOs are $C_m = A_k \exp(2\pi imk/N)$, where A_k is a normalization factor.

At this stage it is illustrative to consider the case of an important molecule, namely, benzene. Let us for simplicity assume that the overlap is negligible. This hypothesis greatly simplifies the mathematics and does not significantly alter physical insights. From the six p_z orbitals of the carbon atoms one can form three bonding and three antibonding MOs. All the bonds are strictly equivalent, giving sixfold symmetry to the molecule. The first bonding MO corresponds to $k = 0$ and $C_m^{(0)} = 1/\sqrt{6}$; the second one is degenerate and corresponds to $k = \pm 1$; and $C_m^{(\pm 1)} = (1/\sqrt{6})\exp(\pm i2\pi m/6)$. The π-electron charge per carbon atom is, of course, $1e$. The energies are $E_0 = \alpha - 2\beta$, $E_{\pm 1} = \alpha - \beta$, $E_{\pm 2} = \alpha + \beta$, and $E_3 = \alpha + 2\beta$. In the ground state the levels 0 and ± 1 are occupied by spin-paired electrons. In the first excited state one electron is transferred from the $k = \pm 1$ level to the $k = \pm 2$ level. In the language of multielectron wave functions, four combinations can be formed with singlet or triplet spin character. To remove this degeneracy it is necessary to consider the electron–electron interaction, which is neglected in the Hückel model.

4. Effect of Substitution on Hydrocarbons

When a heteroatom is substituted onto a carbon atom within a hydrocarbon, or when a substituent group is attached to a carbon in place of a hydrogen, the resulting change of multipole moment causes a change in the absorption spectrum with important consequences on nonlinear optical properties.

The effects of substitution can be quite different depending on whether the molecule is a saturated hydrocarbon or a conjugated one. In general, the interactions of a substituent with the σ bonds are rather mild, whereas they generally affect the π-electronic system more significantly, with consequences felt over all the delocalization range.

In saturated molecules containing only σ electrons, the substitution induces only short-range effects. When an atom or a group of atoms X is attached to a carbon, the C—X bond becomes polar. Depending on the electronegativity of X, electronic charges are displaced away or toward X. The charge on C is therefore modified, and the bond with the next carbon is in turn polarized, although to a lesser extent. This is called the inductive effect, which decreases very rapidly with the distance from X.

In the case of conjugated molecules, a similar inductive effect occurs for the σ electrons of the total molecular electronic charge. For π electrons, the potential at X attracts or repels electronic charges, but alterations can propagate further away than in the case of σ bonds. In addition, X may sustain π orbitals participating in the π MO and, therefore, modify the delocalization via the "mesomeric effect." For π electrons, inductive and mesomeric effects occur simultaneously and are therefore difficult to distinguish. It is only in the case where the substitution induces a weak perturbation in the conjugated molecule that the two effects can be separated. To illustrate this point, let us consider the following simple model. Let Φ_a and Φ_b be the first empty (antibonding) and the last occupied (bonding) orbitals of the conjugated molecule M, and Φ_x the π orbital of the substituent X. To the first order in the perturbation development, the new bonding orbital is $\Psi_b = \Phi_b + [V_{ba}/(E_b - E_a)]\Phi_a + [V_{bx}/(E_b - E_x)]\Phi_x$. The first matrix element of the perturbation potential V mixes states of the unsubstituted conjugated molecule itself, and for this reason the first correction term is associated with the inductive effect. The second matrix element mixes a state of the molecule with a state of the substituent, the corresponding term being associated with the mesomeric effect. The energy gap of the substituted molecule XM, is

$$\Delta = E'_a - E'_b = E_a - E_b + \frac{2|V_{ab}|^2}{E_a - E_b} + \frac{|V_{ax}|^2}{E_a - E_x} - \frac{|V_{bx}|^2}{E_b - E_x}$$

The first term is the gap of the original molecule M. The second term is always positive and tends to increase the gap, but moderately so in view of the large denominator $(E_a - E_b)$. The major part of the effect of the substitution is thus described by the last two terms. If the energy of the oribital Φ_x is low, $E_x < E_b < E_a$, then only $(E_b - E_x)$ is small and the last term dominates. The energy gap of the molecule is then reduced by the substitution. Conversely, if the energy of the orbital is large, $E_b < E_x < E_a$, the last term is positive and the gap is increased. However, in this case new transitions may occur at lower energy, between Φ_x and Φ_a if Φ_x is occupied or between Φ_b and Φ_x if Φ_x is vacant, provided these are symmetry allowed. Although this model is over simplified, it accounts qualitatively for the main features of the effects of substitution on π-electron transitions (Murrel, 1973).

To discuss the consequences of substitution on the charge distribution, let us consider the two important cases of nitrobenzene, NO_2—C_6H_5, and aniline, NH_2—C_6H_5. We model the N—C_6 framework by analogy to a seven-center Hückel chain. We have noted that in the case of an odd number of centers, there exist one orbital, namely, the nonbonding one, for which $E = \alpha$. The sum of the coefficients of the AO in the nonbonding MO attached to one center vanish. This condition, together with the normalization condition, completely determines the coefficients. In the case of aniline there is a lone pair on the nitrogen, one electron participating in the bonding MO, and the remaining one occupying the nonbonding MO. The total charge of the π electrons on the atoms is $1 + \frac{4}{7}$ for N, $1 + \frac{1}{7}$ for the carbons in the ortho and para positions, and 1 for the others. Although the extra electron of the nitrogen has, in the nonbonding MO, a large probability of remaining close to the nitrogen ($4e/7$), because of the delocalization a net negative charge ($3e/7$) has been transferred from the nitrogen to the ring with a preferential localization around the carbons in the para and ortho positions. For this reason the amino group (NH_2) is called a *donor group*. On the other hand, in the case of the nitro group (NO_2), there are no π electrons on the nitrogen because they are attracted by the electronegative oxygens. The π electrons of the ring are attracted toward the NO_2 group, which is called an *acceptor group*.The π charges are, in this case, $1 - \frac{4}{7}$ for N, $1 - \frac{1}{7}$ for the carbon in the para and ortho positions, and 1 for the other carbon atoms. Surprisingly, the π charges in aniline and nitrobenzene calculated using modern self-consistent field methods are in qualitative agreement with this simple model (see Section IV).

The classification of substituent groups into acceptors and donors follows from the generalization of this example. If a substituent group has vacant low-lying π orbitals, it can attract electrons from the conjugated molecule and is classified as an acceptor. The large the electronegativity of the group, the stronger its acceptor character. Examples of acceptors are the NO_2, C≡N, and $CONH_2$ groups. On the other hand, if a substituent group has occupied high-lying π orbitals, it can share some electronic charges with the conjugated molecule and is classified as a donor. Examples of donors are the $N(CH_3)_2$, NH_2, and OH groups. The location of the "extra" negative or positive charges in the para and ortho positions accounts for the directed chemical reactivity of substituted benzenes. In case of a donor substituent, the D—C_6H_5 molecule will react easily with positive ions or electrophilic groups giving ortho or para substitution, while for an acceptor substituent the $A—C_6H_5$ molecule is rather inert toward positive substituents and favors meta substitution of negative ions or electronegative groups.

To characterize more quantitatively the strength and orientation of the π-charge flow, it is customary in chemistry to define the mesomeric dipole moment $\Delta\mu$ as the difference of dipole between the conjugated molecule and the corresponding saturated molecule accommodating the same substituent group: $\Delta\mu_X = \mu(X, \text{conjugated mol}) - \mu(X, \text{saturated mol})$. For instance, for aniline and methylamine for the amino group, $\Delta\mu_{NH_2} = \mu(NH_2—C_6H_5) - \mu(NH_2—CH_3) \approx 1$ D. Since the inductive effects on the σ bonds are of the same nature for the two molecules and decrease rapidly, it is expected that the two molecules will have similar σ contribution to their dipoles so that $\Delta\mu_X$ will depict π-electron behavior only. This is at the origin of the donor ($\Delta\mu_D > 0$) and acceptor ($\Delta\mu_A < 0$) classification and of the definition of a number of scales measuring the "strength" of the substituent group–π-electron interaction. It is interesting to note that for long conjugated molecules the displacement of charges may extend quite far from the substituent and therefore the mesomeric moment scales roughly with the size of the conjugated molecule. It is also worth noting that π electrons cannot be fully characterized solely by the dipole moment but require higher-order multipoles for a more precise description.

In the case of multiple substitution by weakly interacting substituents, their effects on the conjugated molecule are nearly noninterfering and the induced perturbation is then approximately additive. The case of strongly interacting substituents is more interesting, in particular when a strong donor and a strong acceptor are linked to the same conjugated molecule, as in nitroaniline $NO_2—C_6H_4—NH_2$. The donor and the acceptor group are able to exchange charges via the π-electron system of the molecule. We have seen that in a donor monosubstitution the ground state is a mixture of the nonbonding orbital of the donor n_d and a π-orbital of the molecule π_m: $\Psi_0 \approx n_d + \pi_m$, the first excited state being a π orbital of the molecule π_m^*. In an acceptor monosubstitution, the ground state is a π orbital of the molecule π'_m and the first excited state is a mixture of the π orbitals of the acceptor and of the molecule: $\Psi_1 \approx \pi_m'^* + \pi_a^*$. In the case of disubstitution, the ground and first excited states originate from mixtures of the form $\Psi'_0 \approx n_d + \pi_m + \pi'_m$ and $\Psi'_m \approx \pi_m^* + \pi_m'^* + \pi_a^*$, and therefore they keep their character. The first transition corresponds to a charge transfer of the type $n_d \rightarrow \pi_a^*$ with a strong modification of the charge distribution. This transition has a large oscillator strength in the near UV and is usually very sensitive to external perturbations. Going beyond these arguments requires more elaborate theories that are capable of numerical accuracy but will confirm the qualitative trends of these simpler models.

B. Molecular Polarizabilities

1. *Introduction*

In Section IV.A we saw that molecules owing to their complex charge distributions may give rise to a dipole moment $\boldsymbol{\mu}$ when they lack a center of symmetry. If a static field E is applied to a molecule, the electric charges are displaced and the dipole moment is modified. For electric fields small compared to internal ones, the dipole can be expanded in a power

$$\mathbf{p} = \boldsymbol{\mu} + \alpha \cdot \mathbf{E} + \beta : \mathbf{EE} + \gamma : \mathbf{EEE} + \cdots \tag{68a}$$

This expression defines the static molecular polarizability tensors, α, β, and γ. The first one is called the linear polarizability; β and γ are called, respectively, the second- and third-order polarizability or sometimes hyperpolarizabilities. They are polar tensors fully symmetric in the permutation of cartesian indices. Their order of magnitude can be evaluated from simple arguments. The permanent dipole moment is of the order $\mu \approx ed$, where d is the distance between the center of gravity of the positive charges and that of the negative charges, i.e., some angstroms. Accordingly, intramolecular fields are of the order $\mathbf{E}_{\text{int}} \approx e/d^2$. One can expect that the approximation giving the dipole as a power series of $\mathbf{E}$ will diverge when $\mathbf{E} \approx \mathbf{E}_{\text{int}}$. In such a case, all terms in Eq. (68a) have about the same magnitude so that $\alpha \approx d^3 \approx 10^{-23}\ \text{cm}^3$, $\beta \approx d^5/e \approx 10^{-30}$ esu, and $\gamma \approx d^7/e^2 \approx 10^{-37}$ esu. Static polarizabilities were the first to be considered; however, with the event of the laser, large optical fields can be applied to molecules, and it was natural to extend the application range of Eq. (68a) to time-dependent fields. For historical reasons it is customary to use the same notations for frequency-dependent molecular polarizabilities. In the case of monochromatic fields with frequencies such that $\omega = \omega_1 + \omega_2 = \omega_1' + \omega_2' + \omega_3'$, one can write

$$\mathbf{p}(\omega) = \alpha(\omega) \cdot \mathbf{E}(\omega) + \beta(-\omega; \omega_1, \omega_2) : E(\omega_1)E(\omega_2)$$
$$+ \gamma(-\omega; \omega_1', \omega_2', \omega_3') \vdots E(\omega_1')E(\omega_2')E(\omega_3') \cdots \tag{68b}$$

The microscopic tensors $\alpha(\omega)$, $\beta(-\omega; \omega_1, \omega_2)$, and $\gamma(-\omega; \omega_1', \omega_2', \omega_3')$ comply with the general properties of the susceptibility tensor discussed in Section II.

Organic molecules are highly symmetric systems so that a significant number of polarizability components vanish. By application of the results of Section II it is seen immediately that β vanishes for all centrosymmetric molecules and that molecules that are invariant in a rotation of angle $2\pi/n$ around some axis behave as though full symmetry of revolution were applicable for tensors of rank smaller than n. For instance, ammonia (NH_3) has an axial symmetry for α and its β does not vanish, whereas benzene (C_6H_6) has no β and has an axial symmetry for both α and γ. Of course, as for the susceptibilities, invariance under some permutation of indices and dimensionality arguments reduce further the number of independent components.

The linear polarizability $\alpha(\omega)$ is invariant under the permutation of its two indices: it can always be cast in a diagonal form. Irreducible tensor analysis leads to the definition of the mean polarizability as the scalar $\alpha^{(0)} = (\alpha_{11} + \alpha_{22} + \alpha_{33})/3$, which is the only term to be observed in gases. The anisotropy of linear polarizability is described by the $J = 2$ irreducible part with components $\alpha_m^{(2)}$ ($m \leqslant 2$) of the form $\alpha_0^{(2)} = (2\alpha_{33} - \alpha_{22} - \alpha_{11})/\sqrt{6}$. It reduces to a single component $\alpha_0^{(2)} = (2/\sqrt{6})(\alpha_{33} - \alpha_{11})$ for uniaxial molecules (axis of rotation $2\pi/n$ with $n > 2$).

To discuss the cases of β and γ, let us assume for simplicity that Kleinman symmetry is valid so that they are invariant under the permutation of their indices. Using the notations of Eq. (27), they can be expressed as $\beta = \beta^{(1)} + \beta^{(3)}$ and $\gamma = \gamma^{(0)} + \gamma^{(2)} + \gamma^{(4)}$ (Jerphagnon *et al.*, 1978).

The components of the vector $\beta^{(1)}$ are the three mean values $\beta_0^{(1)} = \sqrt{3/5}(\beta_{333} + \beta_{322} + \beta_{311})$, etc. In the case where only one polar axis is determined by symmetry, as for CH_3F or CH_2F_2, the vector $\beta^{(1)}$ has only one component, which is parallel to the permanent dipole. The septor $\beta^{(3)}$ has components of the form $\beta_0^{(3)} = (2\beta_{333} - 3\beta_{322} - 3\beta_{311})/\sqrt{10}$ and $\beta_{\pm 3}^{(3)} = \pm\sqrt{5/4}\,(\beta_{111} \mp 3\beta_{122} \pm i\beta_{112} - 3i\beta_{222})$. Again, if an axis of order $n > 3$ or a polar axis can be defined, only $\beta_0^{(3)}$ will not vanish and will describe the anisotropy of β. Note that a polar molecule with an axial symmetry is characterized by the two parameters $\beta_0^{(1)}$ and $\beta_0^{(3)}$, whereas a one-dimensional molecule (polar long chain) is characterized by a single parameter, i.e., $\beta_{333} \propto \beta_0^{(1)} \propto \beta_0^{(3)}$. In the case of bidimensional molecules, β has four independent components in cartesian and spherical representations, namely, β_{111}, β_{222}, $\beta_{122} = \beta_{212} = \beta_{221}$, $\beta_{211} = \beta_{121} = \beta_{112}$ in the former case and $\beta_{\pm 1}^{(3)}$ and $\beta_{\pm 3}^{(3)}$ in the latter one.

The scalar $\gamma = \frac{1}{5}\Sigma_{i,j}\gamma_{iijj}$ is the mean value of γ, to be measured in gases. The term $\gamma^{(2)}$ depicts the anisotropy of cubic response, similar to that characterized by $\alpha^{(2)}$ in the case of linear response. Its components are of the form $\gamma_0^{(2)} = (2\gamma_{3333} - \gamma_{1111} - \gamma_{2222} + \gamma_{3311} + \gamma_{3322} - 2\gamma_{1122})/\sqrt{7}$. The irreducible part of weight $J = 4$, $\gamma^{(4)}$, expresses a more complex anisotropy.

Before finishing this section, let us note that it is important to make a clear distinction between anisotropy and asymmetry. An isotropic molecule has the same symmetry as that of a sphere and its properties are depicted by purely scalar tensors: $\alpha^{(0)}$, $\gamma^{(0)}$. In an anisotropic molecule, different directions are no longer equivalent but it can still possess a center of symmetry, as for a disc or an ellipsoid, in this case only irreducible tensors of even weight will exist: $\alpha^{(0)}$, $\alpha^{(2)}$, $\gamma^{(0)}$, $\gamma^{(2)}$, and $\gamma^{(4)}$. If in addition there is no center of symmetry, then the molecule is said to be asymmetric, and irreducible tensors of odd weight, $\mu^{(1)}$, $\beta^{(1)}$, and $\beta^{(3)}$ will contribute to its properties.

2. Formal Expression of the Molecular Polarizabilities

Going beyond the symmetry arguments requires the use of the full quantum-mechanical expression of the polarizabilities. The theory of the nonlinear optical response of matter has been discussed in a number of review articles and text books (Armstrong *et al.*, 1962; Butcher and McLean, 1963; Butcher, 1965; Ward, 1965; Flytzanis, 1972, 1975; Yee and Gustafson, 1978; Druet *et al.*, 1979). A review of the subject has been given by Flytzanis (1975). We shall limit our discussion to the principle of the calculation and to the aspects that are important for organic molecules.

The energy spectra and the wave functions of the molecule are assumed to be known: $H_0|A\rangle = \hbar\Omega_A|A\rangle$. The perturbation Hamiltonian that describes the interaction of the molecule with a set of monochromatic field has the usual form $H_I = \Sigma_j(e_j/2m_jc)[\mathbf{A}(\mathbf{r}_j,t)\cdot\mathbf{p}_j + \mathbf{p}_j\cdot\mathbf{A}(\mathbf{r}_jt)] + (e_j^2/m_jc^2)\mathbf{A}^2(\mathbf{r}_j,t)$ where $\mathbf{A}(\mathbf{r}_j,t)$ is the vector potential and $\mathbf{p}_j$ the momentum of the j particle. The interaction Hamiltonian can be written as a multipole expansion by a canonical transformation (Fiutak, 1963).

$$H_I = -(\boldsymbol{\mu}\cdot\mathbf{E} + \mathbf{m}\cdot\mathbf{H} + \mathbf{Q}\cdot\nabla\mathbf{E} + \cdots) \tag{69}$$

where $\boldsymbol{\mu}$, $\mathbf{m}$, and $\mathbf{Q}$ are the electric dipole, magnetic dipole, and electric quadrupole moments of the molecule. For nonmagnetic molecules with dimensions small as compared to the wavelength, it is customary to limit H_I to the first term in the expansion Eq. (69). The polarizabilities obtained from $H_I = \boldsymbol{\mu}\cdot\mathbf{E}$ are called the dipole polarizabilities.

The evolution of the density matrix $\tilde{\rho}$ is governed by the equation

$$\frac{\partial}{\partial t}\tilde{\rho} = \frac{-i}{\hbar}[H_0 + H_I, \tilde{\rho}] \tag{70}$$

When $\tilde{\rho}$ is known then the expectation of the dipole moment is obtained from the trace theorem: $\mathbf{p} = \mathrm{Tr}(\tilde{\rho}\boldsymbol{\mu})$. If the solution of Eq. (70) is obtained by an iteration procedure, $\tilde{\rho}$ is expressed as a series $\tilde{\rho} = \tilde{\rho}^{(0)} + \tilde{\rho}^{(1)} + \tilde{\rho}^{(2)} + \cdots$, where the superscript indicates the order in the perturbation. Correspondingly, $\mathbf{p}$ is obtained in the form $\mathbf{p} = \mathbf{p}^{(0)} + \mathbf{p}^{(1)} + \cdots$. By comparing in this expansion the complex amplitude at the frequency ω with the expression of $\mathbf{p}(\omega)$ one can identify the polarizabilities.

As was mentioned in Section II, the definition of the electric field has some bearing on the expression of the polarizabilities. For a monochromatic wave we use $\mathbf{E}(t) = \mathrm{Re}[\mathbf{E}(\omega)e^{i\omega t}] = \frac{1}{2}\mathbf{E}(\omega)e^{i\omega t} + \text{c.c.}$ corresponding to a dipolar interaction Hamiltonian $H_I = \Sigma_j H_I(\omega_j) + H(-\omega_j) = \Sigma_j[-\frac{1}{2}\boldsymbol{\mu}\cdot\mathbf{E}(\omega_j) + \text{c.c.}]$. As a result, a factor 2^{-n} appears both in the nth-order term of the density matrix and in the expression of the nth-order term of the dipole operator $\mathbf{p}^{(n)} = \mathrm{Re}[\frac{1}{2}\mathbf{p}(\omega)e^{i\omega t} + \text{c.c.}]$. The polarizabilities are defined in terms of the complex field amplitudes $E(\omega)$ and the complex dipole amplitude $\mathbf{p}(\omega)$, leading a $2^{-(n-1)}$ factor in the expression of the nth-order polarizability. Another definition is also used by some authors [for instance, by Butcher (1965)], according to which $\mathbf{E}(t) = \mathbf{E}(\omega)e^{i\omega t} + \text{c.c.}$ The polarizabilities and susceptibilities calculated following this definition have exactly the same form as that obtained with our conventions, but they are larger by a factor 2^{n-1}. This has no importance for α and $\chi^{(1)}$ but changes the values of β and $\chi^{(2)}$ by 2 and those of γ and $\chi^{(3)}$ by 4.

In order to account correctly for the relaxation in the evolution of $\tilde{\rho} = \Sigma_{\mathrm{A,B}} |A\rangle \tilde{\rho}_{\mathrm{AB}} \langle B|$ it is necessary to describe the evolution of both the wave function $|A\rangle$ and it complex conjugate $\langle B|$. This has been analyzed in details by Slichter 1963. Diagramatic techniques have since been developed to carry out the calculation in a systematic manner (Yee and Gustafson, 1978; Druet *et al.*, 1979).

The polarizability consists of a sum of terms, each of which has a numerator formed by a product of dipole moment matrix elements taken between a sequence of state $(G, A, B, \ldots)$ and a denominator formed by a product of monomials including the energy differences between the states corresponding to the numerator matrix elements $(\Omega_{AG}, \Omega_{AB}, \Omega_{CB}, \ldots)$, correlated relaxation parameters $(\Gamma_{AG}, \Gamma_{AB}, \Gamma_{CB}, \ldots)$, and linear combinations of the fields frequencies corresponding to relevant sequence of absorptions or emissions $(\omega'_1, \omega'_1 - \omega'_2, \omega'_1 - \omega'_2 + \omega'_3, \ldots)$. In the expression of the nth-order polarizability there are n monomials in the denominator and $(n + 1)$ matrix elements in the numerator, n of which originate from the n perturbations and one from μ itself. This product gives to the n-order polarizability its character of $(n + 1)$-rank tensor. If the ground state only is populated before any interaction takes place, there are $2^n n!$ diagrams contributing to $\rho^{(n)}$, that is, eight for $\rho^{(2)}$ and 48 for $\rho^{(3)}$. A treatment of the classification of these terms and a discussion of their resonant behavior has been given by Yee and Gustafson (1978), Druet *et al.* (1979), and Prior (1984).

The expressions of the first three polarizabilities are

$$\alpha(\omega) = \sum_A \left(\frac{\langle G|\boldsymbol{\mu}|A\rangle\langle A|\boldsymbol{\mu}|G\rangle}{\hbar(\omega - \Omega_{AG} + i\Gamma_{AG})} + \frac{\langle G|\boldsymbol{\mu}|A\rangle\langle A|\boldsymbol{\mu}|G\rangle}{\hbar(\Omega_{AG} + \omega - i\Gamma_{AG})} \right) \tag{71a}$$

$$\beta(-\omega; \omega_1, \omega_2) = P \sum_{A,B} \frac{\langle G|\boldsymbol{\mu}|A\rangle\langle A|\boldsymbol{\mu}|B\rangle\langle B|\boldsymbol{\mu}|G\rangle}{2\hbar^2(\omega_1 - \Omega_{AG} + i\Gamma_{AG})(\omega - \Omega_{BG} - i\Gamma_{BG})} \tag{71b}$$

$$\gamma(-\omega; \omega'_1, \omega'_2, \omega'_3) = P \sum_{A,B,C} \left(\frac{1}{4\hbar^3}\right) \frac{\langle G|\boldsymbol{\mu}|A\rangle\langle A|\boldsymbol{\mu}|B\rangle\langle B|\boldsymbol{\mu}|C\rangle\langle C|\boldsymbol{\mu}|G\rangle}{(\omega'_1 - \Omega_{AG} + i\Gamma_{AG})(\omega'_1 + \omega'_2 - \Omega_{BG} + i\Gamma_{BG})(\omega - \Omega_{CG} + i\Gamma_{CG})} \tag{71c}$$

The expression of $\alpha(\omega)$ is called the Kramer–Heisenberg formula. In the case of β and γ for simplicity we have written only one term, the symbol P in front of it indicating that all contributions from all diagrams are to be added. An example of the full expression of γ in the case of a four-level system has been given by Bloembergen *et al.* (1978).

The expressions of the polarizabilities are correct provided exact wave functions $|G\rangle$, $|A\rangle$, etc. are used which specify quantum states of both electrons and the nuclei. However, interaction mechanisms of light with electrons and nuclei are different, and it is useful to distinguish their contribution to the polarizabilities (Flytzanis, 1972, 1975). Using the Born–Oppenheimer approximation it is possible to factorize the wave functions according to $|A\rangle = |a(\mathbf{r}_j, \mathbf{q}_l) > |v(\mathbf{q}_l)\rangle$, where the electronic part $|a(\mathbf{r}_j, \mathbf{q}_l)\rangle$ depends on the coordinates of the electrons ($\mathbf{r}_j$) and of the nuclei, which are expressed in terms of the normal modes of vibration ($\mathbf{q}_l$). The nuclei part $|v(\mathbf{q}_l)\rangle$ does not depend on electronic coordinates. The energy of the electrons can be expanded in terms of $\Omega_a(\mathbf{q}_l) = \Omega_a(0) + \Sigma_l \frac{1}{2}\Omega_l^2 q_l^2 + \cdots$ where $\hbar\Omega_a(0)$ is the energy of the electronic state when the nuclei are fixed at their equilibrium position and the Ω_l are the normal modes frequencies. When only the two first terms are retained, the equation of motion of the nuclei is similar to that of a set of harmonic oscillators with energy $\hbar\Omega_v = \Sigma_l (N_l + \frac{1}{2})\hbar\Omega_l$. The differences of energies that appear in the polarizabilities are of the form

$$\Omega_{AB} = \Omega_a(0) - \Omega_b(0) + \Omega_{v'} - \Omega_v \tag{72}$$

The separation between the electronic states is usually much larger than that between vibrational ones $[\Omega_a(0) - \Omega_b(0) \gg \Omega_{v'} - \Omega_v]$. To evaluate Ω_{AB}, two cases have to be distinguished according to whether $a \neq b$ or $a = b$. In the former case $\Omega_{AB} \simeq \Omega_a(0) - \Omega_b(0)$, while in the latter $\Omega_{AB} = \Omega_{v'} - \Omega_v$. The dipole moment appearing in the matrix elements is the sum of two contributions referring to electronic and nuclear coordinates, respectively.

$$\boldsymbol{\mu} = \boldsymbol{\mu}_e + \boldsymbol{\mu}_v = \sum_j e\mathbf{r}_j + \sum_j eZ_j\mathbf{R}_j$$

It is often convenient to expand the ground-state electronic dipole in terms of the normal modes:

$$\langle g|\boldsymbol{\mu}|g\rangle = \boldsymbol{\mu}_0 + \sum_l \boldsymbol{\mu}_l^{(1)} q_l + \cdots$$

The linear polarizability α can then be separated into two terms (Flytzanis, 1975), a purely electronic and a purely vibrational one, $\alpha(\omega) = \alpha_l(\omega) + \alpha_v(\omega)$. The first part originates from the terms for which $a \neq g$, and it has exactly the same form as Eq. (71a) with $\boldsymbol{\mu}$ replaced by $\boldsymbol{\mu}_e$ and all quantum states referring only to the electronic part of the wave function, $|g\rangle$, $|a\rangle$. The vibrational polarizability comes from the terms for which $a = g$ and is

$$\alpha_v(\omega) \approx \sum_l \frac{\boldsymbol{\mu}_l^{(1)}\boldsymbol{\mu}_l^{(1)}}{(\hbar\Omega_l)^2 - (\hbar\omega)^2}$$

The situation is more complex in the case of β and γ because the two types of energy differences appear now in the same term. This situation is unique to nonlinear optics and gives rise to the so-called hybrid contribution (Flytzanis, 1975). When all the states have an electronic part different from the ground state ($a \neq g$), one obtains the purely electronic contributions, β_e and γ_e, which again have the same form as in Eq. (71b) and (71c) with the same changes as for $\alpha_e(\omega)$. When all the states have the same electronic part ($a = g$) one obtains the purely vibrational contributions β_v and γ_v. To understand the meaning of hybrid terms, let us consider one element of the sum making up $\gamma(-\omega'; \omega'_1, -\omega'_2, \omega'_3)$:

$$\gamma_0 = -\left(\frac{1}{4\hbar^3}\right)\frac{\langle G|\boldsymbol{\mu}|A\rangle\langle A|\boldsymbol{\mu}|B\rangle\langle B|\boldsymbol{\mu}|C\rangle\langle C|\boldsymbol{\mu}|G\rangle}{(\omega'_1 - \Omega_{AB} + i\Gamma_{AG})(\omega'_1 - \omega'_2 - \Omega_{BG} + i\Gamma_{BG})(\omega - \Omega_{CG} + i\Gamma_{CG})} \quad (73)$$

Let us consider the case where $a \neq g$, $c \neq g$, but $b = g$ and assume for simplicity that there is only one mode $\Omega_l = \Omega$. The energy denominator is approximately

$$(\omega'_1 - \Omega_{ag} + i\Gamma_{ag})[\omega'_1 - \omega'_2 - (N - N')\Omega + i\Gamma](\omega - \Omega_{cg} + i\Gamma_{cg}) \quad (74)$$

Resonant transitions may occur if $\hbar\omega'_1 = \Omega_{ag}$ or $\hbar\omega = \Omega_{cg}$ as expected and in addition when $(\omega_1 - \omega_2) = (N - N')\Omega$. In the latter case, the difference frequency of the optical fields drives the vibrational mode, which in turn modulates the electronic levels interacting with the other fields. Resonances are observed when the difference $\omega_1 - \omega_2$ is equal to a multiple of the normal mode frequency.

This example shows that the interaction of a combination of frequencies with the vibrational modes of the molecules can induce new resonances. Let us note that in addition the polarizabilities are also temperature dependent, through the mode populations, and that the vibration modes can reveal forbidden electronic transitions, through the breaking of geometrical selection rules by the motion of the nuclei.

Several approaches are possible to exploit further the expressions of the polarizabilities. The most correct one is to try to find the best possible set of wave functions to be entered in the full expressions. This has been attempted for atoms like H or He and also for simple molecules. For more complex molecules it is impossible to produce a complete set of wave functions, and a less ambitious program must be set forth to obtain approximate values for α, β, and γ. In general there are two types of situations for which it is useful to have theoretical values of the polarizabilities. The first one is the discussion of molecular nonlinearities in the transparency range of their spectra, far from any resonance. As we shall see, this is particularly important for device applications. The second one corresponds to the case where one level or a set of levels are close to resonance, for instance, in spectroscopic studies or when some combination of frequencies can approach a particular set of levels. In the latter case, it is possible, by inspection of the full expressions and by use of selection rules, to isolate the contribution of the resonant set of levels one is particularly interested in and to lump the contributions of all the other levels in a dispersionless background. The contribution of the resonant levels is then calculated from the best available wave functions or evaluated from experimentally measured quantities such as oscillator strengths.

The optical polarizabilities far from resonances belong to the "integral properties" of the molecule. They do not depend on details of the energy spectra but on the overall chemical bonding (Hopfield, 1971; Friedel and Lannoo, 1973). They are related to density-of-state-weighted averages of the oscillator strengths, which smooth out a number of details. This explains why the models that can reproduce the general outlines of the density of state and of the charge distribution in the ground state can be successfully applied to the calculation of polarizabilities in the transmission range. To substantiate this argument and to see how the values of the electronic part of the polarizabilities far from the resonance are related to the ground-state charge distribution, let us use the Unsöld approximation to reexpress α and β (Unsöld, 1927). This approximation was first proposed by Unsöld and was subsequently generalized to nonlinear optical effects (Robinson, 1967; Flytzanis and Ducuing, 1968, 1969; O'Hare and Hurst, 1967). Far from resonance, the damping parameters Γ can be neglected and one obtains from Eq. (71a)

$$\boldsymbol{\alpha}(\omega) \approx \sum_{a \neq g} \frac{2e^2\Omega_{ag}}{\hbar^2(\Omega_{ag}^2 - \omega^2)} \langle g|\mathbf{r}|a\rangle\langle a|\mathbf{r}|g\rangle \tag{75}$$

The variations of the matrix element $\langle g|\mathbf{r}|a\rangle$ are very large compared to those of the energies $\hbar\Omega_{ag}$; it is therefore possible to assume that the energy-dependent terms are roughly constant and can be taken out of the summation. Let us consider the element $\alpha_{xx}(\omega)$ and denote by Ω_x the corresponding "mean value" of Ω_{ag}. The expression of $\alpha_{xx}(\omega)$ is

$$\alpha_{xx}(\omega) \approx \frac{2e^2\Omega_x}{\hbar^2(\Omega_x^2 - \omega^2)} \sum_{a \neq g} |\langle g|x|a\rangle|^2$$

Using matrix element algebra $\Sigma_{a \neq g}|\langle g|x|a\rangle|^2 = |\langle g|\hat{x}^2|g\rangle|$, where $\hat{x} = x - \langle g|x|g\rangle$ is the centered coordinate introduced previously, one finds

$$\alpha_{xx}(\omega) \approx \frac{2e^2\Omega_x}{\hbar^2(\Omega_x^2 - \omega^2)} |\langle g|\hat{x}^2|g\rangle|^2 \tag{76}$$

There are many ways for chosing the best "mean" value Ω_x, one being that used by Unsöld, who applied the same approximation to the Thomas–Kuhn sum rule, $\Sigma_{a \neq g}(2m/\hbar^2)\,|\langle g|x|a\rangle|^2\Omega_{ag} = 1$, leading to

$$\frac{1}{\Omega_x} = \frac{2m}{\hbar^2}|\langle g|\hat{x}^2|g\rangle| \tag{77}$$

The linear polarizability depends then only on the quadrupole moment in the ground state: $|\langle g|\hat{x}^2|g\rangle|$. If we assume further that the frequency is small compared to the energy gap ($\omega \ll \Omega_x$), it is found that

$$\alpha_{xx}(\omega \approx 0) \approx \frac{32\pi}{a_0}|\langle g|\hat{x}^2|g\rangle|^2 \tag{78}$$

where $a_0 = \hbar^2/me^2$ is the Bohr radius. The physical interpretation of this result is that the linear polarizability of a molecule measures the "extension" of the electronic distribution. Well-localized distributions concentrated close to the nuclei, such as core orbitals, are difficult to polarize. On the other hand, extended orbitals, such as σ and π valence orbitals, are polarizable. Indeed they contribute to the major part of the refractive index.

When the same method is applied to the nonlinear polarizabilities, similar results are obtained; for instance (Robinson, 1967),

$$\beta_{xyz}(-\omega;\omega_1,\omega_2) \simeq \left(\frac{e^2}{2\hbar^2}\right)\frac{\Omega_x\Omega_y\Omega_z(\Omega_x+\Omega_y+\Omega_z)}{(\Omega_x^2-\omega^2)(\Omega_y^2-\omega_1^2)(\Omega_z^2-\omega_2^2)}\langle g|\hat{x}\hat{y}\hat{z}|g\rangle \tag{79}$$

where β depends directly of the octupole moment in the ground state, $\langle g|\hat{x}\hat{y}\hat{z}|g\rangle$. We have seen in the simple case of a diatomic bond how this moment is linked to the asymmetry of the electronic distribution. While some relation was intuitively expected between the second-order polarizability and the geometric asymmetry of the molecule, the use of the Unsöld approximation establishes clearly how β relates to the octupole moment. This correspondence between polarizabilities and geometry can be generalized further. Far from resonance, the nth-order polarizability can be expressed in a form that involves the semiinvariant moment of order $(n+1)$ and a frequency-dependent factor involving the mean energies $\hbar\Omega_x$ and the energies of the photons of the interacting fields.

3. A Simple System

It is instructive to apply the formalism of the previous section to a simple two-level system, which has been extensively studied (for a review, see Allen and Eberly, 1975) while exact solutions describing its interactions with electromagnetic fields have been found. The aim is to compute the first- and second-order polarizabilities. Usually the two-level system is assumed to have no dipole in the ground state or in the excited state, i.e., $\boldsymbol{\mu}_{aa} = \boldsymbol{\mu}_{gg} = 0$. This corresponds to a centrosymmetric system for which β is forbidden. The density matrix evolution equation of a two-level system with nonvanishing β is

$$\frac{\partial}{\partial t}\tilde{\rho}_{aa} = -\frac{\partial}{\partial t}\tilde{\rho}_{bb} = -\Gamma_{aa}\rho_{aa} - \frac{i}{2\hbar}[(\boldsymbol{\mu}_{ag}\cdot\mathbf{E})\tilde{\rho}_{ga} - \tilde{\rho}_{ag}(\boldsymbol{\mu}_{ga}\cdot\mathbf{E})]$$
$$\frac{\partial}{\partial t}\tilde{\rho}_{ag} = (i\Omega_{ag} + \Gamma_{ag})\tilde{\rho}_{ag} - \frac{i}{2\hbar}[(\boldsymbol{\mu}_{ag}\cdot\mathbf{E})(\tilde{\rho}_{gg} - \tilde{\rho}_{aa}) + (\boldsymbol{\mu}_{aa}\cdot\mathbf{E} - \boldsymbol{\mu}_{gg}\cdot\mathbf{E})\tilde{\rho}_{ag}] \tag{80}$$

The first-order polarizability is simply

$$\alpha(\omega) = \frac{|\mu_{ag}|^2}{\hbar}\left[\frac{1}{(\omega - \Omega_{ag} + i\Gamma_{ag})} + \frac{1}{(\omega + \Omega_{ag} - i\Gamma_{ag})}\right]$$

Four diagrams describe the second-order solution of Eq. (80) in the case of second-harmonic generation. Let us use $V_{ag}(\omega) = -\frac{1}{2}\mu_{ag}\cdot E(\omega)$ and $D(\omega) = (\hbar\omega - \Omega_{ag} + i\Gamma_{ag})$ to shorten the notation. The forms of the resonant contribution to $\rho^{(2)}$ are

$$\rho_{ag}^{(2)}(2\omega) = \frac{[V_{aa}(\omega) - V_{gg}(\omega)]V_{ag}(\omega)}{D(\omega)D(\omega)} \tag{81a}$$

$$\rho_{aa}^{(2)}(2\omega) = \frac{V_{ag}(\omega)V_{ga}(\omega)}{\hbar(\Gamma_{aa} - 2i\omega)}\left[\frac{1}{D(\omega)} + \frac{1}{D^*(-\omega)}\right] \tag{81b}$$

As for the second-order dipole, it can be written as

$$\boldsymbol{\mu}^{(2)} = (\boldsymbol{\mu}_{aa} - \boldsymbol{\mu}_{gg})\rho_{aa}^{(2)} + \boldsymbol{\mu}_{ag}\rho_{ga}^{(2)} + \boldsymbol{\mu}_{ga}\rho_{ag}^{(2)}$$

where we have used the relation $\rho_{aa}^{(n)} + \rho_{gg}^{(n)} = 0$. It is apparent that $\mu^{(2)}$ belongs to the plane of the two vectors $\boldsymbol{\delta} = \boldsymbol{\mu}_{aa} - \boldsymbol{\mu}_{gg}$ and $\mathbf{m} = \boldsymbol{\mu}_{ag}$. Far from resonances, the relaxation vector parameters can be neglected, and the second-order polarizability takes the form

$$\beta_{ijk}(-2\omega;\omega,\omega) \approx \frac{1}{2\hbar^2}\left(\frac{\delta_i m_j m_k}{\Omega_{ag}^2 - \omega^2} + m_i(\delta_j m_k + \delta_k m_j)\frac{\Omega_{ag}^2 + 2\omega^2}{(\Omega_{ag}^2 - \omega^2)(\Omega_{ag}^2 - 4\omega^2)}\right) \tag{82}$$

From this expression it is easy to recognize that β is a bidimensional tensor with only six components (Oudar and Zyss, 1982). In the case where $\mathbf{m}$ and $\boldsymbol{\delta}$ are parallel to the direction $\hat{z}$, then

$$\beta_{zzz}(-2\omega;\omega,\omega) \approx \left(\frac{3\hbar^2}{2m}\right)\frac{W}{(W^2 - \hbar^2\omega^2)(W^2 - 4\hbar^2\omega^2)}f\delta \tag{83}$$

where f is the oscillator strength for the transition and $W = \hbar\Omega_{ag}$. This result is very useful to discuss the nonlinearity of charge-transfer molecules. It shows that β is proportional to the oscillator strength of the transition and to the difference of dipole moment between the excited state and the ground state, δ. It also exhibits resonances when ω or 2ω is close to Ω_{ag}. Far away from resonances in the transparency domain it becomes possible to neglect ω in front of Ω_{ag}, and β becomes proportional to the symmetrized tensor product $\boldsymbol{\delta} \otimes \mathbf{m} \otimes \mathbf{m}$, with only four components as required by Kleinman symmetry:

$$\beta_{ijk}(\omega \approx 0) \approx (1/2\hbar^2\Omega_{ag}^2)(\delta_i m_j m_k + m_i \delta_j m_k + m_i m_j \delta_k)$$

C. Nonlinear Optical Effects in Organic Molecules

1. Introduction

It was shown in Section IV.A that the electronic structure and the optical spectra of conjugated molecules are very different from those of saturated molecules. Naturally, the same trends apply to the linear as well as to the nonlinear optical constants.

It has been recognized for a long time that the linear polarizabilities of σ bonds are almost independent of the molecule with which they participate. In fact, the linear polarizability of saturated molecules is well accounted for by simple additivity rules, as opposed to those of unsaturated molecules, which increase at a faster rate than the number of conjugated electrons (Lefevre and Lefevre, 1965). The contrast between conjugated and saturated molecules is even more apparent in their nonlinear optical behavior. In this section we summarize the principal experimental results obtained for the quadratic response of organic molecules. We also discuss simple models that describe well the basic mechanisms of the origin of these nonlinearities and account quite accurately for the experimental results.

2. Saturated Molecules

The linear and cubic polarizabilities depend on the symmetric part of the molecular wave function, and far from resonance they are directly related to the even-order moments of the electronic charge distribution. Thus, they are not very sensitive to small changes induced by their local environments on strong and well-localized σ bonds. It has been found that for centrosymmetric (Hermann, 1973; Hermann *et al.*, 1973; Hermann and Ducuing, 1974) as well as for noncentrosymmetric molecules (Bigio and Ward, 1974; Ward and Bigio, 1975; Ward and Miller, 1979), α and γ are well described by the sum of bond polarizabilities. It should be mentioned that since the measurements were performed in gases or in liquids, strictly speaking these results apply only to the scalar part, $\alpha^{(0)}$ and $\gamma^{(0)}$. Very little is known about the polarizability anisotropy, i.e., $\alpha^{(2)}$, $\gamma^{(2)}$, and $\gamma^{(4)}$.

The quadratic polarizability β, which depends on the asymmetric part of the wave function and on odd-order moments of the electronic distribution, is on the other hand very sensitive to small deviations from centrosymmetry (Hauchecorne *et al.*, 1971; Finn and Ward, 1974; Bigio and Ward, 1974; Ward and Bigio, 1975; Ward and Miller, 1979). The magnitude of the β of saturated molecules is small, typically in the 10^{-31} to 10^{-30} esu range; thus they are not very useful for applications. However, accurate measurements on small molecules are very important for fundamental purposes, for two reasons. First, they provide the necessary information to check elaborate *ab initio* numerical calculations that can only be performed on small molecules. In addition, they are very useful for calibration of parameters utilized in semiempirical models. In this respect it is worth noting that good agreements between theory and experiments have been obtained in the case of γ and α (Amos and Crispin 1975; Flytzanis, 1975).

The bond additivity scheme works remarkably well for molecules such as CH_3OH or CH_3CH_2OH, but it fails to describe other groups of molecules such as the fluorinated methanes. This is in qualitative agreement with the discussion of Section IV.A. In the heteroatomic bonds, the inductive effect induces a flow of charge toward the most electronegative atom, producing a noncentrosymmetric charge distribution. For small electronegativity difference, the effect does not propagate to other bonds and the additivity scheme works reasonably well; for large electronegativity differences, the inductive effects propagate to surrounding bonds, which can no longer be considered independent.

So far, *ab initio* theories have failed to reproduce the absolute magnitude of β (Bartlett and Purvis, 1979; Purvis and Bartlett, 1981), but semiempirical models have been able to verify trends among related molecules (Hush and Williams, 1972; Dewar *et al.*, 1976; Zyss, 1979a,b,c).

3. Conjugated Molecules

For these molecules, the large delocalization of the π electrons induces cooperative effects and produces very large nonlinear responses. In centrosymmetric molecules the π-electron system by itself presents huge cubic polarizabilities with a strong superlinear dependence in N, the number of conjugated electrons. $\alpha \approx N^3$ and $\gamma \approx N^5$ (Hermann *et al.*, 1973; Hermann and Ducuing, 1974).

However, most often, nonsubstituted molecules are centrosymmetric and the contribution of the delocalized electrons to β vanishes. The quadratic polarizability of conjugated molecules thus arises from the distortion of the π-electron distribution caused by one or more substituent groups. The magnitude and sign of the effects depend on both the nature and the number of substituents as well as on the length of the conjugated system.

Let us start with simple monosubstituted benzenes. They exhibit quadratic polarizabilities more than one order of magnitude larger than saturated molecules of similar dimensions (Levine and Bethea, 1974a,b; Oudar *et al.*, 1977). More importantly, it is found that the sign of the nonlinearity is directly related to the nature (donor or acceptor) of the substituent group (Oudar and Chemla, 1975; Levine, 1975; Chemla *et al.*, 1975). Thus, β is negative for acceptor groups that induce a flow of electrons from the benzene ring toward the substituent group (NO_2, C≡N, etc.) and it is positive for donor groups [OH, NH_2, $(CH_3)_2N$, etc.] that produce an opposite charge flow.

This behavior is well explained by the equivalent internal field model (Oudar and Chemla, 1975; Chemla *et al.*, 1975). In centrosymmetric benzene, the π electrons when polarized by an applied electric field E acquire a dipole moment given by

$$p = \alpha_\pi E + \gamma_\pi E^3 + \cdots \tag{84a}$$

whereas the dipole of the noncentrosymmetric conjugated electrons of the substituted molecule X—C_6H_5 is

$$p = \Delta\mu_X + \alpha E + \beta E^2 + \cdots \tag{84b}$$

where $\Delta\mu_X$ is the mesomeric moment of the substituent group. Since the main effect of the substitution on the π electrons is to create a dipole $\Delta\mu_X$, we can consider that this is equivalent to a static "internal" field E_X such that $E_X = \Delta\mu_X/\alpha_\pi$. The dipole moment of the π electrons in the presence of a laser field E_l and of the substituent group X is given by Eq. (84a) with $E = E_l + E_X$ so that the quadratic polarizability is, after identification of terms of the same order in E,

$$\beta = 3(\gamma_\pi/\alpha_\pi)\,\Delta\mu_X \tag{85}$$

This relation shows that the sign of β is governed by that of $\Delta\mu_X$ and that its magnitude scales both to the strength of the substituent and to the length of the molecule ($\gamma_\pi/\alpha_\pi \approx N^2$).

Good agreement is found with experiment, as shown in Fig. 2 where both experimental and theoretical β_π are plotted against the corresponding $\Delta\mu_X$. The experimental slope, 1.6×10^{-12}, is also in good agreement with the value of $3\gamma_\pi/\alpha_\pi = 2.4 \times 10^{-12}$ for benzene.

The magnitude of the mesomeric moment calculated using the procedure described in Section IV.A entails some uncertainties. More precise evaluation can be deduced from the energy shifts induced by the substituent on the absorption spectrum of the benzene molecule. The determination of experimental spectra together with an analysis sketched similar to that developed in Section IV.A may be used to determine the strength of the perturbation caused

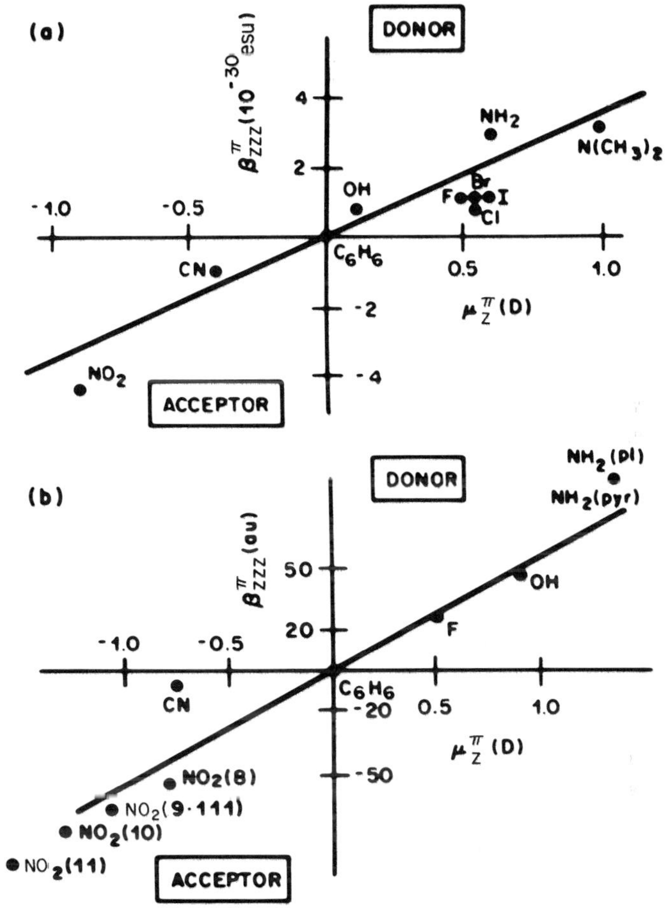

Fig. 2. Relation between the second-order susceptibility of monosubstituted benzene molecules and the moment induced by the substituent group. (a) Experimental versus experimental mesomeric moment (after Oudar and le Person, 1975). (b) Theoretical β^{π}_{zzz} against theoretical μ^{π}_{z} (after Zyss, 1979a,b,c).

by the substituent (Levine, 1975). The effect of the length of the conjugated system has been investigated for substituted stilbenes. The magnitude of β is about one order larger than for benzene, in good agreement with the increased $\gamma_{\pi}/\alpha_{\pi}$, and the right correlation between the sign and nature of the substituent is shown (Oudar, 1977).

A natural step toward the optimization of molecular nonlinearities is to utilize two substituents of opposite nature on the same conjugated system in order to add their effects. Furthermore, the study of disubstituted benzene has an additional interest. Because a number of these molecules crystallize in noncentrosymmetric structures, the direct determination of the molecular polarizabilities is very important for a better understanding of the relations between microscopic and macroscopic responses (see Section V).

As expected, for weakly interacting substituents the quadratic polarizabilities are satisfactorily described by (tensorial) additivity laws (Oudar *et al.*, 1977). Conversely, for strong donor–acceptor interaction the additivity assumption falls short of the experimental values and the discrepancy increases with the degree of conjugation induced by the relative location of the substituents. This is clearly shown in Table II, where experimental data referring to the three (ortho, *o*; meta, *m*; para, *p*) nitroanilines, $NO_2—C_6H_4—NH_2$, are compared to those calculated by additivity rules using the experimental data for nitrobenzene ($NO_2—C_6H_5$) and aniline ($NH_2—C_6H_5$). The origin of the discrepancy resides in the strong intramolecular charge transfer between the donor and the acceptor group that was mentioned in Section IV.A. In the visible range, the optical spectra of these molecules are dominated by an intense charge-transfer transition $n_a \rightarrow \pi_a^*$. In the particular case of nitroanilines, the conjugation depends strongly on the relative positions of the substituents (Lutskii and Gorokhova, 1967). The charge transfer contributes substantially to the first excited state, which has a dipole moment much larger than that of the ground state. The new transition occurs in the range of 3–7 eV with oscillator strengths $f_p \approx 0.51$, $f_m \approx 0.08$, and $f_o \approx 0.14$. In order to account for the contributions other than those related to charge transfer, it has been proposed to decompose the quadratic polarizability into two parts,

$$\beta = \beta_{add} + \beta_{ct} \tag{86}$$

where β_{add} accounts for all transitions except that corresponding to the intramolecular charge transfer and is well described by the sum of the effects caused by individual substituents. The contribution of the intramolecular charge transfer is that of a two-level system formed by the ground state and the

TABLE II

Comparison of the Theoretical Values with Experimental Results[a]

Substance	β_{add} (10^{-30} esu)	W (eV)	f	$\Delta\mu_{ex}$ (D)	β_{ct} (10^{-30} esu)	β_T (10^{-30} esu)	β_x(exp) (10^{-30} esu)
o-NA	2	3.51	0.14	10	10.9	12.7	10.2
m-NA	3.1	3.67	0.08	12.1	4	7	6
p-Na	3.5	3.89	0.51	12.3	19.6	23	34.5
2,4-DNA	2.9				14.9	18	21

[a] Terms: β_{add} is the second-order polarizability calculated from nitrobenzene and aniline; W and f are the energy and oscillator strength of the first excited state with strong charge-transfer contribution; $\Delta\mu_{ex}$ is the difference between excited-state and ground-state dipole moments; β_{ct} is the second-order polarizability induced by charge transfer; and β_T and β_x(exp) are, respectively, theoretical and experimental values.

first excited state. It can be calculated from Eq. (83). Indeed, the nonlinearity duc to the intramolecular charge transfer follows all physical trends discussed in Sections IV.A and IV.B, i.e., proportionality to the oscillator strength of the transition and to the excited-state/ground-state dipole difference and resonance behavior at the fundamental and harmonic frequencies. Excellent quantitative agreement with experiment is obtained when the intramolecular charge transfer is accounted for as shown in Table II.

A simple picture of the physical mechanisms involved in the nonlinearity on the charge transfer can be sketched in terms of Mulliken resonance structures as illustrated in Fig. 3. When an oscillating field is applied to the π electrons of a donor–acceptor molecule, the flow of charge is inhibited in one direction

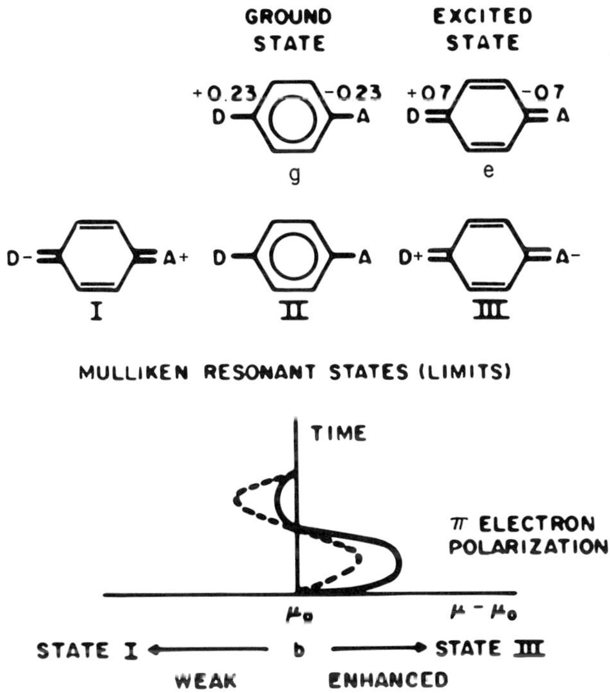

Fig. 3. Simple intuitive model to account for the nonlinearity of conjugated charge-transfer systems. States I, II, and III correspond to extreme situations where one electron is transferred from the acceptor to the donor (case I, extremely unlikely), a neutral state (case II), and a maximal charge transfer state (case III). The actual ground and excited states are essentially combinations of II and III, with little weight on I. The figures on top of D and A correspond to a rough estimate of charge modifications owing to intramolecular interactions as compared to fully screened atoms in a molecule. When polarized parallel to the substitution axis by an optical field, the electronic polarization response will be asymmetrically distorted (solid lines) under the cooperative influence of the donor and acceptor groups, as compared to the symmetric response of an unsubstituted benzene to the electron system (dotted lines). This distortion gives rise to harmonic frequencies in the field radiated by the molecular dipole oscillations.

and enhanced in the other. The charge-transfer molecule acts like an "optical diode" with a very asymmetric response, i.e., a large nonlinearity.

The two-level system description of charge-transfer nonlinearity works remarkably well for even longer molecules. In this case, because the energy of the first excited state decreases with the length of the molecule, the charge-transfer nonlinearity plays a more dominant role. This is clearly demonstrated in Fig. 4 (Oudar, 1977), where the nonlinearity β normalized to the oscillator strength f_{ag} and the dipole moment difference δ are plotted versus the photon energy. The agreement with the theoretical dispersion predicted by Eq. (83)—i.e., the solid line—is excellent and improves as the length of the molecule increases.

This general behavior has been confirmed for a number of charge-transfer molecules, and it can lead to huge nonlinearities (Dulcic and Flytzanis, 1978). However, this is paid at the expense of the transparency range of the molecule.

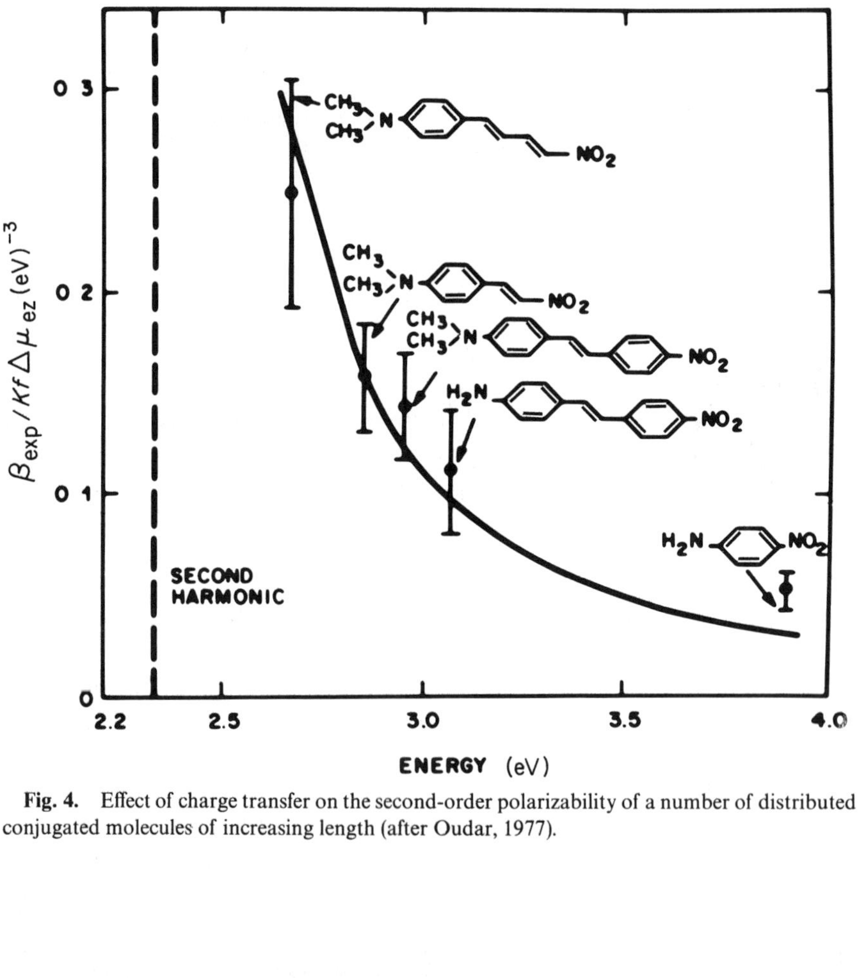

Fig. 4. Effect of charge transfer on the second-order polarizability of a number of distributed conjugated molecules of increasing length (after Oudar, 1977).

TABLE III

Magnitudes of the Nonlinearities of Various Molecules showing Four Steps[a]

Molecule	β (10^{-30} esu)	μ_0 (D)	λ (μm)	Phase	Gap (eV)	Reference
Small Molecules						
$CO(NH_2)_2$	0.45	4.56	1.06	Water	6.2	b
CH_3NO_2	0.06	3.4	1.06	Liquid	4.7	c
Monosubstituted benzenes						
Φ—NO_2	1.11	4.2	1.06	Hexane	4.77	d
Φ—NO_2	2	4.1	1.06	Benzene	4.77	c
Φ—NO_2	2.3	4.2	1.06	Liquid	4.77	e
Charge-transfer distributed benzenes						
(*o*)NO_2—Φ—NH_2	10.2	4.3	1.06	Methanol	3.51	f
(*m*)NO_2—Φ—NH_2	6	4.9	1.06	Methanol	3.67	f
(*p*)NO_2—Φ—NH_2	34.5	6.2	1.06	Methanol	3.89	f
(*p*)NO_2—Φ—NH_2	21.1	7.2	1.06	Molten	3.89	g
Extended charge-transfer molecules						
$N(CH_3)_2$—Φ—C_4H_4—NO_2	650	8.8	1.06	Acetone	2.67	h
CH_3—Φ^*—C_2H_2—Φ—0	1000	8	1.89	DMSO	2.17	i

[a] Terms: Φ stands for an aromatic ring, Φ^* for a nitrogen-containing heterocycle. Dispersion of experimental data for various compounds results from solvent interactions and difference in local field corrections.

[b] Ledoux and Zyss (1982).

[c] Levine and Bethea (1974a and b).

[d] Singer and Garito (1981).

[e] Oudar and Le Person (1975).

[f] Oudar and Chemla (1977).

[g] Levine and Bethea (1976).

[h] Oudar (1977).

[i] Dulcic and Flytzanis (1978).

The rather narrow transparency range of the long charge-transfer molecules may be prohibitive for applications to frequency generation but is less important for applications to electrooptic modulation, which only requires good transparency at the carrier light frequency. To conclude this section, we have collected in Table III the important optical and electrical parameters for a set of molecules, with a scale over four orders of magnitude to show the correlations between molecule structures and nonlinearities.

D. Finite Field Theory of Molecular Susceptibilities

Reliable theoretical predictions of the magnitude of molecular susceptibilities may provide useful guidelines for the synthetic chemist and may orient his work toward previously unexplored species of potential interest. However, the theoretical chemist is confronted with a difficult dilemma when trying to answer such problems: he may restrict his interest to smaller molecules for which full *ab initio* correlated calculations can be undertaken but which have little potential for efficient nonlinear interactions. He may be willing, on the other hand to trade off the absolute accuracy of results for the possibility of dealing with larger molecules of greater interest and then can be satisfied with the use of semiempirical methods, provided tendencies are well accounted for. If a high computational throughput is required in addition, as when scanning portions of the enormous number of unexplored yet potentially interesting "on-the-shelf" available synthetic material, the latter strategy is definitely appropriate. Two approaches are possible: one, refered to as the "sum-over-state" (S-O-S) approach, is dealt with at length in other chapters in this book and will not be detailed here. Starting from full perturbative expansions of molecular susceptibilities in terms of molecular eigenstates, its main purpose is to feed them with accurately computed excited-state energies and transition dipole moments. This method has proved successful quantitatively and is attractive as it allows also for the evaluation of the respective contributions of molecular excited states to the total molecular susceptibility (Morrell and Albrecht, 1979; Lalama and Garito, 1979). Precise calculation of the excited states requires interaction of configurations involving monomolecular ground and excited states, and the number of terms to be computed in the perturbation expansion may end up being critical. However, the theoretical accuracy of the method together with the increasing availability of "number-crunching" computing systems are highly encouraging toward further work in that direction. Another possibility is the use of the so-called "finite-field" perturbation approach, to be detailed hereafter. This approach was developed by Cohen and Roothan (1965) for the calculation of atomic polarizabilities and subsequently extended to that of "medium-scale" molecules (one or two benzene rings) with first-order hyperpolarizabilities (Zyss, 1979a,b,c). The principle of the methods is to diagonalize a molecular Hamiltonian including a static-field dipolar perturbation. In the following, we will refer to a Hartree–Fock (H–F) Hamiltonian, which may be either semiempirical, such as for example in the INDO approximation (Pople and Beveridge, 1970), or *ab initio*. Calling H_0 the unperturbed one-electron H–F Hamiltonian and H the perturbed one, $H = H_0 + \mathbf{r} \cdot \mathbf{E}$, where $\mathbf{r}$ is the one-electron dipolar operator and $\mathbf{E}$ the static perturbing field. Diagonalization of H is undertaken in a wave-function space spanned by atomic orbitals (Slater-type orbitals in the INDO method, or preferably gaussian type orbitals in *ab initio* methods). In the INDO approximations, matrix elements of the perturbation operator $\langle \mu | \mathbf{r} \cdot \mathbf{E} | \lambda \rangle$ are to be considered only for Slater orbitals

$|\mu\rangle$ and $|\lambda\rangle$ attached to the same nucleus. The Hartree–Fock Roothan matrix, which corresponds to operator H in the LCAO formalism, is written $F = F_0 + F'(\mathbf{E})$ where F_0 corresponds to H_0 and $F'(\mathbf{E})\mathbf{r} \cdot \mathbf{E}$. F' has a block diagonal form, each block corresponding to a different atomic nucleus.

$$F'(\mathbf{E}) = \begin{bmatrix} [A] & & & 0 \\ & [B] & & \\ & & [C] & \\ 0 & & & \ddots \end{bmatrix} \tag{87}$$

Full expressions of blocks $[A]$ for outer-valence-shell electrons are to be found in Tables IV and V.

The self-consistent field-dependent operator $F(\mathbf{E})$ has the form $F(\mathbf{E}) = F_s(\mathbf{E}) + F_2$ and $F_1(\mathbf{E}) = \mathbf{F'E} + \mathbf{F}''$, where one- and two-electron contributions are being singled out. $F'(\mathbf{E})$ has already been defined, while $\mathbf{F}''$ is not explicitly field-dependent and corresponds to the kinetic energy operator and electron–nucleus interactions. The two-electron nonlinear contribution to F is given by

$$\langle \mu | F_2 | \nu \rangle = \sum_{\lambda,\theta} [(\mu\nu|\lambda\theta) - \tfrac{1}{2}(\mu\lambda|\nu\theta)] P_{\lambda\theta} \tag{88}$$

where $(\mu\nu|\lambda\theta)$ and $(\mu\lambda|\nu\theta)$ contribute respectively to Coulomb and exchange operators and are given by

$$(\mu\nu|\lambda\theta) = \iint \mu(1)\nu(1) r_{12}^{-1} \lambda(2)\theta(2)\, d\tau_1\, d\tau_2 \tag{89}$$

$$(\mu\lambda|\nu\theta) = \iint \mu(1)\lambda(1) r_{12}^{-1} \nu(2)\theta(2)\, d\tau_1\, d\tau_2 \tag{90}$$

The averaging over orbitals $|\lambda\rangle$ and $|\theta\rangle$ is the LCAO traduction of the mean-field Hartree–Fock hypothesis. Diagonalization of the nonlinear Fock matrix is performed iteratively until self-consistency is achieved. Calling N the current step in the iterative procedure, $F_{(N)}(\mathbf{E}) = F_1(\mathbf{E}) + F_2^{(N-1)}$ where $F_2^{(N-1)}$ is defined from the density matrix $P_{\lambda}^{(N-1)}$ previously calculated at step $N - 1$. The total field-perturbed dipole moment obtained after implementation of INDO approximations takes on the following form:

$$\boldsymbol{\mu}(\mathbf{E}) = \sum_A \left\{ [Z_A - P_{AA}(\mathbf{E})]\mathbf{x}_A - \sum_{\mu \neq \nu} P_{\mu\nu}(\mathbf{E}) \langle \mu | \mathbf{x} | \nu \rangle \right\} \tag{91}$$

$P_{\mu\lambda}(\mathbf{E})$ is the field-dependent matrix representation in the atomic-orbital basis set of the density operator $P = 2\,\Sigma_i|\Psi_i\rangle\langle\Psi_i|$, where one-electron molecular Hartree–Fock orbitals are defined by $|\Psi_i\rangle = \Sigma_\mu c_{i\mu}|\mu\rangle$ (summation over i is over all occupied molecular orbitals, with the factor of 2 accounting for spin degeneracy). The expression of $P_{\mu\nu}$ matrix coefficients is $P_{\mu\nu} = 2\,\Sigma_i c_{i\mu}c_{i\nu}$, while $P_{\mathrm{AA}} = \Sigma_{\mathrm{A}} P_{\mu\nu}$ (summation over all atomic orbitals attached to nucleus A), charge neutrality demands that $\Sigma_{\mathrm{A}} Z_{\mathrm{A}} = \Sigma_{\mathrm{A}} P_{\mathrm{AA}} = 2N$, where $2N$ is the

TABLE IV

Perturbation Block Matrix [A] for Atom A in the Second Row of the Mendeleiev Table[a]

	s	p_x	p_y	p_z
s	$\mathbf{E}\cdot\mathbf{x}_{\mathrm{A}}$	$\frac{5E_x}{\sqrt{3}\,Z_{\mathrm{A}}}$	$\frac{5E_y}{\sqrt{3}\,Z_{\mathrm{A}}}$	$\frac{5E_z}{\sqrt{3}\,Z_{\mathrm{A}}}$
p_x	$\frac{5E_x}{\sqrt{3}\,Z_{\mathrm{A}}}$	$\mathbf{E}\cdot\mathbf{x}_{\mathrm{A}}$	0	0
p_y	$\frac{5E_y}{\sqrt{3}\,Z_{\mathrm{A}}}$	0	$\mathbf{E}\cdot\mathbf{x}_{\mathrm{A}}$	0
p_z	$\frac{5E_z}{\sqrt{3}\,Z_{\mathrm{A}}}$	0	0	$\mathbf{E}\cdot\mathbf{x}_{\mathrm{A}}$

[a] χ_{A} is the coordinate vector of atom A (charge z_{A}).

TABLE V

Perturbation Block Matrix [A] for an Atom in the Third Row of the Mendeleiev Table: $A = 0.39342/Z_{\mathrm{A}}$, $B = 7/2\sqrt{5}\,Z_{\mathrm{A}}$, and $C = \mathbf{E}\cdot\mathbf{x}_{\mathrm{A}}$

STO[a]	3s	p_x	p_y	p_z	d_{z^2}	d_{xz}	d_{yz}	$d_{x^2-y^2}$	d_{xy}
3s	C	xA	yA	zA	0	0	0	0	0
p_x	xA	C	0	0	$\frac{-x}{\sqrt{3}}B$	zB	0	xB	yB
p_y	yA	0	C	0	$\frac{-y}{\sqrt{3}}B$		zB	yB	xB
p_z	zA	0	0	C	$\frac{2}{\sqrt{3}}zB$	xB	yB	0	0
d_{z^2}	0	$\frac{-x}{\sqrt{3}}B$	$\frac{-y}{\sqrt{3}}B$	$\frac{2}{\sqrt{3}}zB$	C	0	0	0	0
d_{xz}	0	zB	0	xB	0	C	0	0	0
d_{yz}	0	0	zB	yD	0	0	C	0	0
$d_{x^2-y^2}$	0	xB	yB	0	0	0	0	C	0
d_{xy}	0	yB	xB	0	0	0	0	0	C

[a] STO: Slater-type orbitals.

total number of valence electrons. However, when singled out for each individual atom, $Z_A - P_{AA}$ allows one to compare the charge surrounding a connected atom in its actual molecular environment to that ($Z_A = P_{AA}$) of the same isolated atom. The field-dependent coefficients $c_{i\mu}(\mathbf{E})$ are obtained by the iterative diagonalization procedure referred to before, and (hyper)polarizabilities of increasing order may be, in principle, obtained by differentiation of $\boldsymbol{\mu}(\mathbf{E})$ with respect to $\mathbf{E}$, as allowed by the Hellman–Feynman theorem (Pople *et al.*, 1968):

$$\alpha_{ij} = \left.\frac{\partial \mu_i}{\partial E_j}\right|_{E=0}, \qquad \beta_{ijk} = \frac{1}{2}\left.\frac{\partial^2 \mu_i}{\partial E_j \partial E_k}\right|_{E=0}, \qquad \gamma_{ijkl} = \frac{1}{3}\left.\frac{\partial^3 \mu_i}{\partial E_j \partial E_k \partial E_l}\right|_{E=0} \tag{92}$$

Using Eq. (91) for the dipole moment, then β_{ijk}, for example, can be expressed in terms of second-order derivatives of the density matrix with respect to the field:

$$\beta_{ijk} = -\frac{1}{2}\sum_{A}\left.\frac{\partial^2 P_{AA}}{\partial E_j \partial E_k}\right|_{E=0} x_A^i - \sum_{A}\sum_{\mu \neq \nu}\left.\frac{\partial^2 P_{\mu\nu}}{\partial E_j \partial E_k}\right|_{E=0} \langle \mu | x^i | \nu \rangle \tag{93}$$

Convergence of the polarizabilities is best ensured by symmetric difference expressions of the dipole derivatives, such as

$$\alpha_{xx} = \frac{\mu_x(E, O, O) - \mu_x(-E, O, O)}{2E} \tag{94}$$

$$\beta_{xxx} = \frac{\mu_x(E, O, O) + \mu_x(E, O, O) - 2\mu_x(O, O, O)}{2E^2} \tag{95}$$

$$\gamma_{xxxx} = \frac{\mu_x(3E, O, O) - \mu_x(-3E, O, O) - 3[\mu_x(E, O, O) - \mu_x(-E, O, O)]}{48E^3} \tag{96}$$

For benzene-like molecules, the order of magnitude of the field being used is 10^{-2} a.u. or 10^7 V cm^{-1}. Calling $\Delta\mu$ the mean fluctuations of the dipole moment at convergence, the error on the nth-order polarizability will be proportional to $\Delta\mu$ and inversely proportional to the nth power of the field E. For the cubic polarizability term γ_{xxxx}, the error is given by $\Delta\gamma_{xxxx} = \Delta\mu/6E^3$, while for β_{zzz} it is $\Delta\beta_{zzz} = 2\,\Delta\mu/E^2$. In practice, the validity of the finite field approach for cubic susceptibilities is limited by the E^{-3} dependence of $\Delta\gamma$, while the method has proved satisfactory for quadratic β susceptibilities. Interesting confirmations and insights on the respective contributions of σ and π electrons to the molecular susceptibilities as well as on the influence of substitution on benzene rings have been derived based on the previous model. Unlike previous models where only π electrons are considered, or where the influence of σ electrons is accounted for by embedding π electrons in an effective medium of higher dielectric constant (Schweig, 1967), the present model deals on an equal footing with both π and σ electrons, which allows for objective comparison of their contributions. It is possible (Zyss, 1979b) to quantitatively distinguish between π-related and σ-related contributions to the overall susceptibility of planar molecules.

Although symmetry considerations straightforwardly show that Fock matrix coefficients are either purely π ($F_{\mu^\pi\nu^\pi}$ where $|\mu^\pi\rangle$ and $|\nu^\pi\rangle$ are π-type atomic orbitals) or purely σ ($F_{\mu^\sigma\nu^\sigma}$ where $|\mu^\sigma\rangle$ and $|\nu^\sigma\rangle$ are σ-type atomic orbitals), mixed $F_{\mu^\pi\nu^\sigma}$ terms being ruled out, self-consistent interaction of σ and π electrons is accounted for: at each step N, both π and σ Fock matrices can be shown to depend on both π and σ density matrixes computed at step $N-1$,

$$F^{(N)} = \begin{bmatrix} \pi \text{ STOs} & \sigma \text{ STOs} \\ |p_x^{\mathrm{B}}\rangle\,|p_x^{\mathrm{A}}\rangle\cdots & |s_x^{\mathrm{A}}\rangle\,|p_y^{\mathrm{A}}\rangle\,|p_z^{\mathrm{A}}\rangle\,|s^{\mathrm{B}}\rangle\cdots \\ F_{\pi\pi}^{(N)}(P_{\pi\pi}^{(N-1)}, P_{\sigma\sigma}^{(N-1)}) & 0 \\ 0 & F_{\sigma\sigma}^{(N)}(P_{\pi\pi}^{(N-1)}, P_{\sigma\sigma}^{(N-1)}) \end{bmatrix} \tag{97}$$

where x refers to a direction perpendicular to the π electron plane. The π and σ contributions to the dipole moments and subsequently to its field derivatives can be defined, provided a similar partition of nuclear charges is introduced to independently ensure π and σ charge neutrality:

$$\langle\mu\rangle = |\mu^\pi\rangle\,|\mu^\sigma\rangle \tag{98}$$

$$|\mu^\pi\rangle = \sum_{\mathrm{A}} Z_{\mathrm{A}}^\pi \mathbf{r}_{\mathrm{A}} - \sum_{j\pi} \langle\Psi_{j\pi}|\mathbf{r}|\Psi_{j\pi}\rangle \tag{99}$$

$$|\mu^\sigma\rangle = \sum_{\mathrm{A}} Z_{\mathrm{A}}^\sigma \mathbf{r}_{\mathrm{A}} - \sum_{j\sigma} \langle\Psi_{j\sigma}|\mathbf{r}|\Psi_{j\sigma}\rangle \tag{100}$$

The $|\Psi_{j\pi}\rangle$ (or $|\Psi_{j\sigma}\rangle$) are π- (or σ) type monomolecular H–F orbitals, Z_{A}^π (or

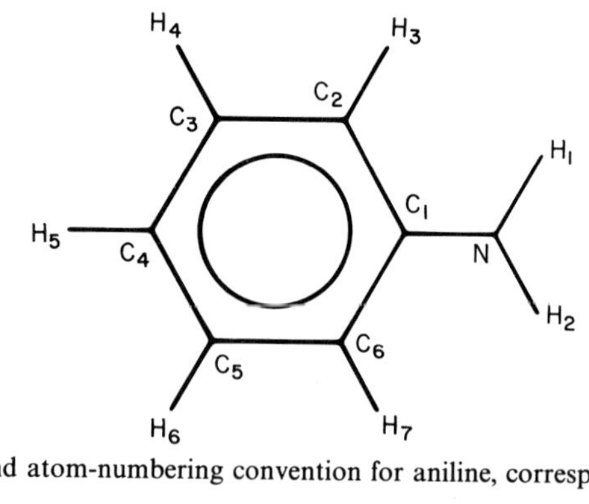

Fig. 5. Geometry and atom-numbering convention for aniline, corresponding to Table VI.

Z^σ_A) being defined as the nearest integer or half-integer to P^π_{AA} (or P^σ_{AA}). It turns out in practice that P^π_{AA} or P^σ_{AA} is close to corresponding Z^π_a or Z^σ_a values. Calling $E(x)$ the function associated with any real value x its closest integer or half-integer value, and remembering that if x and y are close to $E(x)$ and $E(y)$, then $E(x + y) = E(x) + E(y)$, charge neutrality is ensured by

$$\sum_A Z^\pi_A = \sum_A E(P^\pi_{AA}) = E(\sum P^\pi_{AA}) = E(n^\pi) = n^\pi \tag{101}$$

where n^π is the number of π electrons. Identical considerations apply to σ electrons. The point charge distribution for aniline is shown in Table VI, the corresponding geometry being displayed in Fig. 5. The π and σ contributions to molecular susceptibilities may then be defined as

$$\alpha_{ij} = \left.\frac{\partial\langle\mu^\sigma_i\rangle}{\partial E_j}\right|_{E=0} + \left.\frac{\partial\langle\mu^\pi_i\rangle}{\partial E_j}\right|_{E=0} = \alpha^\sigma_{ij} + \alpha^\pi_{ij} \tag{102}$$

$$\beta_{ijk} = \left.\frac{\partial^2\langle\mu^\sigma_i\rangle}{\partial E_j \partial E_k}\right|_{E=0} + \left.\frac{\partial^2\langle\mu^\pi_i\rangle}{\partial E_j \partial E_k}\right|_{E=0} = \beta^\sigma_{ijk} + \beta^\pi_{ijk} \tag{103}$$

Neglecting differential overlap, the susceptibility may be seen to depend on field derivatives of "one-center" density matrix terms, for example;

$$\beta^\pi_{ijk} = -\sum_A \left.\frac{\partial^2 P_{p^A_x p^A_x}(\mathbf{E})}{\partial E_j \partial E_k}\right|_{E=0} r^A_i \tag{104}$$

Figure 2 sums up some results on monosubstituted benzenes, performed along the previously detailed model. As was expected from earlier EFISH experiments and from the internal field description of the effect of substituents

TABLE VI

Point Charge Distribution for Aniline[a]

A	P_{AA}	P^π_{AA}	P^σ_{AA}	Z^π_A	Z^σ_A	Z_A
C_1	3.8217(+)	0.9467(+)	2.8757(+)	1	3	4
$C_2(C_6)$	4.0371(−)	1.071(−)	2.9661(+)	1	3	4
$C_3(C_5)$	3.9529(+)	0.9769(+)	2.976(+)	1	3	4
C_4	4.012(−)	1.0457(−)	2.9645(+)	1	3	4
N	5.2282(−)	1.9108(+)	3.3174(−)	2	3	5
$H_1(H_2)$	0.9275(+)	0	0.9275(+)	0	1	1
$H_3(H_7)$	1.0148(−)	0	1.0148(−)	0	1	1
$H_4(H_6)$	1.0258(−)	0	1.0251(−)	0	1	1
H_5	1.0234(−)	0	1.0234(−)	0	1	1

[a] Symbols (+) and (−) refer respectively to a defect or an excess of charge over the isolated atom.

on the benzene polarizabilities (Oudar and Chemla, 1975; Chemla *et al.*, 1975), the π contribution to β lines up satisfactorily when expressed in terms of the π dipole moment. Remembering that $\beta^{\pi}_{zzz} = 3\gamma^{\pi}_{zzz}\mu^{\pi}_{z}/\alpha^{\pi}_{zz}$ when following the assumptions of the internal field model and taking for α^{π}_{zz} of benzene the independently calculated value of 4×10^{-24} esu yields an average value of γ^{π}_{zzzz} for benzene of 1.56×10^{-36} esu, in satisfactory agreement with Chemla *et al.* (1975), who proposed 1.44×10^{-36} esu. In addition, the expected change of sign for both μ^{π}_{z} and β^{π}_{zzz} coefficients when moving from donor to acceptor substituents is well accounted for, which is not the case when applying simpler Hückel-model assumptions (Cojan, 1977).

Artificial variation of the electronic affinity of the $|2p_x\rangle$ orbitals of oxygen atoms apart from their actual value in nitrobenzene amounts to increasing or diminishing the acceptor strength of the nitro group. The resulting $\beta^{\pi}_{zzz}/\mu^{\pi}_{z}$ ratios for 8, 10, and 11 eV are shown in Fig. 2 to remain fairly constant, as expected from the internal field model. The relative amount of π versus σ contributions to quadratic susceptibilities β is shown in Table VII for a number of benzene donor–acceptor molecules defined in Fig. 6.

Of particular interest is the π (or σ) nonlinearity per π (or σ) electron $\|\beta^{\pi}\|/n^{\pi}$ (or $\|\beta^{\sigma}\|/n^{\sigma}$), which shows the higher nonlinearity, by one order of magnitude at least, of π electrons over that of σ electrons. The ratio $\|\beta^{\pi}\|/\|\beta^{\sigma}\|$ increases from 3.7 for *p*-nitroaniline to 7.8 for 1-*p*-aminophenyl-4-nitrobutadiene ($PNB-NO_2NH_2$), showing that π electrons are all the more prevailing over σ ones as the molecule is more conjugated.

The model developed so far has been dealing with static fields, hence static polarizabilities, while comparison with actual laser-driven nonlinearities requires the introduction of frequency dispersion. This can be done as follows for two-level quantum systems such as the intramolecular change-transfer

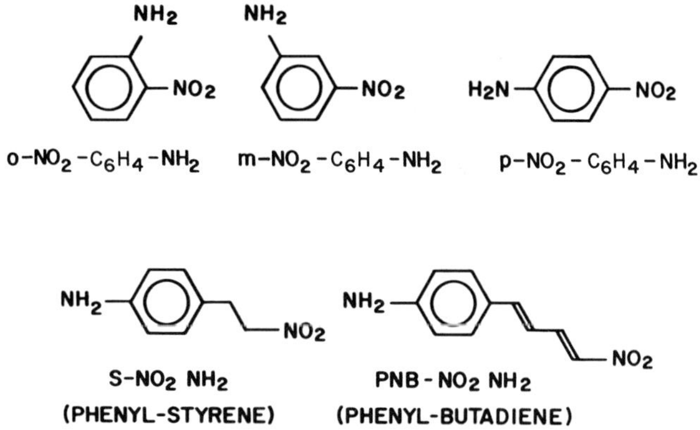

Fig. 6. Structures of various intramolecular charge-transfer conjugated systems for which systematic experimental and theoretical studies have been undertaken.

TABLE VII

Relative Amount of π versus σ Contributions to Second-Order Hyperpolarizability[a]

	o-NO_2—Φ—NH_2	o-NO_2—Φ—NH_2	p-NO_2—Φ—NH_2	S—NO_2NH_2	PNB—NO_2NH_2
$\|\beta^{\pi}\|/\|\beta^{\sigma}\|$	3.32	2.98	3.73	5.36	7.81
$\|\beta^{\pi}\|/n_{\pi}$	7.65	10.03	28	51	98
$\|\beta^{\sigma}\|/n_{\sigma}$	0.69	1.01	2.25	2.77	3.57
n	52	52	52	62	72
n_{π}	12	12	12	14	16
n_{σ}	40	40	40	48	56

[a] Terms: $\|\beta^{\pi}\|$ and $\|\beta^{\sigma}\|$ are the modules of the vector part of the second-order hyperpolarizability of π and σ electrons; n, n_{π}, and n_{σ} are the number of valence, π, and σ electrons of the molecules investigated, respectively. See Fig. 6 for the definitions of molecules referred to here.

molecules of interest in this chapter. Refering to such a system as A—C_6H_4—D, it was shown in Section IV.C that the total nonlinearity β of A—C_6H_4—D could be divided into an additive contribution β_{add} and a charge transfer contribution β_{ct}. This scheme will also apply to the corresponding computed static β (β^0 will refer to static values and β^ω to dispersed values):

$$\beta^0_{A—C_6H_4—D} = \beta^0_{A—C_6H_5} + \beta^0_{D—C_6H_5} + (\beta^0_{A—C_6H_4—D})_{ct} \tag{105}$$

Thus $(\beta^0_{A—\Phi—D})_{ct}$ can be evaluated from Eq. (105), the other β terms therein being calculated. Calling

$$F(W,\omega) = [1 - (2\hbar\omega/W)^2][1 - (\hbar\omega/W)^2]^{-1} \tag{106}$$

the dispersion term for the quadratic susceptibility of a two-level system as given by Eq. (83) and

$$\beta^\omega_{A—C_6H_4—D} = (\beta^0_{A—C_6H_5})F(W_{A—C_6H_5},\omega) + (\beta^0_{D—C_6H_5})F(W_{D—C_6H_5},\omega) + \cdots$$
$$+ (\beta^0_{A—C_6H_4—D} - \beta^0_{A—C_6H_5} - \beta^0_{D—C_6H_5})F(W_{A—C_6H_5—D},\omega) \tag{107}$$

where $W_{A—C_6H_5}$, $W_{D—C_6H_5}$, and $W_{A—C_6H_5—D}$ refer respectively to the two monosubstituted and to the distributed benzene charge-transfer excited-level energy. All static β polarizabilities in Eq. (107) can be calculated by using static finite field schemes, while each must be dispersed independently and subsequently recombined. Table VIII shows the effect of frequency dispersion in relative units with respect to nitrobenzene. Results show a clear increase of static β^0 when increasing the size of the molecule from nitroaniline on to a phenylbutadiene derivative. However, this tendency is much more pronounced for dispersed β^ω: resonant enhancement is then adding its effect to purely static chain-length effects and is stronger at a given laser wavelength for small-gap large systems than for more transparent smaller-size systems.

The introduction of environment-induced perturbations is of importance when dealing with molecular systems in condensed phases. Except for small molecules in the gaseous phase, molecules can hardly be considered as independent units, and the validity limit of an oriented gas description of molecular crystals is reached when considering a strongly hydrogen-bonded system such as that of urea (Zyss and Berthier, 1982). A fairly general model was subsequently developed that may possibly be applied to other interacting systems and refined from a computational point of view. The basic features of the model are the following. Considering for the sake of simplicity a one-dimensional crystal with one molecule per unit cell, such as in Fig. 7, the *in situ* multielectronic molecular Hamiltonian H of molecule M is given by

$$H = H_0 + V - \mathbf{P} \cdot \mathbf{E} \tag{108}$$

where H_0 is the isolated molecule Hamiltonian, V an interaction potential of

TABLE VIII

Static and Optical Theoretical Values for β Compared to Experimental Values[a]

	β^0_{INDO}	$\beta^\omega_{\text{INDO}}$	$\beta^\omega_{\text{exp}}$
Φ—NO_2	1	1	1
p-NO_2—Φ—NH_2	14	16	16
S—NO_2NH_2	34	60	100
PNB—NO_2NH_2	80	234	295

[a] From Oudar (1977), Oudar and Chemla (1977), and Oudar *et al.* (1977). Nitrobenzene is taken as the reference.

molecule M with its nearest neighbors M′ and M″, and $\mathbf{P} \cdot \mathbf{E}$ the dipolar perturbation term; V reflects the interaction of the nucleus and electronic charges of M′ and M″ on molecule M. We assume in the following that only the part of the charge localized around H'' and A', external atoms H-bonded to molecule M, significantly interacts with molecule M. This restriction can be readily lifted and the interaction on M of all neighboring atoms considered in

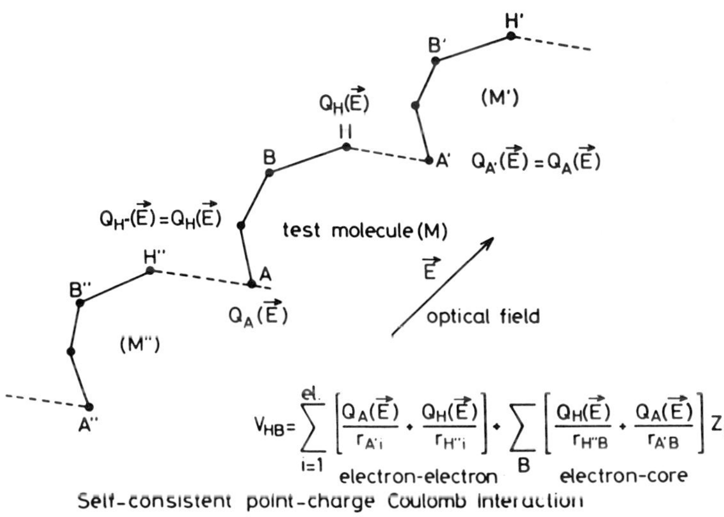

Fig. 7. Simplified one-dimensional crystal with one molecular unit per unit cell. The text molecule M (A ··· BH) is connected to two geometrically equivalent molecular units by hydrogen bonds B—H ··· A′ and B″—H″ ··· A. Whether isolated, or in the presence of a uniform field E, all molecules have identical charge distributions, owing to the translational invariance. As a consequence, the perturbating charges surrounding the text molecule (M) located around A′ and H″ are the same as the charges located around A and H, respectively.

a more general situation. The expression of V is

$$V = \sum_{i=1}^{N} \left(\frac{Q_{A'}(\mathbf{E})}{r_{A'i}} + \frac{Q_{H''}(E)}{r_{H''i}} \right) + \sum_{B=1}^{P} Z_B \left(\frac{Q_{A'}(\mathbf{E})}{r_{A'B}} + \frac{Q_{H''}(E)}{r_{H''B}} \right) \tag{109}$$

where the first summation refers to the N electrons of M and the second to its P atomic nuclei. The translational invariance of the crystal lattice applies to localized charges, so that

$$Q_{A'}(\mathbf{E}) = Q_A(\mathbf{E}) \qquad \text{and} \qquad Q_{H''}(\mathbf{E}) = Q_H(\mathbf{E}) \tag{110}$$

with the resulting expression for V, now called V^{MCF} (for mutually consistent field),

$$V^{\mathrm{MCF}} = Q_A(\mathbf{E}) \left[\sum_{i=1}^{N} \frac{1}{r_{A'i}} + \sum_{B=1}^{P} \frac{Z_B}{r_{A'B}} \right] + Q_H(\mathbf{E}) \left[\sum_{i=1}^{N} \frac{1}{r_{H''i}} + \sum_{B=1}^{P} \frac{Z_B}{r_{H''B}} \right] \tag{111}$$

The procedure for diagonalizing Hamiltonian H will then proceed iteratively as follows, calling N the current step: $Q_A^{N-1}(\mathbf{E})$ and $Q_H^{N-1}(\mathbf{E})$ result from an appropriate localization procedure, such as derived from the Mulliken population analysis (Mulliken, 1955), performed on molecular electronic wave functions calculated at step $N - 1$. These values are used to compute the matrix element of the interaction potential at step N and proceed to the diagonalization of $H^{(N)}$. At convergence, the interacting molecule susceptibilities are obtained by derivation of the dipole moment. An "effective" oriented gas description of the crystal is then applied, starting from interacting molecular units rather than from isolated ones. In the case of anisotropic crystals, corresponding molecules within the same unit cell will in general sustain different electronic charges, as the translational invariance argument clearly does not apply any longer within the same cell. The different relative orientations of the molecules with respect to an externally applied field will result in different polarizations and add to the intra–unit-cell electronic charge anisotropy. This requires some modification on the previously discussed linear crystal model. Figure 8 displays the case of a one-dimensional crystal along axis Z with two molecules per unit cell. It is clear from Fig. 8 that atoms A_1 and A_2 experience different surroundings so that Q_{A_2} differs from Q_{A_1} even in the absence of an external field (intermolecular distances R' and R are chosen to be different). Calling V_{M_1} and V_{M_2} the mutually consistent interaction potentials of molecules M_1 and M_2 in the presence of external field $\mathbf{E}$,

$$V_{M_1} = Q_{H_2'}(\mathbf{E}) \left[\sum_i \frac{1}{r_{H_2'i}} + \sum_{B_1} \frac{1}{r_{H_2'B_1}} \right] + Q_{A_2}(\mathbf{E}) \left[\sum_i^{1} \frac{1}{r_{A_2i}} + \sum_{B_1} \frac{1}{r_{A_2B_1}} \right] \tag{112}$$

$$V_{M_2} = Q_{H_1}(\mathbf{E}) \left[\sum_i \frac{1}{r_{H_1i}} + \sum_{B_2} \frac{1}{r_{H_1B_2}} \right] + Q_{A_1'}(\mathbf{E}) \left[\sum_i \frac{1}{r_{A_1'i}} + \sum_{B_2} \frac{1}{r_{A_1'B_2}} \right] \tag{113}$$

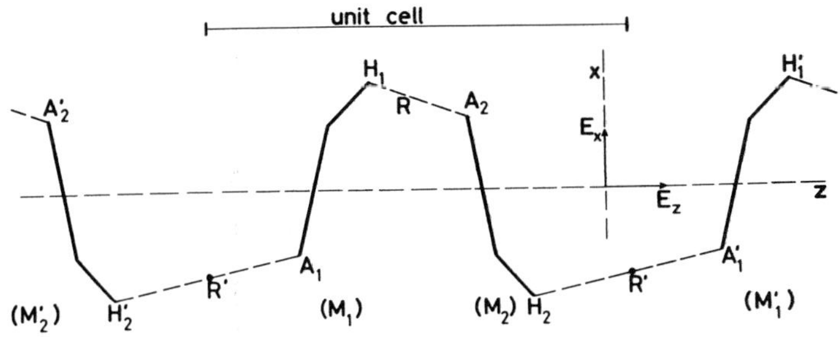

Fig. 8. One-dimensional crystal with two molecules per unit cell. The two molecules are symmetric with respect to the binary axis z and are located in different environments.

Translational invariance remains valid from cell to cell, if not within the same cell, to yield

$$Q_{H_2'}(\mathbf{E}) = Q_{H_2}(\mathbf{E}) \tag{114}$$

$$Q_{A_1'}(\mathbf{E}) = Q_{A_1}(\mathbf{E}) \tag{115}$$

The interacting Hamiltonians $H^{(N)}_{M_1}$ and $H^{(N)}_{M_2}$ must then be diagonalized iteratively and their eigenfunctions and resulting relevant localized charges used to cross-initialize the following step. It was proposed by Otto (1978) to use the complete diffuse coulombic potential to account for external molecular reactions, which does not require point-charge localization approximations.

Table IX shows considerable differences between charge distributions of isolated urea and H-bonded urea both in the presence and in the absence of field along the symmetry axis z of the molecule.

The convergence of dipole and hyperpolarizability coefficients at different interaction steps is reported in Table X and is compared to the isolated urea

TABLE IX

Comparative Population Analysis for Isolated ($\xi\xi$) and Hydrogen-Bonded ($\xi\xi$/H) Urea Molecules Around the Upper Hydrogen of the Amine Groups and the Oxygen Atom

Atom	Field[a]	$Q_{\xi\xi}$	$Q_{\xi\xi/H}$	ΔQ
	O	0.3246	0.4384	11.38×10^{-2}
H_2	F_z	0.3393	0.4554	11.61×10^{-2}
	$-F_z$	0.3096	0.4208	11.12×10^{-2}
O	O	−0.6507	−0.9565	-30.58×10^{-2}
	F_z	−0.6738	−0.9818	30.80×10^{-2}
	$-F_z$	−0.6273	−0.9306	-30.33×10^{-2}

[a] The field intensity is alternatively O, F_z, and $-F_z$ where $F_z = 5 \times 10^{-3}$ in atomic units and is applied along the dipolar axis z of urea.

TABLE X

Convergence of Dipole and (Hyper)Polarizability Coefficients with Iteration Step l and Comparison with Isolated Urea Molecule

Iteration	Isolated	0	1	2	3	4	5	6
μ_z	2.076347	3.084167	3.318160	3.376211	3.390067	3.393389	3.394185	3.394374
α_{xx}	25.09	26.01	26.2	26.3	26.3	26.3	26.3	26.3
α_{yy}	7.77	8.98	10	10	10	10	10	10
α_{zz}	26.8	23.7	29.1	30.5	30.9	20.9	31	31
β_{zzz}	−44	−33.2	−57.6	−68.6	−72.4	−73.6	−74	−74.1
β_{zxx}	23	28.76	30.4	31.16	31.4	31.4	31.5	31.5
β_{zyy}	−3	−1.6	−7.96	−8.7	−8.9	−8.9	−8.9	−8.9

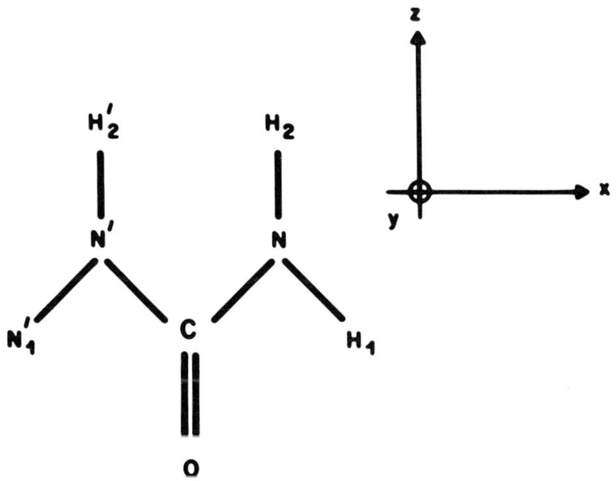

Fig. 9. Conventions of the urea molecule. In the crystal, urea molecules are hydrogen bonded to six neighboring ones, indicating the irrelevance of an "oriented gas" description of crystalline optical properties.

molecule. The term β_{zzz} is negative in both cases, but the influence of the environment enhances by almost a factor of two the value of β_{zzz} (see geometry in Fig. 9). This trend can simply be accounted for (Zyss and Berthier, 1982) by the influence of H-bonding on charge transfer.

V. RELATIONSHIPS BETWEEN MICROSCOPIC AND MACROSCOPIC OPTICAL NONLINEARITIES

A. Introduction

The polarizabilities that were introduced in Section IV characterize isolated molecules. This corresponds quite accurately to the situation encountered in rarefied gases, but in most cases of interest the molecules are embedded in a dense medium, either a liquid or a solid. Then the interactions between molecules are large, and it is necessary to address an old and important problem already mentioned in Section II, that is, relate the optical parameters measured on a macroscopic scale to the microscopic ones. In the case of linear optics, the problem was first considered by Lorentz (1916). In a certain sense the case of organic compounds is simpler to discuss than that of inorganic materials. In organic crystals and liquids the molecules interact through van der Waals forces or hydrogen bonds, which are one or two orders of magnitude weaker than the chemical bonds holding the atoms together within the molecules. Therefore, to some extent, the molecules conserve their individuality in the condensed medium and they can be considered as the microscopic building units of the crystal or the liquid. However, the connection between the microscopic and macroscopic parameters is still difficult to establish for a number of reasons, some of which are given hereafter. Even weak interactions can have important consequences on very sensitive parameters: for instance, it is well known that solvent–solute interaction can cause large shifts in absorption spectra. The state of a molecule in a dense environment can be different from that of the free one. For instance, the relative rotation of two parts of a molecule around a bond, which is allowed for the free molecule, can be hindered in a crystal. Then the geometry and the symmetry of the molecule *in situ* are not the same as those of the free molecules. Finally, the field acting on a molecule in a dense medium is not the applied field but the local field, which is the sum of the applied field, the field generated by the other molecules, and the reaction field of the molecule itself. The exact determination of local field effects is also difficult.

In Section V.B we first recall the nature of cohesive forces in molecular crystals and the resulting geometrical arrangement of molecules in molecular crystals. We then consider the problem of local field effects in linear and nonlinear optical processes. The nonlinear susceptibility tensors of condensed organic matter (fluids and solids) will be expressed in terms of molecular hyperpolarizabilities within the context of an "oriented gas" model. This description, together with a two-dimensional or equivalently two-level description of molecular quadratic susceptibilities (β), is seen to lead to new structural relations between crystal d coefficients, as illustrated by a full treatment of a specific molecular crystal (MAP) in Section V.C. In addition, the problem of inferring molecular susceptibilities from crystal measurements is addressed and limitations are discussed. The optimization of molecular orientations within crystal lattices will be discussed within a theoretical framework set up in Section V.D, while practical strategies tending to actually reach theoretical optimal structures are defined and exemplified for specific efficient nonlinear molecular crystals in Section V.E.

B. Condensed Molecular Matter

Molecules are globally neutral. If their electronic distributions were static, no attraction whatsoever could hold them tight in a condensed phase. In fact, the electronic distribution of a molecule continuously experiences quantum fluctuations, which generate transient fields and induce on neighboring molecules coherent charge fluctuations, which in turn react on the initial molecule. Because of the coherence of the interaction, the averaged potential does not vanish and gives rise to attractive forces.

The first description of the interaction between neutral species was given by London (1930). He showed that the more polarizable electrons (valence π and σ) were responsible for the major part of the attractive interaction. Since they are also responsible for the dispersion of the polarizabilities in the visible domain, he called these forces "dispersion forces." To illustrate the London approach, let us consider two molecules M and M′ which are far enough apart so that the interaction energy is small as compared to the energy of the isolated molecules. The interaction potential can be expanded as a multipole series

$$V = qq'(1/r) + (q\boldsymbol{\mu}' - \boldsymbol{\mu}q')\nabla(1/r) + (\tfrac{1}{3}q'\mathbf{Q} + \tfrac{1}{3}q\mathbf{Q}' - \boldsymbol{\mu}\boldsymbol{\mu}')\nabla\nabla(1/r) + \cdots \quad (116)$$

and the energy of the pair is obtained by a perturbation expansion in V. The first-order correction to the energy is the electrostatic interaction between the two molecules, which has the same form as Eq. (116) but with operators q, $\boldsymbol{\mu}$, $\mathbf{Q}$, etc., replaced by the ground-state permanent charge, dipole, quadrupole, etc. For neutral species it reduces to

$$V_{\text{stat}} \approx -\boldsymbol{\mu}_0 \cdot \boldsymbol{\mu}'_0 \nabla\nabla(1/r) \quad (117)$$

The second-order correction to the theory gives rise to London's dispersion forces,

$$V_{\text{disp}} \approx -A/r^6 \quad (118)$$

where

$$A \approx \sum_{a \neq g} \sum_{a' \neq g'} \frac{|\langle g|\boldsymbol{\mu}|a\rangle|^2 |\langle g'|\boldsymbol{\mu}|a'\rangle|^2}{W_{ag} + W_{a'g'}}$$

Comparing this expression to Eq. (75) written for a static field ($\hbar\omega = 0$), one can perceive some similarity between A and the product of the static linear polarizabilities $\alpha(0) = \Sigma_{a \neq g} 2|\langle g|\mu|a\rangle|^2/W_{ag}$ and $\alpha'(0) = \Sigma_{a \neq g} 2|\langle g'|\mu|a'\rangle|^2/W_{a'g'}$. Using the Unsöld approximation and identifying the fixed energies with the ionization potential I and I', London was able to show that $(W_{ag} + W_{a'g'})^{-1} \approx II'/[(I + I')/W_{ag}W_{a'g'}] + \cdots$, so that the dispersion energy can be approximated by

$$V_{\text{disp}} \simeq -\frac{3}{2}\left(\frac{II'}{I + I'}\right)\frac{\alpha(0)\alpha'(0)}{r^6} \quad (119)$$

The interpretation of this expression is that V_{disp} arises from a dipole–dipole interaction between induced molecular dipoles. At closer distances the electronic distributions start to overlap, leading to strong repulsive forces between the molecules. It is rather difficult to describe the potential arising from processes as complicated as the overlap of charge distribution. It is customary to use empirical formulas to characterize it, such as the Lennard-Jones potential, $\sim +1/r^{12}$ (Lennard-Jones, 1931), and the Buckingham potential, $\sim \exp(-Br)$ (Buckingham, 1938), which are frequently used.

These simple expressions of the interaction potential are well suited for very simple species such as noble gases. For more complicated ones, a more sophisticated approach has to be used, developed by Kitaigorodsky (1961, 1973), called the atom–atom potential method. According to this method, the interaction potential between two molecules is decomposed into a sum of interaction potentials between atoms, which are described following London theory and are assumed to depend only on the atoms and not on the molecule to which they belong. Calculations based on these assumptions are found to be in good agreement with the experimental data for a number of hydrocarbon molecules. The values of the optimized interaction dispersion energies for the C—C, C—H and H—H atoms are, respectively, 4.1×10^{-3} eV, 2.1×10^{-3} eV, and 0.4×10^{-3} eV (Kitaigorodsky, 1973), whereas the mean energy of the covalent bonds between the same atoms ranges from 3.5 to 4.5 eV. This confirms that the intramolecular forces are very weak compared to the chemical bonds. The electron-density distribution of a free molecule and that of the same molecule in a crystal bounded by dispersion forces are very similar. In fact, it is possible to describe the electron density in such a crystal as that of the free molecules separated by narrow spatial gaps almost free of electrons. The arrangement of the molecules in such a crystal is therefore mostly determined by packing considerations. It depends strongly on the shape and size of the molecules. The main distinction between the liquid and the molecular solid is the regular and permanent orientation of the molecules in the crystal compared to the more random and nonpermanent one in the liquid. Let us note that in a liquid, because of the dense packing and the intramolecular forces, neighboring molecules tend to assume preferential positions relative to each other. They form loosely bonded and nonpermanent quasi-crystalline groups in which some motions are hindered.

There are other intermolecular bonds with energies in the range 8×10^{-2} eV to 26×10^{-2} eV, intermediate between intramolecular bonds and van der Waals bonds. They usually involve a hydrogen atom that is normally bound to an electronegative atom such as N, O, Cl, or F and interacts strongly by electrostatic forces to another electronegative atom. This type of interaction is called hydrogen bonding and provides cohesion to ice and urea crystals, for instance. It is also involved in the bonding of highly associative liquids, where it can favor the formation of dimers. Although molecules bonded by hydrogen bonds retain their individualities, the perturbation due to the hydrogen bond is far from being negligible. In particular, it is responsible for the large dielectric constant of liquids such as H_2O and CH_3OH, and may result in large modifications of the nonlinear polarizabilities, as shown for urea in Section IV.D.

1. Local Field Effects

Strictly speaking, the electromagnetic theory applies to distributions of charges, currents, and electromagnetic fields all *in vacuo*. Therefore, microscopic fields acting on a microscopic volume element are the sum of external fields plus fields generated by all charges and currents of the system. If some charges are located within the volume element, it is necessary to subtract the own reaction fields of these charges. Again, it is a formidable task to perform such calculation and to describe material systems in the presence of macroscopic electromagnetic fields it is customary to use Eq. (1). The quantities appearing in these equations are the macroscopic fields applied to the material system, **E** and **B**. The other variables are defined as functions of the electric dipole **P**, the magnetic dipole **M**, the electric quadrupole **Q** densities, etc., which are related to their microscopic counterparts by averaging over "volume small on the macroscopic scale and large compared to the atomic dimensions" (Lorentz, 1916). The exact definition of these averages raises a number of questions that have been addressed correctly only recently. It has been shown that the statistical ensemble average, assuming a rather coarse distribution in real space, can be used to relate microscopic and macroscopic Maxwell equations (Mazur and Nijboer, 1953; DeGroot, 1969). A more satisfactory approach has been proposed by Robinson (1971, 1973), based on the intuitive understanding that for a given problem the spatial Fourier components of the fields are irrelevant beyond some limiting wavelength. He introduced a formal mathematical procedure to eliminate the high spatial frequency components, which he called "truncation," and proved rigorously that the truncated variables obey the macroscopic Maxwell equations and correspond to definitions of **P**, **M**, **Q**,... which avoids all the usual mathematical difficulties. One appealing aspect of this approach is that the cut-off spatial frequency depends on the specific problem one wishes to solve. For instance, in the case of optics, the limiting spatial frequency is of the order of a few hundred angstroms, whereas in X-ray diffraction it is of the order of one or a fraction of an angstrom. Let us note that other formal treatments of this problem in the Bloch formalism have been proposed (Adler, 1962; Pine, 1965; Wiser, 1963).

In practice the local field $\xi(\omega)$ is often written as a function of the applied field $\mathbf{E}(\omega)$ and of the dipole density $\mathbf{P}(\omega)$ as

$$\xi(\omega) = \mathbf{E}(\omega) + 4\pi L\mathbf{P}(\omega) \tag{120}$$

where L is a dimensionless rank-two tensor. The derivation of Eq. (120) relies on several assumptions. The first one is that the dipole approximation satisfactorily describes the interactions among molecules. This is true in rarefied gases where the intermolecular distance is large compared to molecular dimensions. In solid and liquids where these two quantities are of the same order of magnitude, the dipole approximation is very dubious. Fortunately, when one is only interested in dipolar polarization effects, the effects of higher-multipole fields are greatly reduced by symmetry cancellation. The other assumption used to derive Eq. (120) is that the polarization within the dense medium is uniform. Using this hypothesis it is possible to transform the summation over a small volume surrounding a molecule into a surface integral. The shape of the small cavity is determined by the environment of the molecule. In liquids or cubic crystals it is taken as spherical by symmetry, as considered first by Lorentz, but in general it must be ellipsoidal. As long as the dimensions of the cavity are small compared to the optical wavelength, the electric field can be calculated by standard electrostatic methods (Bottcher, 1952). It is found that for a uniform applied field, the field in the cavity is uniform, too. It does not depend on the size of the cavity but on its shape. The components of L are the demagnetization factors

$$L_{\mathrm{I}} = \frac{a_1 a_2 a_3}{2} \int_0^{\infty} \frac{ds}{(s + a_{\mathrm{I}}^2)[(s + a_1^2)(s + a_2^2)(s + a_3^2)]^{1/2}} \tag{121}$$

where a_1, a_2, and a_3 are the semiaxes of the ellipsoidal cavity. The demagnetization factors are related by $L_1 + L_2 + L_3 = 1$. For a spherical cavity, by symmetry $L_1 = L_2 = L_3 = \frac{1}{3}$; for a very long cylinder, $L_1 = L_2 = \frac{1}{2}$, $L_3 = 0$; and for a flat one, $L_1 = L_2 = 0$, $L_3 = 1$.

In the case of a linear dielectric, with N nonpolar molecules per unit volume, the induced polarization is

$$P(\omega) = N\alpha(\omega) \cdot \xi(\omega) = \chi^{(1)}(\omega) \cdot \mathbf{E}(\omega)$$

To express the local field corrections in a closed form, it is convenient to introduce the tensor $f(\omega)$, given by

$$f(\omega)[1 - 4\pi L\alpha(\omega)] = 1$$

in terms of which it is possible to write

$$\chi^{(1)} = N\alpha(\omega) f(\omega) \tag{122}$$

and

$$\xi(\omega) = f(\omega) \cdot \mathbf{E}(\omega) \tag{123}$$

It is important to notice that owing to the high symmetry of $\alpha(\omega)$ and L, in a number of cases the tensors $\alpha(\omega)$, L, and $f(\omega)$ commute. It is then possible to treat them as scalar quantities. The components of $f(\omega)$ are called the local field factors. They are

$$f_{\mathrm{I}}(\omega) = 1 + (\varepsilon_{\omega} - 1)L_{\mathrm{I}} \tag{124a}$$

For a cubic crystal or an isotropic liquid ($L = \frac{1}{3}$), they reduce to the well-known Lorentz correction factor

$$f(\omega) = (\varepsilon_\omega + 2)/3 \tag{124b}$$

So far we have only considered induced dipoles; however, external fields, especially static fields, can also orient dipoles. The field exerted on one molecule by the other molecules contains one component, the reaction field, that depends on the dipole of the molecule itself and therefore cannot contribute to its own orientation (Onsager, 1936). The local field factors describing this configuration are known as the Onsager correction factors, and their expression is

$$f(0) = \varepsilon_0(\varepsilon_\omega + 2)/(2\varepsilon_0 + \varepsilon_\omega) \tag{124c}$$

The distinction between induced dipole and induced rotation is particularly important to explain why polar liquids do not exhibit spontaneous polarization.

If we consider a dense nonlinear medium, the microscopic dipole of a molecule is given by Eq. (68b) where the fields acting on the molecule are the local fields of Eq. (120):

$$\mathbf{p}(\omega) = \alpha \cdot \xi(\omega) + \beta : \xi(\omega_1)\xi(\omega_2) + \gamma \vdots \xi(\omega'_1)\xi(\omega'_2)\xi(\omega'_3) + \cdots \tag{125}$$

The polarization density is the sum over the unit volume of the molecular dipoles, or more rigorously one of the averages or truncations discussed above. It can be split into its linear and its nonlinear parts,

$$\begin{aligned} \mathbf{P}(\omega) &= N\mathbf{p}(\omega) = \mathbf{P}_\mathrm{L}(\omega) + \mathbf{P}_\mathrm{NL}(\omega) \\ &= N\alpha \cdot \xi(\omega) + N[\beta : \xi(\omega_1)\xi(\omega_2) + \gamma \vdots \cdots] \end{aligned} \tag{126}$$

Assuming for simplicity that tensors α, L, and f commute, one obtains for the nonlinear susceptibilities the expressions

$$\chi^{(2)}(-\omega; \omega_1, \omega_2) = f(\omega)f(\omega_1)f(\omega_2)N\beta(-\omega; \omega_1, \omega_2) \tag{127a}$$

$$\chi^{(3)}(-\omega; \omega'_1, \omega'_2, \omega'_3) = f(\omega)f(\omega'_1)f(\omega'_2)f(\omega'_3)N\gamma(-\omega; \omega'_1, \omega'_2, \omega'_3) \tag{127b}$$

It is interesting to note the appearance of an extra local field factor in $\chi^{(2)}$ and $\chi^{(3)}$ as compared to $\chi^{(1)}$. This is a general result in the expression of nonlinear susceptibilities where one local field factor is present for each frequency, i.e., $(n + 1)$ local field factors in $\chi^{(n)}$.

2. Macroscopic and Microscopic Nonlinearities in Organic Crystals

In this section we discuss the relationships between the molecular polarizabilities β and γ and the crystalline susceptibilities $\chi^{(2)}$ and $\chi^{(3)}$. To simplify the notations, we do not specify the frequency dependence when possible, remembering that they should be the same for the two couples $\beta, \chi^{(2)}$ and $\gamma, \chi^{(3)}$. We shall detail the formulas for the tensor β and $\chi^{(2)}$, the corresponding ones for γ and $\chi^{(3)}$ being straightforward extensions.

The crystallographic axes are labeled with capital indices (I, J, K). According to the crystal symmetry, there are N_g equivalent positions in the unit cell occupied by identical molecules, which are characterized *in situ* (which may be different from that of the free molecule) by the same second-order polarizability β. The equivalent positions are labeled by the index $s = 1, 2, \ldots, N_g$. To each molecule is attached a set of molecular axes (i, j, k), their choice reflecting the molecular symmetry, and the polarizabilities are naturally expressed in this set of axes. The angle between the crystallographic axis I and the microscopic axis i_s is denoted $\theta_{Ii}^{(s)}$. The transformation of the tensor β from the microscopic axis to the macroscopic ones is given by Eq. (26).

$$\beta_{IJK} = \sum_{ijk} \cos\theta_{Ii}^{(s)} \cos\theta_{Jj}^{(s)} \cos\theta_{Kk}^{(s)} \beta_{ijk} \tag{128}$$

If one assumes that the intermolecular forces are weak compared to the intramolecular ones, it is possible to write each element of the susceptibility tensor as (Chemla *et al.*, 1975)

$$\chi_{IJK}^{(2)}(-\omega; \omega_1, \omega_2) = \frac{1}{V} f_I(\omega) f_J(\omega_1) f_K(\omega_2) \times \sum_s \sum_{ijk} \cos\theta_{Ii}^{(s)} \cos\theta_{Jj}^{(s)} \cos\theta_{Kk}^{(s)} \beta_{ijk}(-\omega; \omega_1, \omega_2) \tag{129}$$

Here V is the volume of the unit cell. In some complicated structures, molecules can occupy nonequivalent sites, and there can be several species. Then the expression of $\chi^{(2)}$ has the same form as Eq. (129), with, however, additional summation over the nonequivalent subsets. It is convenient to introduce a new tensor, the unit cell nonlinearity per molecule, b_{IJK} (Zyss and Oudar, 1982),

$$\begin{aligned} d_{IJK} &= N f_I(\omega) f_J(\omega_1) f_K(\omega_2) b_{IJK} \\ b_{IJK} &= \frac{1}{N_g} \sum_{ijk} \left(\sum_{s=1}^{N_g} \cos\theta_{Ii}^{(s)} \cos\theta_{Jj}^{(s)} \cos\theta_{Kk}^{(s)} \right) \beta_{ijk} \end{aligned} \tag{130}$$

where N_g is the number of equivalent positions in the unit cell [for example, $N_g(622) = 12$ or $N_g(222) = 4$] and N is the number of molecules in the crystal unit volume.

The actual number of molecules in the unit cell n, when not equal to N_g for a given crystal, is either a submultiple or a multiple of N_g. The former case happens when molecular-symmetry operations eventually coincide with crystal-symmetry operations, thus leaving molecules invariant with respect to some of the crystal point group symmetry operations. For example, in the urea crystal (group $\bar{4}2m$), $n = 2$, while $N_g = 8$. The redundancy in the summation is compensated by dividing by N_g instead of n. The integer N_g/n is then a sort of degeneracy factor accounting for possible coincidences between molecular and crystalline symmetries. The latter case happens when the crystal unit cell accommodates two or more sets of nonequivalent molecular positions. Equations (130) can then be generalized to

$$d_{IJK} = f_I(\omega) f_J(\omega_1) f_K(\omega_2) \sum_{t=1}^{P} N_t b^t_{IJK} \tag{131}$$

$$b^t_{IJK} = \frac{1}{N_g} \sum_{ijk} \sum_{s=1}^{N_g} \cos\theta^{(s)}_{Ii_t} \cos\theta^{(s)}_{Jj_t} \cos\theta^{(s)}_{Kk_t} \beta_{ijk} \tag{132}$$

where t is any of the P unequivalent positions and N_t is the actual number of molecules in one of the p nonequivalent subsets. Local molecular frames are fixed with respect to the molecules, so that β_{ijk} in Eq. (132) is not dependent on either s or t. Equation (132) is then formally identical to Eq. (130), and we shall not distinguish, in the following, between b^t_{IJK} and b_{IJK}. This rather formal procedure has the practical advantage of leading to a general (only group-dependent) expression of b_{IJK}, referring to the lowest-order *nonlinearity of the unit cell per molecule*. This expression is independent of the actual number of molecules in the unit cell. Similar considerations can be used for connecting "linear" quantities such as the index of refraction (or dielectric susceptibility tensor) and the molecular polarizability α. The indices of refraction n_{L} are related, within the principal dielectric axis reference frame, to a diagonal tensor a_{LL} representing the average linear polarizability of the crystal per molecule by the Lorenz–Lorentz relation (Bottcher, 1973):

$$a_{\mathrm{LL}} = \frac{3}{4\pi N} \cdot \frac{n_{\mathrm{L}}^2 - 1}{n_{\mathrm{L}}^2 + 2} \tag{133}$$

where a_{IJ} is in turn related to α_{ij} within the oriented gas model by

$$a_{IJ} = \frac{1}{N_g} \sum_{i,j} \sum_{s=1}^{N_g} (\cos\theta^{(s)}_{Ii} \cos\theta^{(s)}_{Jj}) \alpha_{ij} \tag{134}$$

These considerations will find applications in Section V.C. To analyze specific situations, one may also resort to the irreducible tensor decomposition technique so as to avoid the rather tedious implementation of Eq. (129) (Chemla *et al.*, 1975).

In the transparency domain where the polarizability tensors and susceptibility tensors are invariant in the permutation of their cartesian indices, the decomposition in irreducible parts is particularly simple: β (as well as $\chi^{(2)}$ and b) is the sum of a vector and a septor, $\beta = \beta^{(1)} + \beta^{(3)}$. In the spherical representation, the transformation from the microscopic axis to the crystallographic axis is determined by the Wigner matrix $D^{(J)}_{Mn}(\Psi,\Theta,\Phi)$ and is independent for the two irreducible components that *do not mix*, conversely to the cartesian ones. The two analogs of Eq. (130) are

$$b_1^{\mathrm{M}} = \frac{1}{V}\sum_{m=-1}^{1}\left[\sum_s D^{(1)}_{m\mathrm{M}}(\Psi^{(s)},\Theta^{(s)},\Phi^{(s)})\right]\beta_1^m \tag{135a}$$

$$b_3^{\mathrm{M}} = \frac{1}{V}\sum_{m=-3}^{3}\left[\sum_s D^{(3)}_{m\mathrm{M}}(\Psi^{(s)},\Theta^{(s)},\Phi^{(s)})\right]\beta_3^m \tag{135b}$$

To illustrate how this representation leads to simpler calculations, let us apply Eq. (135a) to the example of point group $mm2$. First, we remark that if the molecules are not polar ($\beta^{(1)} = 0$) the vector part of b and $\chi^{(2)}$ vanishes, although it is otherwise allowed by crystal symmetry.

If the molecules are polar, we can choose the microscopic axis so that $\beta^{(1)}$ lies along $\hat{z}$; then $\beta_1^{\pm 1} = 0$ and $\beta_1^0 = \sqrt{\tfrac{3}{5}}(\beta_{zzz} + \beta_{zyy} + \beta_{zxx})$, leading to

$$b_1^0 = \sqrt{\tfrac{3}{5}}(b_{zzz} + b_{zyy} + b_{zxx}) = (4/V)\cos\theta\beta_1^0 \tag{136}$$

Along the same lines, the expression of the other irreducible components can be written using the tabulated $D^{(J)}_{m\mathrm{M}}$ (Buckmaster *et al.*, 1972). In the particular example of the crystallographic point group $mm2$, $b^{(3)}$ has only two components, $b_3^0 = (2b_{zzz} - 3b_{zxx} - 3b_{zyy})\sqrt{10}$ and $b_3^{\pm 2} = \sqrt{3/4}(b_{zxx} - b_{zyy})$. To develop Eq. (135b) it is necessary to consider the specific orientations of the molecule (Chemla *et al.*, 1975).

3. Macroscopic and Microscopic Nonlinearities in Fluids

A number of measurements aiming at the determination of molecular polarizabilities are performed in fluids, gases, solutions, or molten solids. For a fluid under no perturbation, the molecular environment is nonpermanent: even if in some liquids the relative orientation of neighboring molecules momentarily reproduces some aspects of the crystalline phase, on the average there is no macroscopic orientation. The only nonvanishing component of the macroscopic susceptibilities, therefore, are scalar or pseudo-scalar, for the even- and odd-rank tensors, respectively. This gives for the first three nonlinearities

$$\chi^{(1)}(\omega) = f(\omega)N\alpha^{(0)}(\omega) \tag{137a}$$

$$\chi^{(2)}(-\omega;\omega_1,\omega_2) = f(\omega)f(\omega_1)f(\omega_2)N\beta^{(0)}(-\omega;\omega_1,\omega_2) \tag{137b}$$

$$\chi^{(3)}(-\omega;\omega_1',\omega_2',\omega_3') = f(\omega)f(\omega_1')f(\omega_2')f(\omega_3')N_\gamma^{(0)}(-\omega;\omega_1',\omega_2',\omega_3') \tag{137c}$$

where the local factors for optical frequencies $f(\omega)$ have the form of Eq. (124a). The scalars $\alpha^{(0)} = \frac{1}{3}\alpha_{ij}\delta_{ij}$ and $\gamma^{(0)} = \frac{1}{5}\gamma_{ijkl}\delta_{ij}\delta_{kl}$ and the pseudo-scalar $\beta^{(0)} =$

$\frac{1}{6}e_{ijk}\beta_{kji}$ have been defined in Section III. The pseudo-scalar $\beta^{(0)}$ does not vanish only if the frequencies ω, ω_1, and ω_2 are all different. It has the same symmetry as the parameter describing the rotatory power in noncentrosymmetric fluids [Giordmaine (1965)]. The average polarizability $\alpha^{(0)}$ is obtained by measuring the refractive index and is related to the dielectric constant by the Clausius–Mossotti equation $(\varepsilon_\omega - 1)/(\varepsilon_\omega + 2) = (4\pi/3)N\alpha$, i.e., Eq. (133). The mean third-order polarizability $\gamma^{(0)}$ can be determined from four-wave mixing or third-harmonic experiments.

Orientation of the molecules in a fluid can be achieved by application of a static field $E(0)$. The energy of a molecule in the static local field $\xi(0)$ is given by

$$W = W_0 - \tilde{\mu} \cdot \xi(0) - \tfrac{1}{2}\tilde{\alpha} : \xi(0)\xi(0) - \tfrac{1}{3}\tilde{\beta} \vdots \xi(0)\xi(0)\xi(0) - \cdots \tag{138}$$

where in this section to distinguish the frequency-dependent and the static polarizabilities we denote the static dipole and static linear, quadratic, and cubic polarizabilities by $\tilde{\mu}$, $\tilde{\alpha}$, $\tilde{\beta}$, $\tilde{\gamma}$. Since they are fully symmetric tensors, their decomposition in irreducible parts is very simple:

$$\tilde{\alpha} = \tilde{\alpha}^{(0)} + \tilde{\alpha}^{(2)}, \qquad \tilde{\beta} = \tilde{\beta}^{(1)} + \tilde{\beta}^{(3)}, \qquad \tilde{\gamma} = \tilde{\gamma}^{(0)} + \tilde{\gamma}^{(2)} + \tilde{\gamma}^{(4)}$$

The orientation of the molecule is specified by the direction of its total static dipole $\mathbf{p}(0) = \partial W/\partial\xi(0)$, which is field-dependent in the molecular frame, while other parameters describing the orientation of the molecule around $\mathbf{p}(0)$ are not relevant. They are uniformly distributed among all molecules with a given orientation of $\mathbf{p}(0)$. For any physical property characterized by a tensor T, the thermodynamical average measured in the laboratory frame is

$$T_{IJ\cdots L} = \sum_{ijk} \langle \cos\theta_{Ii} \cos\theta_{Jj} \cdots \cos\theta_{Ll} \rangle\, T_{ij\cdots l} \tag{139a}$$

where the average of a parameter $X(\Omega)$ depending on the molecular orientation (Ω) is

$$\langle X \rangle = \int X(\Omega) \exp[-W(\Omega)/kT]\, d\Omega \bigg/ \int \exp[-W(\Omega)/kT]\, d\Omega \tag{140}$$

The expression of thermodynamical average for cartesian tensors has been discussed by Kielich (1969) and by Cyvin *et al.* (1965). As the rank of the tensor increases, it rapidly becomes untractable, while the use of irreducible tensors leads to a more convenient formalism and indeed to an exact analytical solution (Chemla and Bonneville, 1978). The thermodynamical average for a spherical tensor is

$$T^J_{\mathrm{M}} = \sum_m \langle D^{(J)}_{m\mathrm{M}}(\Psi, \Theta, \Phi) \rangle\, T^J_m \tag{139b}$$

If we choose the laboratory frame in such a way that $\hat{z}$ is directed along the direction of the static field $\xi(0)$, it is possible to express the energy of the molecule as

$$W(\Psi\Theta) = \sum_J (-1)^m C_m^J(\Theta, \Psi) S_{-m}^J[\xi(0)]$$

where $C_m^J(\Theta, \Psi)$ are the Racah reduced spherical harmonics that contain all the angular information. The parameter S describes the static field–molecular polarizability coupling and has the following form:

$$S_{-m}^J[\xi(0)] = \sum_n (-1)^{(n-J)/2} N(n, J) \kappa_m^{(n,J)} \frac{\xi_{(0)}^n}{n!}$$

where $\kappa_m^{(n,J)}$ is the irreducible component (J, m) of the polarizability tensor of order n and $N(n, J)$ is a normalization factor (Jerphagnon *et al.*, 1978). Using this expression in Eq. 139b it is found that

$$\langle D_{mM}^{(J)}(\Psi, \Theta, \Phi)\rangle = (-1)^m \delta_{MO} \langle C_m^J(\Theta, \Psi)\rangle \tag{141}$$

The averages of the reduced spherical harmonics are called the generalized Langevin functions, because if only the terms linear in $\xi(0)$ are kept [i.e., $W = W_0 - \tilde{\mu} \cdot \tilde{\xi}(0)$], the first of these functions reduces to the well-known Langevin function:

$$\langle \cos \Theta \rangle = \tilde{\mu} \cdot \xi(0)/3kT \tag{142}$$

Equation (141) can be cast in a form that gives an exact expression for $\langle D_{mM}^{(J)}\rangle$ valid to any order in $\xi(0)$. Some care is required when dealing with the exponential in Eq. (140). When W is expressed by Eq. (138), all terms of the same order in $\xi(0)$ must be collected together. This introduces the effective temperature-dependent static polarizabilities,

$$\tilde{\alpha}(T) = \tilde{\alpha} + \tilde{\mu}\tilde{\mu}/kT$$
$$\tilde{\beta}(T) = \tilde{\beta} + \tilde{\mu}\tilde{\alpha}/kT + \tilde{\mu}\tilde{\mu}\tilde{\mu}/(kT)^2$$
$$\vdots$$

This correction is important and should not be neglected. For instance, using the estimates $\tilde{\mu} \approx 10^{-18}$ esu, $\tilde{\alpha} \approx 10^{-23}$ esu, $\tilde{\beta} \approx 10^{-30}$ esu, and $kT \approx 4 \times 10^{-14}$ erg at room temperature, it is found that $\tilde{\mu}^2/kT \approx 2.5 \times 10^{-23}$ esu, $\tilde{\mu}\tilde{\alpha}/kT \approx 2.5 \times 10^{-28}$ esu, and $\tilde{\mu}^3/(kT)^2 \approx 610^{-28}$ esu. Conversely, in most cases it is possible to identify the direction of the total static dipole $p(0)$ with that of the permanent dipole $\tilde{\mu}$ [for $\xi(0) = 3 \times 10^4$ V/cm = 100 esu, $\tilde{\mu} \approx 10^3 \times \tilde{\alpha}\xi(0) \approx 10^8 \times \tilde{\beta}\xi^2(0) \approx 10^{12} \times \tilde{\gamma}\xi^3(0)$].

Let us consider as an example the first-order term in the expression of the second-order susceptibility induced by a static field measured in EFISH experiments. The average of the vector part of $\beta(-2\omega; \omega, \omega)$ is given by

Eqs. (139b) and (141) written for $J = 1$ and $M = 0$, i.e., the analog of Eq. (136),

$$b_1^0 = N\langle\cos\Theta\rangle\beta_1^0 = N\frac{\tilde{\mu}\cdot\xi(0)}{3kT}\beta_1^0 \tag{143}$$

To the same order in $\xi(0)$, the macroscopic nonlinear susceptibility has a vector part but no septor:

$$b_1^0 = \sqrt{\frac{3}{5}}(b_{ZZZ} + b_{ZYY} + b_{ZXX}) \tag{144a}$$

$$b_3^0 = \frac{1}{\sqrt{10}}(2b_{ZZZ} - 3b_{ZYY} - 3b_{ZXX}) = 0 \tag{144b}$$

$$b_3^{\pm 2} = \sqrt{\frac{3}{4}}(b_{ZXX} - b_{ZYY}) = 0 \tag{144c}$$

These relations imply that $b_{ZXX} = b_{ZYY} = \frac{1}{3}b_{ZZZ}$ and $b_1^0 = \sqrt{3/5}(\frac{5}{3}b_{ZZZ})$, so that the relation between the cartesian components of b and the non-normalized vector part of β, which is $\beta_z = \beta_{zzz} + \beta_{zyy} + \beta_{zxxx}$ (this definition is the one most encountered in the literature; we use it in order to obtain an expression involving the same quantities as those commonly appearing in the general literature), is

$$b_{ZZZ} = -3b_{ZXX} = N\frac{\tilde{\mu}\cdot\tilde{\xi}(0)}{5kT}\beta_z \tag{145a}$$

The macroscopic susceptibility tensor is

$$\chi_{ZZZ}^{(2)} = -3\chi_{ZXX}^{(2)} = Nf(2\omega)f^2(\omega)f(0)\frac{\tilde{\mu}\cdot\xi(0)}{5kT}\beta_z \tag{145b}$$

Let us notice that a nonlinear polarization at 2ω arises also through the third-order nonlinearity,

$$\chi_{ZZZZ}^{(3)} = Nf(2\omega)f^2(\omega)f(0)\gamma^{(0)} \tag{146}$$

where $\gamma^{(0)}$ is the scalar part of the cubic polarizability. The total nonlinear polarization at frequency 2ω that contributes to EFISH is

$$P_Z^{\mathrm{NL}}(2\omega) = Nf(2\omega)f^2(\omega)f(0)\left[\gamma^{(0)} + \frac{\tilde{\mu}\beta_z}{5kT}\right]E(\omega)E(\omega)E(0) \tag{147}$$

It is worth noticing that the combination of parameters in the brackets appears in other effects that are also induced by a static field orientation, for instance, in induced birefringence (Chemla and Bonneville, 1978; Bonneville and Chemla, 1978).

C. Oriented Gas Description of an Intramolecular Charge-Transfer Crystal: MAP

We shall now combine two types of considerations, the two-level description of intramolecular charge transfer and the oriented gas description of molecular crystals.

As was shown in Section IV, the quadratic susceptibility of quantum two-level systems is rigorously two-dimensional (in the geometric sense) and is a third-order cartesian tensor defined within the $(\boldsymbol{\delta}, \mathbf{m})$ plane, where $\boldsymbol{\delta}$ is the difference between excited and ground state dipole moments of the molecule and $\mathbf{m}$ is the molecular transition dipole vector. The term $\boldsymbol{\delta}$ must meet structural symmetry requirements and will be, for example, directed along the 2 axis if the molecular symmetry point group is *mm*2 as in *p*-nitroaniline. It will then be directed along *a* as shown in Fig. 10. However symmetry requirements

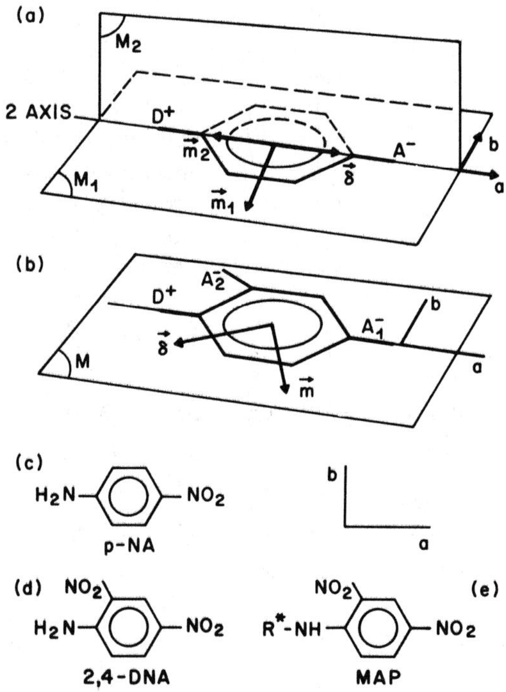

Fig. 10. (a) Planar donor–acceptor intermolecular charge-transfer molecule with *mm*2 symmetry. (b) Planar charge-transfer molecule with mirror (*m*) symmetry and two acceptors A_1 and A_2 interacting with donor D and examples of such structures. (c) *p*-Nitroaniline, (d) 2,4-dinitroaniline, and (e) MAP (Oudar and Hierle, 1977), where R* stands for an aliphatic chiral group (see Section V.E).

for **m** are more subtle. For example, in the case of molecules like *p*-nitroaniline, **m** can in principle be either along a or perpendicular to a, within the molecular plane (a, b) depending on the respective parities of ground and excited state wavefunctions with respect to mirror M_2 symmetry. However, in most cases (Liptay, 1973), the transition dipole moment is directed along a. Such transitions involve essentially an electronic excitation from a nonbonding π-donor orbital to a vacant π-antibonding acceptor one ($n_D \rightarrow \pi_A^*$). As a result, the charge transfer contribution to the linear and nonlinear polarizabilities of this type of molecules is clearly a *one-dimensional property*: that is, the only nonzero tensor components of these polarizabilities are the ones that involve only the a subindex (α_{aa}, β_{aaa}, etc.). This feature is, in fact, fully confirmed by the results of detailed quantum-mechanical calculations on *p*-nitroaniline (Lalama and Garito, 1979; Morrell and Albrecht, 1979).

In more complex cases, a single charge-transfer axis cannot be readily defined. Representative examples of such molecules among the nitroaniline family are *o*- or *m*-nitroaniline, 2,4-dinitroaniline, etc. For all these molecules, the relative positions of the substituent radicals lower the molecular symmetry, leaving only a mirror plane, parallel to the aromatic ring. The same arguments as above apply to this restricted symmetry, with the result that the ground and excited state dipole moments are within this plane, as well as the charge-transfer transition dipoles. As a consequence, charge-transfer contribution to the linear and nonlinear polarizabilities of these molecules has an intrinsic *two-dimensional* character. In other words, if c is the axis perpendicular to the molecular plane, the polarizability tensor components that involve the c subindex are zero. This two dimensional character leaves only four independent coefficients to the β tensor (assuming Kleinman symmetry relations), namely, β_{aaa}, β_{bbb}, β_{baa}, and β_{abb}. Results of finite-field INDO calculations on 2,4-dinitroaniline in Table XI show the pronounced two-dimensional character of β.

The same feature is not valid for α, since the two-dimensional π-electron contribution is less prevailing for α than for β. Inspection of the in-plane components shows that β_{aaa} is significantly larger than the other coefficients, while α is not far from being isotropic in the molecular plane. Having defined the two-dimensional nature of the β tensor of charge-transfer molecules well, we may proceed to insert that description within the oriented gas model. We shall focus on the specific case of MAP (Oudar and Hierle, 1977), a monoclinic crystal ($P2_1$) with two molecules in the unit cell interchanged by a twofold rotation around the binary axis. The structure of MAP is shown in Fig. 10, and more details about the molecular and crystalline properties of that compound are to be found in Sections V.E and VI.B. MAP being very similar to planar 2,4-dinitroaniline, a two-dimensional description of its quadratic polarizability is relevant, and the relation between the four $P2_1$ permitted b

TABLE XI

Results of INDO Calculations on 2,4-Dinitroaniline (2,4-DNA)

Dipole moment (in Debye units)	μ_a	−4.74
	μ_b	−4.27
Linear polarizability (in 10^{-24} esu)	α_{aa}	23.7
	α_{ab}	0.1
	α_{bb}	19.8
	α_{cc}	4.0
Lowest-order hyperpolarizability (in 10^{-30} esu)	β_{aaa}	−6.33
	β_{baa}	0.34
	β_{abb}	2.07
	β_{bbb}	−0.43
	β_{acc}	0.08
	β_{bcc}	0.09

coefficients and four β coefficients goes as follows:

$$\begin{aligned}
b_{YXX} &= \cos\theta_{Yx}\cos^2\theta_{Xx}\beta_{xxx} + \cos\theta_{Yy}\cos^2\theta_{Xy}\beta_{yyy} + \cos\theta_{Xx}(\cos\theta_{yY}\cos\theta_{Xx} \\
&\quad + 2\cos\theta_{Yx}\cos\theta_{Xy})\beta_{yxx} + \cos\theta_{Xy}(\cos\theta_{Yx}\cos\theta_{Xy} \\
&\quad + 2\cos\theta_{Yy}\cos\theta_{Xx})\beta_{xyy} \\
b_{YYY} &= \cos^3\theta_{Yx}\beta_{xxx} + \cos^3\theta_{Yy}\beta_{yyy} \\
&\quad + 3\cos\theta_{Yy}\cos\theta_{Yx}(\cos\theta_{yx}\beta_{yxx} + \cos\theta_{Yy}\beta_{xyy}) \\
b_{YZZ} &= \cos\theta_{yz}\cos^2\theta_{Zx}\beta_{xxx} + \cos\theta_{Yy}\cos^2\theta_{Zy}\beta_{yyy} + \cos\theta_{ZX}(\cos\theta_{Yy} \\
&\quad + 2\cos\theta_{Yx}\cos\theta_{Zy})\beta_{yxx} + \cos\theta_{Zy}(\cos\theta_{Yx}\cos\theta_{Zy} \qquad (148a) \\
&\quad + 2\cos\theta_{Yy}\cos\theta_{Zx})\beta_{xyy} \\
b_{YZX} &= \cos\theta_{Yx}\cos\theta_{Zx}\cos\theta_{Xx}\beta_{xxx} + \cos\theta_{Yy}\cos\theta_{Zy}\cos\theta_{Xy}\beta_{yyy} \\
&\quad + (\cos\theta_{Yy}\cos\theta_{Zx}\cos\theta_{Xx} + \cos\theta_{Yx}\cos\theta_{Zy}\cos\theta_{Xx} \\
&\quad + \cos\theta_{Yx}\cos\theta_{Zx}\cos\theta_{Xy})\beta_{yxx} + (\cos\theta_{Yx}\cos\theta_{Zy}\cos\theta_{Xy} \\
&\quad + \cos\theta_{Yy}\cos\theta_{Zx}\cos\theta_{Xy} + \cos\theta_{Yy}\cos\theta_{Zy}\cos\theta_{Xx})\beta_{xyy}
\end{aligned}$$

where X_i, Y_j, Z_k are the cosine components of the crystallographic reference frame unit factors expressed in the molecular reference frame unit vectors. However tedious to manipulate, Eqs. (148a) appear highly appropriate if β molecular coefficients are to be inferred from d or b coefficients (note that a factor of 2 that would normally appear when d is expressed in terms of β as there are two molecules per unit cell does not show up when b is prefered to d

as b is "averaged" over equivalent positions in the unit cell). In fact, the relation between b and β is linear, and the 4×4 "crystallographic" matrix connecting b to β is just inverted to yield the desired β coefficients. This method for finding individual β coefficients is highly desirable, as solution EFISH measurements will provide only a linear combination of β coefficients of Eq. (145a) while in the solid state the four b coefficients can be independently measured and hence the four β. The situation of MAP is particularly illustrative of that approach, as point-group symmetry arguments at the crystallographic level and chemical arguments at the molecular level (i.e., relevance of the quantum two-level description and henceforth of an anisotropic two-dimensional β tensor) converge to yield the same number of coefficients for b and β, namely, four. Feeding Eqs. (148a) with the proper crystallographic data for MAP (Knossow *et al.*, 1976) demonstrates a singularity that prevents the inverse straightforward expression of β coefficients in terms of b. The rather tedious calculations of the analytic expression of the determinant in terms of literal variables $\cos\theta_{Xi}$, $\cos\theta_{Yj}$, and $\cos\theta_{Zk}$ is required if we wish to further analyze the singularity of the associated linear system, as it is not clear in the absence of further investigation whether the singularity is accidentally linked to the precise numerical values in the specific case of the MAP structure or whether it is more generally linked to the basic geometric features of point group 2. This analysis can be more readily performed within the reference frame shown in Fig. 11 and defined as follows: since only the Y binary crystal axis is imposed by symmetry, the X axis can be arbitrarily chosen in the plane perpendicular to Y. We also choose it to be in any of the two molecular planes, so that $X = x$ is a common axis for macroscopic and microscopic reference frames, and call α the rotation angle around this common axis that transforms y into Y. Considering the two molecules in the unit cell, axis $X = x$ is simply the intersection of the corresponding molecular planes, and 2α is the angle between these planes. With this choice, Eqs. (148a) take a much simpler form:

$$\begin{aligned} b_{YXX} &= (\cos\alpha)\beta_{yxx}, & b_{YYY} &= (\cos^3\alpha)\beta_{yyy} \\ b_{YZZ} &= (\cos\alpha\sin^2\alpha)\beta_{yyy}, & b_{XYZ} &= -(\sin\alpha\cos\alpha)\beta_{xyy} \end{aligned} \tag{148b}$$

The particular frame now clearly evidence the essential (i.e., nonaccidental) singularity of the linear system of equations: the tensor b does not depend on β_{xxx}, which also means that this coefficient cannot be reached from second-harmonic generation data in the crystal. In addition, b_{yyy} and b_{yzz} are proportional. In other words, the two-dimensional assumption implies a new structural relation between two components of b, namely, $b_{YZZ}/b_{YYY} = \tan^2\alpha$. This is a predictive relation that can readily be tested from experimental nonlinear and crystallographic data.

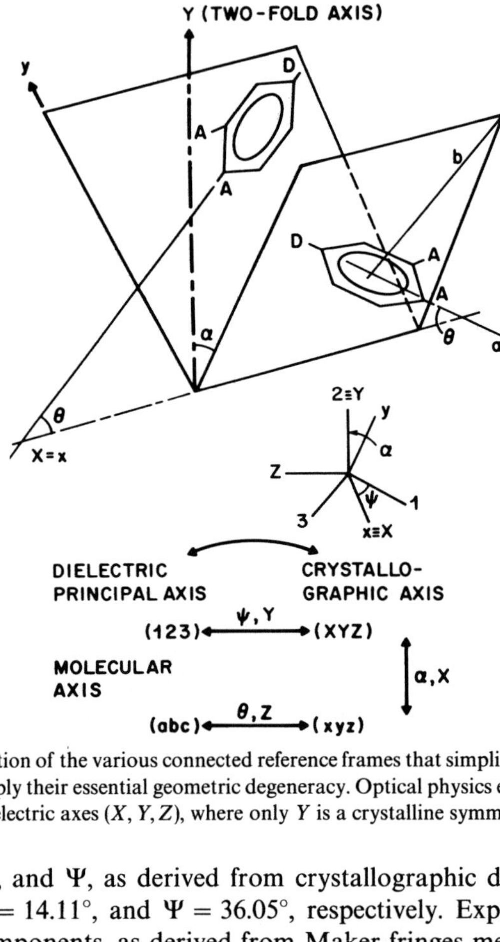

Fig. 11. Definition of the various connected reference frames that simplify Eqs. (148a)–(168b) and help show simply their essential geometric degeneracy. Optical physics experiments will refer to the principal dielectric axes (X, Y, Z), where only Y is a crystalline symmetry axis.

Angles α, Θ, and Ψ, as derived from crystallographic data analysis, are $\alpha = 36.89°$, $\Theta = 14.11°$, and $\Psi = 36.05°$, respectively. Experimental values of b tensor components, as derived from Maker fringes measurements of d coefficients of MAP (see Section VI.B), are given in Table XII. The experimental value of b_{YZZ}/b_{YYY} is 0.54, versus 0.56 for $\tan^2\alpha$. This satisfactory agreement validates the various models introduced so far for describing quadratic nonlinear phenomena in charge-transfer molecular organic crystals. Additional considerations on the cancellation of $d_{213}^2/d_{211}d_{233} - 1$ (Oudar and Zyss, 1982) have shown the cancellation of this quantity to be a relevant criterion as to the possible one-dimensional nature of the molecular charge transfer. In view of this criterion, a one-dimensional nonlinearity can be discarded in the case of MAP, as could be inferred from the respective para and ortho positions of the substituents, which can be refered to the respective contributions of the ortho and para charge-transfer processes taking place in MAP.

TABLE XII

Components of the b and β Tensors of MAP, Derived from SHG Experimental Results (in Crystal and Solution)[a]

b_{YXX}	b_{YYY}	b_{YZZ}	b_{YXZ}	β_{xxx}	β_{xyy}	β_{yyy}	β_{yxx}	β_{aaa}	β_{baa}	β_{abb}	β_{bbb}
−1.62	−3.59	−1.93	−1.12	−28.5	2.3	−6.8	−2.0	−27.1	5.6	−0.48	−7.7

[a] Units are 10^{-30} esu.

It is of interest to compare experimental crystal inferred β-coefficient values, as from Table XII, to EFISH measurement (Oudar and Chemla, 1977) of *p*- and *o*-nitroaniline molecules. A vectorial addition scheme on measured $\boldsymbol{\beta}$ values of these two molecules, assuming in both cases a one-dimensional β tensor directed from the acceptor on to the donor nitrogen, yields $(\boldsymbol{\beta})_a = -40 \times 10^{-30}$ esu and $(\boldsymbol{\beta})_b = -8.8 \times 10^{-30}$ esu to be compared to $(\boldsymbol{\beta})_a = -27.6 \times 10^{-30}$ esu and $(\boldsymbol{\beta})_b = -2.2 \times 10^{-30}$ esu for MAP. There seems to be much more anisotropy in favor of the para process for MAP than could be induced from this single additive scheme. In addition, the norm of $\boldsymbol{\beta}$, i.e., $(\boldsymbol{\beta} \cdot \boldsymbol{\beta})^{1/2}$, is reduced by almost a factor of two from the "additive" molecule onto actual MAP. In fact, as far as vectorial quantities are involved, MAP behaves very much like *p*-nitroaniline, with the difference of an additional smaller transverse β component making the β tensor two-dimensional. There remains to be solved the problem of finding β coefficients in terms of measured b coefficients, a task that requires additional data in view of the singularity of the linear crystallographic system of Eq. (148b). This is provided by solution EFISH measurement, which yields

$$\beta_\mu = (\beta_{xxx} + \beta_{xyy})\cos(\Theta + \eta) + (\beta_{yxx} + \beta_{yyy})\sin(\Theta + \eta) \qquad (149)$$

where $\eta = 42°$, obtained from INDO–Hartree–Fock calculations on (2,4-DNA), is the angle between the molecular dipole moment and axis a (the "para" charge transfer axis). Values of β tensor coefficients expressed in the *abc* reference frame are given in Table XII, showing as expected from simpler arguments, a negative prevailing para-oriented (i.e., along a) component $(\boldsymbol{\beta})_a = \beta_{aaa} + \beta_{abb}$ of the $\boldsymbol{\beta}$ tensor. The $(\boldsymbol{\beta})_b$ component (mostly "ortho" related) is by one order of magnitude smaller but nonnegligible, thus confirming the essential two-dimensional nature of the $\boldsymbol{\beta}$ tensor and related charge-transfer phenomenon: $(\boldsymbol{\beta})_b - \beta_{bbb} + \beta_{baa} \simeq (\boldsymbol{\beta})_a/10$. Both components are negative.

Similar considerations can be fruitfully applied as well to the linear properties of the molecule and of the crystal and thus can establish, within the framework already discussed, a connection between the crystal index of refraction and the molecular polarizability. It is, however, not *a priori* justified to describe the polarizability properties of the molecule by a two-dimensional

α tensor as was done for β because the linear polarizability contains non-negligible σ contributions that may involve out-of-plane bonds while the hyperpolarizability mainly originates from in-plane π-electron contributions. Both crystal and molecular tensors have four components, and the linear system connecting them can be readily inverted here to yield

$$\alpha_{xx} = a_{XX}, \qquad a_{yy} = [\cos^2(\alpha)\, a_{YY} - \sin^2(\alpha)\, a_{ZZ}] \cos 2\alpha$$
$$a_{xy} = a_{XZ}/\sin\alpha, \qquad a_{zz} = [\cos^2(\alpha)\, a_{ZZ} - \sin^2(\alpha)\, a_{YY}] \cos 2\alpha \tag{150}$$

The a coefficients (i.e., crystalline polarizability per molecule as defined in Section V.B) are in turn connected by the Lorenz–Lorentz relations to the indices of refraction, and Sellmeir are expressions derived:

$$a_{\mathrm{LL}}(\lambda) = a_{\mathrm{LL}}^{(0)} + a_{\mathrm{LL}}^{(1)} \frac{\lambda_{\mathrm{L}}^2}{\lambda^2 - \lambda_{\mathrm{L}}^2} - a_{\mathrm{LL}}^{(2)} \lambda^2$$

Results are found in Table XIII. They show that the α tensor is nearly isotropic in the plane xy (i.e., $\alpha_{xy} \ll \alpha_{xx} \approx \alpha_{yy}$). This implies that $a_{XY} \simeq 0$ and subsequently that X and Z are principal dielectric axes. This result is all the more striking as it ascribes dielectric significance to axes X and Z, which were defined purely structurally as the intersection of the two symmetry-related molecular planes. This correspondence goes more intimately into the packing geometry than the classical identification of the crystal symmetry axis with principal dielectric axis.

TABLE XIII

Linear Polarizability Results Obtained from the Reference Indices of MAP Crystals[a]

Sellmeir term	(0)	(1)	(2)	λ_L (μm)
a_{XX}	32.5	5.3	0.42	0.377
a_{YY}	24.9	2.6	0.19	0.413
a_{ZZ}	22.0	1.3	0.20	0.406
a_{XZ}	~0	~0	~0	
α_{xx}	32.5	5.3		
α_{xy}	~0	~0		
α_{yy}	28.6	4.3		
α_{xx}	18.3	−0.4		
α_{aa}	32.3	5.2		
α_{ab}	−0.9	−0.24		
α_{bb}	28.8	4.4		
α_{cc}	18.3	−0.4		

[a] Units for $a^{(0)}$ and $a^{(1)}$ are 10^{-24} esu and for $a^{(2)}$ are 10^{-16} esu.

An additional two-dimensional hypothesis on the α tensor would lead to a structural relation, very similar to that satisfied by the b coefficients, namely, $a_{ZZ}/a_{YY} = \tan^2\alpha$. This relation is satisfactorily met by $a^{(1)}$ ($a^{(1)}_{ZZ}/a^{(1)}_{YY} \approx 0.5$, versus $\tan^2\alpha = 0.56$), which governs the dispersion of the first electronic charge transfer absorption band in the UV. However, this is not true for the nonresonant $a^{(0)}$ term (and thus for a), which virtually contains contributions from the out-of-plane higher excited states that involve the chiral radical.

We have defined a convenient framework for discussing from now on the problem of crystallographic optimization: i.e., for a given molecular β susceptibility, find the molecular orientation in the crystal lattice that will lead to the highest b (or d) tensor coefficients. Our aim is to go beyond the classical negative statement on centrosymmetry when $\chi^{(2)}$ processes are concerned and to quantitatively estimate the cost of the crystallographic projection factors so as to show "optimal" structures.

Before discussing this problem in general in Section V.D, we shall illustrate, in the case of MAP, the gain on nonlinear efficiency that can be expected when moving from the actual MAP structure on to an "optimized" one. This is shown in Fig. 12, where the two unit-cell molecules are consistently rotated within their mean planes at fixed $\alpha = 36.89°$, from the actual $\Theta = 14.11°$, on to an arbitrary Θ value. The β_{yyy} (or β_{yxx}) coefficients can be readily connected to b_{YYY} (or b_{YXX}) by Eq. (148b). It is found that $\Theta = 14.11°$ is far from optimum, and that one order of magnitude could be gained on β_{yyy} or β_{yxx} if equivalent molecules could be packed at "better" angles. Finally, when optimizing on both α and Θ angles it is found that the theoretical optimal value for d_{YXX} is 6.7

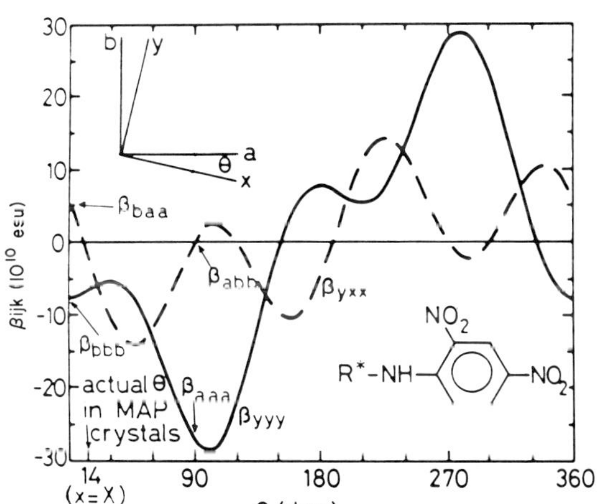

Fig. 12. Variation of the molecular β_{ijk} components upon rotation of MAP molecules in the *ab* plane. The numerical values are those of Table XII.

times larger than the phase-matchable d_{21} coefficient in actual MAP crystals, while the ratio would be of 12 for the non–phase-matchable d_{YYY} coefficient. As will be discussed in Section V.E, NPP (Zyss *et al.*, 1984) exemplifies an optimal monoclinic lattice for the phase-matchable d_{21} coefficient while MNA (Levine *et al.*, 1979) exhibits an optimal structure for the d_{22} coefficient. The molecular unit in both cases are *p*-nitroaniline derivatives and can be considered to have equivalent molecular polarizabilities. The former crystal is ideally adapted to optical parametric applications, while the latter is well tailored toward electro-optic applications.

D. From Noncentrosymmetric to Optimized Lattices

As was previously demonstrated in this section, the mere exclusion of centrosymmetric structures is not a precise enough criterion to help select relevant crystalline structures for quadratic nonlinear optics: some noncentrosymmetric structures may depart only slightly from a corresponding centrosymmetric one. A more subtle situation is illustrated in Fig. 13, where, for the sake of clarity, a two-dimensional crystal with two molecules per unit

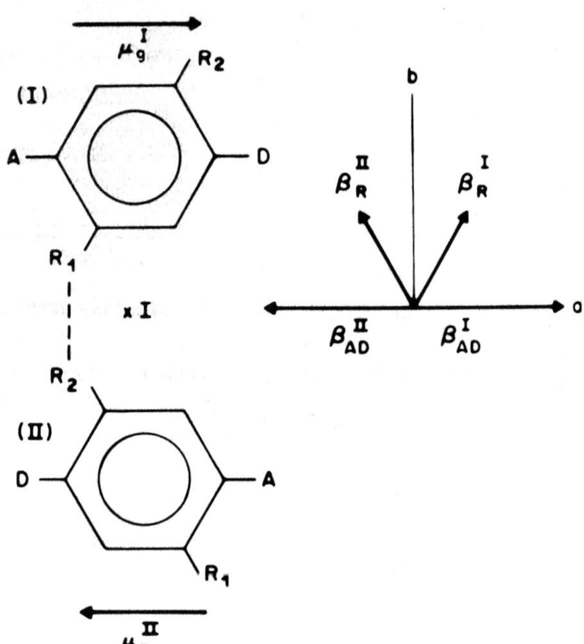

Fig. 13. Hypothetical noncentrosymmetric arrangement of molecules, showing mutual cancellation of the prevailing charge-transfer-related contribution to the molecular nonlinearity.

cell is being discussed. The molecular unit is a benzene ring with four different substitutents, A, D, R_1, and R_2, A and D being electron-accepting and electron-donating groups, respectively, interacting in para (1,4) positions via the benzene ring while R_1 and R_2 are two aliphatic substituents making no significant contributions to the dipole, the nonlinearity, or any other π-electron-related properties of the molecule. If the ground-state dipole, essentially parallel to the donor–acceptor charge-transfer axis, has a significant value, it may drive the dipoles of molecules **(I)** and **(II)** in the frequently met head-to-tail configuration of Fig. 13. If, in addition, groups R_1 and R_2 contain some chemical features, such as hydrogen-bonding atoms, endowing them with mutual affinity, the final structure may end up resembling that of Fig. 13. Point I is not an inversion center for the full molecule, but may be viewed as a "partial" inversion center for that portion of the molecule that does not include R_1 and R_2 or for the sole π-electron system of the molecule. Assuming geometric additivity of the nonlinearities of molecules $R_1{-}C_6H_4{-}R_2$ and $A{-}C_6H_4{-}D$, tilted with respect to each other by an angle of 60° or equivalently, additivity of σ and π contribution to the total molecular nonlinearity leads to

$$b_{bbb} = \beta_R(3\sqrt{3}/8), \qquad b_{aaa} = 0$$

$$b_{baa} = \beta_R(\sqrt{3}/8), \qquad b_{abb} = \tfrac{3}{8}\beta_R$$

where β_R is the supposedly one-dimensional weak nonlinearity of $R_1{-}C_6H_4{-}R_2$, while the major part of the molecular nonlinearity, β_{AD}, corresponding to the π charge-transfer-related contribution, does not show up. In fact, the lattice geometry of Fig. 13 cancels out the major part of the molecular susceptibility while the weakest σ contribution remains as the only molecular contribution to the crystal susceptibility tensor.

This theoretical situation is encountered in practice in urea-like crystals, as can be seen from the simplified lattice description of Fig. 14. The actual crystalline packing of urea is not favorable, as it cancels out the major β_{yyy} (y parallel to the dipole of urea) molecular term while the only molecular term to contribute is the weaker β_{yxx} term (Zyss and Oudar, 1982; Morrell and Albrecht, 1979) with $\beta_{yxx} \approx \beta_{yyy}/3$. For the $\bar{4}2m$ group, the general expression of the only Kleinman-allowed coefficient is $b_{XYZ} = \cos\alpha \sin 2\phi(1 + \Lambda \sin^2 \alpha)\beta_{yxx}/2$, where $\Lambda = -\beta_{yyy}/\beta_{xxx} \approx -3$, α is the angle of the dipole with the 4 axis Z, and ϕ is the angle with crystal axis X of the projection of the dipole on the (001) plane. The optimal value of b_{xyz} is still found for $\phi = \pi/4$, but a better α value, allowing for the optimal combination of "longitudinal" β_{yyy} and "transverse" β_{yxx^3} molecular contributions, would be given by $\cos\alpha = [(1 + \Lambda)/3\Lambda]^{1/2}$ and $\alpha = 46.6°$, finally raising d_{XYZ} by a factor of 1.56 above the actual urea value. The nonlinear efficiency of this theoretical material would be 2.5 times above that of urea with a comparable transparency.

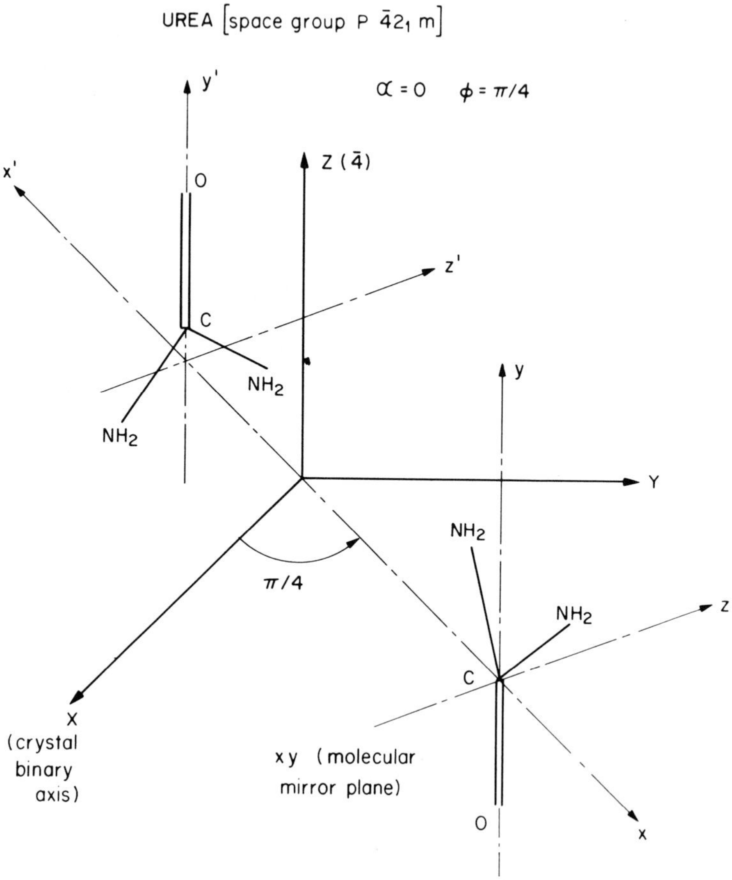

Fig. 14. The small number of related molecules (two instead of a standard $Z = 8$) in the $\bar{4}2m$ unit cell of the urea crystal is a consequence of the coincidence of molecular and crystalline symmetry axis. Optimization of a (with ϕ remaining equal to $\pi/4$) would eventually increase the crystalline nonlinearity.

It was very much along the same considerations that we have shown in the previous section that most of the crystal nonlinearity of MAP originated from the weaker "ortho" molecular charge-transfer-related nonlinearity at the expense of the higher "para-related" one.

The approach used in Section V.C for the specific case of the MAP crystal can be generalized for all crystal point groups (Zyss and Oudar, 1982) and leads to a precise geometric reference definition of the best molecular orientation for phase-matched second-harmonic generation. Such a configuration is defined by four angular parameters (see Fig. 15) Θ, Ψ, λ, and ρ: Θ and Ψ position the supposedly one-dimensional non-linear molecule (axis μ)

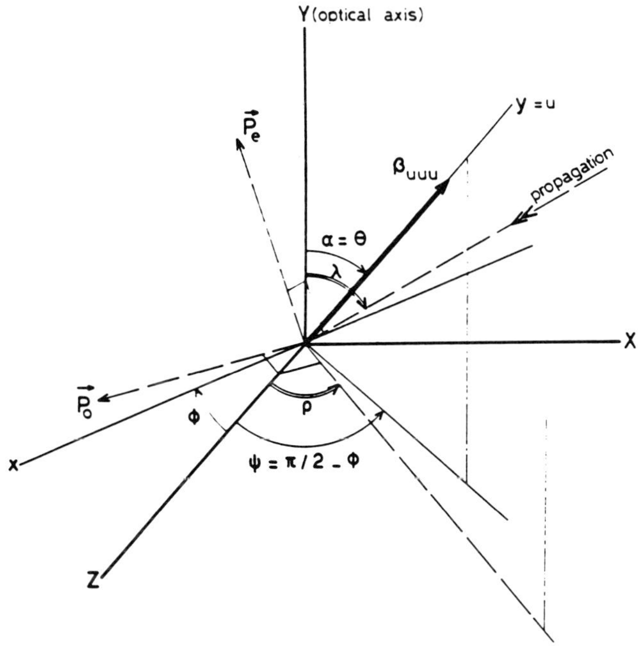

Fig. 15. Definition of the angular parameters involved in the determination of the best molecular orientation for phase-matched second-harmonic generation.

with respect to the crystal reference frame, while λ and ρ define the propagation direction of the interacting beams. Table XIV displays the optimal phase-matched nonlinearities for all Kleinman-allowed crystal point groups together with the corresponding molecule-propagation orientations. The most favorable groups are those having the highest maximum effective crystalline nonlinearity per molecule (i.e., $b_{\text{eff}}^{\text{max}}$). From that point of view, biaxial groups (with the exception of 222) that correspond to the least symmetrical lattices are to be preferred. It turns out that, following systematic exploration from crystallographic data files, these groups are the most frequently encountered among organic structures, while the situation is reversed for inorganic materials, where more symmetric (i.e., less favorable) structures are more frequent. The optimal configurations for biaxial groups, including the "best" four groups of Table XIV are found in Fig. 16.

It is worth generalizing further to other crystal point groups the structural relations already obtained for MAP in the case of a single crystalline twofold axis. The following situations are to be encountered: if the number of macroscopic coefficients $p(g)$ allowed by the crystal symmetry group g is greater than four (number of molecular coefficients in a two-dimensional model of the nonlinearity), the linear relation such as established in Sec-

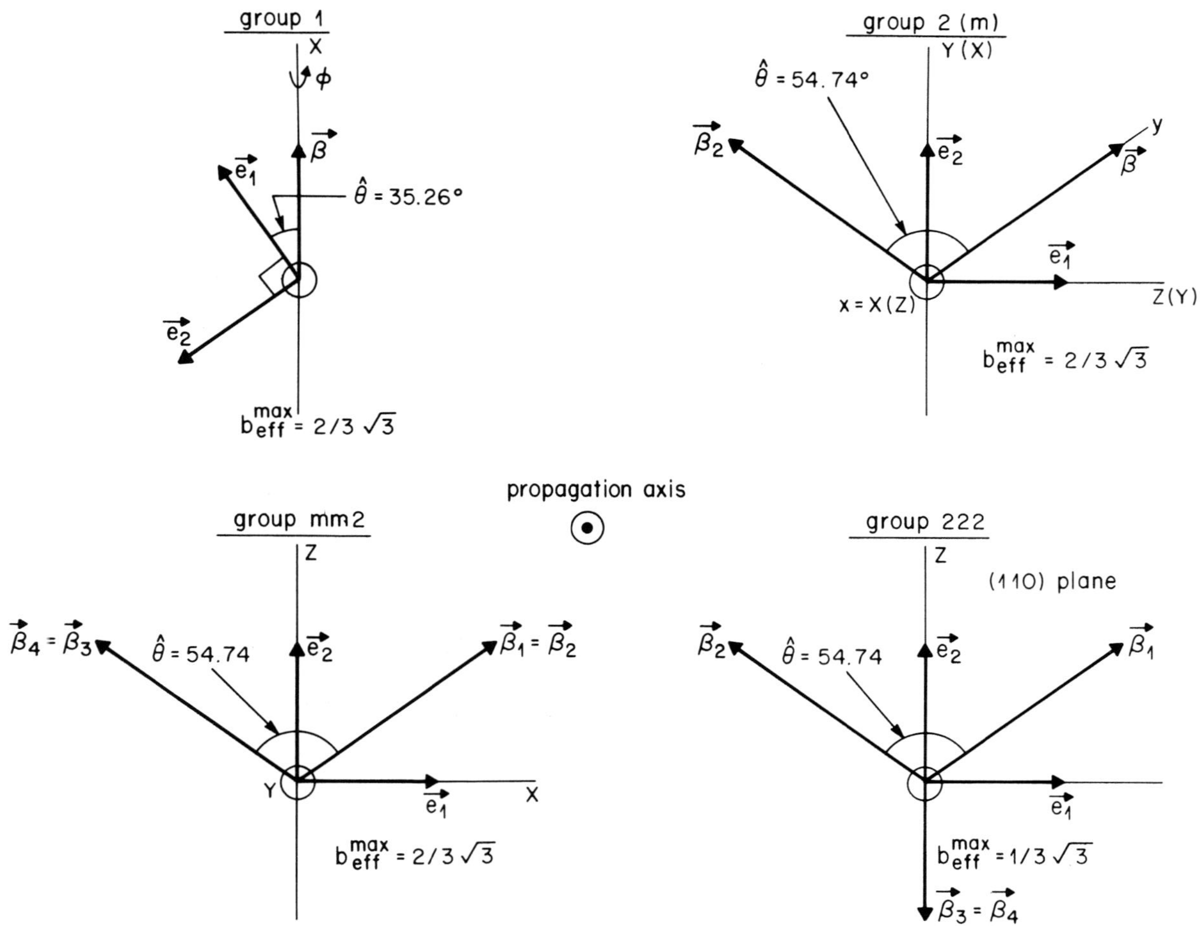

Fig. 16. Optimal molecular orientations (one-dimensional case) and phase-matching configurations for biaxial crystal groups.

TABLE XIV

Optimal Effective Phase Matched Nonlinearities for all Kleinman-Allowed Crystal Point Groups[a]

Group	NL coefficient	Θ	Ψ	λ	ρ	$b_{\text{eff}}^{\max}$
1, 2, m, $mm2$		See text and Fig. 16				$\frac{2}{3}\sqrt{3} \cong 0.38$
$\overline{62}m$	$d_{22}, \frac{1}{4}$	90°	60°	0°(1, 2)	0°(1), 30°(2)	$\frac{1}{4}$(1, 2)
$\bar{6}$, 3	$d_{22}, \frac{1}{4}$	90°		0°(1, 2)	0°(1), 30°(2)	$\frac{1}{4}$(1, 2)
3m	$d_{22}, \frac{1}{4}$	90°	30°	0°(1, 2)	0°(1), 30°(2)	$\frac{1}{4}$(1, 2)
32	$d_{11}, \frac{1}{4}$	90°	60°	0°(1, 2)	30°(1), 0°(2)	$\frac{1}{4}$(1, 2)
222		See text and Fig. 16				$\frac{1}{3}\sqrt{3}$
6mm, 6, 4mm, 4	$d_{15}, \frac{1}{3}\sqrt{3}$	54.74°		90°(2)		$\frac{1}{3}\sqrt{3}$
4m, 23, 43m	$d_{14}, \frac{1}{3}\sqrt{3}$	54.74°	45°	45°(1), 90°(2)	0°(1), 45°(2)	$\frac{1}{3}\sqrt{3}$
$\bar{4}$	0					0

[a] Between parentheses is the type of phase matching, following Midwinter (Midwinter and Warner, 1965), corresponding to an angular parameter, or an optimal value of b_{eff}; $(d_{22}, \frac{1}{4})$ for $\bar{6}2m$ group, for example, means that the optimal phase-matching configuration involves a d_{22} coefficient with $b_{222}^{\max} = \beta_{uuu}/4$. The $\bar{4}$ group shows no possibility of phase-matched interactions.

TABLE XV

Expressions of the Macroscopic Nonlinear Susceptibilities b_{IJK} in Terms of the Four Two-Dimensional Molecular Hyperpolarizability Coefficients β_{ijk} for Lower Crystal Symmetry[a]

	1	2	m	$mm2$	222
b_{XXX}	β_{xxx}	0	β_{xxx}	0	0
b_{YYY}	β_{yyy}	$\beta_{yyy}\cos^3\alpha$	0	0	0
b_{ZZZ}	0(*)	$-\beta_{yyy}\sin^3\alpha$	$\beta_{yyy}\cos^3\alpha$	0	
b_{XYY}	β_{xyy}	0	$\beta_{yyx}\cos^2\alpha$	0	0
b_{YXX}	β_{yxx}	$\beta_{yxx}\cos\alpha$	0	0	0
b_{YZZ}	0(*)	$\beta_{yyy}\cos\alpha\sin^2\alpha$	0	0	0
b_{ZYY}	0(*)	0	$-\beta_{yyy}\cos^2\alpha\sin\alpha$	$-\beta_{xyy}(\sin^2\Phi\sin 2\alpha)/2$ $+\beta_{yxx}\sin^2\Phi\cos\alpha$ $+\beta_{yyy}\cos^2\Phi\cos\alpha\sin^2\alpha$	0
b_{XZZ}	0(*)	0	$\beta_{xyy}\sin^2\alpha$	0	0
b_{ZXX}	0(*)	$-\beta_{yxx}\sin\alpha$	$\beta_{xyy}(\sin 2\Phi\sin 2\alpha)/2$ $+\beta_{yxx}\cos^2\Phi\cos\alpha$	0 $+\beta_{yyy}\sin^2\Phi\cos\alpha\sin^2\alpha$	
b_{XYZ}	0(*)	$-\beta_{xyy}\cos\alpha\sin\alpha$	0	0	$\sin 2\Phi\cos\alpha(\beta_{yxx}$ $-\beta_{yyy}\sin^2\alpha)$ $-\beta_{xyy}\cos 2\Phi\sin 2\alpha$

[a] Definitions of reference frames XYZ, xy, and angles α and ϕ are to be found in Figs. 17 and 18. A zero value of coefficient d_{IJK} is a consequence of both symmetry considerations and Kleinman relations, regardless of the molecular properties. Starred 0 (i.e., 0(*)) for group 1 points out that the canceling out of the corresponding coefficients is of a different origin: it is a consequence of the reduced dimensionality of the molecular system and is to be viewed as a structural relation similar to those occuring in groups m, 2, and 3.

tion V.C in a special case will lead to, at least, $p(g) - 4$ relations between the macroscopic coefficients. Such is the case for groups $1[p(1) = 10]$ and $m[p(m) = 6]$. In addition, nontrivial structural relations may occur even for symmetry groups with $p(g) \leqslant 4$. In the case of group 2, with $p(2) = 4$, one such structural relation has been exhibited: in adequately defined crystalline and molecular reference frames, one out of the four molecular coefficients was found to be missing in the linear relations connecting b and β.

The set of linear equations connecting b to β can be written in matrix notation: $b = A\beta$, where A is a structural crystallographic matrix [four lines, $p(g)$ columns] relating the crystal tensor b [one column, $p(g)$ lines] to the molecular tensor β (one column, four lines). If $\rho(A)$ is the rank of A—or the dimension of the largest non singular square submatrix of A, with $\rho \leqq \mathrm{Inf}[4, p(g)]$—then the number S of structural relations connecting macroscopic tensor coefficients is $S = p(g) - \rho(A)$. In more practical terms, a straightforward way to find $p(g)$ and exhibit the S structural relations is just to try to express β in function of b. This procedure corresponds to the problem of inducing molecular coefficients from crystal tensor coefficients.

The general expression of b (or β) tensor coefficients in terms of β (or b) tensor coefficients are to be found in Tables XV and XVI, respectively, for low-symmetry groups and in Tables XVII and XVIII, respectively, for higher-symmetry groups in Figs. 17 and 18.

Four point groups yield the unusual structural relations previously evidenced in MAP.

Group 1 $(p = 10, \rho = 4, S = 6)$: In this group with no symmetry constraint to fix any of the crystal axes we are free to make molecular and crystalline axis coincide, leading to $b_{IJK} = \beta_{ijk}$ for all permutations of $IJK = ijk$. The six obvious structural structural relationships directly display the influence of the two-dimensional assumption on the crystal property by

$$d_{ZZZ} = d_{ZXX} = d_{ZYY} = d_{ZZY} = d_{ZZX} = d_{XYZ} = 0$$

Group 2 $(p = 4, \rho = 3, S = 1)$: That analyzed in detail in Section V.C. One structural relation is obtained,

$$\tan^2 \alpha = b_{YZZ}/b_{YYY}$$

Group m $(p = 6, \rho = 4, S = 2)$: Two structural relations exist,

$$\tan^2 \alpha = b_{XZZ}/b_{XYY} = b_{ZZZ}/b_{ZYY}$$

Group 3 $(p = 4, \rho = 3, S = 1)$: One structural relation exists,

$$2d_{YYY} + 3d_{ZXX} \tan \alpha = 2d_{ZZZ} \tan^3 \alpha$$

TABLE XVI

Expressions of the Two-Dimensional Hyperpolarizabilities β_{ijk} in Terms of the Macroscopic Nonlinear Susceptibilities for Low-Symmetry Crystals[a]

	1	2	m	$mm2$
β_{xxx}	b_{XXX}	Abs	XXX	Abs
β_{yyy}	b_{YYY}	$b_{YYY}/\cos^3\alpha = b_{YZZ}/\cos\alpha\sin^2\alpha$	$-b_{ZZZ}/\sin^3\alpha = -b_{YYZ}/\cos^2\alpha\sin\alpha$	$b_{ZZZ}/\cos^3\alpha$
β_{xyy}	b_{XYY}	$-b_{XYZ}/\sin\alpha\cos\alpha$	$b_{XZZ}/\sin^2\alpha = b_{YYX}/\cos^2\alpha$	$(2b_{ZZZ}tg^2\alpha/tg2\Phi$ $+ b_{XXZ}tg\Phi - b_{YYZ}/tg\Phi)\sin 2\alpha$
β_{yxx}	b_{YXX}	$b_{XXY}/\cos\alpha$	$-b_{XXZ}/\sin\alpha$	$(b_{XXZ} + b_{YYZ} - b_{ZZZ}tg^2\alpha)/\cos\alpha$

[a] "Abs" refers to the absence of the corresponding β coefficient in the expression of tensor d for the relevant symmetry group. The structural relations between macroscopic coefficients appear in this table as equalities (examples: $b_{YYY}/\cos^3\alpha = b_{ZZZ}/\cos\alpha\sin^2\alpha$ for group 2, etc.).

TABLE XVII

Expressions of the b_{III} Macroscopic Nonlinear Susceptibilities in Terms of the Four Two-Dimensional Molecular Hyperpolarizability Coefficients β_{ijk} for Higher Crystal Symmetry[a]

	b_{XXX}	b_{YYY}	b_{ZZZ}
3	$(\beta_{xxx} - 3\beta_{xyy}\sin^2\alpha)/4$	$\frac{1}{4}(\beta_{yyy}\sin^2\alpha - 3\beta_{yxx})\sin\alpha$	$\beta_{yyy}\cos^3\alpha$
32	$\cos\Phi(\cos^2\Phi - 3\sin^2\Phi)(\beta_{xx} - 3\beta_{xyy}\sin^2\alpha)/4$ $+ \sin\Phi(3\cos^2\Phi - \sin^2\Phi)(\beta_{yyy}\sin^2\alpha - 3\beta_{yxx})/4$	0	0
$3m$	0	$\frac{1}{4}[\sin\Phi(\sin^2\Phi - 3\cos^2\Phi)(\beta_{xxx} - 3\beta_{xyy}\sin^2\alpha)$ $+ \cos\Phi(\cos^2\Phi - 3\sin^2\Phi)(\beta_{yyy}\sin^2\alpha - 3\beta_{yxx})\sin\alpha]$	Idem 3
4, $4mm$, 6, $6mm$	0	0	0
$\bar{4}$	0	0	0
$\bar{4}2m$	0	0	0
$\bar{6}$	Idem 3	Idem 3	0
$\bar{6}2m$	0	$\frac{1}{4}[\cos\Phi(\cos^2\Phi - 3\sin^2\Phi)(\beta_{xxx} - 3\beta_{xyy}\sin^2\alpha)$ $+ \sin\Phi(\sin^2\Phi - 3\cos^2\Phi)(\beta_{yyy}\sin^2\alpha - 3\beta_{yxx})\sin\alpha]$	0
23, $\bar{4}3m$	0	0	0

	b_{ZXX}	b_{XYZ}
3	$(\beta_{yxx} + \beta_{yyy}\sin^2\alpha)\cos\alpha/2$	0
32	0	0
$3m$	Idem 3	0
4, $4mm$, 6, $6mm$	Idem 3	0
$\bar{4}$	$(\beta_{yxx} - \beta_{yyy}\sin^2\alpha)\cos\alpha/2$	$\beta_{xyy}\sin\alpha\cos\alpha$
$\bar{4}2m$	0	$\cos\alpha[\beta_{xyy}\cos 2\Phi\sin\alpha$ $+ \sin 2\Phi(\beta_{yxx} - \beta_{yyy}\sin^2\alpha)/2]$
$\bar{6}$	0	0
$\bar{6}2m$	0	0
23, $\bar{4}3m$	0	$-\beta_{yyy}\cos\alpha\sin^2\alpha\sin\Phi\cos\Phi$ $-\beta_{xyy}\sin\alpha\cos\alpha\cos 2\Phi$ $-\beta_{yxx}\sin\Phi\cos\Phi\cos\alpha$

[a] "Idem 3 (or any other group)" means that the corresponding coefficient is identical to that of group 3.

TABLE XVIII

Expressions of Two Dimensional Hyperpolarizabilities β_{ijk} in Terms of the Macroscopic Nonlinear Susceptibilities for Low-Symmetry Crystals

	β_{xxx}	β_{yyy}	β_{xyy}	β_{yxx}
3, 3*m*	Pres[a]	$d_{ZZZ}/\cos^3\alpha$	Pres	$(2b_{ZXX} - b_{ZZZ}tg^2\alpha)/\cos\alpha$
32, $\bar{6}$, $\bar{6}2m$	Pres	Pres	Pres	Pres
4, 4*mm*, 6, 6*mm*	Abs	Idem 3	Abs	Idem 3
$\bar{4}$	Abs	Pres	$d_{XYZ}/\sin\alpha\cos\alpha$	Pres
$\bar{4}2m$, 23, $\bar{4}3m$	Abs	Pres	Pres	Pres

[a] "Pres" refers to a molecular coefficient that is present in the expression of *b* for the relevant group but that cannot be expressed in terms of the macroscopic coefficients.

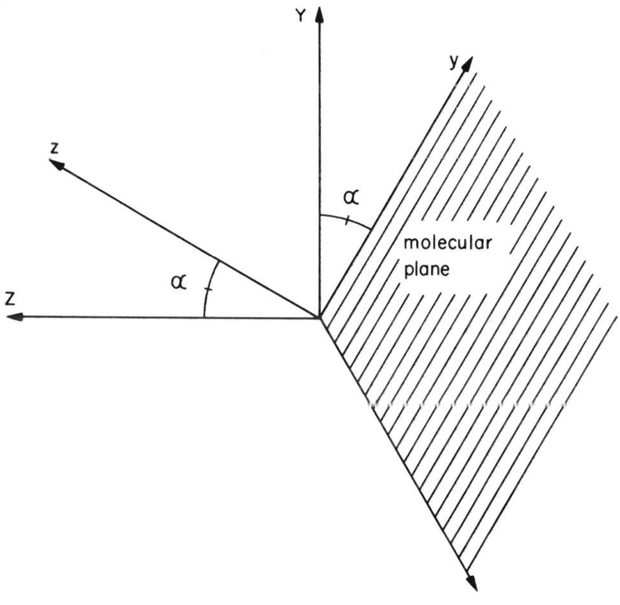

Fig. 17. Definition of macroscopic crystalline and microscopic molecular axis for point groups 2, *m*, 3, 4, $\bar{4}$, 6, and $\bar{6}$. Group 2: Y is conventionally along the symmetry axis; $X = x$ is the intersection of one of the molecular planes with (010) plane (both molecular planes intersect along the same direction by symmetry); y is the projection of Y on one of the molecular planes. The angle between the two equivalent molecular planes is 2α. Group *m*: Y is conventionally perpendicular to the mirror plane, $X = x$ being the intersection of the mirror plane with the molecular planes (y, as for group 2). The angle between the two equivalent molecular planes is then $\pi - 2a$. Group 3: for this group Y and Z have to be permuted on the drawing, Z being conventionally the threefold axis. Groups 4 and $\bar{4}$: Z is the fourfold axis, and X is the intersection of any molecular plane with the (001) plane. Other conventions are identical to that of group 2. Group 6: conventions in Fig. 3 of Zyss and Oudar (1982). Group $\bar{6}$: straightforward extensions of the definitions given by group 3, Z being now the inverse sixfold axis.

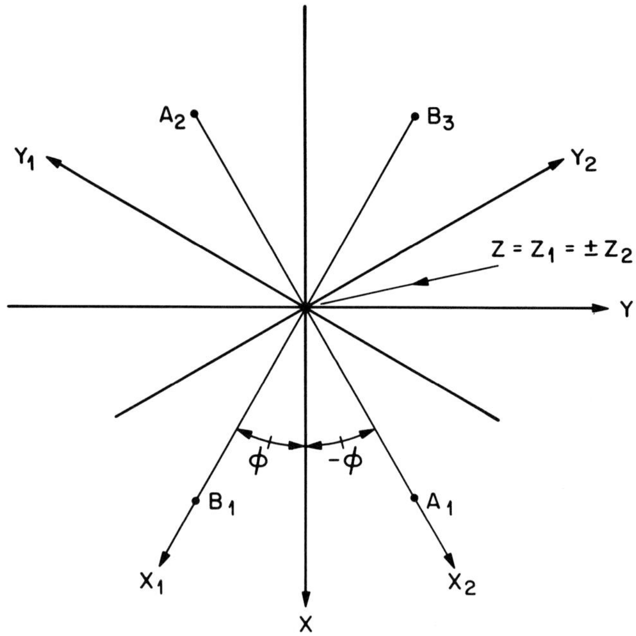

Fig. 18. Axis conventions for groups *mm*2, 222, 32, 3*m*, 4*mm*, $\bar{4}2m$, 6*mm*, $\bar{6}2m$, 23, and $\bar{4}3m$. Group *mm*2: Planes of equivalent molecules A_2 and B_2 intersect the crystal plane XY along X_1 (angle ϕ between X_1 and the mirror XZ). The locations of A_1 and B_1 are similar, but with angle ϕ. Two parameters, ϕ and a (already defined for group 2 as the half angle between molecular planes of A_1 and B_1 or of A_2 and B_2) suffice then to fully determinate the orientation of the molecular planes in the crystallogrphic reference frame. Group 222: same notions apply with X and Y being then the twofold axis, the only difference being $Z = Z_1 = -Z_2$ (instead of $Z = Z_1 = Z_2$ for group *mm*2). Group 32: X is along any of the three twofold axis perpendicular to the threefold axis Z. Here also $Z = Z_1 = -Z_2$. Group 3*mm*: ZX is any of the three mirror planes $Z = Z_1 = Z_2$. Group 4*mm*: Z is the fourfold axis and ZX is one of the two mirror planes. Otherwise, the same definitions as for *mm*2 apply here: $Z = Z_1 = Z_2$. Group $\bar{4}2m$: Z is the inverse fourfold axis and X a perpendicular twofold axis. Group 6*mm*: Z is the sixfold axis and ZY is one of the three mirror planes. Group $\bar{6}2m$: Z is the inverse sixfold axis and ZY is a mirror plane. Cubic groups 23 and $\bar{4}3m$: Z is, respectively, the main twofold axis or inverse fourfold axis, following crystallographic conventions.

An attempt to generalize this approach (Zyss and Oudar, 1982) to the electro-optic tensor *r*, also following an oriented gas description and assuming a two-dimensional molecular electro-optic tensor, leads to similar structural relations between crystalline *r* coefficients, which remain to be checked experimentally. Possible departures from such expected structural relations would be linked to the relative contribution to the electro-optic tensor of out-of-plane versus in-plane Raman modes. It is expected that the high anisotropy of planar charge-transfer molecules will also reflect on their vibrational properties and promote in-plane vibrations with respect to out-of-plane ones.

E. Examples of Efficient Organic Crystal Structures

The classification of crystal structures proposed in Section V.C is useful in as much as it offers a quantitative estimation of how much an actual crystallographic structure differs from the theoretical optimum. However, much remains to be done in order to actually reach the optimum. This is by no means a straightforward task, because there exists no such accurate and predictive relationship between molecular and crystallographic structures, in contrast to the one that exists between molecular structures and (hyper)polarizabilities. However, we know from experience that crystallographic structures are very sensitive to seemingly minor changes on the molecular structure. In view of the diversity of modifications readily available through organic synthesis from a basic initial nonlinear molecular structure, a variety of crystallographic structures will be scanned in practice. The "statistical" approach, provided one has sufficient synthesis through-put available, is likely to help get close to or even reach the true crystallographic optimum. Consideration should be paid to the cross-influence of polarizability- or transparency-related features and crystallographic-related ones. One might, at worst, although starting from a relevant nonlinear molecule, end up with an ideal crystallographic orientation but degraded electronic and optical properties. However, this rather blind statistical approach can be guided along some more predictive lines if we introduce in the selection process some of our knowledge, however limited, of the structural origins of intermolecular forces. This will be illustrated in the following, based on three different organic crystalline structures that exhibit higher crystalline nonlinearities and were conceived along three different strategies for optimization of crystal structure.

1. Chirality and Charge Transfer: Methyl-(2,4-dinitrophenyl)-aminopropanoate (MAP)

Molecules are said to have chiral character whenever their sole symmetry elements, if any, are symmetry axes. An example of such molecules is shown in Fig. 19; with a molecule containing a carbon atom linked to four different substituent groups. It can be easily checked that no such chiral molecule can be superimposed on its image through a mirror or inversion symmetry operation. In fact, chiral molecules have two forms, one being the image of the other through a joint inversion or mirror symmetry operation, and these are respectively termed left-handed and right-handed enantiomers. Optically pure materials—i.e., those composed of only left-handed or right-handed molecules—cannot therefore be centrosymmetric or allow mirror planes as symmetry elements. Such compounds are, in fact, readily available in nature (amino acids, sugars, and alkaloids are well known to be single enantiomers

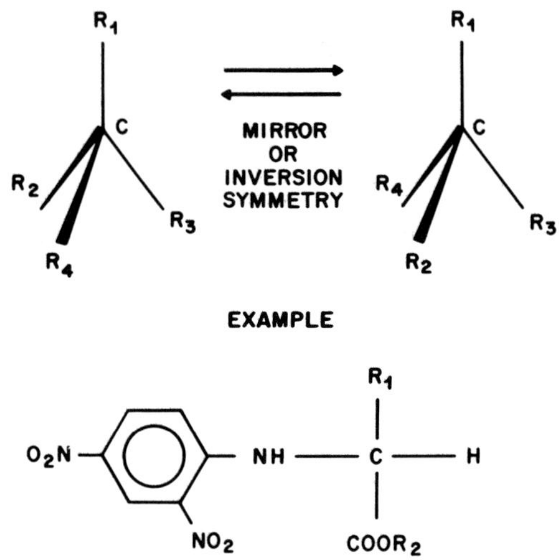

Fig. 19. Example of a family of chiral structures derived from 2,4-dinitroaniline by inclusion of an asymmetric carbon in the amine group. An symmetrically substituted carbon is a single chiral system: mirror or inversion symmetry will exchange left-handed and right-handed enantiomers. The MAP crystal, grown from purely left-handed optical isomers, is noncentrosymmetric.

only). One absolute way to ensure the absence of a center of symmetry in a crystalline structure is thus to grow a crystal made of purely, say, left-handed molecules, possibly natural amino acid derivatives. It should, however, be noted that a racemic crystal, composed of an equal mixture of left-handed and right-handed molecules, may well be noncentrosymmetric but must then have a mirror among its symmetry elements. This was demonstrated with racemic malic acid, which was shown to generate second-harmonic radiation (Betzler *et al.*, 1978). The chiral character brought to a given molecule, say, by addition of a chiral group must interfere only weakly or not at all with the desired additional property of interest. The idea that subsequently led to MAP (Oudar and Hierle, 1977) was to start from a classical reportedly highly nonlinear molecule, 2,4-dinitroaniline, and then, by introducing some chiral character into it, ensure that the resulting pure enantiomeric crystal would not be centrosymmetric. As was discussed at length in Section V.B, the resulting structure is not necessarily very much off a centrosymmetric one, as this strategy depends more on the symmetry of the interaction forces rather than on their intensity. This situation would compare closely to that encountered in spectroscopy when pure symmetry arguments will help predict the existence of a given transition line but not its intensity. The nonlinearity of 2,4-dinitroaniline, measured by EFISH in solution, is 21×10^{-30} esu, of the same order as that of *p*-nitroaniline, which is 34.5×10^{-30} esu (Oudar and Chemla,

TABLE XIX

SHG Powder Test Results for a Series of Chiral 2,4-Dinitroaniline Analogs (MAP Corresponds to $R^1 = R_2 = CH_3$)

	R_2			
R_1	H	CH_3	C_2H_5	$CH(CH_3)2$
CH_3	225	1500	3000	100
$CH(CH_3)_2$	5	450		2500
$CH—CH_2—(CH_3)_2$	210			

1977). However, *p*-nitroaniline as well as 2,4-dinitroaniline crystals have centrosymmetric crystalline structures (Wyckoff, 1969; Sidding *et al*., 1959). Relying on the fact that natural amino acids are left-handed, a number of esters of aromatic amino acids were prepared, all resulting in potentially second-harmonic–generating noncentrosymmetric structures, which were subsequently sorted out by the powder test. Results are shown in Table XIX; the molecular structure refers to that in Fig. 19.

A good trade-off, taking into account stability, simplicity of the synthesis scheme, lower molecular weight (to favor a more dense packing of the active units in the crystal), and powder SHG efficiency, turns out to be $R_1 = R_2 = CH_3$, eventually leading to MAP. EFISH measurements show that the modification of 2,4-dinitroaniline on to MAP has not proved disruptive to the molecular β tensor (Oudar and Chemla, 1977); for 2,4-dinitroaniline $\beta = 21 \times 10^{-30}$ esu, and for MAP, $\beta = 22 \times 10^{-30}$ esu (projection of the vector part of β along the respective dipolar axis). MAP was initially grown in solution, but difficulties subsequently led to developing a Stockbärger–Bridgman technique adapted to the melt temperature of MAP, $T_m = 79 \pm 1°C$. MAP is monoclinic and belongs to space group $P2_1$ with lattice parameters $a = 6.829$ Å, $b = 11.121$ Å, $c = 8.116$ Å, and $\beta = 95.59°$ (Knossow *et al*., 1976) The comparison of the actual structure of MAP with that of the corresponding theoretical optimal structure was detail in Section V.C, with the striking result that almost one order of magnitude could be gained in nonlinear efficiency with molecules comparable to MAP but ideally oriented (Oudar and Zyss, 1982).

2. Vanishing Ground-State Dipole Moment: 3-Methyl-4-nitropyridine-1-oxide (POM)

Important contributors to the crystal lattice structure are electrostatic dipole–dipole interaction forces. Before discussing the possibility of a vanishing molecular dipole moment essentially leading to a near cancellation of the electrostatic interaction between molecules, it should be noted that a small dipole and a high first-order molecular hyperpolarizability are not incompatible requirements, contrary to general belief. The two-level model already discussed in Section IV.B, when used to describe first-order nonlinear processes in charge-transfer molecules, leads to a proportionality relation between the first-order hyperpolarizability and the difference between the excited charge-transfer and ground-state dipole moments, i.e., Eq. (83). Therefore, a high value of the excited state dipole moment by means of an important charge-transfer process of a noncentrosymmetric molecule is compatible with a small ground-state dipole value to finally ensure a high value of β. Although the simplest examples of molecules with zero ground-state dipole moments are centrosymmetric molecules, these would have no molecular first-order hyperpolarizability and subsequently a vanishing crystal nonlinearity. One needs in fact a nonlinear molecule with a nonzero value of an odd-order multipole, of higher order than the dipole. POM (Zyss *et al.*, 1981) fulfills these requirements. The difficulty of finding noncentrosymmetric conjugated charge-transfer molecules of the donor–acceptor type with low ground-state dipole moments originates from the already significant amount of charge transfer generally present in the ground state. This is evidenced by a simplified two-level model of the charge transfer where we assume σ, π charge separation (see illustration in Fig. 20). Isolating those electrons actually taking part in the charge transfer (π electrons) from irrelevant ones (σ electrons), we call $|\phi_\pi^-\rangle$ the wavefunction of π electrons in the limiting case of no charge-transfer interactions (shielding the donor from the acceptor in some way), $|\phi_\pi^+\rangle$ the wavefunction of the same π electrons in the opposite limiting case or "ionic" case when one full electronic charge gets transfered from the donor on to the acceptor site, and $|\phi_\sigma\rangle$ the σ-electron wavefunctions (unchanged at first order under the excitation). We may express $|\Psi^g\rangle$ (or $|\Psi^e\rangle$); the total π–σ wave function of the actual ground state (or excited state), as

$$|\Psi^g\rangle = A[(\alpha|\phi_\pi^-\rangle + \beta|\phi_\pi^+\rangle)\otimes|\phi_\sigma\rangle] = A(\phi_\pi^g \otimes \phi_\sigma) \tag{151}$$

$$|\Psi^e\rangle = A[(\alpha'|\phi_\pi^-\rangle + \beta'|\phi_\pi^+\rangle)\otimes|\phi_\sigma\rangle] = A(\phi_\pi^e \otimes \phi_\sigma) \tag{152}$$

where A stands for the electron antisymmetrization operator (we deal here with closed-shell molecular systems with an even number of electrons).

Orthogonality and normalization of $|\Psi_g\rangle$ and $|\Psi_e\rangle$, and neglect, for simplification of the $\langle\phi_\pi^+|\phi_\pi^-\rangle$ overlap yield $\alpha' = \beta^*$, $\beta' = -\alpha^*$, $\alpha\alpha^* + \beta\beta^* = 1$, and $(\Psi_e) = A[(\beta^*|\phi_\pi^-\rangle - \alpha^*|\phi_\pi^+\rangle)\otimes|\phi_\sigma\rangle]$. The dipole moment operator can then, as justified in Section IV.D, be divided into in π and σ contributions for the electronic as well as nucleus-related parts, $\mu = \mu_\sigma + \mu_\pi + \mu_\sigma^n + \mu_\pi^n$, where μ_n is the additional nucleus contribution $\langle\mu_\sigma\rangle = \langle\phi_\sigma|\mu_\sigma|\phi_\sigma\rangle$ for both

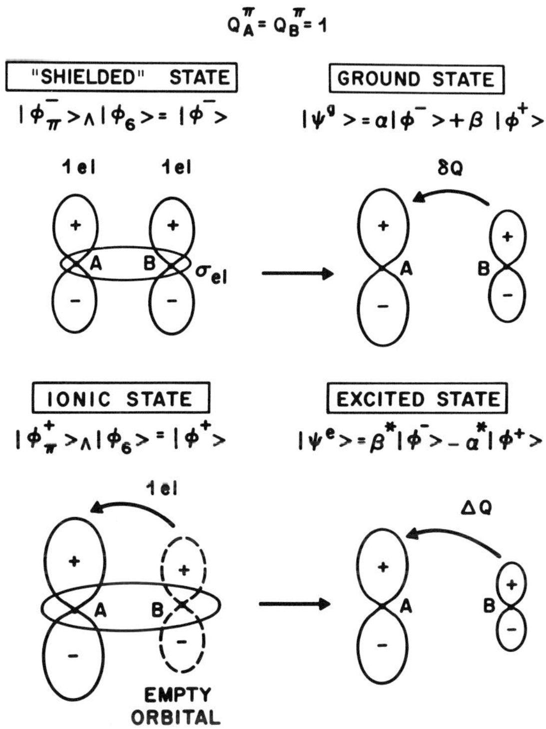

Fig. 20. Orbital definitions for a single two-level description of charge transfer involving idealized shielded and ionic states. See text for precise definitions; A = acceptor and B = donor

fundamental and excited charge transfer states, while

$$\langle \mu_\pi^{\rm g} \rangle = |\alpha|^2 \langle \phi_\pi^- |x| \phi_\pi^- \rangle + |\beta|^2 \langle \phi_\pi^+ |x| \phi_\pi^+ \rangle \tag{153}$$

$$\langle \mu_\pi^{\rm e} \rangle = |\beta|^2 \langle \phi_\pi^- |x| \phi_\pi^- \rangle + |\alpha|^2 \langle \phi_\pi^+ |x| \phi_\pi^+ \rangle \tag{154}$$

where we have again neglected $\langle \phi_\pi^- |x| \phi_\pi^+ \rangle$. From the definitions of $|\phi_-^\pi\rangle$ ("shielded" charge transfer state) and $|\phi_+^\pi\rangle$ (ionic state) comes the expectation value of the corresponding dipole operators:

$$0 = \langle \phi_-^\mu |x| \phi_-^\mu \rangle + \mu_\pi^{\rm n} \tag{155}$$

$$R_{\rm AB} = \langle \phi_+^\mu |x| \phi_+^\mu \rangle + \mu_\pi^{\rm n} \tag{156}$$

the former are the noninteracting π electrons spherically centered around the corresponding π nuclear charges, and the latter correspond to one full charge transfer from site A to site B. Remembering that $|\alpha|^2 + |\beta|^2 = 1$. the final expressions of $\langle \mu^g \rangle$ and $\langle \mu^e \rangle$ are

$$\langle \mu^g \rangle = \langle \mu^\sigma \rangle + \langle \mu^g_\pi \rangle + \mu^n_\pi + \mu^n_\sigma = (1 - |\alpha|^2) R_{AB} + \mu^t_\sigma$$

$$\langle \mu^e \rangle = \langle \mu^\sigma \rangle + \langle \mu^e_\pi \rangle + \mu^n_\mu + \mu^n_\sigma = |\alpha|^2 R_{AB} + \mu^t_\sigma$$

$$\langle \Psi^g | \mu | \Psi^e \rangle = -|\alpha\beta| R_{AB} + \langle \mu^\sigma \rangle$$

$$\langle \Delta\mu \rangle = \langle \mu^e \rangle - \langle \mu^g \rangle = (2|\alpha|^2 - 1) R_{AB}$$

where μ^t_σ, the global (electron–nuclear) σ wavefunction, is given by $\langle \mu_\sigma \rangle + \mu^n_\sigma$. Taking $|\alpha|^2 = 0.75$, $R_{AB} = 5$ Å yields (α and β are assumed to hold the same phase with no loss of generality)

$$\langle \mu^g \rangle - \mu^t_\sigma = 3.17 \text{ D}$$

$$\langle \mu^e \rangle - \mu^t_\sigma = 9.52 \text{ D}$$

These orders of magnitude are typical of moderately strong conjugated change-transfer molecules such as singly substituted benzenes. The charge-transfer contribution to the ground state, i.e., 3.17 D, is nonnegligible and will significantly contribute to the total dipole moment $\langle \mu^g \rangle$ unless otherwise counter-balanced by μ^t_σ, the σ dipole of the molecule or more generally of that part of the molecule not involved in the charge transfer process. In the latter case, i.e., $\mu^t_\sigma = (|\alpha|^2 - 1) R_{AB}$, the total ground-state dipole moment is seen to vanish.

The criterion for such cancellation can be stated as follows: when the charge-transfer dipole present in the ground state is balanced by the pole of other non-charge transfer participating charges (the corresponding nuclear dipoles being included in these definitions), the total ground-state dipole moment vanishes, although the molecule may still have a high dipole transition probability toward a highly dipolar excited state as required for nonlinear activity.

The cancellation of dipole–dipole interactions is of great potential interest for the following reason: simple electrostatic arguments show that strong dipoles have a natural tendency to form dimers of antiparallel molecules or clusters with subsequent negative consequences. First, the resulting crystal structure will have a strong probability to display centrosymmetry or near centrosymmetry, thus leading to null or weak geometrical factors when building up the crystal nonlinearity from the individual molecular contributions. Second, the occurrence of molecular clusters certainly has a disrupting influence on the regular crystal growth process from a solution. The rate of occurrence of such clusters might be reduced when molecules interact less, which is the case when they have weaker or no dipoles. The good quality of POM crystals as grown in solution is most likely related to this feature.

Cancellation of dipole–dipole interaction is in fact not to be viewed as an absolute guarantee against centrosymmetry of the crystal structure, but rather as a favorable preliminary starting basis toward the prevention of centrosymmetry through action on the molecular structure: the absence of otherwise prevailing electrostatic interactions increases the sensitivity of the crystal structure to perturbative actions such as changing or adding a substituent. Let us now proceed to show how the electronic properties of the pyridine 1-oxide molecule and of some of its derivatives combine the previously defined molecular requirements. In addition, pyridine 1-oxide will be shown to display some original and potentially interesting charge transfer features.

The heterocyclic molecule C_6H_5N gives rise to a highly polarized semipolar (or coordination) bond N^+—O^- upon conversion to the *N*-oxide (see Fig. 21), which has a dipole moment of 4.24 D (Linton, 1940), indicating considerable back donation of charge into the pyridine ring. Since then; this polarized bond has been extensively studied both theoretically and experimentally (Minkin *et*

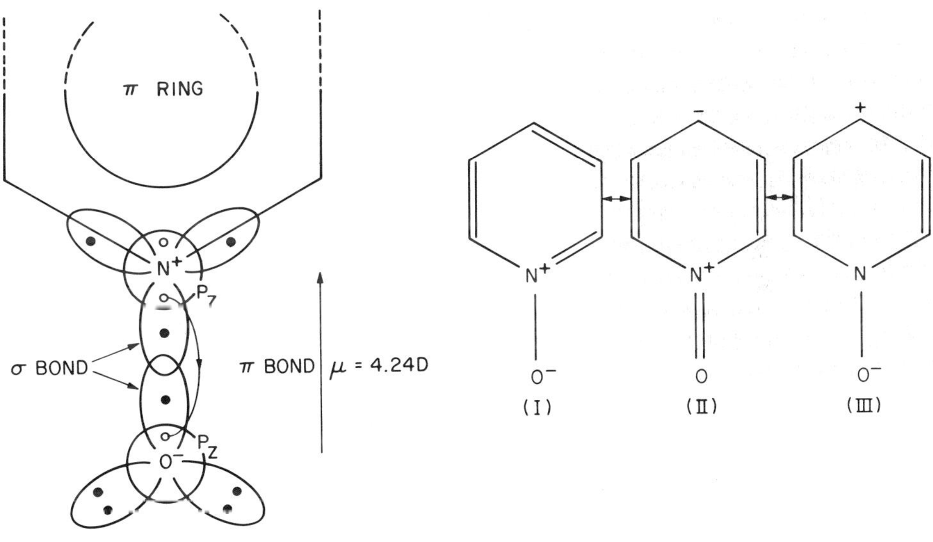

(a) (b)

Fig. 21. (a) Simplified valence bond picture of the electronic population of the *N*-oxide bond in the pyridine 1-oxide molecule ground state. The nitrogen atom has five valence electrons, three of them engaged in σ-type sp^2 bonds. One of the remaining p_z electrons contributes to the 6π molecular orbitals of the aromatic ring; the other gives rise in connection with the lone p_z electron of oxygen to a highly polarized π bond. The rather vacant p_z orbital of nitrogen and the relatively more occupied p_z of oxygen qualify the former as a potential acceptor and the latter as a potential donor when included in a charge transfer molecule. (b) The resonant structures of pyridine 1-oxide responsible for both the actual chemical feasibility and physical interest of donor or acceptor 4-substituents are labeled **I**, **II**, and **III**.

al., 1970; Borisov, 1975). By the resonance effect, the *N*-oxide group can behave as an electron acceptor as well as an electron donor (see Fig. 22), in addition to being an electron-inductive acceptor. The presence of the nitrogen atom within the ring certainly enhances the conjugation of the coordination bond with the aromatic π electrons.

By studying dipole moments of different para-substituted pyridine 1-oxides, it was shown (Katrizky *et al.*, 1957) that a surfeit as well as a deficit of electrons can be created at the 4 position, thus asserting the effective contribution of both resonating structures **II** and **III** of Fig. 21: an electron-withdrawing group ($\delta\mu < 0$) ($\delta\mu$ stands here for the variation of the mesomeric moment) or acceptor group ($\delta\mu > 0$) in the 4 position will respectively reduce or increase

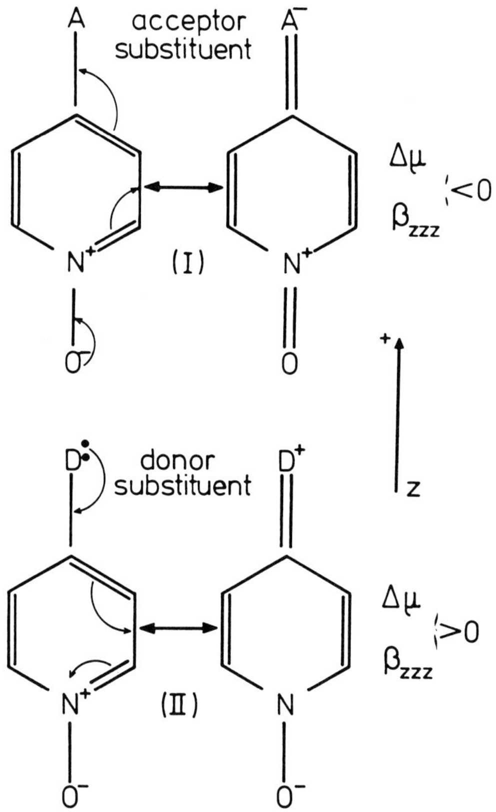

Fig. 22. Charge-transfer mechanism in excited states of substituted pyridine 1-oxide (case I, acceptor in the 4 position; case II, donor in the 4 position); $\Delta\mu$ is the difference between the excited and ground states dipole moments; β_{zzz} is a component of the molecular first-order hyperpolarizability tensor. In case I, the oxygen of the *N*-oxide bond behaves as a donor, while in case II, the nitrogen acts as an acceptor.

the mesomeric moment of the substituted molecules with respect to that of the pyridine 1-oxide molecule. It has been shown (Borisov, 1975) that this effect is felt more strongly for pyridine 1-oxide than for pyridine or identically substituted benzene. This strong push–pull property has an essential chemical consequence: it accounts for the equally easy synthesis of para derivatives of pyridine 1-oxide with donor as well as acceptor groups. Ochiai and Ishikawa (1942) were the first to realize that the electronic properties of pyridine 1-oxide might facilitate electrophilic substitution on the pyridine ring, and Ochiai thus discovered the easy nitration of pyridine 1-oxide in the 4 position, leading to 4-nitropyridine 1-oxide. Both types of molecules appear to be appealing candidates for nonlinear optics. They display charge transfer properties as shown in Fig. 22. In terms of a simplified two-level system, the excited state of **I** has a negative δ while **II** has a positive one. The experimental evidence of a prevailing charge transfer state in the lower excited-state spectrum of 4-nitropyridine 1-oxide has been brought about by Yamakawa *et al.* (1974) who have reported a strong intensity band (346.3 nm in ether) corresponding to a large electronic transfer from the *N*-oxide group oxygen to the nitro group via the pyridine ring. At this level we may conclude that a number of substituted pyridine 1-oxides, being highly conjugated charge transfer heterocyclic molecules, meet our basic molecular requirements. Besides, the push–pull property of the *N*-oxide bond adds some versability to the molecular engineering process previously developed for homocyclic aromatic derivatives.

Relying on the fundamental understanding of the charge transfer mechanism in acceptor as well as donor 4-substituents, we may from now on focus on 4-nitro-substituted and related molecules because of their small total dipole moment. 4-Nitropyridine 1-oxide meets this requirement, as can be justified by a simple additivity argument, as the dipoles of nitrobenzene (4.01 D) and of pyridine 1-oxide (4.24 D) nearly balance each other (Minkin *et al.*, 1970). This is supported by other experimental results (Katrizky *et al.*, 1957). In other words, the electronic affinity of the single oxygen of pyridine 1-oxide suffices to balance the affinity of the two oxygens of the nitro group because of the privileged location of the relaying nitrogen of the N—O bond inside the ring as opposed to the outer position of the nitro group's nitrogen. As was proved experimentally (Yamakawa *et al.*, 1974) and supported by our previous arguments, the nitro group cancels out the total dipole moment and, in the same place, gives rise to a strong charge transfer. However, 4-nitropyridine 1-oxide happens to crystallize with a center of symmetry in space group P_{mna} (D_{2h}^{7}) (Shiro *et al.*, 1977). We may now use the near cancellation of dipole–dipole interactions in the crystal, which makes the structure more sensitive to additional substitutions, as was stated before. The "ideal" substituent to 4-nitropyridine 1-oxide should have the following properties: first, it should not interfere destructively with the basic charge transfer mechanism from the *N*-oxide group to the nitro group, which accounts for the expected high optical nonlinearity; and second, it should remove the centrosymmetry of the 4-nitropyridine 1-oxide crystal. In view of these requirements, a number of substituted 4-nitropyridine 1-oxide molecules were prepared and passed the SHG powder test (see Chapter II-3). Out of these, three molecules, with zero, one, and two methyl groups respectively, are worth a more detailed inspection (see Table XX) to illustrate the respective influences of electronic-related structural features and crystallographic-related ones on the ultimately desired nonlinear property. The spectra of these molecules have been extensively studied (Yamakawa *et al.*, 1974). The strong charge transfer band already mentioned for nitropyridine oxide is almost unmodified by the substitution of a methyl group in position 3, as can be derived by comparison of the respective absorption spectra. However, removal of the charge-transfer character from the dimethyl compound is caused by the steric hindrance of the two methyl groups raising the nitro group by an angle of 56° above the aromatic plane and thus preventing full occurrence of the charge transfer. It may be concluded therefrom that a single methyl group in position 3 helps remove centrosymmetry. This discussion is summed up in Table XX. It should also be noted (Zyss *et al.*, 1981) that bromine, chlorine, and methyl groups (substituents of similar size) in the 3 position all give rise to noncentrosymmetric structures. A similar point has been reported (Levine *et al.*, 1979) for nitroaniline derivatives. Direct-current–induced second-harmonic generation in the liquid phase is hardly feasible when the dipole moment of the solute molecule is negligible. Therefore, this type of systematic measurement is difficult to apply to 4-nitropyridine 1-oxide and a number of its derivatives. However, a dimethylamino group in the 4-position gives rise to a high dipole moment (Boulton *et al.*, 1966) and, being a donor group, will promote the acceptor potential of the *N*-oxide bond. In addition, as the acceptor strength of the nitro group compares with the donor strength of the dimethylamino group, the measured value of the molecule's SHG tensor β of 4-dimethylaminopyridine 1-oxide should be of the same order of magnitude as that of 4-nitropyridine 1-oxide, which in turn should not be very different from that of POM.

The heterocyclic molecules **I** and **II** that have been studied are shown in Fig. 23, along with their aromatic counterparts (**I′** and **II′**) (Oudar, 1977). Results and comparisons are to be found in Table XXI. Results for **I**, **II**, **I′**, and **II′** are consistent, although molecule **I** has a β value smaller than that of **I′**. (The higher gap of **I**, as compared to **I′**, would favor transparency in the transparency versus efficiency trade-off encountered in the optimization of nonlinear molecular crystals.) Molecules **II** and **II′** have similar

TABLE XX

Molecular Electronic Properties, Crystalline Symmetry Requirements, and Interest for Crystal Growth[a]

Compound	Molecular property	Crystal symmetry	Potential interest for crystal growth
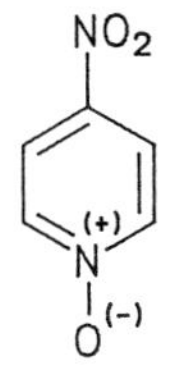	Strong C.T. (+)	P_{mna}, centrosymmetric (−)	None
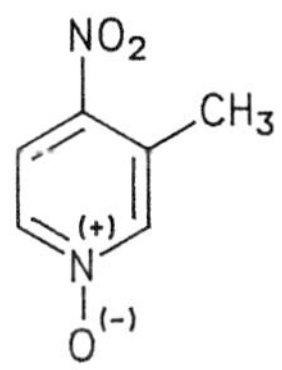	Strong C.T. (+)	$P_{2_12_12_1}$, noncentrosymmetric (+)	Good
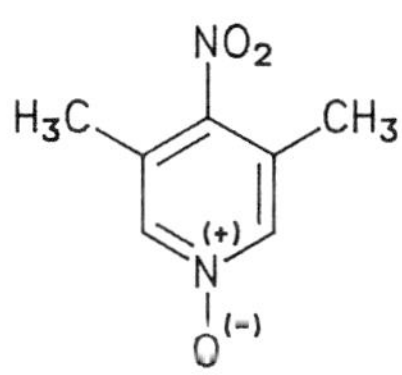	Weak C.T. (−)	$P_{4_12_12_1}$, noncentrosymmetric, forbidden by Kleinman symmetry (−) P_{bca}, centrosymmetric ( )	None

[a] Molecular electronic properties and crystalline symmetry requirements must be simultaneously met to qualify a crystal for efficient three-wave nonlinear interactions. This is clearly the case for POM. Plus or minus signs refer to a positive or negative characteristic of the molecule or crystal.

TABLE XXI

Direct-Current-Induced Second-Harmonic Generation Results for the Vector Part of the Molecular First-Order Hyperpolarizability of Heterocyclic Molecules **I** and **II** of Fig. 23

Molecule	Dipole moment (Debye)	Solvent for β measurement	C.T. band (eV)	β (10^{-30} esu)
I (**I**′)	6.76 (6.93)	Chloroform	4.49 (3.89)	5 (34.5)
II (**II**′)	6.9 (7.42)	Chloroform	3.1 (2.95)	500 (450)

[a] Between parentheses are results from Oudar (1977) for molecules **I** and **II**′ of Fig. 23. The dipole moment of **I** is from Katrizky *et al.* (1957). The dipole moment of **II** is from Boulton, *et al.* (1966) and use of an additivity scheme. All measurements were done using a now-classical dc-induced second-harmonic generation experimental setup with a YAG:Nd^{3+} laser (1.06 μm as a fundamental source; β is the vector part of the hyperpolarizability tensor projected along the dipolar axis of the molecule).

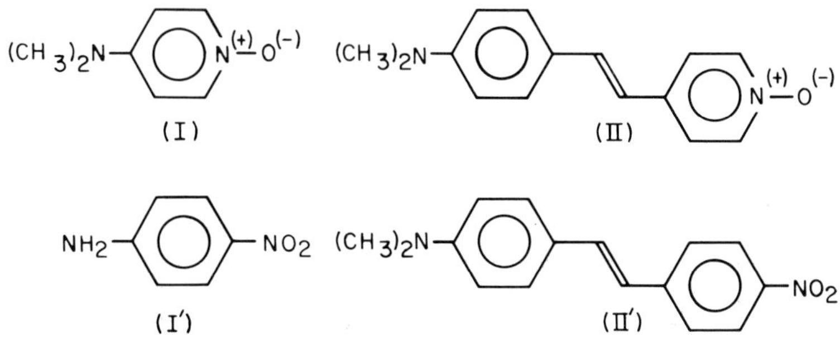

Fig. 23. 4-Dimethylaminopyridine 1-oxide (**I**) and 4-(4′-dimethylaminostyryl)-pyridine 1-oxide (**II**) are examples of pyridine 1-oxide derivatives with a donor group in the 4 position leading to a high value of the dipole moments and acceptor–donor charge transfer. Molecules **I′** and **II′** are the nearest aromatic counterparts of **I** and **II** with available data on the molecular SHG tensor.

efficiencies. The increase of β from a molecule with a single aromatic ring to a molecule with two conjugated rings is also noteworthy in the case of heterocyclic molecules. POM crystals have been extensively studied for their interesting nonlinear properties, based on the previously described combination of a weakly dipolar ground-state charge distribution, together with the required charge-transfer-related spectral properties. Extensive crystal growth investigations have evidenced the suitability of this structure for solution growth (Hierle *et al.*, 1984). It is of interest to position the POM crystallographic structure with respect to optimal structures previously proposed. The space group of POM is $P2_12_12_1$ (Shiro *et al.*, 1977) with four molecules per unit cell and unit cell parameters $a = 21.359$ Å, $b = 6.111$ Å, and $c = 5.132$ Å. The directional nature of the charge transfer from the *N*-oxide to the nitro group in the para position makes a one-dimensional model relevant: the single d coefficient for point group 222, assuming valid, Kleinman relations, is

$$d_{XYZ} = \frac{4}{V} f_x^{2\omega} f_y^{\omega} f_z^{\omega} (\sin\Phi\cos\Phi\cos\Theta\sin^2\Theta)\beta_{uuu}$$

where u is the charge-transfer axis, V is the unit cell volume, f_X, f_Y, and f_Z are Lorentz–Lorentz local field factors, and Φ and Θ are angular parameters defined in Fig. 15, where X, Y, and Z are the three binary axes. The optimal value of the angular factor is 0.192, corresponding to $\Theta = 54.7°$ and $\Phi = 45°$. The actual value of that factor, corresponding to $\Theta = 58.6°$ and $\Phi = 30.3°$, goes down to 0.1005. An increase by a factor of four in the nonlinear figure of merit of the material is thus to be expected if proper structural molecular modifications are introduced so as to reach the group optimum without alteration of the molecular susceptibility. The N-oxide/nitro charge-transfer system thus deserves further exploration, especially if one is ready to trade off the large transparency of POM for more molecular efficiency and if one allows for a greater conjugated system than pyridine to connect the donor and acceptor groups.

3. Chirality and Hydrogen-Bonding: N-(4-Nitrophenyl)-(L)-prolinol (NPP)

As discussed in previous parts of this section, chirality, although guaranteeing with absolute certainty a noncentrosymmetric structure, may not ensure a significant departure from centrosymmetry. Dipolar interaction forces will tend to favor the head-to-tail antiparallel "dimerization" of molecules, and chirality may not significantly modify that tendency. The strategy that eventually led to POM consisted in decreasing the negative influence of these forces by setting up a molecular structure with a vanishing ground-state dipole. In the present case, a symmetric strategy was explored where dipolar forces are to be surpassed by the effect of intermolecular hydrogen bonds. Hydrogen bonds are more energetic by one or two orders of magnitude than van der Waals and dipole–dipole potentials, while being only a little more than one order of magnitude below intramolecular bond energies. Furthermore, hydrogen bonds are, as opposed to dipolar potentials, perfectly neutral as to the centro- or noncentrosymmetric nature of the resulting crystal structure. The combination of chirality and hydrogen-bonding potential can thus promote a greater number of interesting structures than chirality can by itself.

The structure of NPP (Zyss *et al.*, 1984) is shown in Fig. 24. The molecule is seen to be equivalent, from the electronic point of view, to a *p*-nitroaniline structure, the nitro group acting in both cases as an electron acceptor while the nonbonding lone-pair orbital of the amino nitrogen gives equivalent donor character to the prolinol group of NPP as for the dimethylamino group of dimethylaminonitrostilbene (DANS). It is thus expected that both species will display similar nonlinear properties. The crystal structure of NPP is monclinic, as is that of MAP, the most frequently encountered class for this type of compounds, with parameters $a = 5.261$ Å, $b = 14.908$ Å, $c = 7.185$ Å, and $\beta = 105.18°$. The smaller values of a and c show the pronounced planar packing parallel to b, and as actually demonstrated in Fig. 25, the planar NPP molecules are stacked in the (101) plane, which also turns out to coincide with the average molecular plane. This particular arrangement suggests that the (101) plane is a possible cleavage plane. Within this plane, the structure is characterized by infinite hydrogen-bond chains along the binary axis. The hydrogen bond connects the two oxygen of the nitro group of one molecule to the hydroxyl oxygen of the next one. It turns out that, contrary to urea, the molecular "chains" have the same mutual polar orientation, which is a most

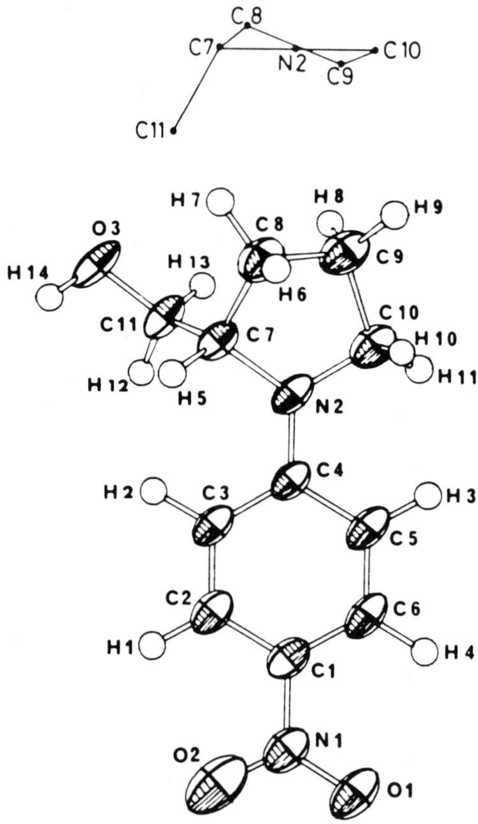

Fig. 24. Molecule of NPP projected on the aromatic plane, the thermal ellipsoids being at the 50% probability level (hydrogen atoms excepted).

favorable factor. The hydrogen-bond length (2.86 Å) corresponds to a bonding of medium intensity. Highly relevant to nonlinear optimization is the value of the Θ angle between the charge-transfer axis connecting the two nitrogen atoms of the NPP molecule and the binary axis: $\Theta = 58.6°$. The optimal angle $\Theta = 54.7°$ (Zyss and Oudar, 1982) is thus almost reached ensuring quasi-optimization of the d_{YXX} coefficient, where Y coincides with the binary axis b and X is the perpendicular to Y in the (101) plane (X is obviously a principal dielectric axis as is Z, perpendicular to the mean molecular or cleavage plane $\langle 10\bar{1} \rangle$). The packing of NPP owing to the higher inclination angle Θ is more favorable than that of MNA (Levine *et al.*, 1979; Lipscomb *et al.*, 1981), as far as the optimization of the phase-matchable d_{21} coefficient is concerned, while MNA has a higher electro-optic r_{11} coefficient. This will be discussed in more detail in Section VI.B.

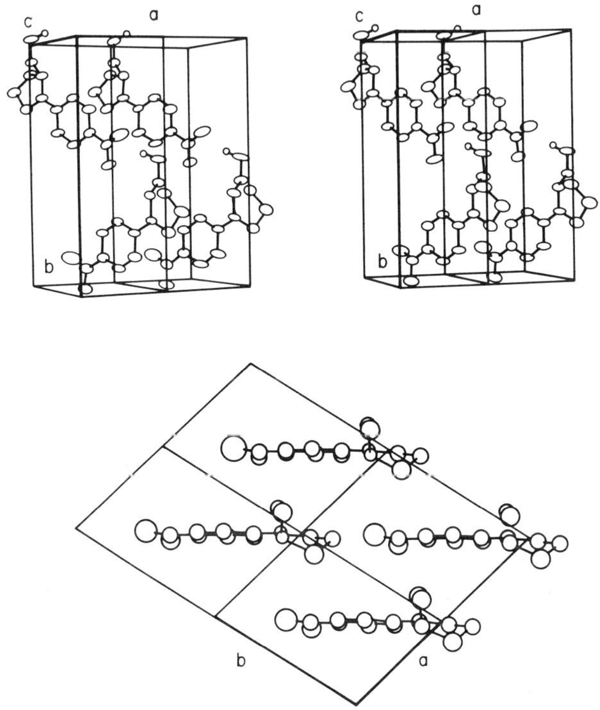

Fig. 25. Projection along binary 2_1 axis *b* showing a phase packing of the molecule in the structure suggests that the (101) plane is a possible cleavage plane. Within this plane, the structure is characterized by infinite hydrogen-bond chains along the binary axis. The hydrogen bond connects the two oxygens of the nitro group of one molecule to the alcohol group of the next one.

VI. NONLINEAR ORGANIC CRYSTALS: FROM BULK TO WAVEGUIDED DEMONSTRATION

A. Device-Related Considerations

The actual use of organic crystals in nonlinear devices, either in bulk or in waveguides, is dependent upon two kinds of requirements.

The first requirement relates to the mechanical properties, long-term chemical stability, capacity to be processed through various basic operations such as cutting, polishing, and electrode deposition, and resistance to various radiations including daylight illumination. To our knowledge, no systematic studies have been reported for nonlinear crystals of the type described here. However, it has been observed that crystals like MAP and POM could be cut and polished, provided that adequate but not necessarily very sophisticated tools were available: cutting can be satisfactorily performed using a nylon or silk thread saw in contact with a proper solvent. MAP can be kept in daylight for years, while cut and polished faces of POM tend to degrade superficially in a matter of weeks when exposed to the atmosphere. This problem can readily be solved by keeping POM samples in glycerine-filled cells where it is safe indefinitely. A more convenient packaging technique, involving the immersion of the crystal in a transparent resin, is presently being developed. The increasing industrial interest in nonlinear organics will probably spur in the near future more systematic studies that were not required at an earlier pioneering stage.

The second requirement is directly linked to the function expected from the nonlinear device and relates to such considerations as transparency, nonlinear figure of merit, dielectric constant, damage threshold, and the possibility of phase-matching. These cannot be discussed in general, as each material with its own specific features will more or less meet the particular external constraints imposed upon it. The main characteristics of the incoming beams, with characteristics ranging potentially from high-power laser pumps to weak fluorescence signals to be detected will be prominent. However, we shall develop in more detail here the various phase-matching configurations that can be encountered and their relevance in terms of device operation. Phase matching is an essential prerequisite to the efficient operation of nonlinear crystals in parametric interactions. In its absence, a fraction of the crystal length smaller than the coherence length will be effectively contributing to the outgoing emission, which will correspond, in general, at most, to a few tens of microwatts. In the phase-matched case, the coherence length goes to infinity and other limiting factors need to be considered. It is worthwhile revisiting some classical concepts such as "noncritical phase matching" in view of both the specific requirements of some advanced short-pulse nonlinear experiments and the potential of organics to meet these requirements. A generic type of experiment is described in Fig. 26: a nonlinear material can be used to detect weak infrared signals by up-conversion, parametric amplification, or difference mixing. In realistic situations, which would, for example, correspond to time-resolved study of near-IR semiconductor fluorescence, one wishes to collect as much of the weak infrared signal as possible both angularly and spectrally. In other words, the nonlinear material that is going to process the signal should remain phase-matched or close to phase-matched for the largest angular and spectral mismatch factors $\Delta\theta$ and $\Delta\lambda$ in order to up-convert or amplify as much as possible of the initial signal. When time resolution is expected, optimization of the spectral acceptance $\Delta\lambda^{\max}$ is essential, as incoming short pulses would otherwise be time-expanded at the possible expense of the desired time-related information.

Referring back to Eq. (38); the second-harmonic intensity is related in the plane-wave approximation to $\operatorname{sinc}^2(\Delta kl/2)$ where l is the interaction length

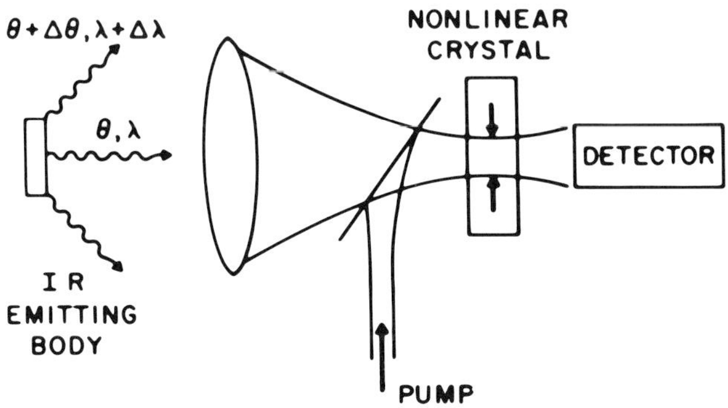

Fig. 26. A nonlinear crystal may be used for up-converting or amplifying (respectively by sum or difference mixing with a strong pump beam) a weak IR signal such as the luminescence of an excited semiconductor. In the general case discussed here, the signal is assumed to be weak, broadband, and nondirectional. This will result in attempting to increase θ, the collecting angle, and lowering the IR beam area in the nonlinear crystal so as to optimize the nonlinear yield. However, focusing on the crystal is limited by the phase-matching angular acceptance angle. When subpicosecond time resolution is desired, the spectral acceptance of the phase-matching conditions will reveal an additional limiting factor.

and $\Delta k = k_{2\omega} - 2k_\omega = (4\pi/\lambda)(n_{2\omega} - n_\omega)$ is the wave-vector mismatch factor; n_ω and $n_{2\omega}$ are wavelength dispersed indices of refraction and depend also on the orientation (θ), with respect to a crystal-attached framework, of the phase-matched colinear beam wave vectors. The λ and θ "aperture" conditions, using λ_0 and θ_0 as the wavelength and angle for exact phase-matching, become

$$-1 < \Delta k(\theta, \lambda) l < 1 \tag{157}$$

A first-order Taylor expansion of $\Delta k(\theta, \lambda)$, close to the phase-matched $\Delta k(\theta_0, \lambda_0) = 0$, yields

$$-1/l < \Delta\theta \frac{\partial}{\partial\theta} \Delta k \bigg|_{\theta_0, \lambda_0} + \Delta\lambda \frac{\partial}{\partial\lambda} \Delta k \bigg|_{\theta_0, \lambda_0} < 1/l \tag{158}$$

where $\Delta\theta = \theta - \theta_0$ and $\Delta\lambda = \lambda - \lambda_0$. In the simple case of a monochromatic wave ($\Delta\lambda = 0$), the aperture of the phase-matching cone is given by

$$|\Delta\theta| < \left(l \frac{\partial}{\partial\theta} \Delta k \bigg|_{\theta_0, \lambda_0} \right)^{-1} \tag{159}$$

where $\Delta\theta$ is maximal when $(\partial/\partial\theta)\Delta k|_{\theta_0, \lambda_0} = 0$. In type I or type II phase matching, in principal planes, one of the beams is extraordinary (i.e., θ-dependent) and the other is ordinary (i.e., non$-\theta$-dependent); the previous condition is then equivalent to $(\partial/\partial\theta) n^e(\theta, \lambda) = 0$, leading to $\theta = 0$ or $\pi/2$, remembering that, owing to its elliptic shape, n^2 is extremal when propagation takes place along the ellipse axis (i.e., the dielectric axis). This would also apply to type II phase matching. This situation corresponds to the well-known, noncritical phase-matching configuration and entails the double benefit of an optimal aperture cone and no walk-off. Remembering that the so-called phase-matching curve is given by $\Delta k(\theta, \lambda) = 0$, then $(\partial/\partial\theta)\Delta k|_{\theta_0} = 0$ expresses that the phase-matching curve has a horizontal slope at θ_0, λ_0 (see Fig. 27). In order to get an estimate of the aperture cone in noncritical phase-matching conditions, Δk must be further expanded to second order in $\Delta\theta$:

$$\Delta\theta < \sqrt{2}\left(\left| l \frac{\partial^2}{\partial\theta^2}\Delta k \right|_{\theta_0, \lambda_0}\right)^{-1/2} \tag{160}$$

The second-order derivative of Δk with respect to θ relates to the inverse of the curvature radius at θ_0, λ_0 of the phase-matching curve. It is clear that this curvature radius becomes the next limiting factor: the larger it is, the wider is the angular domain over which the phase-matching curve will closely approach its $\lambda = \lambda_0$ horizontal target. Let us turn next to broadband (or ultrashort) well-collimated signals, a situation symmetric to the previous one, where $\Delta\theta = 0$. Permutation of θ and λ provides the spectral acceptance of the phase matching,

$$\Delta\lambda < \left(l \frac{\partial}{\partial\lambda}\Delta k|_{\theta_0, \lambda_0}\right)^{-1} \tag{161}$$

where $\Delta\lambda$ is maximal for $(\partial/\partial\lambda)\Delta k|_{\theta_0,\lambda_0} = 0$, which corresponds to a vertical slope at θ_0, λ_0 for the phase-matching curve in the θ, λ plane (see Fig. 27). This situation, which one would be wrongly tempted to call "critical" in opposition to its classically termed "noncritical" counterpart, is in fact favorably meets spectral acceptance requirements. A more suitable denomination should be *θ-noncritical* for the former case and *λ-noncritical* for the latter one, this terminology explicitly indicates their respective advantages. The second-order considerations apply as well in the λ-noncritical configuration, leading to a phase-matching bandwidth ultimately limited by the curvature of the phase-matching curve at θ_0, λ_0:

$$\Delta\lambda < \sqrt{2}\left(\left| l \frac{\partial^2}{\partial\lambda^2}\Delta k \right|_{\theta_0, \lambda_0}\right)^{-1/2} \tag{162}$$

In the general case of broadband noncollimated beams, bandwidth $\Delta\lambda$ and aperture angle $\Delta\theta$ of the nonlinear crystal are related and given by

$$\left| \Delta\theta \frac{\partial}{\partial\theta}\Delta k \right|_{\theta_0, \lambda_0} + \left| \Delta\lambda \frac{\partial}{\partial\lambda}\Delta k \right|_{\theta_0, \lambda_0} < l^{-1} \tag{163}$$

In general, the angular departure from θ_0, accounted for here by one parameter $\Delta\theta$ for the sake of simplicity, is to be actually defined by two

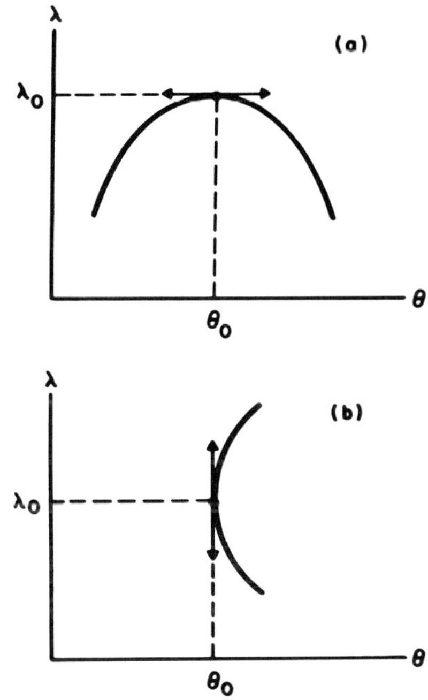

Fig. 27. Phase-matching curves with (a) "θ-noncritical" phase-matching conditions and (b) "λ-noncritical" phase-matching conditions fulfilled at λ_0, θ_0.

parameters, $\Delta\phi$ and $\Delta\psi$. In a uniaxial crystal, $\Delta\phi$ would refer to an angular mismatch within the plane defined by the propagation vector and the optical axis and $\Delta\psi$ to an angular mismatch off the same plane. The corresponding partial derivatives of Δk with respect to ϕ and ψ may be very different in an anisotropic medium, resulting in an elliptic acceptance cone. The acceptance anisotropy leads to practical design considerations on the focusing system when nonlinear efficiency is critically at stake for weak-signal detection. The occurrence of quasi-λ-noncritical phase-matching conditions for a fundamental wave in the 1.5-μm range seems to be an intrinsic feature for organic crystals (see also Section VI.B) as illustrated in Fig. 28 for a type I, $e^{\omega} + e^{\omega} \rightarrow 0^{2\omega}$, configuration. The condition $(\partial/\partial\lambda)\Delta k|_{\theta_0,\lambda_0} = 0$ is then equivalent to

$$\frac{1}{2}\frac{dn_0}{d\lambda}\bigg|_{\lambda_0/2,\theta_0} = \frac{\partial n^e}{\partial\lambda}\bigg|_{\lambda_0,\theta_0} \tag{164}$$

which states that the slope at $\lambda_0/2$ of the harmonic ordinary wave is twice that at λ_0 of the fundamental one. For $\lambda_0/2$ within the visible range, the influence of the charge-transfer excited levels is strongly felt ($\lambda_0/2$ is closer to the UV pole of the Sellmeir expression of the index), giving to the index a strongly

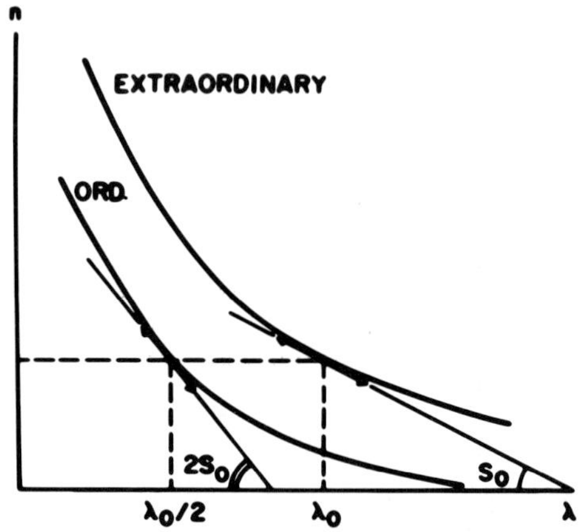

Fig. 28. Occurrence of λ-noncritical phase matching at $\lambda = \lambda_0$ evidenced in an $e^{\omega} + e^{\omega} \to 0^{2\omega}$ type I phase-matching configuration by the value of the slope of the harmonic ordinary index dispersion curve twice that for the extraordinary fundamental curve. Such a situation is realistic for organic crystals in the near infrared due to compensation effects between IR and UV resonances.

ascending outlook, while at λ_0, the compensating influence of the closer IR poles decreases the index variations.

The next relevant criterion when short-pulse nonlinear interactions are of interest is the matching of group velocities (Akhmanov *et al.*, 1975). In the second-harmonic generation case, the useful crystal interaction length will be limited to

$$l(u_{2\omega}^{-1} - u_{\omega}^{-2}) \leqslant \tau \tag{165}$$

where u_ω(or $u_{2\omega}$) refers to the fundamental (or harmonic) wave group velocity while τ is the fundamental wave duration. Beyond l, the signals will cease to significantly overlap and no harmonic can be further coherently generated. In what follows, we shall remain, within a principal dielectric plane. The Fresnel equation that related $\mathbf{k}$ to ω can be expressed in both cartesian and polar coordinates:

$$F(k_x, k_y, \omega) = G(k, \theta, \omega) = 0 \tag{166}$$

Calling $\mathbf{u}$ and $\mathbf{u}_1$ the classical polar coordinate framework unit vectors ($\mathbf{u} = \cos\theta\mathbf{x} + \sin\theta\mathbf{y}$, $\mathbf{u}_1 = -\sin\theta\mathbf{x} + \cos\theta\mathbf{y}$) and remembering that $G = k - [n(\omega,\theta)\omega]/c$,

$$u_\omega^X = \frac{\partial\omega}{\partial k_x} = -\frac{\partial F/\partial k_x}{\partial F/\partial\omega} = \frac{\cos\theta\,\partial G/\partial k - [(\sin\theta)/k]\partial G/\partial\theta}{\partial G/\partial\omega} \tag{167}$$

$$u_\omega^Y = \frac{\partial\omega}{\partial k_y} = -\frac{\partial F/\partial k_y}{\partial F/\partial\omega} = \frac{\sin\theta\,\partial G/\partial k + [(\cos\theta)/k]\partial G/\partial\theta}{\partial G/\partial\omega} \tag{168}$$

leading to

$$\mathbf{u}_\omega = \left(\frac{\partial G}{\partial \omega}\right)^{-1}\left[\mathbf{u} + \frac{1}{k}\left(\frac{\partial G}{\partial \theta}\right)\mathbf{u}_1\right] = c\frac{1/n(\partial n/\partial \theta)\mathbf{u}_1 - \mathbf{u}}{n + \omega(\partial n/\partial \omega)} \tag{169}$$

where

$$\frac{\partial G}{\partial \omega} = -\frac{1}{c}\left(n + \omega\frac{\partial n}{\partial \omega}\right) \quad \text{and} \quad \frac{\partial G}{\partial \theta} = -\left(\frac{\omega}{c}\right)\frac{\partial n}{\partial \theta} \tag{170}$$

Using Eq. (10), $\mathbf{u}_\omega$ can be written as

$$\mathbf{u}_\omega = \frac{c}{n + \omega(\partial n/\partial \omega)}(\tan \rho \mathbf{u}_1 - \mathbf{u}_2) \tag{171}$$

where $\|\mathbf{u}_\omega\| = c/\cos \rho[n + \omega(\partial n/\partial \omega)]$. In a λ-noncritical phase-matching configuration of type I previously discussed ($e^\omega + e^\omega \to 0^{2\omega}$),

$$n_\omega^{\mathrm{e}} = n_{2\omega}^{0} \quad \text{and} \quad 2\frac{\partial n^{\mathrm{e}}}{\partial \omega}\bigg|_{\omega} = \frac{\partial n^{0}}{\partial \omega}\bigg|_{2\omega} \tag{172}$$

Consequently, $\partial G/\partial \omega$ has the same value for both fundamental-extraordinary and harmonic-ordinary waves, so that

$$(\mathbf{u}_{2\omega}^{0} - \mathbf{u}_{\omega}^{\mathrm{e}}) \cdot \mathbf{u} = 0$$

In a λ-noncritical configuration, the group velocities are thus matched along the wave-vector direction: the harmonic group velocity is then along **u** as there is no walk-off for an ordinary wave ($\partial G/\partial \theta = 0$). The fundamental group velocity is departing by the walk-off angle ρ from the wave-vector direction, but its component along that direction is equal to that of the ordinary harmonic: a partial group velocity matching is thus achieved (see Fig. 29). In a

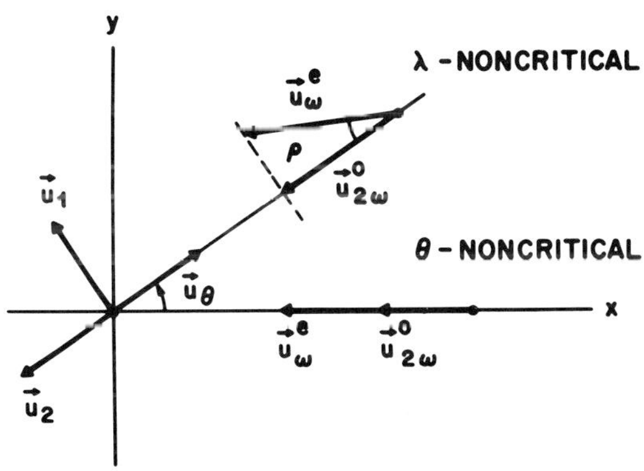

Fig. 29. A λ-noncritical phase-matching configuration entails group velocity matching along the propagation direction which is a θ-noncritical phase-matching configuration; group velocities are parallel to the relevant dielectric axis, but their intensities are mismatched.

θ-noncritical configuration, both group velocities are parallel to the wave-vector or principal dielectric axis direction but the moduli are not matched.

B. Assessments of Important Nonlinear Crystals

1. Introduction

The measurement of precise experimental linear and nonlinear data for newly synthesized organic crystals may prove a long and arduous task. It is, however, essential for both fundamental and application-related developments: such data are necessary to compare various concepts and models with actual organic structures and therefore possibly improve and refine their validity. This was demonstrated in Section V.C, where analysis of MAP quadratic nonlinear coefficient values in terms of an oriented gas of two-dimensional molecules confirmed the relevance of the model. When it comes to applications, each entailing its own highly specific requirements, precise optical data will help the potential user select the best material. It is not possible to decide on a limited number of such data as the needs of users vary considerably, essentially following the characteristics of one or various laser sources for interaction in the material, such as pulse duration, repetition rate, peak and average power, wavelength of interest, as well as efficiency requirements. However, very basic and universally required data are the UV–Vis–IR transmission spectrum, index of refraction in the visible and IR ranges, nonlinear coefficients, second-harmonic and parametric phase-matching curves, estimates of damage thresholds, and possibly various figures of merit at different time scales (ideally in the nanosecond, picosecond, and femtosecond regimes). Single crystals that are being discussed here in these terms were already introduced and described at both molecular and crystalline levels in Section V.E, namely, MAP, POM, and NPP.

MAP was originally grown in ethyl acetate–n-hexane solutions (Oudar and Hierle, 1977), but the Stockbarger–Bridgman method was developed subsequently and proved more readily capable of yielding high-quality single-crystal samples of a few cubic centimeters. The absorption is shown in Fig. 30. Oudar and Hierle have gathered a substantial amount of data on MAP, to be referred to hereafter. The indices of refraction, measured by the deviation minimum method for properly cut prisms, were measured in the visible range and at 1.064 and 1.348 μm. The coefficients of the following Sellmeir equation were fitted to the experimental index values:

$$n^2 = A + \frac{B\lambda^2}{\lambda^2 - C} - D\lambda^2 \tag{173}$$

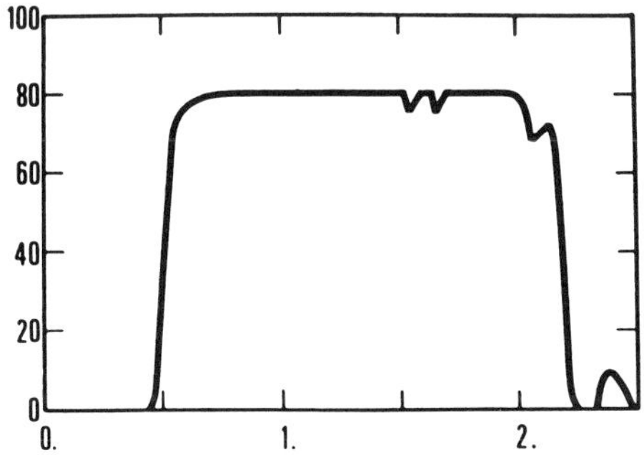

Fig. 30. Optical transmission of a MAP sample. [From Oudar and Hierle (1977).]

Results are given in Table XXII. For the point group 2 there are four independent coefficients, $d_{21} = d_{26}$, $d_{23} = d_{34}$, $d_{25} = d_{36} = d_{14}$, and d_{22}, when the Kleinman's symmetry relations are taken into account.

The values of the coefficients are given in Table XXIII, the higher ones being roughly 40 times larger than the d_{11} coefficient of quartz.

The calculated phase-matching curves for SHG and the experimental phase-matching curve for parametric processes (Oudar and Sigelles, 1977) may be found in Figs. 31 and 32, respectively. At 1.06 μm, reference to the phase-matching curves in Fig. 31 indicates that two phase-matching configurations are possible, of type I and type II, respectively, with detailed characteristics in Table XXIV.

Using the type I phase-matching configuration, the efficiency of MAP rises to 15 times that of $LiNbO_3$. SHG efficiencies of the order of 30% over samples of 1 mm thickness were measured using a 700-kW peak power 10-nsec YAG laser at 1.064 μm. The calculated parametric gain with a pump wavelength of 0.532 μm and 750 MV cm^{-2} intensity is $\Gamma = 14$ cm^{-1}.

TABLE XXII

Refractive Indices at 1.064 and 0.532 μm and the Coefficients of a Least-Squares Fit in Eq. (173)

		n_X	n_Y	n_Z
Wavelengths	1.064	1.5078	1.5991	1.8439
(μm)	0.532	1.5568	1.7100	2.0353
Sellmeier	A	2.1713	2.3100	2.7523
coefficients	B	0.10305	0.2258	0.6079
	C	0.16951	0.17988	0.1606
	D	0.1667	0.01886	0.05361

TABLE XXIII

Characteristics of the Different SHG Phase-Matching Configurations of MAP

Curve number (Fig. 31)	Type	Plane of propagation	Polarizations $E_{2\omega}$	E_ω	E_ω	Useful range of effective NLO coefficients (10^{-9} esu)
1	II	*YZ*	*X*	*YZ*	*X*	19–21
2	II	*ZX*	*ZX*	*Y*	*ZX*	21–29
3	I	*ZX*	*ZX*	*Y*	*Y*	0
4	I	*ZX*	*Y*	*ZX*	*ZX*	36–40
5	II	*ZX*	*Y*	*ZX*	*Y*	0
6	II	*XY*	*XY*	*Z*	*XY*	0–13

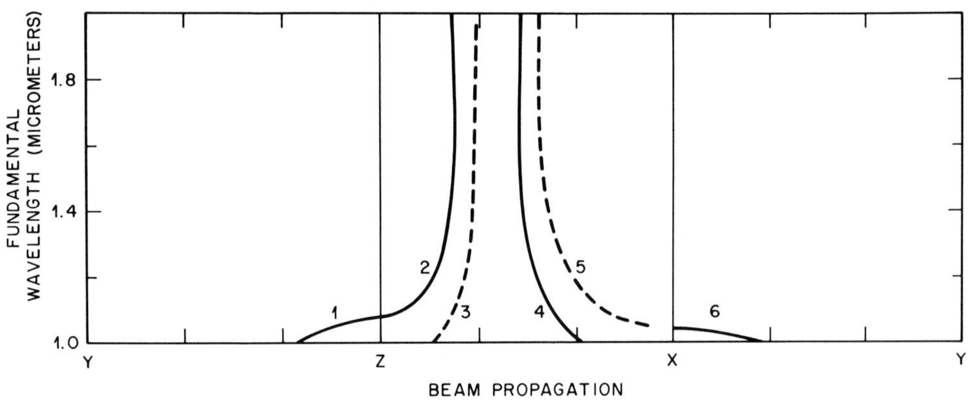

Fig. 31. Calculated phase-matching curves for SHG in the principal planes. Curve numbers refer to Table XXIII.

TABLE XXIV

Parameters for SHG Phase Matching at 1.06 μm[a]

	Beam direction: Calculated	Observed	d^2/n^3 Relative in $LiNbO_3$	$tg(\rho)$
Type I	$\phi = -2°$	$-2.2 \pm 1°$	15	0.20
	$\phi = -72°$	$-71 \pm 1°$	0.8	0.20
Type II	$\phi_z = 10.3°$	$11 \pm 1°$	6	0.042

[a] The angle between the wavevectors and the orthogonal direction within the ⟨010⟩ plane to the *a* crystallographic axis is ϕ.

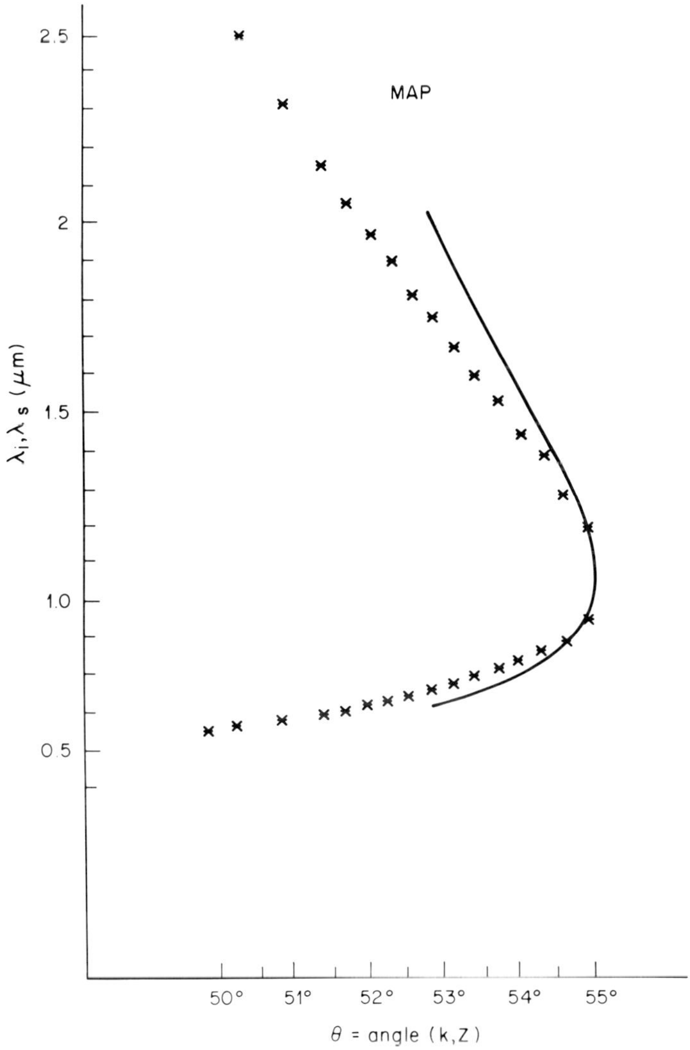

Fig. 32. Phase-matching curve of MAP obtained by parametric fluorescene. The pump wavelength is 532 nm. Experimental points are indicated by an *. The corresponding theoretical curve is the solid line; θ is the angle between the collinear wavevectors and the X dielectric axis (see Fig. 11). At degeneracy, the parametric process depicted here is the inverse of the type I SHG process at 1.06 μm in curve 4 of Fig. 31. [From Oudar and Sigelles (1977).]

2. POM

The growth of POM is accomplished by slowly decreasing the temperature of slightly supersaturated solutions in various possible solvents such as acetonitrile or methylacetate, in the presence of properly cut seeds. Kinetic studies of the various thermal or concentration-related parameters underlying the growth process have been performed in a number of solvents (Hierle *et al.*, 1984). The optical properties of POM have been and are still extensively studied in the nanosecond (Zyss *et al.*, 1981), picosecond (Zyss *et al.*, 1985), and femtosecond pulse duration regimes (Ledoux *et al.*, 1986). Indices of refraction were measured by the minimum deviation method, both in the visible range and at 1.06 μm, and are found in Table XXV, while subsequently better Sellmeir coefficient (same expression as for MAP) are in Table XXVI.

Since POM is orthorhombic ($P2_1,2_1,2_1$ space group), directions *a*, *b*, and c refer unambiguously to crystallographic symmetry-related orientations (Shiro *et al.*, 1977), which was not the case for monoclinic MAP, where further

TABLE XXV

Values of Refractive Indices as Functions of Wavelength (Experimental Points)[a]

λ (μm)	n_a	n_b	n_c	n_c^*
1.0632	1.663	1.829	1.625	
0.644	1.709	1.915	1.637	$1.642 \pm 5 \times 10^{-3}$
0.579	1.728	1.953	1.648	1.646 ± 10^{-3}
0.546	1.742	1.981	1.656	$1.65 \pm 6 \times 10^{-3}$
0.532	1.750	1.997	1.660	—
0.509	1.766	2.028	1.668	$1.65 \pm 1.2 \times 10^{-2}$
0.480	1.793	2.082	1.682	$1.66 \pm 2 \times 10^{-2}$
0.468	1.809	2.114	1.69	—
0.435	—	—	1.717	—

[a] Where n_c^* stands for an additional measurement of the lowest index n_c by means of an Abbe refractometer.

TABLE XXVI

Sellmeir Coefficients of POM

	n_a	n_b	n_c
A	2.4315	2.5521	2.4529
B	0.3556	0.7962	0.1641
C	0.1276	0.1289	0.128
D	0.0579	0.0941	0.0

reference to principal dielectric axes had to be made. The following ordering appears from Table XXVI, showing a strong anisotropy and subsequent birefringence selecting direction *b* for plane *ac*:

$$n_b \gg n_a > n_c$$

POM has only one nonvanishing *d* coefficient when Kleinman relations are taken into account:

$$d = d_{14} = d_{25} = d_{36} \tag{174}$$

The nonlinear coefficient *d* was independently measured by the Maker fringes methods and by observation of second-harmonic generation from a phase-matched orientation of a 1.06-μm YAG laser. Both methods agree on a value of $d = (23 \pm 3)\ 10^{-9}$ esu. Calculated SHG phase-matching curves are displayed in Fig. 33, with corresponding polarization conditions and a range of useful nonlinear coefficients in Table XXVII. The rather vertical slope of curve 1 (i.e., $d\theta/d\lambda \approx 0$) corresponds to the lambda-noncritical phase-matching discussed in Section VI.A. The situation "λ-noncritical" behavior meet here for branch 1 in the near-IR was also met for branches 2 and 4 of MAP in the corresponding spectral range (see Fig. 31) and will be shown for NPP. It was ascribed in Section VI.A to a general feature of this class of compounds where the index of refraction of the fundamental IR wave is weakly dispersed owing to the compensation of UV charge-transfer-state related dispersion by the IR vibrational dispersion, while the index of refraction of harmonic close to UV resonances than to IR ones experiences stronger dispersion. The practical advantages that can be derived from λ-noncritical phase-matching conditions were discussed in Section VI.A. The relative efficiencies of the two symmetry permitted phase-matched configurations are in Table XXVIII. The parametric

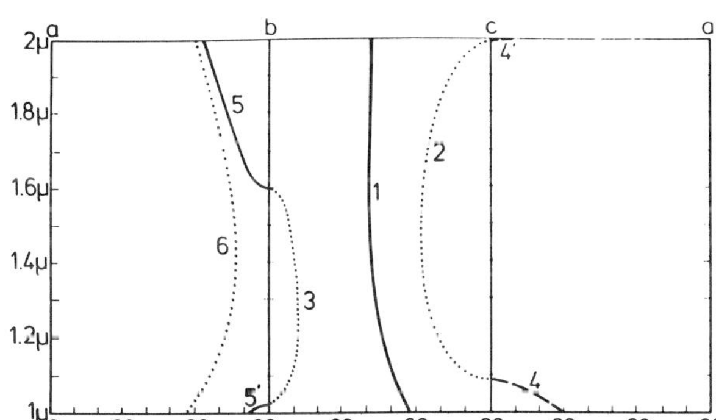

Fig. 33. Second-harmonic phase-matching curves in the principal planes as a function of wavelength. Dotted lines represent symmetry-forbidden interactions. Dashed and full lines stand for type II and type I phase-matching, respectively.

TABLE XXVII

Phase-Matching Conditions in the Different Situations Displayed in Fig. 33

Curve number	Phase-matching type	Propagation plane	Polarizations $E^{2\omega}$	E^{ω}	E^{ω}	d_{eff}	ρ_{max} (rad)	Range of useful nonlinear optical (NLO) coefficients (10^{-9} esu)
1	I	*bc*	*a*	*bc*	*bc*	$d \sin(2\theta + 2\rho)$	0.116	15–20
2	II	*bc*	*a*	*a*	*bc*	0	—	0
3	I	*bc*	*bc*	*a*	*a*	0	—	0
4 (4′)	II	*ac*	*ac*	*b*	*ac*	$d \sin(2\theta + \rho)$	−0.0524	0–16.5
5 (5′)	I	*ab*	*c*	*ab*	*ab*	$d \sin(2\theta + 2\rho)$	−0.0585	0–17
6	II	*ab*	*c*	*c*	*ab*	0	—	0

gain phase-matching curve together with the second-harmonic generation phase-matching curve (branch 1 of Fig. 33) were actually measured in the near IR (Zyss *et al.*, 1985) using the IR emission of an $LiIO_3$ parametric emitter pumped by a mode-locked frequency-doubled YAG laser in a configuration allowing for direct comparison between a second $LiIO_3$ reference crystal and POM.

The maximum power density sustained at 0.532 μm by POM is 50 MV cm^{-2}, with 15 nsec laser pulse duration, versus a damage threshold higher than 270 MV cm^{-2} with 7-psec laser pulses.

The phase matching of POM has been experimentally studied in the three processes:

(1) sum frequency recombination of idler and signal waves back onto the pump: $\omega_i + \omega_s + \omega_p$ (process I)
(2) doubling of the idler frequency: $\omega_i + \omega_i + 2\omega_i$ (process II)
(3) doubling of the signal frequency: $\omega_s + \omega_s + 2\omega_s$ (process III)

TABLE XXVIII

Selected SHG Phase-Matching Data for the Two Available Configurations at 1.06 μm (Fundamental Wavelength)

Type of phase matching	Beam direction	d_{eff}^2/n^3 Absolute (10^{-18} esu)	d_{eff}^2/n^3 Relative to $LiNbO_3$	$tg\rho$
Type I (curve of Fig. 33)	$\theta = 54°18'20''$ (with *b*)	73	3.3	0.11
Type II (curve 4 of Fig. 33)	$\theta = 12°48'$ (with *c*)	18.7	0.82	−0.0250

Here, ω_i and ω_s refer to the frequency of the idler and signal emitted by the $LiIO_3$ parametric emitter pumped by the second harmonic of the YAG laser at ω_p.

The three resulting experimental curves, I, II, and III, corresponding to the phase-matching conditions for processes I, II, and III are compared with theoretical curves in Fig. 34. POM crystals can be used for frequency doubling of typical semiconductor laser wavelengths (1.55 or 1.3 μm). When pumping at $\lambda_p = 0.53$ μm, it is possible, as can be seen from curve I in Fig. 34, to frequency-convert an infrared beam at $\lambda = 0.85$ μm into $\lambda = 1.4$ μm. The three following conversion processes are of interest for communication systems and corre-

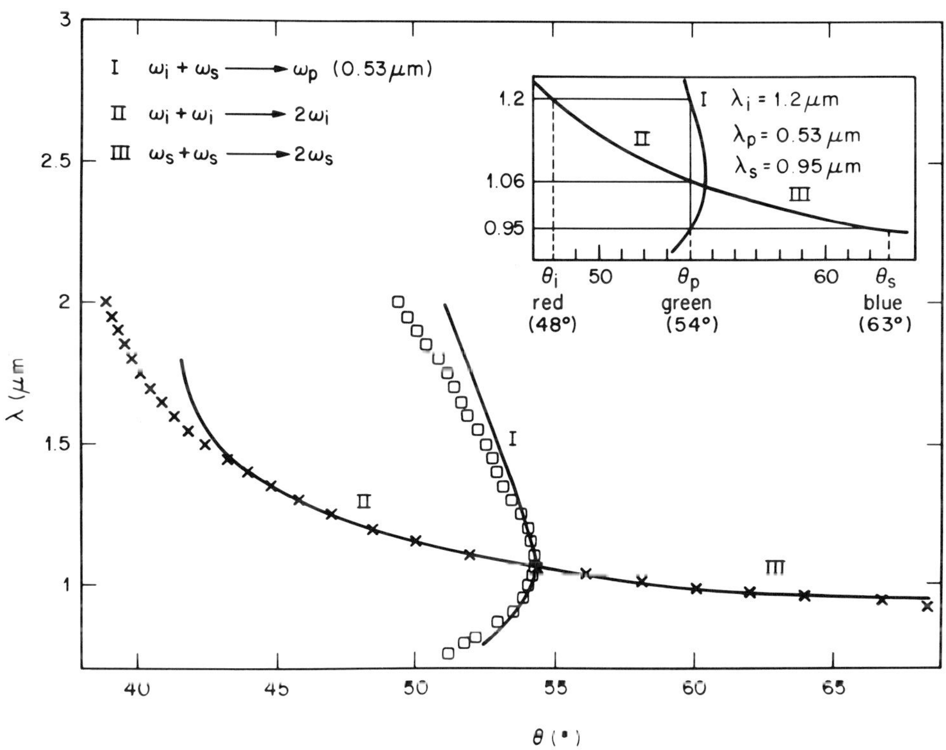

Fig. 34. Different experimental phase-matching curves in POM crystal. I, Sum frequency generation phase-matching curve. II, Idler-wave frequency-doubling phase-matching curve. III, Signal-wave frequency-doubling phase-matching curve. The inset exemplifies these interactions at idler and signal wavelengths chosen so as to generate three visible colors (red, green, and blue). Experimental points are indicated by **x** for SHG and squares for the parametric process, while theoretical curves are solid lines.

spond to phase-matching and transparency requirements satisfied by POM:

$$0.85\ \mu m \rightarrow 1.3\ \mu m \quad \text{(i)}$$

$$0.85\ \mu m \rightarrow 1.55\ \mu m \quad \text{(ii)}$$

$$1.3\ \mu m \rightarrow 1.55\ \mu m \quad \text{(iii)}$$

The pump wavelengths required for such conversions are 0.514 μm, 0.549 μm, and 0.707 μm, respectively, with phase-matching angles 55.7°, 51.1°, and 43.9°, respectively, as shown in Fig. 34. At 0.707 μm, the theoretical parametric phase-matching curve is almost vertical, a situation that can be taken advantage of, in principle, for processing broad-band optical signals (such as corresponding to subpicosecond pulses). For pump wavelength above 0.7 μm, the concavity of the phase-matching curve is inverted, as evidenced by the curve at $\lambda_p = 0.9\ \mu m$ in Fig. 35. In addition, as shown in the insert of Fig. 34, a POM crystal, on which a parametric emitter is shining, may perform as a three-color device (red, green, and blue): the visible emissions can be either simultaneous, when the IR fundamental beams are focused, or alternate, by monitoring the incidence angle on the POM crystal.

Parametric amplification was obtained by difference frequency mixing between the pump beam at 0.53 μm and the idler and signal beam delivered by a $LiIO_3$ parametric generator. The experimental gain, obtained from the curve I_s^a/I_s^i (where I_s^a and I_s^i are respectively the amplified and incoming beam intensities), is decreased by saturation effects from the theoretical value.

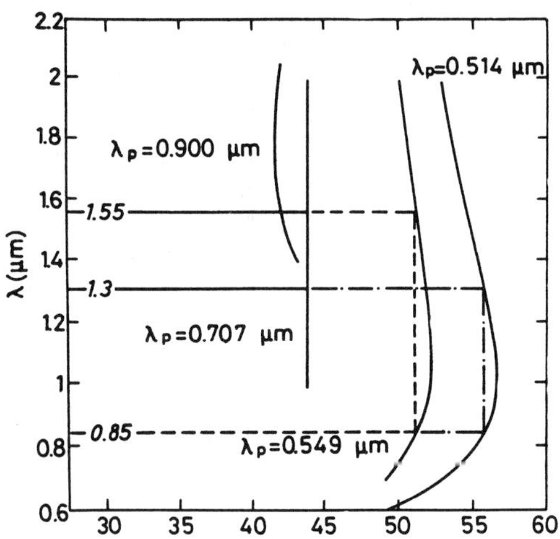

Fig. 35. Calculated phase-matching curves in POM crystal at various pump wavelengths. Possible conversions or amplification schemes corresponding to semiconductor laser emission (in italic numerals) are outlined.

Comparable parametric gains are being obtained with a 25-mm $LiIO_3$ crystal and an 8-mm POM crystal, leading to an experimental value of $\Gamma = 8\ cm^{-1}$ for POM. Parametric emission was achieved in a POM crystal by use of a high-intensity (200–260 MW cm^{-2}) pump beam at $\lambda_p = 0.532\ \mu m$, the crystal being oriented close to degeneracy. Plot of the emitted infrared beam intensity I_{IR}^{e}, in terms of the intensity of the incident pump beam intensity I_p, are shown in Fig. 36. For $I_p = 250\ MW\ cm^{-2}$, $\Gamma = 5.2\ cm^{-1}$ has been measured for $LiIO_3$ and $\Gamma = 14\ cm^{-1}$ for POM, which is consistent with the previously reported value of 10 cm^{-1} at 130 MW cm^{-2}. The emission intensity is of the order of a few hundred kilowatts per square centimeter at degeneracy (1.06 μm).

3. *NPP*

NPP is a monoclinic ($P2_1$) crystal with two molecules per unit cell, and its most appealing structural feature is the near optimal angle (58.6°) of molecular transition dipoles with respect to the binary symmetry axis (see Fig. 37). The nonlinear coefficients of NPP, in view of its highly planar structure where molecular planes are parallel to the crystallographic cleavage plane $\langle 001 \rangle$, are essentially d_{22} and d_{21}. All other coefficients involving the Z out-of plane axis may be assumed to be negligible. In addition to the modified Bridgman–Stockbarger technique described by Badan *et al.* in Chapter II-4 and used to grow bulk single-crystalline NPP crystals of a few cubic centimeters in volume, a thin-film growth technique has been developed (Ledoux *et al.*, 1985). NPP powder is melt ($T_m = 115.9°C$) between glass or silica slides until there remains just a single seed to act as a nucleation center during the subsequent slow cooling stage. Full recrystallization of samples of 10 mm^2 area and 11 μm thickness is achieved in a matter of a week for a drop to about 12°C below the melting point. The d_{21} and d_{22} coefficients were measured on this type of sample by transverse second-harmonic generation in a geometry where the interaction length l is small or negligible as compared to the coherence length. The nonlinear coefficients of NPP were estimated to be $d_{21} = 200 \pm 30 \times 10^{-9}$ esu and $d_{22} = 70 \pm 10 \times 10^{-9}$ esu. The very high value of d_{21} is being essentially involved in further phase-matched second-harmonic generation experiments in the bulk.

Time-resolved second-harmonic and parametric interaction experiments in the femtosecond regime have been performed in NPP (Ledoux *et al.*, 1985). In nonresonant parametric processes, group velocity mismatch is the major cause of temporal broadening. Second-harmonic generation phase-matching curves in planes ZX and ZY are shown in Fig. 38, demonstrating a θ-noncritical phase-matching configuration ($d\lambda/d\theta \approx 0$) at $\lambda = 1.15\ \mu m$. When getting away from the Z axis in the ZX plane and consequently increasing the wavelength, phase-matching curves exhibit a λ-noncritical type of shape (i.e.,

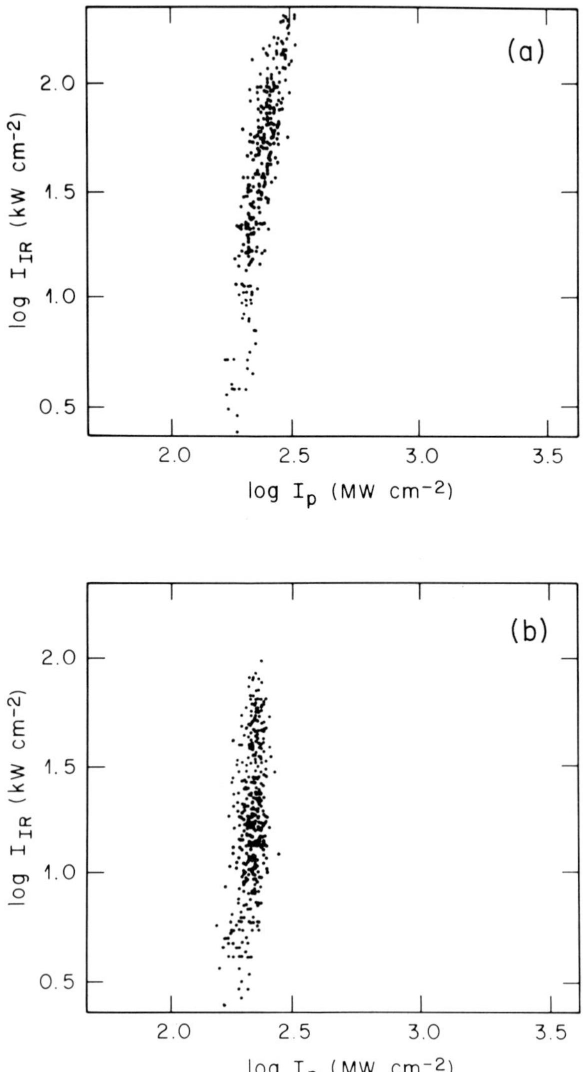

Fig. 36. Parametric emission curves in (a) an 8-mm-thick POM crystal and (b) 25-mm-thick $LiIO_3$ crystal near degeneracy, indicating similar efficiencies.

$d\theta/d\lambda \approx 0$). The parametric phase-matching curve is shown in Fig. 39. Below 1.24 μm, the parametric amplification process competes with a strong Raman-gain emission, which dominates in the 800 nm region. The possibility of a resonant Raman gain proportional to the pump intensity interacting with a nonresonant parametric gain proportional to the pump field could help

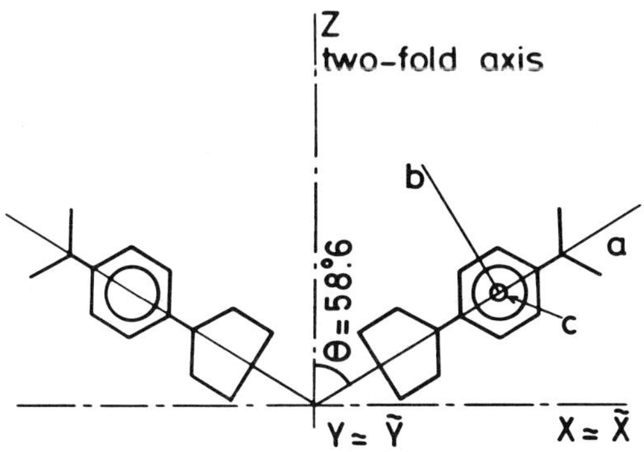

Fig. 37. In cleaved NPP samples, molecules are stacked parallel to the cleavage plane YX, Z being normal to the cleavage plane. Optimal d_{eff} coefficients are involved when propagating perpendicular to the molecular planes (i.e., cleavage plane) along Z. For phase-matched processes, the best polarization configuration corresponds to infrared beams along Y so as to involve the optimized d_{21} coefficient.

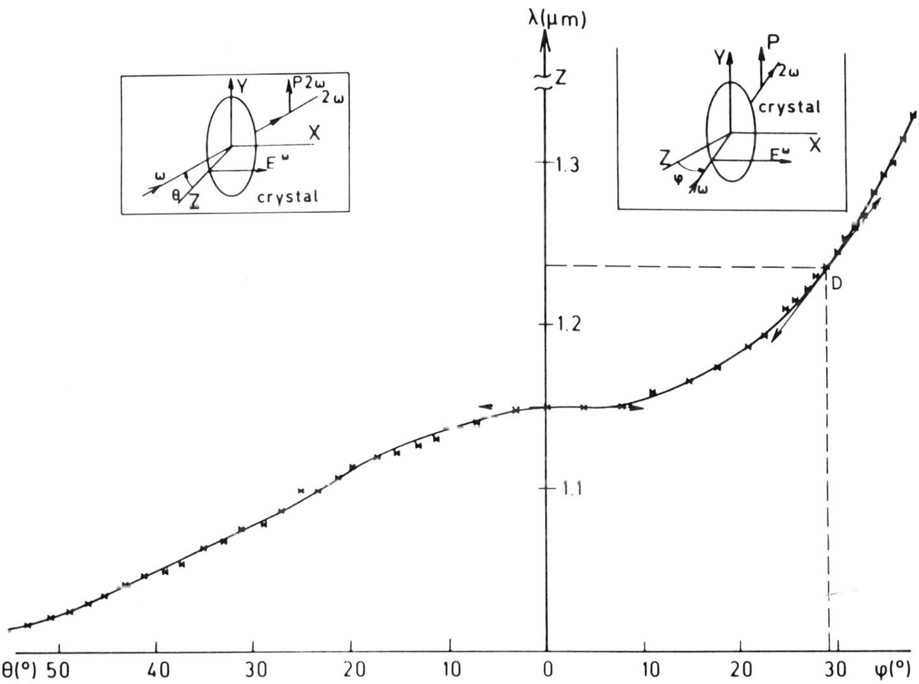

Fig. 38. Second-harmonic phase-matching curves of NPP in the ZX and ZY plane, showing a room-temperature noncritical phase-matching configuration at 1.15 μm. The point corresponding to the 620-mm second-harmonic generation is also shown.

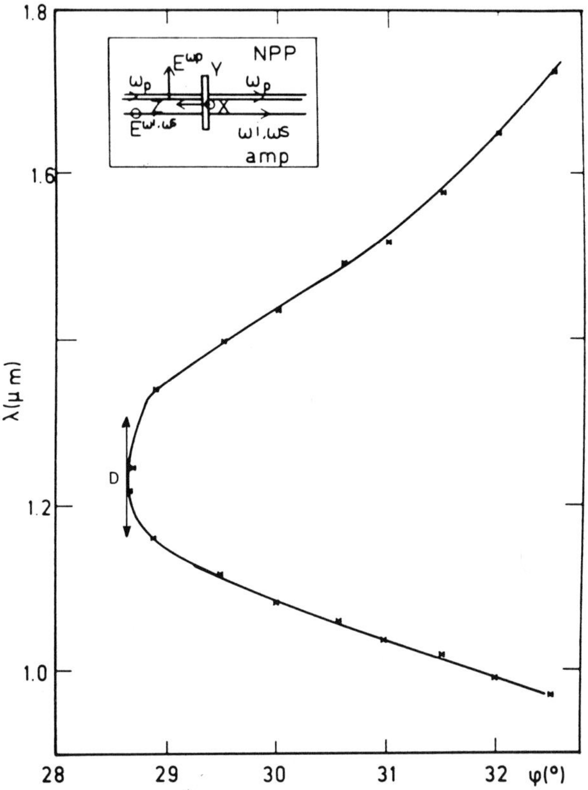

Fig. 39. Parametric gain phase-matching curve in the *ZX* plane: θ is defined with respect to axis *Z* in the *ZX* plane.

directly calibrate $\chi^{(2)}$ third-order tensors with respect to Raman cross-section related to the fourth-order tensor $(\partial\chi^{(1)}/\partial Q)^2$. In addition to parametric amplification, emission has been observed with a large bandwidth at degeneracy, extending from 1 to 1.4 μm. This can be accounted for by the large curvature of the parametric matching curve apart from degeneracy. An unsaturated gain of the order of 10^5 has been measured on a 1.3-mm-thick sample with a beam radius of 1 mm^2 and an energy of 1 μJ per pulse. This corresponds to an effective nonlinear coefficient d_{eff} for NPP, surpassing that of $LiNbO_3$ by one order of magnitude. The time profile of the signal after amplification has been studied by cross-correlation techniques for both POM and NPP samples. The cross-correlation traces are shown in Fig. 40. The time-expanded amplified signals show durations that can be estimated to be at most of the order of 200 psec, versus the initial 100 psec of both pump and IR continuum. As was previously mentioned, this expansion can be ascribed

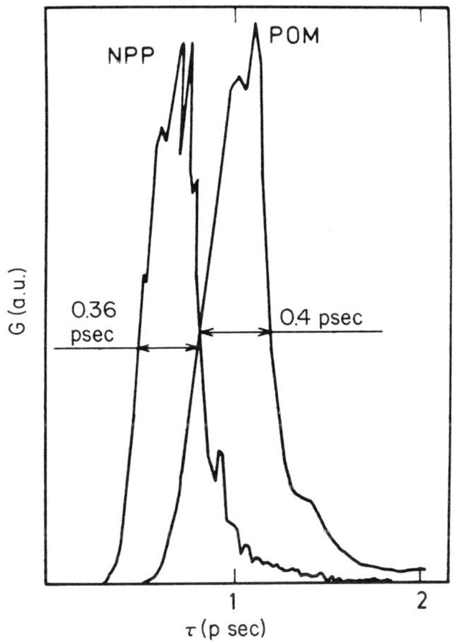

Fig. 40. Cross-correlation trace of the pump and parametric signal emitted by a 1.4-mm-thick NPP sample and 4-mm-thick POM sample at degeneracy ($\lambda = 1.24\ \mu m$).

essentially to group velocity mismatch and is compatible with the use of parametric amplification for time-resolved spectroscopy in semiconductors in the subpicosecond range. A new type of spectroscopy, tagged "PASS" after "parametric amplification sampling spectroscopy," has been developed therefrom and used to time-resolve, below the subpicosecond time scale, the IR luminescence of dyes or III–V semiconductor quantum wells (Hulin *et al.*, to be published).

C. Waveguided Nonlinear Optics

The optimization of nonlinear devices depends on the ability to grow nonlinear materials in structures allowing for waveguided interaction. A number of crucial advantages result from waveguided optics, which we shall briefly recall. However, the comparatively limited amount of earlier decisive experiments as opposed to the significantly higher amount of speculation accounts for the practical difficulties involved with meeting the stringent requirements of integrated nonlinear optics, namely, low absorption loss at the operating wavelength, limited volume and surface scattering defects, and maximum orientational ordering, or at least homogeneous ordering, for quadratic nonlinear interactions.

Before getting into the comparison of the respective performances of waveguiding and bulk configurations toward nonlinear optical device operation and more specifically into the advantages of highly birefringent nonlinear organic waveguides over more classical structures, it is necessary to recall the essential features of linear waveguides and dwell more than is usually done on birefringence-related features. Figure 41 shows the illustrative example of a nonlaterally bounded planar waveguide of thickness t, with medium 1 on the upper side ($Z > 0$) and medium 3 as the substrate ($Z < -t$). All quantities, such as index of refraction n and dielectric constant ε, corresponding to medium 1 (or medium 3) will be accordingly indexed and refered to as n_1 or ε_1 (or n_3 or ε_3), while n and ε will refer to the waveguiding medium.

The waveguide will be assumed to be of single-crystalline nature and birefringent, a situation to be encountered in the case of organic waveguides made of the materials previously discussed. Classical derivations of waveguide propagation properties (Yariv, 1975) generally do not make room for birefringence, and we shall demonstrate in this section that this is a necessary condition to be able to phase match propagation modes of the same order at different frequencies. We shall also assume that the waveguide is homogeneous in space so that the dielectric properties do not depend on space variables, and media 1 and 3 will be considered as isotropic. These may be air for medium 1 and a medium with zero or negligible birefringence as compared to that of the guide for medium 3, such as fused silica. The following index ordering will be considered:

$$n > n_3 > n_1$$

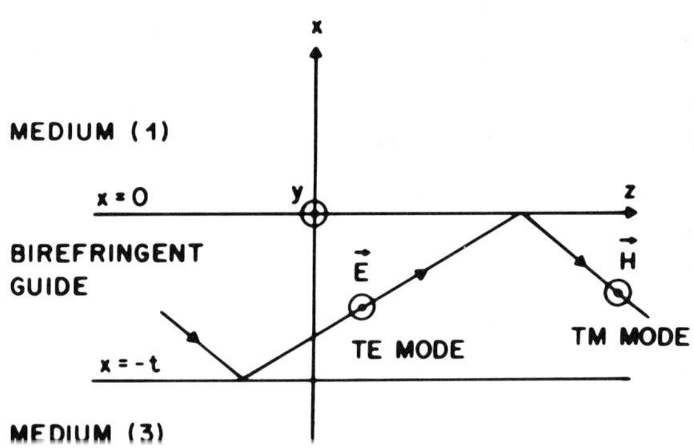

Fig. 41. Reference geometry for a nonlaterally bounded asymmetric plane of thickness t bordered by a cladding (medium 1) and a substrate (medium 3) of lower indices. The theory of TE and TM mode propagation is being developed for birefringent media in view of application to organic molecular crystalline guiding material.

where n refers to the smallest of the three principal indexes of refraction of the guidc, $n - \mathrm{Inf}(n_X, n_Y, n_Z)$, with X, Y, and Z bcing thc principal diclectric axcs. We shall, in the following derivation only consider, as shown in Fig. 41, propagation in a principal dielectric plane, say XZ, and a guide geometry to where the planar interface's normal coincides with a principal axis, say X. Maxwell equations at frequency ω are written, as in Section II, as

$$\nabla \times \mathbf{E} = -\mu_0 i\omega \mathbf{H}, \qquad \nabla \times \mathbf{H} = i\omega \boldsymbol{\varepsilon} \mathbf{E} \tag{175}$$

where $\boldsymbol{\varepsilon}$ refers to the dielectric tensor of the waveguide. It is possible to linearly decompose any vector field, such as $\mathbf{E}$ and $\mathbf{H}$ in transverse and longitudinal components, respectively, orthogonal and parallel to the propagation plane XZ:

$$\mathbf{E} = \mathbf{E}_t + \mathbf{E}_l \qquad \text{and} \qquad \mathbf{H} = \mathbf{H}_t + \mathbf{E}_l \tag{176}$$

Then, very much as classical reflection and transmission through a planar homogeneous boundary, this allows independent consideration of in- and out-of-plane polarized components of the fields, and the wave propagation problem can also be separated.

$$\nabla \times \mathbf{E}_t + \nabla \times \mathbf{E}_l = -\mu_0 i\omega(\mathbf{H}_t + \mathbf{H}_l) \tag{177}$$

$$\nabla \times \mathbf{H}_t + \nabla \times \mathbf{H}_l = i\omega \boldsymbol{\varepsilon} \mathbf{E}_t + i\omega \boldsymbol{\varepsilon} \mathbf{E}_l \tag{178}$$

It can be easily shown, as follows, that $\nabla \times \mathbf{E}_t$ (or $\nabla \times \mathbf{E}_l$) is longitudinal (or transverse), the same holding true for the $\mathbf{H}$ field. This result is a straightforward consequence of the invariance of the problem with respect to position variable Y whereby $\mathbf{H} = \mathbf{H}(x, z)$ and $\mathbf{E} = \mathbf{E}(x, z)$, yielding

$$\nabla \times \mathbf{E}_t = -\frac{\partial E_y}{\partial z}\mathbf{X} + \frac{\partial E_y}{\partial x}\mathbf{Z} \tag{179}$$

$$\nabla \times \mathbf{E}_l = \left(\frac{\partial E_x}{\partial z} - \frac{\partial E_z}{\partial x}\right)\mathbf{Y} \tag{180}$$

with similar relations for $\mathbf{H}$. Thus, each Maxwell equation can be geometrically decoupled in two independent equations, namely,

$$\nabla \times \mathbf{E}_t = -\mu_0 i\omega \mathbf{H}_l, \qquad \nabla \times \mathbf{H}_l = i\omega \boldsymbol{\varepsilon} \mathbf{E}_t \tag{181}$$

and

$$\nabla \times \mathbf{E} = \mu_0 i\omega \mathbf{H}_t, \qquad \nabla \times \mathbf{H}_t = i\omega \boldsymbol{\varepsilon}^{l} \mathbf{E}_l \tag{182}$$

Note that this separation holds true inasmuch as the geometry of the guide allows for the proper breakdown of tensor $\boldsymbol{\varepsilon}$ into transverse and longitudinal parts, i.e.,

$$\boldsymbol{\varepsilon} = \boldsymbol{\varepsilon}^{\mathrm{t}} + \boldsymbol{\varepsilon}^{\mathrm{l}} \tag{183}$$

where

$$\boldsymbol{\varepsilon} = \begin{bmatrix} 0 & 0 & 0 \\ 0 & \varepsilon_Y & 0 \\ 0 & 0 & 0 \end{bmatrix}, \qquad \boldsymbol{\varepsilon}^{\mathrm{l}} = \begin{bmatrix} \varepsilon_X & 0 & 0 \\ 0 & 0 & 0 \\ 0 & 0 & \varepsilon_Z \end{bmatrix}.$$

Such is also the case in the slightly generalized situation, where the propagation plane would be rotated around dielectric axis Y by an arbitrary angle, the dielectric tensor being still decomposed into $\boldsymbol{\varepsilon}^{\mathrm{t}}$ and $\boldsymbol{\varepsilon}^{\mathrm{l}}$ with:

$$\boldsymbol{\varepsilon}^{\mathrm{l}} = \begin{bmatrix} -(\varepsilon_X \sin^2\theta + \varepsilon_Z \cos^2\theta) & 0 & \sin\theta\cos\theta(\varepsilon_X - \varepsilon_Z) \\ 0 & 0 & 0 \\ \sin\theta\cos\theta(\varepsilon_X - \varepsilon_Z) & 0 & -(\varepsilon_X \cos^2\theta + \varepsilon_Z \sin^2\theta) \end{bmatrix} \tag{184}$$

We shall only detail the $\theta = 0$ or $\pi/2$ case, as depicted in Fig. 41, the situation where θ differs from the two values bearing no essential difference from what follows. It should, however, be noted that the general situation where the crystal dielectric axes are fully random with respect to the guide geometry (i.e., defined by θ and another angular parameter ϕ) does not allow for separation of Maxwell equations as in Eq. (182). This rather involved situation results from a geometric configuration where the walk-off angle rotates the $\mathbf{E}$ vector out of the propagation plane: when the propagation plane is a principal dielectric plane, the walk-off angle might be nonzero as θ differs from 0 or $\pi/2$ but is still keeping the field polarization in plane, i.e., longitudinally oriented. Elimination of $\mathbf{H}_{\mathrm{l}}$ (or $\mathbf{H}_{\mathrm{t}}$) from Eq. (181) [or Eq. (182)] yields the following pair of independent equations:

$$\nabla \times \nabla \times \mathbf{E}_{\mathrm{t}} - \mu_0\omega^2\varepsilon^{\mathrm{t}}\mathbf{E}_{\mathrm{t}} = 0 \tag{185}$$

$$\nabla \times \nabla \times \mathbf{E}_{\mathrm{l}} - \mu_0\omega^2\boldsymbol{\varepsilon}^{\mathrm{l}}\mathbf{E}_{\mathrm{l}} = 0 \tag{186}$$

where the ε^{t} notation accounts for the scalar nature of tensor $\boldsymbol{\varepsilon}^{\mathrm{t}}$. The existence of these two equations together with the related proper boundary continuity conditions on the fields, i.e., continuity of the $\mathbf{E}$ field projection on the interface planes, demonstrates the breakdown of the field determination problem into independent longitudinal and transverse component determination. These two field geometries are classically refered to as TE and TM modes. All fields are taken as waves propagating in the z direction with a purely x-dependent amplitude profile, which gives for a TM mode

$$\mathbf{E} = [E_x(x)\mathbf{X} + E_z(x)\mathbf{Z}]e^{i(\omega t - \beta Z)} \tag{187}$$

$$\mathbf{H} = H_y(x)e^{i(\omega t - \beta Z)}\mathbf{Y} \tag{188}$$

and for the TE mode

$$\mathbf{E} = E_y(x)e^{i(\omega t - \beta Z)}\mathbf{Y} \tag{189}$$

$$\mathbf{H} = [H_x(x)\mathbf{X} + H_z(x)\mathbf{Z}]e^{i(\omega t - \beta Z)} \tag{190}$$

Equation (185) simplifies to Eq. (191), remembering that $\varepsilon_0\mu_0\omega^2 = k_\omega^2$, and that $\varepsilon = \varepsilon_0 n^2$:

$$\frac{d^2}{dx^2}E_y + (k_\omega n_Y^2 - \beta^2)E_y = 0 \tag{191}$$

corresponding to the TE mode. Similar equations are to be written in medium 1 (or medium 3) where n_y is to be replaced by n_1 (or n_3). The TM mode can be more easily derived from elimination of $\mathbf{E}_t$ from Eq. (182), to yield a purely $\mathbf{H}_t$-dependent equation, rather than from manipulation of Eq. (187):

$$\frac{d^2H_y}{dx^2} + (k_\omega^2 n_Z^2 - \beta^2 n_Z^2/n_X^2)H_y = 0 \tag{192}$$

Similar equations may be written for the isotropic media 1 and 3. The anisotropy of the guiding material shows up in the form of the n_Z^2/n_X^2 factor weighing the β propagation factor in the TM-mode propagation equation. For the detailed derivation and discussion of the solutions of Eqs. (191) and (192) we refer to a classical treatise (Yariv and Yeh, 1984). We use the same notation, however, adapted in the TM case to account for birefringence.

For the TE mode, the $E_y(x)$ field distribution is given by

$$Ce^{-qx} \qquad \text{for} \qquad x \geqslant 0 \tag{193}$$

$$C\left(\cos hx - \frac{q}{h}\sin hx\right) \qquad \text{for} \qquad -t \leqslant x \leqslant 0 \tag{194}$$

$$C\left(\cos ht + \frac{q}{h}\sin ht\right)e^{p(x+1)} \qquad \text{for} \qquad x \leqslant -t \tag{195}$$

with

$$h = (n_Y^2 k_\omega^2 - \beta^2)^{1/2} \tag{196}$$

$$q = (\beta^2 - n_1^2 k_\omega^2)^{1/2} \tag{197}$$

$$p = (\beta^2 - n_3^2 k_\omega^2)^{1/2} \tag{198}$$

Here C is the constant that can be related to the energy flow of the wave, and β is assumed to give h, p, and q positive real values for confined modes; in other words, $n_1 < n_3 < \beta/k_\omega < n_Y\beta/k_\omega = n_{\text{eff}}$, which is sometimes referred to as the effective index of the propagation mode.

As for the TM mode, the $H_y(x)$ field distribution is given by

$$-C\frac{\bar{h}}{\bar{q}}e^{-qx} \qquad \text{for} \qquad x > 0 \tag{199}$$

$$C\left[-\frac{\bar{h}}{\bar{q}}\cos\bar{h}x + \sin\bar{h}x\right] \qquad \text{for} \qquad -t < x < 0 \tag{200}$$

$$-C\left[\frac{\bar{h}}{\bar{q}}\cos\bar{h}t + \sin\bar{h}t\right]e^{p(x+t)} \qquad \text{for} \qquad x < -t \tag{201}$$

where q and p have been defined previously and $\bar{q} = qn_Z^2/n_1^2$ and $\bar{h} = (k_\omega^2 n_Z^2 - \beta^2 n_Z^2/n_X^2)^{1/2}$. The continuity relations at interfaces $x=0$ and $x=-t$ for E_y and H_z in the TE mode and H_y and E_z in the TM mode lead to conditions on h, $\bar{h}$, p, $\bar{p}$, q, and $\bar{q}$ [$\bar{p}$ defined as $\mathbf{p}(n_Z^2/n_3^2)$].

$$\tan ht = \frac{p+q}{h(1-pq/h^2)} \qquad \text{TE mode:} \tag{202}$$

$$\tan\bar{h}t = \frac{\bar{p}+\bar{q}}{h(1-pq/\bar{h}^2)} \qquad \text{TM mode:} \tag{203}$$

Expression (203) accounts for the guide birefringence through the definition of parameter $\bar{h}$. These equations can be simply interpreted physically and lead straightforwardly to the so-called "mode quantization." Defining ϕ_1 and ϕ_3 by $\tan\phi_1 = q/h$ and $\tan\phi_3 = p/h$, Eq. (202) can be further simplified to

$$\tan ht = \tan(\phi_1 + \phi_3) \tag{204}$$

or

$$ht = \phi_1 + \phi_3 + m\pi \tag{205}$$

where m is an integer, and ϕ_1 (and ϕ_3) can be simply interpreted as the classical dephasing factor at the reflection of the wave at interface guide/medium 1 (and guide/medium 3), in a classical ray description of wave guiding, equivalent to the Maxwellian point of view adopted here. A similar interpretation would apply to TM modes and Eq. (205) where ϕ_1 (and ϕ_3) would have to be replaced by $\bar{\phi}_1$ (and $\bar{\phi}_3$), given by $\tan\bar{\phi}_1 = \bar{q}/\bar{h}$ (and $\tan\phi_3 = \bar{p}/\bar{h}$). Equation (205), or its equivalent in the TM case, indicates that for a given wavelength and wave vector β there exists a corresponding discrete infinity of permitted thickness values t, indexed by the integer m. In practice, t is an explicit function of β, ω, and m, and p, q, h, $\bar{p}$, and $\bar{q}$ are explicit functions of β and indirectly of ω through the indexes that can be defined by Sellmeir dispersion functions.

The exponential evanescence of the field distribution in media 1 and 3 for both TE and TM modes corresponds to confinement of the energy in the guide. The m parameter can be simply related to the shape of the confined field distribution within the guide: ϕ_1 and ϕ_3 can be taken between 0 and $\pi/2$, as p, q, and h are positive. It follows that $0 < \phi_1 + \phi_3 < \pi$ and

$$\begin{array}{lll} 0 < ht < \pi & \text{for} & m = 0 \\ \pi < ht < 2n & \text{for} & m = 1 \\ \vdots & & \\ m\pi < ht < (m+1)\pi & \text{for} & m \end{array} \tag{206}$$

According to Eq. (194) or (200), the squared field distribution in the guide has an sinusoidal shape of periodicity π/h or $\pi/\bar{h}$. As obviously derived from Eq. (206), $m = 0$ then corresponds to a single lobe with no zero in the energy distribution profile within the guide, $m = 1$ to two lobes and one zero, etc. The TE (or TM) mode of order m is refered to as TE^m (or TM^m), the TE^0 (or TM^0) modes exhibiting no oscillatory transverse energy distribution as opposed to higher-order modes. This feature will be of crucial importance in the evaluation to follow of various possible nonlinear coupling schemes in waveguides. Typical mode dispersion curves, relating β or n_{eff} to t, are shown in Fig. 42. The nonlinear propagation equations can be derived exactly as in Section III except for the need to take into account the transverse x dependence of the fields involved as opposed to the simplified x-independent plane wave formalism. A number of excellent articles have reviewed this problem, and we shall concentrate, in the following, on the relevant physical parameters for future nonlinear device optimization. The classical nonlinear propagation equation for bulk as well as waveguided interactions, say for a $TE^m_{2\omega}$ polarized mode and a second-harmonic generation process, is written as

$$\nabla \times \nabla \times \mathbf{E}^m_{2\omega} - \frac{4\omega^2}{c^2}\varepsilon_{2\omega}\mathbf{E}^m_{2\omega} = \frac{16\pi\omega^2}{c^2}\mathbf{P}^{\text{NL},m}_{2\omega} \tag{207}$$

where $\mathbf{P}^{\text{NL},m}_{2\omega}$ is the nonlinear polarization at 2ω generated by TE^m_ω or TM^m_ω mode via the nonlinear tensor $\chi^{(2)}$,

$$\mathbf{P}^{\text{NL},m}_{2\omega} = \chi^{(2)} : \mathbf{E}^m_\omega \otimes \mathbf{E}^m_\omega \tag{208}$$

Assuming that the previously described quantized eigenmode scheme is not significantly perturbated by the nonlinear conversion process, the energy will have to flow from the TE^m_ω (or TM^m_ω) mode on to the TE^m_ω mode. This will be achieved with reasonable efficiency when and only when the source term of Eq. (207) governing the energy transmission from ω to 2ω displays sufficient transverse spatial overlap with the $TE^n_{2\omega}$ mode of interest at the left-hand side of that equation. It is therefore natural that the nonlinear conversion rate should depend on a so-called overlap integral I_R defined as

$$I_\text{R} = \int_\Sigma E^n_{2\omega} P^{\text{NL},m}_{2\omega}\, d\sigma \tag{209}$$

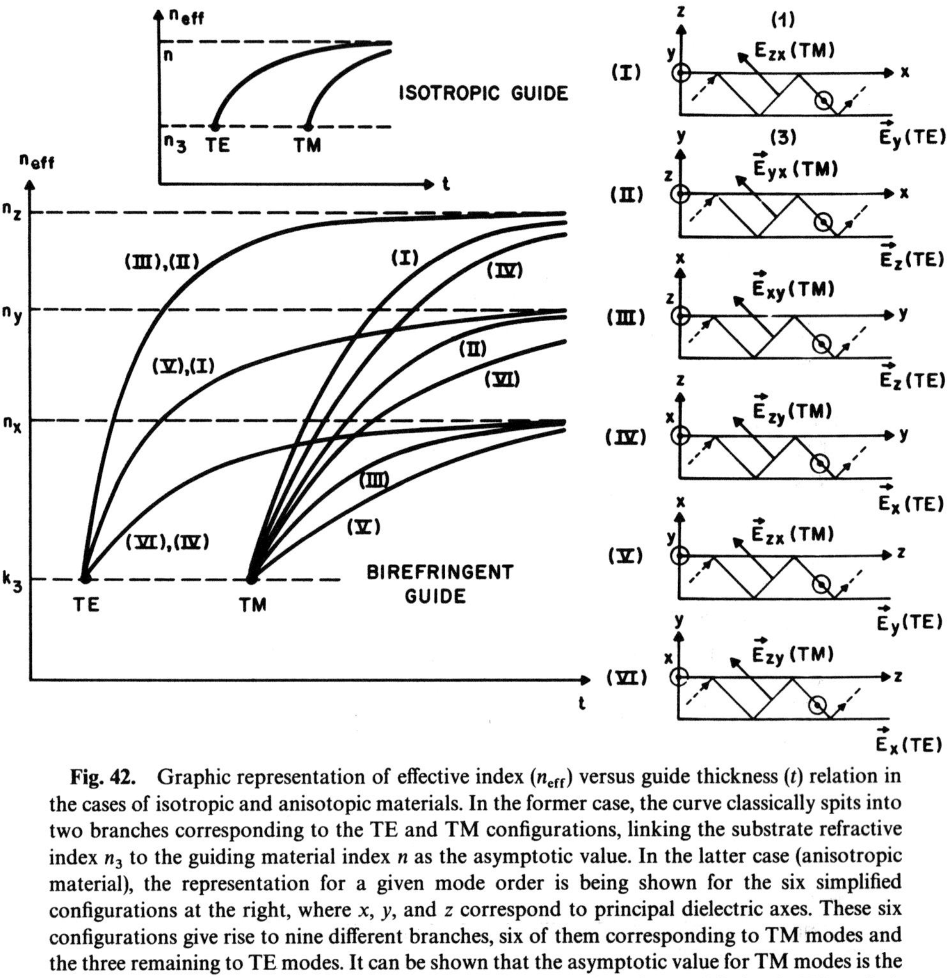

Fig. 42. Graphic representation of effective index (n_{eff}) versus guide thickness (t) relation in the cases of isotropic and anisotopic materials. In the former case, the curve classically spits into two branches corresponding to the TE and TM configurations, linking the substrate refractive index n_3 to the guiding material index n as the asymptotic value. In the latter case (anisotropic material), the representation for a given mode order is being shown for the six simplified configurations at the right, where x, y, and z correspond to principal dielectric axes. These six configurations give rise to nine different branches, six of them corresponding to TM modes and the three remaining to TE modes. It can be shown that the asymptotic value for TM modes is the index of refraction in the direction perpendicular to the planar interfaces, while for the TE mode, it is obviously that of the index of refraction in the direction perpendicular to the propagation plane. The result for TM modes can be physically accounted for by remembering that a TM mode is the superposition of a standing wave (Fabry–Perot mode), polarized parallel to the interface planes within the propagation plane, and of a traveling wave, polarized perpendicular to the interface. Only the latter contributes to the effective index.

where integration is performed over the guide transverse cross section Σ. In the simpler guide geometry of interest here, this factor reduces to

$$I_R \propto \int_{-t}^{0} [E_{\omega}^{m}(x)]^2 E_{2\omega}^{n}(x)\, dx \tag{210}$$

Simple symmetry arguments rule out the coupling of energy into an odd-order harmonic mode. Furthermore, optimization of the I_R integral is better achieved with nonoscillating or weakly oscillating field energy profiles, which tend to favor low-order interacting modes. The ideal combination in SHG, parametric, or higher-order interactions would correspond to $n = m = 0$ single-lobe transverse field profiles.

The phase-matching condition that imposes equal phase vectors for the nonlinear polarization source of Eq. (207) and the 2ω "free" guided mode will be required in order to achieve efficient coupling of the waves along the full interaction length. This can be written as $\beta^n_{2\omega} = 2\beta^m_{\omega}$, where β for guided waves plays the part of k for plane waves in bulk interactions. The analogy ceases when it comes to actually fulfilling this condition. The plane-wave phase-matching condition is ensured, at a given frequency ω, by acting on the continuous angular variable θ, positioning the crystal with respect to the interacting beam's propagation as was amply detailed in Section III. The guide thickness t can be considered as playing a very similar role to that of θ, remembering that, for each mode m, at a given frequency ω, the guide thickness is an explicit function of the mode wave vector β (the reverse also holds true, but β is not an explicit function of t). In other words, $t = f^m(\beta, \omega)$ or $t = g^n(\beta, \omega)$ with f^m (or g^n) corresponding to TE^m (or TM^n) mode dispersions as in Eq. (205). The phase-matching condition relating to TM^m_{ω} and a $TE^n_{2\omega}$ mode is written as

$$t = f^n(2\beta, 2\omega) = g^m(\beta, \omega) \tag{211}$$

with f^n (or g^m) corresponding to TE^n (or TM^m) mode dispersions. This condition is formally similar to that of plane waves in bulk interactions where t is traded for θ, β for k, and where f^n or g^m become the f and g function, defined as

$$k = \frac{n(\omega, \theta)\omega}{c} \Rightarrow \theta = f(k, \omega)$$

with f (or g) corresponding to an ordinary (or extraordinary) angular dependence of n. In this case, the phase-matching condition is

$$\theta = f(2k, 2\omega) = g(k, \omega) \tag{212}$$

In bulk crystals, birefringency is useful in splitting the otherwise isotropic index dispersion into ordinary and extraordinary curves, which is necessary for compensation of the index dispersion from ω to 2ω in the absence of intermediary resonances. In waveguided optics, there already exists in isotropic media a splitting of the $\beta(t)$ curve at given ω or of the $\beta(\omega)$ curve at given t owing to the different relative orientations of TE and TM modes with respect to the propagation plane, for very much the same reasons that make

Fresnel reflection or transmission factors across the boundary between two isotropic media differ. However, when one is aiming at interactions between the zero-order mode, where the overlap integral is maximized at either $t = f^0(2\beta, 2\omega) = g^0(\beta, \omega)$ or $t = g^0(2\beta, 2\omega) = f^0(\beta, \omega)$, the weak departure between TE and TM modes does not make room for satisfaction of either of these conditions. Of course, matching a TE^0_ω (or TM^0_ω) mode with a $TE^0_{2\omega}$ (or $TM^0_{2\omega}$) mode is ruled out by the monotonous dependence of β/k_ω with respect to ω for a given t, similar to that of $n(\omega)$ for a bulk material. In addition to the aforementioned TE and TM dispersion curve-limited splitting in isotropic media, birefringency may introduce an additional significant distance between the TE and TM curves as shown in Fig. 42. In the insert of Fig. 42, the isotropic dispersion for a given integer n exhibits two TE^m or TM^m branches, starting from the respective "cut-off" thicknesses in the (n_{eff}, t) plane, to asymptotically reach at higher t the n values of the isotropic guide refractive index. As shown in that same figure, the two isotropic branches split, when birefringence is accounted for in $3 + 2 \times 3 = 9$ branches, the increase corresponding to the six nonequivalent guide-propagation geometries in the right-hand portion of Fig. 42. Let us examine in more detail case V, for which detailed expressions of the TE and TM dispersion relations were previously given. The effective index of the TE mode goes from n_3 to n_Y as imposed by P [Eq. (198)] and h [Eq. (196)] positive. The effective index of the TM mode goes from n_3 to n_X as imposed by p and h positive. Notice on the figure that couples of distinct branches relate identical TM cut-off point and n_{eff} asymptotic values, changing from one to six the number of TM branches when moving from an isotropic on to a birefringent medium. However, for TE modes there is only a single branch connecting the TE cut-off index to a given n_{eff} asymptote. This changes from one to only three the number of TE branches when moving from an isotropic to a birefringent medium. This feature can be simply accounted for by inspection of Eq. (202) for the TE mode and Eq. (203) for the TM mode. In the TE case, p, q, and h do not make reference to the refractive index along the propagation direction, which makes dispersion curves IV and VI, I and V, and II and III identical. However, dispersion relation Eq. (204) contains the propagation direction index via h, and configurations III and V, I and IV, and II and VI cannot be considered equivalent as far as TM modes are concerned.

Returning to the phase-matching problem in birefringent waveguides, an illuminating framework for discussion is offered by the three-dimensional $(n_{eff} = \beta/k_\omega, \omega, t)$ space, which leads naturally to comparison with the bulk situation depicted within the (k, ω, θ) space. The bulk index surface is given in Fig. 43, and the insert refers to the transform of the three (ellipse–circle) couples classically depicting in polar coordinates (n, θ) the angular dependence of the extraordinary and ordinary indexes in the three principal planes.

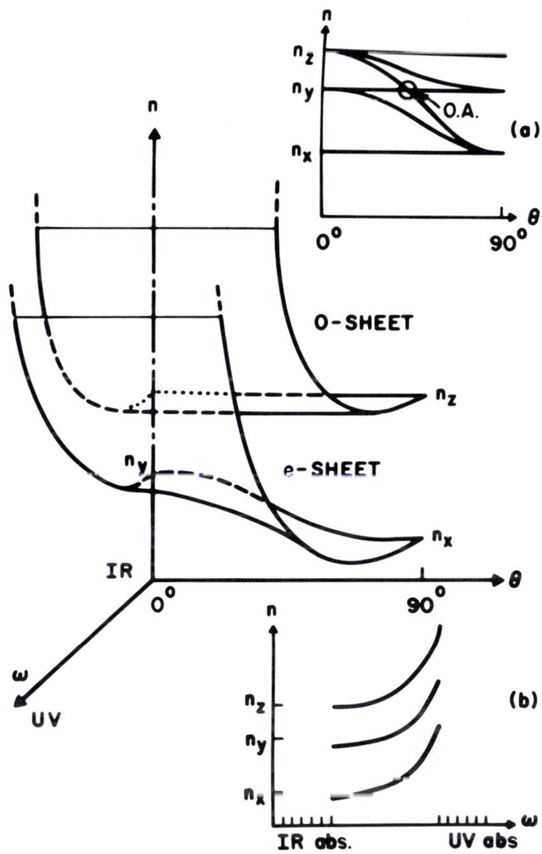

Fig. 43. Three-dimensional representation of the index surface in the (n, θ, ω) space. The constant-ω section (a) of this surface corresponds to the classical ellipse–circle couple in polar coordinates $n(\theta)$, while the constant-θ section (b) corresponds to the usual index dispersion curves.

Through that transform, defining $Y = n$, $X = \theta$, the circle turns $n = n_Z$ into a straight line $Y = n_Z$ and the ellipse $(n^2 \cos^2 \theta / n_Y^2) + (n^2 \sin^2 \theta / n_X^2) = 1$ into a periodic curve:

$$Y = [(\cos^2 X / n_2^2) + (\sin^2 X / n_1^2)]^{-1/2} \tag{213}$$

The relevant curves are shown in insert (a) of Fig. 43. The $n = n(\theta, \omega)$ surface appears to be a six-sheet manifold Σ_σ composed of three couples of ordinary–extraordinary sheets. Projection on the n, ω plane of the intersections of these sheets with planes $\theta = 0$ or $\theta = \pi/2$ yields the three classical Sellmeir dispersion curves of the principal indices, exemplified in insert (b). The $\omega(\theta)$ phase-matching curves can then be defined as follows. Calling A_2^ω the affinity of axis ω and ratio 2, and $A_2^\omega(\Sigma_\sigma)$ the transform of Σ_σ through A_2^ω, the phase matching-curve C_σ can be defined as

$$C_\sigma = A_0^n\left[\sum_\sigma \cap A_2^\omega\left(\sum_\sigma\right)\right] \tag{214}$$

where A_0^n is the orthogonal projection on the ω, θ plane. The curve C_σ has in principle at most 36 branches, corresponding to the maximal number of intersections of two six-sheet manifolds, although they may not all be permitted in view of the symmetry-imposed limitations on the coupling configurations.

In a very similar way, the $n_{\text{eff}}^m = n_{\text{eff}}^m(t, \omega)$ surface for a given m-order mode divides into two TE and TM surfaces, each in turn formed of respectively three and six sheets, as shown in Fig. 44. In this figure, the sole three-sheet TM^m manifold is represented for the sake of simplicity with just the embryo of the corresponding six-sheet TM^m manifold. As it decreases to smaller values,

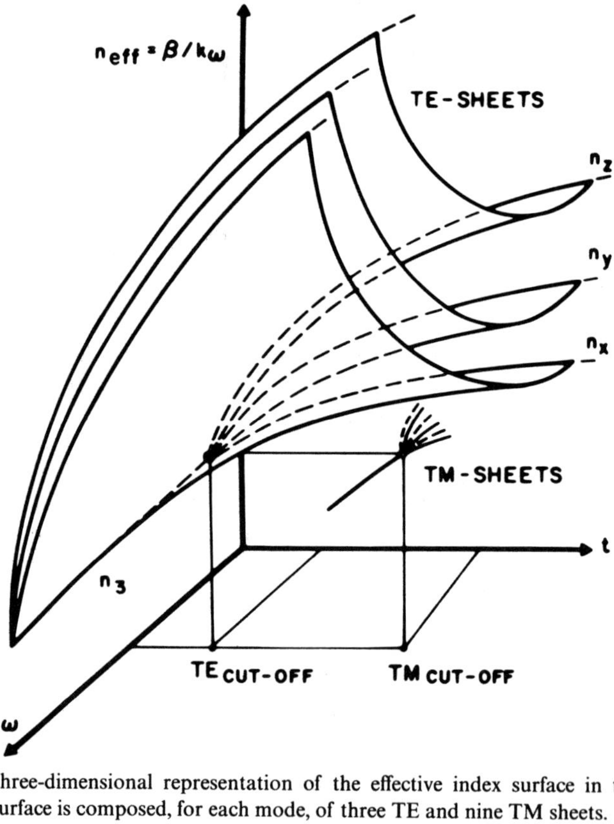

Fig. 44. Three-dimensional representation of the effective index surface in the $(n_{\text{eff}}, \omega, t)$ surface. The surface is composed, for each mode, of three TE and nine TM sheets.

the three sheets join along the cut-off values n_3 of the effective n_3 index, corresponding to the isotropic substrate index. It has been assumed here for the sake of clarity that the substrate index is weakly dispersive as compared to that of the guide. For higher t value, sections of the sheets with constant-t planes resemble more and more the Sellmeir dispersion curves. Calling Σ_{9m} the $9m$-sheet manifold representing all modes up to order m, the phase-matching curve can be defined in the (ω, t) plane by

$$C_{9m} = A_0^{n_{\text{eff}}}[\textstyle\sum_{9m}] \cap A_2^{\omega}(\textstyle\sum_{9m})] \tag{215}$$

Let us now consider the notions of group velocity and group velocity matching for a given mode. This notion is relevant for efficient waveguided short-pulse interactions, which will require matching of the group velocities $\partial\omega/\partial\beta$ of the interacting waves. At given thickness t_0 and given frequency ω_0, u is related to the slope of the intersection of the $t = t_0$ plane with the relevant sheet of manifold Σ_{9m} at $\omega = \omega_0$. By differentiation of Eq. (205) with $dt = 0$, one gets for a TE mode

$$t\,dh = d\phi_1 + d\phi_3 = \frac{h\,dp - p\,dh}{h^2 + p^2} + \frac{h\,dq - q\,dh}{h^2 + q^2} \tag{216}$$

Similarly, for a TM mode,

$$t\,dh = d\bar{\phi}_1 + d\bar{\phi}_3 = \frac{\bar{h}\,d\bar{p} - \bar{p}\,d\bar{h}}{\bar{h}^2 + \bar{p}^2} + \frac{\bar{h}\,d\bar{q} - \bar{q}\,d\bar{h}}{\bar{h}^2 + \bar{q}^2} \tag{217}$$

Remembering the expressions of p, $\bar{p}$, q, $\bar{q}$, h, and $\bar{h}$ given previously,

$$\begin{aligned} \frac{\partial p}{\partial\beta} &= \frac{\beta}{p}, & \frac{\partial\bar{p}}{\partial\beta} &= \left(\frac{n_Z^4}{n_3^4}\right)\frac{\beta}{\bar{p}} \\ \frac{\partial q}{\partial\beta} &= \frac{\beta}{q}, & \frac{\partial\bar{q}}{\partial\beta} &= \left(\frac{n_Z^4}{n_1^4}\right)\frac{\beta}{\bar{q}} \\ \frac{\partial h}{\partial\beta} &= \frac{\beta}{h}, & \frac{\partial\bar{h}}{\partial\beta} &= \left(\frac{n_Z^2}{n_X^2}\right)\frac{\beta}{\bar{h}} \end{aligned} \tag{218}$$

Since the bulk group velocity is $(u_i)^{-1} = (1/c)[\omega(\partial n/\partial\omega) + n]$,

$$\begin{aligned} \frac{\partial p}{\partial\omega} &= \frac{-n_3 k_\omega}{p u_3}, & \frac{\partial p}{\partial\omega} &= -\left(\frac{n_Z^4}{n_3^3}\right)\frac{k_\omega}{\bar{p}u_3} + \frac{2p}{k_\omega}\left(\frac{n}{n_Z u_Z} - \frac{1}{n_3 u_3}\right) \\ \frac{\partial q}{\partial\omega} &= \frac{-n_1 k_\omega}{q u_1}, & \frac{\partial\bar{q}}{\partial\omega} &= -\left(\frac{n_Z^4}{n_1^3}\right)\frac{k_\omega}{\bar{q}u_1} + \frac{2\bar{q}}{k_\omega}\left(\frac{1}{n_Z u_Z} - \frac{1}{n_1 u_1}\right) \\ \frac{\partial h}{\partial w} &= \frac{n_Y k_\omega}{h u_Y}, & \frac{\partial\bar{h}}{\partial\omega} &= \frac{n_Z^2 k_\omega}{n_x \bar{h} u_x} + \frac{\bar{h}}{k_\omega}\left(\frac{1}{n_Z u_Z} - \frac{1}{n_X u_X}\right) \end{aligned} \tag{219}$$

Substitution in Eq. (216) [or Eq. (217)] or the partial derivatives of p, q, and h (or $\bar{p}$, $\bar{q}$, and $\bar{h}$) as detailed above and subsequent extraction of $\partial\omega/\partial\beta$ yields, for a TE mode such as in Fig. 41,

$$\left.\frac{\partial\omega}{\partial\beta}\right|_{\mathrm{TE}} = -\left(\frac{\beta}{k_\omega}\right)\frac{t + p^{-1} + q^{-1}}{t\dfrac{n_Y}{u_Y} + \dfrac{p^{-1}}{p^2 + h^2}\left(\dfrac{n_3}{u_3}h^2 + \dfrac{n_Y}{u_Y}p^2\right) + \dfrac{q^{-1}}{q^2 + h^2}\left(\dfrac{n^1}{u_1}h^2 + \dfrac{n_Y}{u_Y}q^2\right)} \tag{220}$$

For the TM mode of Fig. 41 the expression is more complicated:

$$\left.\frac{\partial\omega}{\partial\beta}\right|_{\mathrm{TM}} = -\left(\frac{\beta}{k_\omega}\right)$$

$$\times\frac{t + \bar{p}^{-1}\dfrac{\bar{p}^2 + \alpha_1\bar{h}^2}{\bar{p}^2 + \bar{h}^2} + \bar{q}^1\dfrac{\bar{q}^2 + \alpha_2\bar{h}^2}{\bar{q}^2 + \bar{h}^2}}{\alpha t + \bar{h}^2\delta_1 + \dfrac{\bar{p}^{-1}}{\bar{p}^2 + \bar{h}^2}(\alpha\bar{p}^2 + \beta_1\bar{h}^2 + \delta_2\bar{p}^2\bar{h}^2) + \dfrac{\bar{q}^{-1}}{\bar{q}^2 + \bar{h}^2}(\alpha\bar{q}^2 + \beta_2\bar{h}^2 + \delta_3\bar{q}^2\bar{h}^2)} \tag{221}$$

where the α, β, and δ coefficients are given by

$$\alpha_1 = \frac{n_Z n_X}{n_3^2}, \qquad \alpha_2 = \frac{n_Z n_X}{n_1^2}, \qquad \alpha = \frac{n_X}{u_X} \tag{222}$$

$$\delta_1 = \left(\frac{1}{k_\omega}\right)\frac{n_X^2}{n_Z^2}\left(\frac{1}{n_Z u_Z} - \frac{1}{n_X u_X}\right) \tag{223}$$

$$\delta_2 = \left(\frac{1}{k_\omega^2}\right)\frac{n_X^2}{n_Z^2}\left(\frac{1}{n_3 u_3} - \frac{1}{n_X u_X} - \frac{1}{n_Z u_Z}\right) \tag{224}$$

$$\delta_3 = \left(\frac{1}{k_\omega^2}\right)\frac{n_X^2}{n_Z^2}\left(\frac{1}{n_1 u_1} - \frac{1}{n_X u_X} - \frac{1}{n_Z u_Z}\right) \tag{225}$$

$$\beta_1 = \frac{n_X^2 n_Z^2}{n_3^3 u_3}, \qquad \beta_2 = \frac{n_X^2 n_Z^2}{n_1^3 u_1} \tag{226}$$

It may be checked that for identical isotropic media 1–3 and guide media (i.e., $n = n_X = n_Z = n_1 = n_3$ and $u = u_X = u_Z = u_1 = u_3$), Eq. (221) simplifies as expected to

$$\frac{\partial\omega}{\partial\beta} = \left(\frac{\beta}{k_\omega}\right)\frac{u}{n} = n \tag{227}$$

Let us now discuss, in the case of SHG, the advantage provided by confinement in waveguide structures over the classical bulk interaction configurations. In the case of bulk interaction, the conversion rate $\eta_b = P^{2\omega}/P^\omega$ is given by

$$\eta_b = Kd^2L^2(P^{(\omega)}/\pi\omega_0^2)f(\Delta kL/2) \tag{228}$$

where $K = 2\omega^2 n^{-3}\ (\mu_0/\varepsilon_0)^{3/2}$, n and ω are the index of refraction and wavelength of the pump beam, d is the effective nonlinear coefficient (i.e., the combination of nonlinear tensor coefficients weighted by angular projection factors, as addressed by the couple of interacting transverse nonlinear polarizations), $P^{(\omega)}$ is the total fundamental power, ω_0 is the radius of the fundamental gaussian beam at its waist, and $f(x) = \sin^2(x)/x^2$ reflects the mismatch between fundamental and harmonic wavevectors, $\Delta k = k_{2\omega} - 2k_\omega$, L being the crystalline length involved in the interaction.

This expression assumes no pump depletion, and $L < b_0$, where b_0 is the confocal parameter of the gaussian beam ($b_0 = 2\pi n\omega_0^2/\lambda$), or, stated differently that the transverse section of the gaussian beam remains essentially constant over L and equal to the beam waist area $\pi\omega_0^2$. Unfortunately, it is impossible, owing to the very nature of a gaussian beam, to maintain a constant power density $P^\omega/\pi\omega_0^2$ over the crystal interaction length indefinitely, although TEM_{00} gaussian beams display minimal diffraction. It can be shown that the optimal conversion rate is obtained in bulk when the crystal length L is of the order of b_0 (namely, $L = 5.68\ b_0$), yielding the following more realistic value for η_b:

$$\eta_b^{conf} = 2K\lambda^{-1}d^2LP^\omega f(\Delta kL/2) \tag{229}$$

Expression (229) can be seen to be in sharp contrast with Eq. (230), corresponding to the conversion rate η_{wg} in a planar waveguide (Azema *et al.*, 1979) of thickness a:

$$\eta_{wg} = Kd^2L^2P^\omega a^{-1}I_R f(\Delta kL/2) \tag{230}$$

where I_R is the overlap integral between nonlinear polarization and harmonic field transverse distributions over the guide cross-sectional area, already referred to, and which can be assumed to hold the optimal value of $1/a$. The following condensed expressions of η_b^{conf} and η_{wg} point out differences:

$$\eta_b^{conf} \approx \lambda^{-1}LP^\omega \tag{231}$$

$$\eta_{wg} \approx a^{-2}L^2P^\omega \tag{232}$$

Expression (231) demonstrates a bulk SHG rate dependent on the total power (and not power density) and proportional to the interaction length. Expression (232) shows the guided SHG rate to be proportional to the power density (P^ω/a^2) and quadratic in L.

In Eqs. (231) and (232) the proportionality constants are of the same order and, further assuming waveguide dimensions of the order of the wavelength (i.e., $a \approx \lambda$), one ends up with the following ratio: $\eta_{wg}/\eta_b^{conf} \approx L/\lambda$. In the case where $L = 1$ cm and $\lambda = 1\ \mu$m, this ratio, highly in favor of the waveguide configuration, would be, in theory, of the order of 10^4. Of course, this very high ratio is somewhat unrealistic in view of the aforementioned assumption of no pump depletion. In practice, a ratio of the order of 10^2 can be reasonably expected when moving from bulk to guides, all other parameters such as pump power or interaction length being kept constant. To summarize, the main advantage of waveguides rests in the possibility of maintaining, over theoretically infinite dimensions in the case of a perfectly guiding medium, optical power transversely confined over dimensions of the order of the wavelength. This is, of course, not possible for bulk interactions, where the stronger the focusing, the smaller is the confocal parameter b_0 beyond which the linear L dependence will show up and consequently reduce the conversion efficiency. For illustration, a continuous pump power of 100 mW, corresponding to powerful (however, currently available) GaAlAs semiconductor laser infrared emission, corresponds to a power density of the order of 10 MW cm^{-2} when confined along a guide of average cross-sectional area 1 μm × 1 μm. This power density comes close to the optical damage threshold of $LiNbO_3$ and partially accounts for the demand for new materials, which are able to sustain higher power densities.

Phase-matching conditions in waveguides have been previously discussed in general. However, when aiming for efficient conversion, one should ensure a maximal overlap integral I_R in addition to the phase-matching condition. TM_ω^0 (or TE_ω^0) to $TE_0^{2\omega}$ (or $TM_0^{2\omega}$) matching, however ideal from the transverse overlap optimization point of view, cannot, in general, be implemented in nonbirefrigent or weakly birefringent media, as the amount of (TM^ω, TE^ω) anisotropic shift* is in general too small to compensate for the effective index dispersion. Various solutions have been invented to bypass that drawback, such as covering the nonlinear waveguide by a linear one of higher index and thus favorably distorting the interaction field energy profiles at ω and 2ω. Another possibility is to grow a linear, purely guiding layer over a nonlinear substrate, the nonlinear interaction taking place between substrate modes of identical symmetries. In the latter configuration, a trade-off has to be found between the transverse spatial extension of the nonlinear substrate mode, so that a significant part of the fields may be able to overlap in a nonlinear environment, and the power density, which goes opposite but is crucial in the optimization process. In the following, we shall examine a number of configurations where waveguides are made of single-crystalline

* As already noted there is a shift between TE and TM mode even in the absence of birefringence, owing to the anisotropy of the interface reflection.

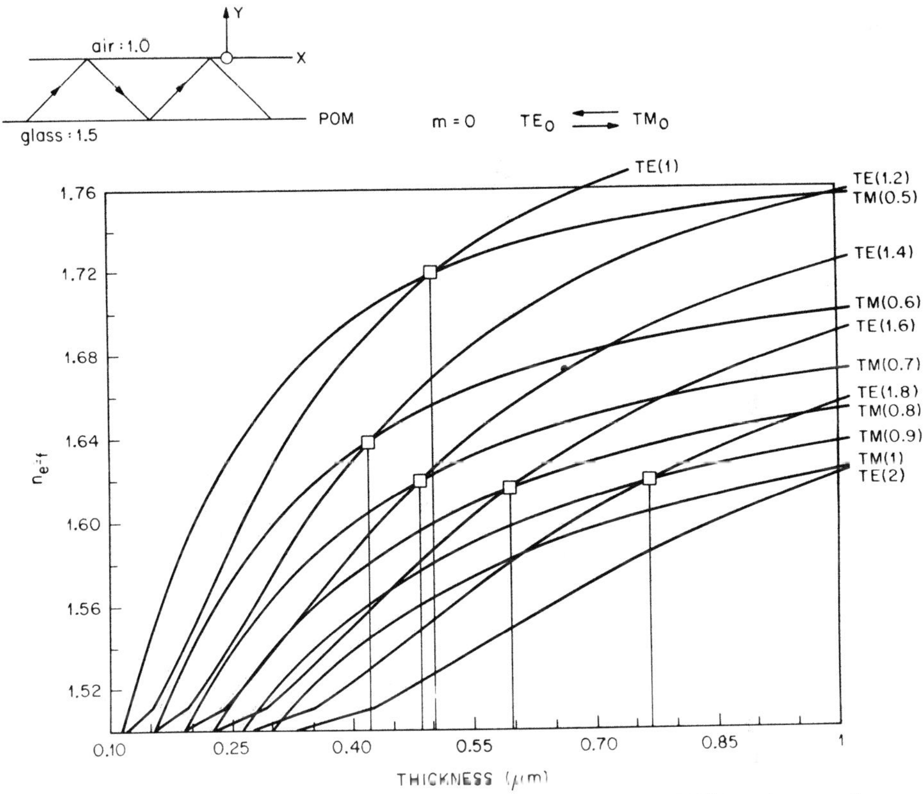

Fig. 45. Dependence of the effective index of POM for TE_0^{ω} and $TM_0^{2\omega}$ modes at various wavelengths (configuration is inset). The crossing points correspond to the phase-matching thickness.

highly nonlinear organic materials such as POM and MAP, discussed in Section V.B. In Fig. 45, TE_0^{ω} effective index dependence on guide thickness has been drawn at various wavelengths λ, ranging from 0.5 to 2 μm. The material is MAP supposedly grown on a glass substrate of index 1.5. The geometry of the guide is shown in the insert of Fig. 45, and the absence of crossing points shows the impossibility of phase matching in this type of configuration, which does not allow for birefringency-related compensation of the index dispersion. On the contrary, as shown in Figs. 46 and 47, relative to POM, phase-matching within a given mode ($m = 0$ in Fig. 46 and $m = 1$ in Fig. 47) is permitted when nonlinear interaction is taking place between TE_{ω}^{m} and $TM_{2\omega}^{m}$ modes. Birefringency is seen to be sufficient to allow for compensation of the effective index dispersion within the same m order for both fundamental and harmonic waves. Figure 48 shows the dependence of the phase-matched thickness for

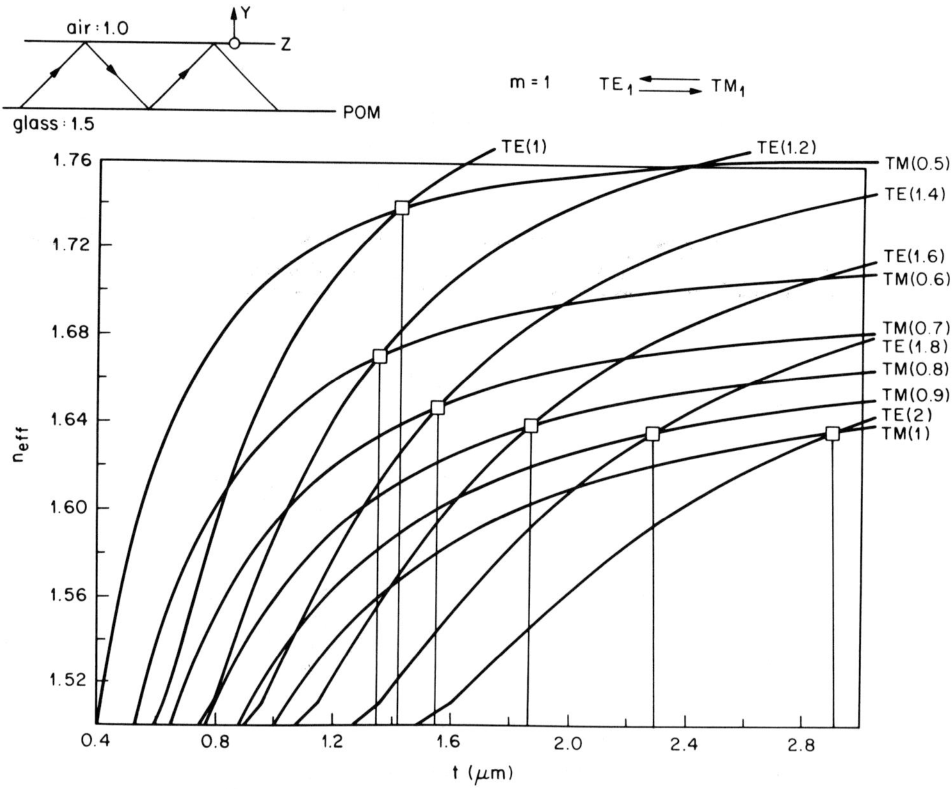

Fig. 46. Same as in Fig. 45 for TE_1^{ω} and $TM_1^{2\omega}$ modes.

various wavelengths in the $m = 0$ matching case. The minimum of t in the vicinity of $\lambda = 1.2$ μm makes a precise definition of t noncritical.

In the following, we discuss briefly a new concept in nonlinear waveguiding that makes use of both spectrally and spatially resonant nonlinear configurations. The spectrally resonant situation corresponds to transverse Fabry–Perot or waveguided modes, where the mode quantization phenomenon may be considered as corresponding to some geometry-induced resonances in the propagation equations. The spectrally resonant configuration corresponds to single or possibly multiple coincidences between excited states of the nonlinear molecules and relevant combinations of the interacting beam wavelength: this field, known as nonlinear spectroscopy, has been well explored for quadratic and mainly cubic nonlinearities. Although spatial and spectroscopic resonances have been conclusively explored independently, their combination (Zyss, 1985) has not been considered and will be briefly discussed. We shall start with a brief reminder of surface SHG effects.

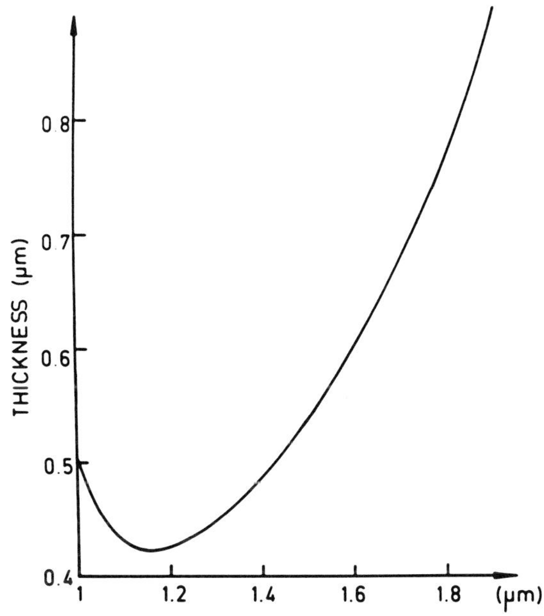

Fig. 47. Phase-matched thickness dependence on the fundamental wavelength for $TE_0^\omega \rightarrow TM_0^{2\omega}$ interaction in POM (configuration in the box of Fig. 45).

When irradiating a nonlinear surface at incidence angle θ with a coherent beam of intensity $I(\omega)$ at frequency ω, coherent second-harmonic emission can be detected at angle θ (corresponding to specular reflection of the beam at ω) with intensity $I(2\omega)$ given by the following expression (Heinz *et al.*, 1983):

$$I(2\omega) = \frac{32\pi^3\omega^2 \sec^2\theta}{c^3\varepsilon(\omega)\varepsilon^{1/2}(2\omega)} \left| \mathbf{e}(2\omega) \cdot \chi_s^{(2)}(-2\omega; \omega, \omega) : \mathbf{e}(\omega)\mathbf{e}(\omega) \right|^2 I^2(\omega) \quad (233)$$

where $\varepsilon(\omega)$ [or $\varepsilon(2\omega)$] is the fundamental (or harmonic) dielectric constant of the medium through which the fundamental (or harmonic) waves propagate, and $\mathbf{e}(\omega)$ [or $\mathbf{e}(2\omega)$] is the fundamental (or harmonic) unit polarization vector, and $\chi_s^{(2)}$ is the surface second-harmonic tensor including local field corrections. This effect is illustrated in Fig. 48. If we suppose that the upper medium has no significant dispersion, it can be clearly understood why the harmonic emission will peak in a direction parallel to that of the specular reflection of the fundamental pump beam: the ω beam (wave vector $\mathbf{k}_\omega$) excites linearly (alternatively, nonlinearly) a "forced" polarization wave at ω (alternatively, 2ω) with wave vector parallel to the substrate given by $k_\omega \sin\theta_\omega$ (alternatively; $2k_\omega \sin\theta_\omega$). The ω and 2ω polarization waves can be viewed as gratings of respective periodicity $T_\omega = 2\pi/(k_\omega \sin\theta_\omega)$ and $\pi/(k_\omega \sin\theta_\omega) = T_\omega/2$. These gratings will, in turn, radiate back the reflected fundamental and the emitted

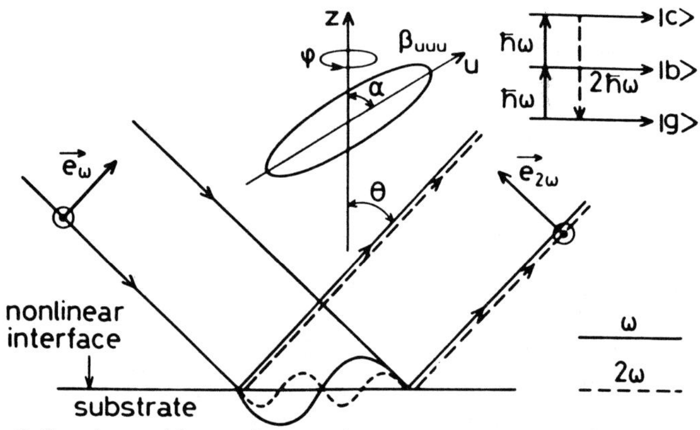

Fig. 48. Reflected second-harmonic generation experiment. When shined on at incident angle θ^{ω} by a fundamental beam of finite transverse extension, nonlinear dipoles covering the substrate surface are polarized at frequency ω (or 2ω) with spatial wave vector $k^{\omega} \sin \theta^{\omega}$ (or $2k^{\omega} \sin k^{\omega}$). The 2ω surface polarization radiates in turn in direction $\theta^{2\omega}$ given by $k^{2\omega} \sin \theta^{2\omega} = 2k^{\omega} \sin \theta^{\omega}$, which, when neglecting the dispersion of the propagation medium, yields $\theta^{\omega} = \theta^{2\omega} = \theta$. Assuming a two-dimensional phase with a random azimuthal distribution ϕ and an average α angle of dipoles with respect to the surface normal, α and the molecular higher hyperpolarizability coefficient β_{uuu} may then be inferred from the intensity and polarization state of the emitted second harmonic (Heinz *et al.*, 1983; Chen *et al.*, 1981). In addition, tuning of ω (and 2ω) to resonance with excited electronic states of the adsorbed molecules may further enhance the sensitivity of this technique (Heinz *et al.*, 1982), which can then be referred to as nonlinear spectroscopic ellipsometry. This is shown in the insert, where a three-level molecular system is shown to be simultaneously resonant at ω and $2\omega^{m}$ (Marowski *et al.*, 1985).

harmonic in directions that are "phase-matched" to the respective emitting molecular gratings:

$$k_{\omega} \sin \theta_{\omega} = k_{\omega} \sin \theta_{\omega}^{\mathrm{R}} \tag{234}$$

and

$$2k_{\omega} \sin \theta_{\omega} = k_{2\omega} \sin \theta_{2\omega} \tag{235}$$

where $\theta_{\omega}^{\mathrm{R}}$, the reflected angle, equals θ_{ω}, and where $\theta_{2\omega}$ the harmonic emission angle, is given by

$$n_{\omega} \sin \theta_{\omega} = n_{2\omega} \sin \theta_{2\omega} \tag{236}$$

Hence $\theta_{2\omega} = \theta_{\omega}$ when the index of dispersion is negligible. In the case of a surface covered with a monomolecular layer composed of molecular units with a random azimuthal angular distribution and a peaked angular distribution of the polar axis with respect to the surface normal (α being the latter angle), $\chi_{s}^{(2)}$ can be traced back to its molecular origin β, the deposited molecular first-order hyperpolarizability, by a simple geometric summation scheme (the so-called oriented gas model, which reduces here to be a two-dimensional oriented gas). The surface can then be considered as optically uniaxial with optical axis perpendicular to the surface. The only nonvanishing component of $\chi_s^{(2)}$ (the other being deduced by Kleinman symmetry arguments) is then

$$[\chi_s^{(2)}]_\perp = \tfrac{1}{2} N_s < \cos\alpha \sin^2\alpha > \beta \qquad (237)$$

$$[\chi_s^{(2)}]_{\perp\perp\perp} = N_s < \cos^3\alpha > \beta \qquad (238)$$

where it has been further assumed that the molecule is one-dimensional resulting in a single molecular tensorial coefficient $\beta = \beta_{uuu}$, with u along the polar axis, and N_s is the molecular coverage density. It is thus possible, by a nonlinear ellipsometric measurement, to measure the peak value α_0 of the α distribution, assuming, as done by Heinz *et al.* (1983) and Chen *et al.* (1981), that $\langle\alpha\rangle = \alpha_0$, as well as β. In a typical experiment, Heinz *et al.* thus measured the polar orientation of *p*-nitrobenzoic acid absorbed at air–silica and ethanol–silica interfaces. In a similar configuration, Rasing *et al.* (1985) has studied monomolecular layers of sodium dodecylnaphthalene sulfonate [CH_3—$(CH_2)_{11}$—$C_{10}H_6$—SO_3Na, where $C_{10}H_6$ stands for a naphthalene structure] at the air–water interface assuming an orientational model as in Eqs. (237) and (238) (i.e., azimuthal isotropy of long chain axis and peaked α distribution). When increasing the surface pressure π, analysis of the reflected SHG at the air–water interface along the lines detailed above shows a limiting high-pressure value of $\alpha \approx 30°$ at π above 5 mN m^{-1}. This experiment quantitatively supports the well-accepted notion of molecules tending to adopt an upright position at higher compression as induced from the classical π/σ pressure isotherms. Surface second-harmonic generation can be further enhanced by tuning the fundamental and/or harmonic wavelength to resonance with excited electronic levels of the monomolecular layers. This was first demonstrated by Heinz *et al.* (1982) by adjustment of the emission wavelength of an Nd^{3+}. YAG laser pumped dye to half the energy of the $S_0 \rightarrow S_2$ electronic transition of rhodamine dyes adsorbed on fused silica substrates (30,000 cm^{-1} for rhodamine 110 and 28,900 cm^{-1} for rhodamine 6G). Doubly resonant excitation was achieved on a slab covered with Nile blue by simultaneous tuning of ω (near 600 nm) and 2ω with $S_0 \rightarrow S_1$ and $S_0 \rightarrow S_2$ transitions of the dye, respectively. Neglecting nonresonant terms and maintaining the validity of the perturbational sum-over-state description, $\chi_s^{(2)}$ is seen to have both ω- and 2ω-resonant denominators (Marowski *et al.*, 1985),

$$\chi_{s,ijk}^{(2)} \propto \frac{\langle g|x_i|c\rangle}{(W_c - 2h\omega + i\Gamma_{ac})} \times \frac{\langle g|x_j|b\rangle\langle b|x_k|c\rangle|c\rangle + \langle g|x_k|b\rangle\langle b|x_j|c\rangle}{(W_b - h\omega + i\Gamma_{ab})} \qquad (239)$$

where b, W_b and c, W_c correspond to the eigenstates and energies of resonant S_2 and S_0 excited states, respectively, and g to the ground state. The doubly resonant scheme is illustrated in Fig. 48, and Marowski reports a second-harmonic generation efficiency of the order of 20% of that obtained, under similar illumination conditions, from a 1-mm-thick phase-matched KDP crystal.

Although full spatial resonance, such as obtained in waveguides and resulting in eigenmode guiding, has not been considered in relation to nonlinear surface phenomena, interferometric enhancement has been developed in the quadratic case by Reider *et al.* (1983), using a technique reminiscent of that used previously by Kajzar *et al.* (1983) for cubic effects $\chi^{(3)}$ in Langmuir–Blodgett deposited layers of polydiacetylenes (see Chapter II,5). Enhancement by a factor of 10^2 to 10^4 has thus been reported, as compared to the second-harmonic signal from a single surface reflection, when irradiating a quartz slab coated by rhodamine 110 on both sides: the enhancement originates, at proper angular incidence, from multiple, constructively interfering, nonlinear reflections at the slab-coating interface. Let us now consider the geometry of Fig. 49 where a classical waveguide is topped by nonlinear layers that can be modelled by a thin nonlinear interface between the guide and medium 1 (the air, for example). Let us adopt, for the sake of simplicity, a ray-optics point of view, which will ease the extension from simple surface SHG on to the situation depicted here. The fundamental wave will excite a surface harmonic polarization wave, which will in turn radiate back into the guide, provided it is transparent at both ω and 2ω. The case where the waveguide would not be transparent at 2ω can be treated by considering a second waveguide on top of the first one, both being separated by the nonlinear surface, and with adequate spectral transparency at 2ω for the second one. The discussion to follow needs only minor adaptations to deal with this case. Returning to the configuration in Fig. 49, the back-radiated second-harmonic will have to correspond to a propagation eigenmode if one wants efficient second-harmonic emission. This leads to the following phase-matching conditions:

$$n_\omega \sin\theta_\omega = n_{2\omega}\sin\theta_{2\omega} \tag{240}$$

$$\bar{h}^{\omega} t = \bar{\phi}_1^{\omega} + \bar{\phi}_3^{\omega} + m\pi \tag{241}$$

$$\bar{h}^{2\omega} t = \bar{\phi}_1^{2\omega} + \bar{\phi}_3^{2\omega} + n\pi \tag{242}$$

Equation (240) corresponds to the condition of Eq. (236) in surface SHG, while Eq. (241) [or Eq. (242)] shows that the fundamental (or harmonic) waves as an eigenmode of order m (or n). The situation depicted here corresponds to TM modes, more likely to be of interest as the molecules tend to adopt a statistically upright position with respect to the substrate. TE mode polarization

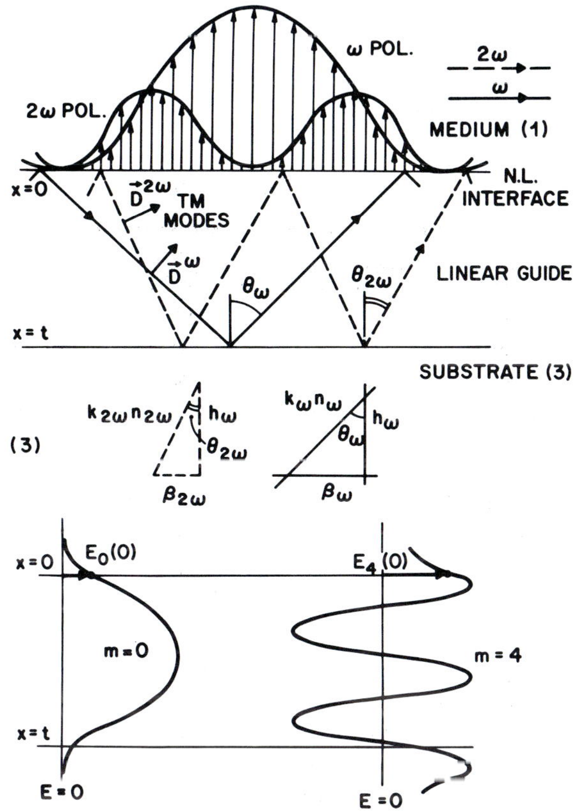

Fig. 49. Planar waveguide topped by a nonlinear layer: optical waves are confined in the guide and interact in the nonlinear layer. Phase matching is ensured when the phase periodicity of the linear molecular polarization wave is twice that of the harmonic one. The efficiency of the process depends only on the product of the interacting fields at $x = 0$, which allows for the interaction of waves belonging to different modes.

would generally be orthogonal to the mean molecular dipole orientations, thus cancelling the $\mathbf{E} \cdot \boldsymbol{\mu}_{01}$ terms ($\boldsymbol{\mu}_{01}$ transition dipole). Definitions of ϕ_1, ϕ_3, and $\bar{h}$ were given previously as functions of β, ω, and relevant indices of refraction at ω and 2ω. Restating Eq. (240) in the equivalent form, $\beta_{2\omega} = 2\beta_\omega$, Eqs. (241) and (242) correspond to the classical phase-matching conditions: i.e., for given m, n, and ω (possibly imposed by some resonant conditions on the molecular species), the proper guide thickness t can be extracted after elimination of $\beta = \beta_\omega = \beta_{2\omega}/2$ from Eqs. (241) and (242). The major difference when comparing this with the more classical SHG conversion process in waveguides is that no overlap integral is weighting the efficiency. What matters is just the "scalar" factor α_R (as opposed to the overlap "integral" I_R)

of the ω and 2ω waves at $x = 0$ given by

$$\alpha_R = [E_\omega^m(0)]^2 \times E_{2\omega}^n(0) \tag{243}$$

This factor does not vanish for simple symmetry arguments; furthermore, there is no reason to systematically favor lower-order modes, inasmuch as α_R is concerned, as was done in the classical case to optimize I_R. As depicted at the bottom of Fig. 49, proper thickness–index combinations may well lead to $E^4(0) \geqslant E^0(0)$, and an $n = 4$ mode can then be prefered to an $m = 0$ mode. Birefringency is therefore not required for the guiding material, as was previously found relevant for the $m = n = 0$ mode matching. However, lower-order modes may still be prefered on other grounds: lower-order modes correspond to a larger θ angle, thus favoring the geometric selection factor $\mathbf{E} \cdot \boldsymbol{\mu}_{01} \approx E\mu_{01} \sin\theta$ (in the general case of upright molecules), which governs the efficiency of field–molecule interactions.

Addressing such a device at resonance(s) of the nonlinear species raises the problem of subsequent absorption. However, the nonlinear effect involved, as opposed to a bulk configuration, takes place only at an interface, which should be lest costly than in the case of actual propagation in the absorbing medium of space. A full discussion of the best trade-off requires modernization within the Maxwellian formalism, with a nonlinear layer of thickness τ with $\tau \ll t$, covering the waveguide of thickness t (Zyss, 1985).

In conclusion, a number of interesting possibilities are opened up by organic materials based waveguiding devices. Some of them are listed in Fig. 50, of which two, (a) and (c), have been specifically addressed here. A review on organic materials for nonlinear integrated optics with more details on material technology has been given by Zyss (1985). Other chapters of this book address the crystallization and layer-deposition technique more specifically than is done here. Configuration (d), single-crystalline nonlinear organic cored capillaries or hollow fibers, may open new interesting directions in view of obvious compatibility with optical fiber systems. Efforts are currently undertaken to grow NPP or *N*-(4-nitrophenyl)-*N*-methylaminoacetonitrile (NPAN) within capillaries of compatible dimensions with single-mode operation. Once insertion loss problems are solved, this type of devices might more readily find its way in a low-loss single-mode optical fiber system than planar waveguides would tend toward parametric amplification, electo-optic modulation, or directional coupling.

VII. CONCLUSION

The subtle and intimate interplay between structural and nonlinear optical properties of organics, considered at both the molecular and the crystalline levels, has been amply evidenced: the increase of basic knowledge derived

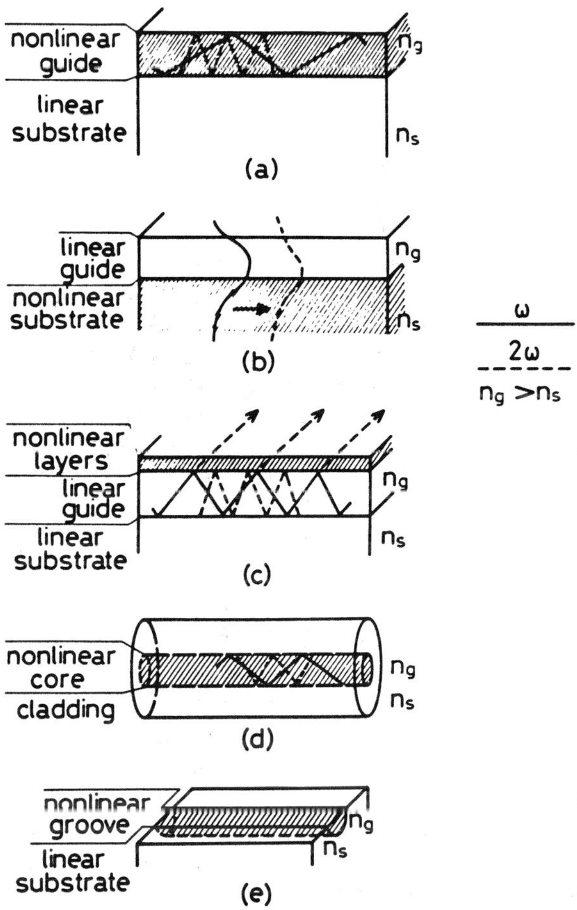

Fig. 50. Examples of waveguiding nonlinear structures. (a) Propagation of both harmonic and fundamental waves together with harmonic conversion in a nonlinear waveguide topping a linear substrate. (b) Linear waveguiding structure topping a nonlinear substrate. Both harmonic and fundamental are guided, but the nonlinear conversion, from the evanescent tail of the fundamental on to that of the harmonic, takes place in the nonlinear bulky substrate. (c) A very thin layer (or monolayer) is deposited on top of a linear waveguide. Nonlinear conversion takes place at reflections of the fundamental at the linear–nonlinear interface. For a proper choice of the guide index of refraction and of the fundamental coupled-in eigenmode (i.e., angle of reflection in a ray description of the waveguiding phenomenon), the harmonic generated at the interface can be coupled out and may give raise to interferometric enhancement. It may also be coupled into a propagating mode, provided a phase-matching condition is satisfied. In the case of resonant or multiresonant nonlinear interactions, this kind of structure may well be the ideal one, as the high adsorption entailed rules out significant interfaction length if one wants to profit from the resonant enhancement. (d) Waveguiding hollow fiber or capillary with a nonlinear core. The situation is similar to that in (a). Another possibility, although less favorable, is to guide the harmonic in the cladding (so-called "radiation modes"). (e) Nonlinear groove carved in a linear substrate.

therefrom should, in itself, suffice to justify and reward the already considerable investment in the field. In fact, the original avenues opened up by a cross-fertilizing approach blurring the borderlines of physics and chemistry are just starting to be fully exploited: powerful "number-crunching" computers coupled with crystallographic data banks and running ever more reliable light–matter interaction routines are in the process of supplying previously scarce organic nonlinear optical data. One may realistically envision a time when organic chemists, whatever the precise end of their work, will manipulate molecular hyperpolarizabilities as any other general-purpose molecular data and with no less familiarity than dipole moments, infrared oscillator strength, dielectric constant, or rotatory power. In fact, simple common-sense arguments suffice to point-out the necessity of going beyond such classically available data when one wishes to account, with sufficient insight, for molecular optical properties, very much like higher-order multipoles are needed when an accurate description of the electronic charge density is at stake. In fact, a comprehensive description of molecular behavior in an electromagnetic environment would require a field- and vibration-dependent wave-function amplitude $|\Psi(\mathbf{x}, \mathbf{Q}, \mathbf{E}, \mathbf{H})|^2$ or, equivalently, of its nth-order moments ($n \geqslant 1$),

$$\mu^{(n)}(\mathbf{Q}, \mathbf{E}, \mathbf{H}) \propto \int x^n |\Psi(\bar{x}, \bar{Q}, \bar{E}, \bar{H})|^2 \, d^3x$$

The $\mu^{(n)}$ terms can be accounted for by their successive derivatives with respect to $\mathbf{Q}$ and $\mathbf{E}$, namely $\partial\mu^{(n)}/\partial Q^i \, \partial E^j$, disregarding here for the sake of simplicity magnetic-field-related effects. Some of these "susceptibilities" are quite familiar and belong to the common "cultural" background of chemists, such as polarizabilities and related indices of refraction ($n = 1, i = 0, j = 1$: $\propto \partial\mu^{(1)}/\partial E$), infrared oscillator strength ($n = 1, i = 1, j = 0$: $\propto \partial\mu^{(1)}/\partial Q$), Raman cross section ($n = 1, i = 1, j = 1$: $\propto \partial\alpha/\partial Q$), classical multipoles ($n = 2, i = 0, j = 0$: $\propto$ quadrupole), etc. Except for the Raman cross section proportional to $\partial^2\mu^{(1)}/(\partial Q \, \partial E)$, these quantities all remain within the border of linear approximations and are useful insofar as this restriction is valid. Molecular hyperpolarizabilities such as β ($n = 1, i = 0, j = 2$) or γ ($n = 1, i = 0, j = 3$) are next on the list after the polarizability tensor to help refine classical molecular emission and absorption phenomena and related models. While an increasing amount of theoretical and experimental data on molecular nonlinear properties is being gathered and made available, the data should be fruitfully referred to whenever any physicochemical property of interest, not necessarily in the field of nonlinear optics itself, is found sensitive to electronic delocalization, structural asymmetry, charge transfer, the electronic nature of UV-excited state, etc. In very much the same way, few scientists are interested in dipole moments as such, but a majority are considering them as essential information to be used in connection with other complementary data, in order to account for a host of interesting phenomena such as interactions in fluids, viscosity, lattice cohesion, adsorption on surfaces and wetting, chemical reactivity, etc. In the study of molecule–solvent interactions in solutions, some insight may be derived from the absorption spectrum shift in relation to the polarity of both constituents. Additional results for the dispersion and concentration dependence of the solution nonlinearity and possible deviations from an additive solvent–solute model would fruitfully complement the linear data. This is also obvious when adsorption of molecular species is being investigated: surface second-harmonic generation has proved to be a powerful detection and characterization tool. Similarly, local field corrections have been shown to increasingly weight macroscopic susceptibilities when higher-order nonlinear effects are considered: comparison between macroscopic and microscopic susceptibilities, possibly measured in different phases, would certainly be helpful in refining current views on the dielectric theory of matter. The molecular-engineering approach followed here is reasonably predictive at the molecular level, where simple reference models such as based on the two-level system or an equivalent internal field can be confidently addressed to yield simple trends to which properly calibrated quantum-mechanical models of every desired level of sophistication may add further numerical accuracy. This favorable situation is in striking constant with that prevailing at the crystal level. It is impossible to reliably predict the crystalline structure of a molecular crystal from the sole knowledge of the molecular structure except in very simple cases. Even in the case of a crystal built of enantiomeric chiral molecules, the geometrically imposed absence of an inversion center does not guarantee a significant departure from centrosymmetry. This results in a costly dead end for the major part of the synthetic chemist's work, however promising at the molecular level. In practice, a number of techniques aimed at "forcing" the alignment of molecules in polar lattices, out of their natural tendency to form centrosymmetric structures, are being actively explored: the Langmuir–Blodgett technology, the polar orientation of nonlinear liquid crystalline structures by application of a poling field in the mesophase and subsequent freezing in the glassy phase, and inclusion compounds all resort to cohesive interactions of various natures that prevail over dipolar interactions. The cost entailed by the dilution of the active nonlinear unit and the statistically random nature of the molecular ordering may be willingly affordable in cases where there exists a possibility of acting on the lattice orientation. The possibility of mixing organic with inorganic structures may also be claimed as a way to mechanically harden organic structures while keeping the benefit of their unique structural flexibility and polarizability. In fact, going further along this way, boron–oxygen structures mimicking aromatic compounds have been recently proposed and shown to rival urea for UV nonlinear applications (Chen *et al.*, 1984). Besides these considerations, organic nonlinear materials are nonetheless expected to lead to a practical breakthrough in the field of optical materials and optical applications, which demands that a number of stringent technology related requirements be met (Section VI).

It is often and rightly claimed by research chemists attracted in the field that the number of molecular structures actually investigated for nonlinear optics is negligible as compared to the full potential of organic synthesis. As a consequence, major advances should still be ahead of us, and *p*-nitroaniline or pyridine-like structures should certainly not be the ultimate in molecular engineering. Nevertheless, such compounds as NPP or MNA, surpassing lithium niobate by one to two orders of magnitude in nonlinear efficiency, are well worth a concentration of efforts following the example of urea, while this should not be at the expense of a more advanced synthesis work. The technological efford required to take these materials from laboratory measurements on to "real-life" device operation, either in bulk or thin-film configurations, is bound to significantly surpass that invested in the previous molecular engineering phase. Of paramount importance, when referring to bulk applications such as optical frequency doublers and parametric amplifiers, is the necessity to improve, if not invent, polishing and packaging techniques, as surface scattering is presently one of the major obstacles against the observation of cavity-enhanced parametric effects. The damage threshold of nonlinear organics under continuous or high repetition rate (i.e., megahertz or above) laser irradiation, unlike the case of low repetition rate experiments allowing for thermal relaxation, has not been well assessed. Satisfactory behavior under such conditions would considerably enlarge the potential of organics for applications in view of the increasing number of continuous laser source-driven systems. Waveguided interactions of short-pulse (i.e., picosecond or subpicosecond) laser sources seem to offer bright prospects for application owing to the combined benefit of power confinement and high gain: such applications are in keeping with current trends in optical communication toward higher bit rates and tighter integration.

As far as applications are concerned, materials exhibiting parameteric gains surpassing by two orders of magnitude that of lithium niobate are bound to raise interest sufficient to generate the industrial growth and processing technologies required for applications, namely, purification, cutting, polishing, packaging, thin-film deposition, etc. However, there obviously is still a long pioneering way to go from laboratory experiments on to marketable devices. Parallel to it, the investigation of nonlinear properties of organic matter, regardless of applications, will go on developing irreversibly as a major source of structural and spectroscopic insights on molecular electronic phenomena.

Developments in the future will proceed along these two avenues, where equally fascinating challenges are to be met in technology as well as in basic science.

REFERENCES

Adler, S. L. (1962). *Phys. Rev.* **126,** 413.

Agranovitch, V. M., and Ginzburg, V. L. (1966). "Spatial Dispersion in Crystal Optics and the Theory of Excitons." Wiley (Interscience), New York.

Akhmanov, S. A., Kovrygin, A. I., and Sukhorukov, A. P. (1975). *In* "Quantun Electronics: A Treatise" Ed. H. Rabin and C. L. Tang, eds.), Vol. 1, Part. B, p. 476. Academic Press, New York.

Allen, L., and Eberly, J. H. (1975). "Optical Resonance and Two Level Atoms." Wiley (Interscience), New York.

Amos, A. T., and Crispin, R. J. (1975). *J. Chem. Phys.* **63,** 1890.

Armstrong, J. A., Bloembergen, N., Ducuing, J., and Pershan, P. S. (1962). *Phys. Rev.* **127,** 1918.

Asby, R. (1969a), *Opto-electronics (London)* **1,** 165.

Asby, R. (1969b), *Phys. Rev.* **187,** 1062, 1070.

Azema, A. Botineau, J., Gires, F., and Saissy, A. (1979). *SPIE Proc.* **213,** 26.

Bartlett, R. J., and Purvis, G. D. (1979). *Phys. Rev. A* **20,** 1313.

Bergman, J. G., Crane, J. R., Levine, B. F., and Bethea, C. G. (1972). *Appl. Phys. Lett.* **20,** 21.

Betzler, K., Hesse, H., and Loose, P. V. (1978). *J. Mol. Struct.* **47,** 393.

Bigio, I. J., and Ward, J. F. (1974). *Phys. Rev. A* **9,** 35.

Bjorkholm, J. E. (1966). *Phys. Rev.* **142,** 128.

Bloembergen, N. (1965). "Nonlinear Optics." Benjamin, New York.

Bloembergen, N., Logan, H., and Lynch, R. T. (1978). *Indian J. Pure Appl. Phys.* **16,** 151.

Bonneville, R., and Chemla, D. S. (1978). *Phys. Rev. A.* **17,** 2046.

Borde, C. J. (1979). *J. Phys. (France)* **60,** 819.

Borisov, E. V. (1975). *J. Phys. Chem.* **49,** 1560.

Born, M., and Wolf, E. (1965). "Principles of Optics." Macmillan, New York.

Bottcher, C. J. (1952). "Theory of Electric Polarization." Elsevier, Amsterdam.

Bottcher, C. J. F. (1973). "Theory of Dielectric Polarization." Elsevier, Amsterdam.

Boulton, A. J., Glover, G. M., Hely Hutchinson, M., Katritzky, A. R., Short, D. J., and Sutton, L. E. (1966). *J. Chem. Soc. B* p. 822.

Boyd, G. D., and Kleinman, D. A. (1968). *J. Appl. Phys.* **39,** 3597.

Buckingham, R. A. (1938). *Proc. R. Soc. London, Ser. A* **168,** 264.

Buckmaster, M. A., Chatterjee R., and Shing, Y. H. (1973). *Phys. Status Solidi A* **13,** 9.

Butcher, P. N. (1965). "Nonlinear Optical Phenomena." Ohio State Univ. Press, Columbus.

Butcher, P. N., and McLean, T. P. (1963). *Proc. Phys. Soc., London* **81,** 219.

Byer, R. L. (1975). *In* "Quantum Electronics: A Treatise" (H. Rabin and C. L. Tang eds.), Vol. I, Part B, p. 588. Academic Press, New York.

Carenco, A., Jerphagnon, J., and Perigaud, A. (1977). *J. Chem. Phys.* **66,** 3806.

Chemla, D. S., and Bonneville, R. (1978). *J. Chem. Phys.* **68,** 2214.

Chemla, D. S., Oudar, J. L., and Jerphagnon, J. (1975). *Phys. Rev. B* **12,** 4534.

Chen, C. K., Heinz, T. F., Ricard, D., and Shen, Y. R. (1981). *Phys. Rev. Lett.* **46,** 15.

Chen, C.-T., Wu, B.-C., You, G.-M., Jiang, A.-D., and Huang Y. C. (1984). *Proc. Int. Quantum Electr. Congr., 13th, 1984.*

Cohen, H. D., and Roothan, C. C. J. (1965). *J. Chem. Phys.* **43,** 534.

Cojan, C. (1977). Ph.D. Thesis, p. 114. University of Paris-Sud.

Coulson, C. A. (1979), "Valence." Oxford Univ. Press, London and New York.
Cyvin, S. J., Rauch, J. E., and Decius, J. C. (1965). *J. Chem. Phys.* **43,** 4083.
Davydov, B. L., Derkacheva, L. D., Dunina, V. V., Zhabotinskii, M. E., Zolin, V. F., Koreneva, L. G., and Samokhina, M. A. (1970). *JETP Lett. (Engl. Transl.)* **12,** 16.
Davydov, B. L., Derkacheva, L. D., Dunina, V. V., Zhabotinskii, M. E., Zolin, V. F., Koreneva, L. G., and Samokhina, M. A. (1971). *Opt. Spectrosc.* **30,** 274.
DeGroot, S. R. (1969). "The Maxwell Equations." 1969 North-Holland Publ., Amsterdam.
Dewar, M. J. S., Suck, S. H., Weiner, P. K., and Bergman, J. G. (1976). *Chem. Phys. Lett.* **38,** 228.
Donaldson, W. R., and Tang, C. L. (1984). *Appl. Phys. Lett.* **44,** 25.
Druet, S. A. J., Taran, J.-P. E., and Borde, C. J. (1979). *J. Phys. (France)* **40,** 819–840.
Dulcic, A., and Flytzanis, C. (1978). *Opt. Commun.* **25,** 402.
Finn, R. S., and Ward, J. F. (1974). *J. Chem. Phys.* **60,** 454.
Fiutak, J. (1963). *Can. J. Phys.* **41.** 12.
Flytzanis, C. (1972). *Phys. Rev.* **178,** 1218.
Flytzanis, C. (1975). *In* "Quantum Electronics: A Treatise" (H. Rabin and C. L. Tang, eds.), Vol. 1, Part A, Academic Press, New York.
Flytzanis, C., and Ducuing, J. (1968). *Phys. Rev. Lett. A.* **26,** 315.
Flytzanis, C., and Ducuing, J. (1969). *Phys. Rev.* **178,** 1218.
Friedel, J., and Lannoo, M. (1973). *J. Phys. (Orsay, Fr.)* **34,** 115, 483.
Giordmaine, J. A. (1962). *Phys. Rev. Lett.* **8,** 21.
Giordmaine, J. A. (1965). *Phys. Rev. A* **138,** 1599.
Giordmaine, J. A., and Rentzepis, P. M. (1967). *J. Chem. Phys.* **64,** 215.
Gott, J. R. (1971). *J. Phys. B* **4,** 116.
Halbout, J. M., Blit, S., Donaldson, W., and Tang, C. L. (1979). *IEEE J. Quantum Electron.* **QE-15,** 1176.
Halbout, J. M., Sarhangi, A., and Tang, C. L. (1980). *Appl. Phys. Lett.* **37,** 864.
Harriss, E., Oshman, M. K., and Byer, R. L. (1967). *Phys. Rev. Lett.* **18,** 732.
Hauchecorne, G., Kerherve, F., and Mayer, G. (1971). *J. Phys.* **32,** 47.
Heilbronner, E., and Bock, H. (1970). "Das HMO-Modell und seine Anwendung." Verlag Chemie, Weineim.
Heilmeir, G. H., Ockman, N., Braunstein, R., and Kramer, D. A. (1964). *Appl. Phys. Lett.* **5,** 229.
Heinz, T., Chen, C., Ricard, D., and Shen, Y. (1982). *Phys. Rev. Lett.* **48,** 478.
Heinz, T., Tom, H., and Shen, Y. (1983). *Phys. Rev. A* **28,** 1883.
Hermann, J. P. (1973). *Opt. Commum.* **9,** 74.
Hermann, J. P. (1976b). Ph.D. diss., University of Paris-Sub (Orsay).
Hermann, J. P., and Ducuing, J. (1974). *J. Appl. Phys.* **45,** 5100.
Hermann, J. P., Ricard, D., and Ducuing, J. (1973). *Appl. Phys. Lett.* **23,** 178.
Hierle, R., Badan, J., and Zyss, J. (1984). *J. Cryst. Growth* **69,** 545.
Hochstrasser, R. M. (1973). *Acc. Chem. Res.* **6,** 263.
Hopfield, J. J. (1971). *In* "The Physics of Optoelectronic Materials" W. A. Albers, Jr., ed.), p. 41 Plenum, New York.
Hulin, D., Migus, A., Antonetti, A., Ledoux, I., Badan, J., Oudar, J. L., and Zyss, J. *Appl. Phys. Lett.* (to be published).
Hush, N. S., and Williams, M. L. (1972). *Theor. Chim. Acta.* **25,** 346.
Jerphagnon, J. (1970a). *Phys. Rev. B* **2,** 1091.
Jerphagnon, J. (1970b). *Appl. Phys. Lett.* **16,** 298.
Jerphagnon, J. (1971). *IEEE J. Quantum Electron.* **QE-7,** 42.
Jerphagnon, J., and Kurtz, S. K. (1970a). *J. Appl. Phys.* **41,** 1667.
Jerphagnon, J., and Kurtz, S. K. (1970b). *Phys. Rev. B* **1,** 1739.
Jerphagnon, J., Chemla, D. S., and Bonneville, R. (1978). *Adv. Phys.* **27,** 609.

Kajzar, F., Messier, J., Zyss, J., and Ledoux, I. (1983). *Opt. Commun.* **45,** 133.
Kajzar, F., Ledoux, I., and Zyss, J. (1985). *Phys. Rev. A* (to be published).
Katritzky, A. R., Randall, E. W., and Sutton, L. E. (1957). *J. Chem. Soc.* p. 1769.
Kielich, S. (1969). *IEEE J. Quantum Electron.* **QE-5,** 562.
Kitaigorodsky, A. I. (1961). "Organic Chemical Cristallography." Consultants Bureau, New York.
Kitaigorodsky, A. I. (1973). "Molecular Crystals and Molecules" (Engl. Transl.). Academic Press, New York.
Kleinman, D. A. (1962). *Phys. Rev.* **126,** 1977.
Kleinman, D. A., Askin, A., and Boyd, G. D. (1966). *Phys. Rev.* **145,** 338.
Knossow, M., Mauguen, and Y., de Rango, C. (1976). *Cryst. Struct. Commun.* **5,** 723.
Koreneva, L. G., Zolin, V. F., and Davydov, B. L. (1975). *In* "Molecular Crystals in Nonlinear Optics." Nauka, Moscow.
Kurtz, S. K. (1975). *In* "Quantum Electronics: A Treatise" (H. Rabin and C. L. Tang, eds.), Vol. 1, Part A. Academic Press, New York.
Kurtz, S. K., and Perry, T. T. (1968). *J. Appl. Phys.* **39,** 3798.
Kurtz, S. K., Jerphagnon, J., and Choy, M. (1979). *In* "Landolt-Bornstein Tables" New Ser., Group III, p. 456. (H. Heleweg and A. M. Heleweg, eds.), Springer-Verlag, Berlin and New York.
Lalama, S. J., and Garito, A. F. (1979). *Phys. Rev. A* **20,** 1179.
Ledoux, I., and Zyss, J. (1982). *Chem. Phys.* **73,** 203.
Ledoux, I., Josse, D., Vidakovic, P., and Zyss, J. (1985). *Opt. Eng.* **25,** 202.
Ledoux, I., Zyss, J., Migus, A., Etchepare, J., and Grillon, G., Antonetti, A. (1986). *Appld. Phys. Lett.*
Lefevre, C. G., and Lefevre, R. J. W. (1965). *Adv. Phys. Org. Chem.* **3,** 48.
Lennard-Jones, J. E. (1931). *Proc. Cambridge Philos. Soc.* **27,** 469.
Levine, B. F. (1975). *J. Chem. Phys.* **63,** 115.
Levine, B. F. (1976). *Chem. Phys. Lett.* **37,** 516.
Levine, B. F., and Bethea, C. G. (1974) *Appl. Phys. Lett.* **24,** 445.
Levine, B. F., and Bethea, C. G. (1974). *J. Chem. Phys.* **60,** 3856.
Levine, B. F., and Bethea, C. G. (1975). *J. Chem. Phys.* **63,** 2666.
Levine, B. F., and Bethea, C. G. (1976). *J. Chem. Phys.* **65,** 2439.
Levine, B. F., and Bethea, C. G. (1977). *J. Chem. Phys.* **66,** 1070.
Levine, B. F., Bethea, C. G., Thurmond, C. D., Lynch, R. T., and Bernstein, J. L. (1979). *J. Appl. Phys.* **50,** 2523.
Linton, E. P. (1940). *J. Am. Chem. Soc.* **62,** 1945.
Lipscomb, G. F., Garito, A. F., and Narang, R. S. (1981). *J. Chem. Phys.* **75,** 1509.
Liptay, W. (1973). *In* "Excited States" (E. C. Lim, ed.), Vol. 1, p. 129. Academic Press, New York.
London, F. (1930). *Z. Phys. Chem.* **11,** 222.
Lorentz, H. A. (1916). "The Theory of Electrons." Teubner Lepzig (reprinted by Dover, New York, 1951).
Lutskii, A. E., and Gorokhova, N. I. (1967). *Opt. Spectrosc. (Engl. Transl.)* **27,** 917.
Maker, P. D., Terhune, R. W., Nisenoff, M., and Savage, C. M. (1962). *Phys. Rev. Lett.* **8,** 21.
Manley, J. M., and Rowe, H. E. (1959). *Proc. IRE* **47,** 2115.
Marowski, G., Gierulski, A., and Dick, B. (1985). *Opt. Commun.* **52,** 339.
Mazur, P., and Nijboer, B. R. A. (1953). *Physica (Amsterdam)* **19,** 971.
Meredith, G. R., Buchalter, B., and Hanzlik, C. (1983). *J. Chem. Phys.* **78,** 1533.
Midwinter, J. E., and Warner, J. (1965). *J. Appl. Phys.* **16,** 1135.
Minkin, V. I., Osipov, O. A., and Zhdanov, Y. A. (1970). "Dipole Moments in Organic Chemistry." Plenum, New York.
Morrell, J. A., and Albrecht, A. C. (1979). *Chem. Phys. Lett.* **64,** 46.

Mulliken, R. S. (1955). *J. Chem. Phys.* **23,** 1833.
Murrel, J. N. (1973). "The Theory of Electronic Spectra of Organic Molecules." Wiley, New York.
Nayar, B. K. (1983). *ACS Symp. Ser.* **233,** 153.
Nayar, B. K. (1986). *In* "Nonlinear Optics: Materials and Devices" (C. Flytzanis and J. L. Oudar, eds.), p. 142. Springer-Verlag, Berlin and New York.
Nicoud, J. F., private communication.
Ochiai, E., and Ishikawa, M. O. (1942). *Proc. Jpn. Acad.* **18,** 561.
O'Hare, J. M., and Hurst, R. P. (1967). *J. Chem. Phys.* **46,** 2356.
Onsager, L. (1936). *J. Am. Chem. Soc.* **58,** 1456.
Otto, P. (1978). *Chem. Phys.* **33,** 407.
Oudar, J. L. (1977). *J. Chem. Phys.* **67,** 446.
Oudar, J. L., and Chemla, D. S. (1975). *Opt. Commun.* **13,** 10.
Oudar, J. L., and Chemla, D. S. (1977). *J. Chem. Phys.* **66,** 2664.
Oudar, J. L., and Hierle, R. (1977). *J. Appl. Phys.* **48,** 2699.
Oudar, J. L., and Le Person, H. (1975). *Opt. Commun.* **15,** 258.
Oudar, J. L., and Sigelles, M. (1977). CNET Internal Report (unpublished).
Oudar, J. L., and Zyss, J. (1982). *Phys. Rev. A* **26,** 2076.
Oudar, J. L., Chemla, D. S., and Batifol, E. (1977). *J. Chem. Phys.* **67,** 1626.
Orlov, R. Y. (1966). *Sov. Phys.—Crystallogr.* (*Engl. Transl.*) **11,** 410.
Pauling, L. (1960). "Nature of the Chemical Bond." Cornell Univ. Press, Ithaca, New York.
Pershan, P. S. (1963). *Phys. Rev.* **130,** 919.
Pine, A. S. (1965). *Phys. Rev. A* **139,** 901.
Pople, J. A., and Beveridge, D. L. (1970). "Approximate Molecular Orbital Theory." McGraw-Hill, New York.
Pople, J. A., McIver, J. M., and Ostlund, N. S. (1968). *J. Chem. Phys.* **49,** 2960.
Prior, Y. (1984). *IEEE J. Quantum Electron.* **QE-20,** 37.
Purvis, G. D., and Bartlett, R. J. (1981). *Phys. Rev. A* **23,** 1594.
Rasing, T., Shen, Y. R., Mahn Won Kim, Valint, P., and Bock, J. (1985). *Phys. Rev. A* **31,** 537.
Reider, G. A., Schmidt, A. J., and Marowski, G. (1983). *Opt. Commun.* **47,** 223.
Rentzepis, P. M., and Pao, Y. H. (1964). *Appl. Phys. Lett.* **5,** 156.
Robinson, F. N. H. (1967). *Bell Syst. Tech. J.* **46,** 913.
Robinson, F. N. H. (1971). *Physica* (*Amsterdam*) **54,** 329.
Robinson, F. N. H. (1973). "Macroscopic Electromagnetism." Pergamon, Oxford.
Schweig, A. (1967). *Chem. Phys. Lett.* **1,** 195.
Shen, R. Y. (1968). *Phys. Rev.* **167,** 818.
Shiro, M., Yamakawa, M., and Kubota, T. (1977). *Acta Crystallogr. Sect. B* **B33,** 1549.
Sidding, K. M., Saraogi, J., and Ali, S. (1959). *Acta Crystallogr.* **12,** 248.
Singer, K. D., and Garito, A. F. (1981) *J. Chem. Phys.* **75,** 3572.
Singh, S. (1971). *In* "Handbook of Laser" (R. J. Pressley, ed.), p. 489. Chem. Rubber Publ. Co., Cleveland, Ohio.
Slichter, C. P. (1963). "Principles of Magnetic Resonance." Harper, New York.
Twieg, R. D., and Dirck, C. (1986). To be published.
Unsöld, A. (1927). *Z. Phys.* **43,** 388.
Vidakovic, P., Badan, J., Hierle, R., and Zyss, J. (1984). *Proc. Int. Quantum Electron. Congr., 137b, 1984.*
Ward, J. F. (1965). *Rev. Mod. Phys.* **37,** 1.
Ward, J. F., and Bigio, I. J. (1975). *Phys. Rev. A* **11,** 60.
Ward, J. F., and Miller, C. K. (1979). *Phys. Rev. A* **19,** 826.
Warner, J. (1975). *In* "Quantum Electronics: A Treatise" (H. Rabin and C. L. Tang, eds.), Vol. 1, Part B, p. 703. Academic Press, New York.

Wiser, N. (1963). *Phys. Rev.* **129,** 62.
Wyckoff, R. W. (1969). "Crystal Structures," Vol. 6, Part 1, p. 120. Wiley, New York.
Yamakawa, M., Kubota, T., Ezumi, K., and Mizuno, Y. (1974). *Spectrochim. Acta* **30,** 2103.
Yariv, A. (1975). "Quantum Electronics." Wiley, New York.
Yariv, A., and Yeh, P. (1984). "Optical Waves in Crystals." Wiley, New York.
Yee, T. K., and Gustafson, T. K. (1978). *Phys. Rev. A* **18,** 1597.
Zyss, J. (1979a). *J. Chem. Phys.* **70,** 3333.
Zyss, J. (1979b). *J. Chem. Phys.* **70,** 3341.
Zyss, J. (1979c). *J. Chem. Phys.* **71,** 909.
Zyss, J. (1985). *J. Mol. Electron.* **1,** 25.
Zyss, J., and Berthier, G. (1982). *J. Chem. Phys.* **77,** 3635.
Zyss, J., and Oudar, J. L. (1982). *Phys. Rev. A* **26,** 2028.
Zyss, J., Chemla, D. S., and Nicoud, J. F. (1981). *J. Chem. Phys.* **74,** 4800.
Zyss, J., Nicoud, J. F., and Coquillay, M. (1984). *J. Chem. Phys.* **81,** 4160.
Zyss, J., Ledoux, I., Hierle, R., Raj, R., Oudar, J. L. (1985). *IEEE J. Quantum Electron.* **QE21,** 1286.

Chapter II-2

Molecular Hyperpolarizabilities of Organic Materials

DAVID PUGH

Department of Pure and Applied Chemistry
University of Strathclyde
Glasgow GI 1X1, United Kingdom

JOHN O. MORLEY

Imperial Chemical Industries Organic Division
Manchester M93DA, United Kingdom

I. INTRODUCTION

The molecular polarizability and hyperpolarizability tensors relate the dipole moment induced in an isolated molecule to the applied fields:

$$p_i = \alpha_{ij}E_j + \beta_{ijk}E_jE_k + \gamma_{ijkl}E_jE_kE_l + \cdots \tag{1}$$

where p_i is the component of the induced dipole moment in the coordinate direction i, α_{ij} is the polarizability tensor, β_{ijk}, γ_{ijkl}, etc. are the hyperpolarizability tensors, and E_j is a component of the electric field acting on the molecule.

The expression parallels that for the macroscopic polarization of a material,

$$P_i = \varepsilon_0[\chi_{ij}^{(1)}E_j + \chi_{ijk}^{(2)}E_jE_k + \chi_{ijkl}^{(3)}E_jE_kE_l + \cdots] \tag{2}$$

In both equations the tensors are frequency dependent and should be written as functions of the frequency of the fields and the polarization produced. For example, $\beta_{ijk}(-2\omega, \omega, \omega)$ is the third-rank tensor describing the second-order electro-optic process in which a polarization of frequency 2ω is produced as a result of the action of fields of frequency ω.

For inorganic materials such as semiconductors (GaAs, CdS) or ionic crystals ($LiNbO_4$, KDP), Eq. (1) is not of much practical importance since individual molecular properties cannot be closely related to those of covalently or ionically bound crystals. However, it has been established that certain types of organic molecule have very high values of β_{ijk} and, when they occur in noncentrosymmetric crystal structures, of the related susceptibility $\chi_{ijk}^{(2)}$. In the solid state these compounds form molecular crystals in which the molecules interact through weak intermolecular forces (dispersion, dipole–dipole, etc.) and retain their individual identity to a high degree. The structure of the individual molecules is much more complicated than the structure of the atomic or ionic units in inorganic materials. The study of the molecular tensors of Eq. (1) is therefore pertinent to the identification of materials that have large nonlinear polarizabilities. The relationship between the tensors of Eq. (1) and the susceptibilities of Eq. (2) is discussed in Chapter III-1.

There are two general methods for the calculation of molecular β_{ijk} tensors. In the first, the molecular ground state is recalculated directly in the presence of a finite perturbing electric field. The new ground-state wave function, as a function of field, implicitly contains the information expressed in Eq. (1), provided the perturbing field is sufficiently general, and the β_{ijk} tensor is evaluated via the Hellmann–Feynman theorem. The method has been used extensively by Zyss and co-workers (see Chapter III-1).

The second method, the excited-state perturbation expansion or sum over states method, makes use of the fact that the unperturbed molecular eigenfunctions provide a complete set of functions in terms of which the perturbed state can be expanded. The choice of these functions enables readily available experimental information relating to the excited states of the molecule to be included implicitly in the calculation as a calibration. The method is described in detail in Sections III and IV. In Section V the theoretical and experimental results for 4-nitroaniline (PNA), one of the most extensively studied compounds, are described in detail, while in Section VI the results for other molecules are briefly reviewed. While the results of calculations on large molecules by this method are not of a high order of absolute accuracy, it is at present the only method that can be routinely applied to them.

The sum over states method is also the basis for the qualitative understanding of the magnitude of nonlinear effects. In perturbation theory, states that have similar energies mix most effectively and therefore low-lying excited states can modify the ground state substantially and lead to high polarizabilities in all orders. High values of β_{ijk} in organic molecules are usually associated with the systems of delocalized π orbitals in aromatic compounds. Such structures have excited electronic states of low energy that often give rise to near ultraviolet (UV) or visible spectra. There are additional criteria, relating to the matrix elements of the perturbation expansion, that must also be satisfied to produce large values of β, and if these conditions are met then a substantial contribution to the hyperpolarizability will result. The origin of the large hyperpolarizability is consequently electronic. This is also true of semiconductors, where electronic transitions across the band gap are the dominant virtual excitations. In ionic compounds there are additional large contributions, due to displacement of ions, which are present at low frequencies, below the microwave, but are not found in experiments at infrared or optical frequencies. The ionic displacements are also associated with large piezoelectric effects in the same crystals. The frequency dispersion of β_{ijk} in organic molecules should take place mainly at optical frequencies and be susceptible to calculation in terms of electronic states.

The measurement of β_{ijk} for an isolated molecule in the vapor phase is not usually possible because the vapor pressure of the substances is small at ordinary temperatures. Therefore, solution and liquid measurements of second-harmonic generation (SHG) are the main source of information on β_{ijk} and the theory underlying the interpretation of these measurements is described in the next section.

II. ELECTRIC-FIELD-INDUCED SHG IN LIQUIDS AND SOLUTIONS

The random orientation of molecules in liquids or solutions will result in average cancellation of even-order polarization effects, even though the molecules themselves have no center of symmetry. There is no macroscopic permanent polarization in a liquid of polar molecules (no zero-order effect) and no SHG from the combined second-order terms of Eq. (2). It is important to be able to measure the electro-optic constants of molecules in liquid or solution, since the cooperative effects of interactions are likely to be much less important than in the crystalline state. For example, in the sum over states method the excited states of the liquid, although somewhat changed from those of the vapor, are still identifiable as being excited states of one molecule, whereas in a regular crystal lattice their formal description might have to be in terms of a delocalized exciton band. In molecular crystals, delocalized crystal effects are generally of minor importance, as shown by the data in Table I, where a change from the gas phase to the crystalline state for a series of polar organic materials produces only a relatively small movement in the electronic transition energy. Indeed, a far greater movement of the transition energies of these materials occurs in the change from nonpolar solvents such as hexane to protonic solvents such as methanol where intermolecular hydrogen bonding often stabilizes the excited state.

It is also the case that measurements on oriented crystals are difficult and time consuming and that solution data relating to the approximate properties of one molecule are a major source of information for preselecting crystalline materials.

In attempting to isolate a measure of the molecular hyperpolarizability from liquid or solution studies, there is a close parallel with earlier work aimed at evaluating the dipole moments and linear polarizabilities of individual molecules from measurements of the macroscopic dielectric constant and refractive index [see, for example, Bottcher (1952)]. The same corrections for internal field effects are used in the analysis that extracts information about β_{ijk} from measurements of SHG. To establish the technique for dealing with internal fields we examine both linear and nonlinear effects.

If an external field $E_Z(\omega)$ is applied to a liquid, the first-order macroscopic polarization is

$$P_Z(\omega) = \varepsilon_0 \chi^{(1)}_{(\omega)} E_Z(\omega) \tag{3}$$

where in a liquid $\chi(\omega)$ is a scalar. On a microscopic scale the local field $E^{\mathrm{L}}_Z(\omega)$

TABLE I

Lowest Electronic Transition Energies (nm) of Organic Materials in Different States[a]

Molecule	Reference	Gas phase	Solid phase	Solution spectra	Solvent
1,4-Diaminoanthraquinone	Sheppard and Newsome (1942)	492	540	575	C_6H_{12}
				590	CH_3OH
Indigo	Sheppard and Newsome (1942)	544	650	600	$C_2H_4Cl_2$
				605	CH_3OH
Merocyanine	Sheppard and Newsome (1942)	470	520	510	C_6H_{12}
				558	CH_3OH
Phthalocyanine	Edwards and Gouterman (1970); Lyons *et al.* (1960)	686	719	699	$C_{10}H_9Cl$

[a] The merocyanine described is 3-ethyl-2-[2-(3-oxo-2,3-dihydrobenzo[*b*]thien-2-ylidine)-ethylidene]benzthiazole. The solid-state spectra were obtained from solid films on glass, except for the phthalocyanine value, which was obtained for a single crystal (polarized light in ac plane).

acting on a molecule is the sum of $E_Z(\omega)$ and an internal field $E_Z^I(\omega)$ that results from the polarization of the medium surrounding the reference molecule. The polarization and internal field are linearly proportional to $E_Z(\omega)$. The internal field at any frequency is related to the component of the polarization at that frequency by a geometrical factor g, determined in the simpler theories by the shape of the cavity assumed to surround the reference molecule ($g = \frac{1}{3}$ for a spherical cavity). Writing

$$E_Z^I(\omega) = gP_Z(\omega)/\varepsilon_0 \tag{4}$$

and

$$E_Z^L(\omega) = f_\omega E_Z(\omega) \tag{5}$$

gives for the case of linear polarization

$$E_Z^L = E_Z + E_Z^I = (1 + g\chi^{(1)})E_Z = f_\omega E_Z$$

where Eqs. (3) and (4) have been used. Then

$$g = (f_\omega - 1)/\chi^{(1)}_{(\omega)} \tag{6}$$

Setting $g = \frac{1}{3}$ and the dielectric constant $\varepsilon(\omega) = \chi^{(1)}(\omega) + 1$ gives the Clausius–Mossotti relation for the internal field factor,

$$f_\omega = [\varepsilon(\omega) + 2]/3 \tag{7}$$

Onsager's method (Onsager, 1936), which is applicable to concentrated polar liquids, provided there is no association, gives

$$f_\omega = (\varepsilon_\infty + 2)\varepsilon_\omega/(\varepsilon_\infty + 2\varepsilon_\omega) \tag{8}$$

where ε_∞ is the high-frequency value of $\varepsilon(\omega)$. At high frequencies, where no reorientation of permanent dipoles in the field is possible, Maxwell's relation

$$n_r^2(\omega) = \varepsilon(\omega)$$

between the refractive index and dielectric constant holds. In this limit, at frequencies not too near an absorption band, both Clausius–Mossotti and Onsager factors reduce to the Lorenz–Lorentz factor, $(n_r^2 + 2)/3$.

The macroscopic polarization produced by a field in the Z direction is obtained by adding the contributions of individual molecules, so that

$$P_Z = N\langle w(\mu_Z + \alpha_{ZZ}E_Z^L)\rangle \tag{9}$$

where w is the angular distribution function of the molecules, the angle brackets represent an average over all molecules, and N is the number of molecules per unit volume. The molecules are partially oriented by the field and, to first order in E_Z^L,

$$w = \exp(\mu_Z E_Z^L/kT)/(p \cdot f) \tag{10}$$

where the orientational partition function

$$(p \cdot f) = 1 + \tfrac{1}{6}(\mu_Z E_Z^{\mathrm{L}}/kT)^2 + \cdots = 1$$

to the order of the effect considered. Expanding the exponential and again retaining only linear terms in E_Z gives

$$P_Z = N\langle \mu_Z^2 E_Z^{\mathrm{L}}/kT + \alpha_{ZZ} E_Z^{\mathrm{L}} \rangle \tag{11}$$

or

$$P_Z = N f_\omega \bar{\alpha} E_Z$$

where Eq. (5) has been used and

$$\bar{\alpha} = \langle \alpha_{ZZ} \rangle + \langle \mu_Z^2 \rangle / kT \tag{12}$$

is the effective microscopic polarizability. The angular averages are now over a random distribution, since the Boltzmann weighting factor has been treated explicitly. For the permanent dipoles,

$$\langle \mu_Z^2 \rangle = \tfrac{1}{3}\mu^2 \tag{13}$$

In the expression $\langle \alpha_{ZZ} \rangle$ the subscripts refer to the laboratory coordinate system in which E_Z is fixed. In order to relate the result to a specific combination of components of the molecular tensor, α_{ZZ} must be transformed to a coordinate system fixed in a particular molecule. If the transformation matrix between the laboratory system (X, Y, Z) and the molecular system (x, y, z) is a_{iI} then the second-rank tensor transforms as the product of two vectors and

$$\langle \alpha_{ZZ} \rangle = \langle a_{iZ} a_{jZ} \alpha_{ij} \rangle = \langle a_{iZ} a_{jZ} \rangle \alpha_{ij} \tag{14}$$

since α_{ij} is now a fixed value, specified with respect to molecular axes. The evaluation of averages over products of direction cosines has been comprehensively treated by Cyvin *et al.* (1965) and also by Andrews (1980). The a_{iI} terms transform as coordinates, and even powers of any quantity must occur for nonzero average values. Thus

$$\begin{aligned}\langle \alpha_{ZZ} \rangle &= \langle a_{iZ} a_{jZ} \rangle \alpha_{ij} = \langle a_{iZ} \alpha_{iZ} \rangle \alpha_{ii} \\ &= \left(\tfrac{1}{2} \int_0^\pi \cos^2\theta \sin\theta \, d\theta \right) \alpha_{ii} \\ &= \tfrac{1}{3}\alpha_{ii} = \tfrac{1}{3}(\alpha_{xx} + \alpha_{yy} + \alpha_{zz}) = \alpha \end{aligned} \tag{15}$$

where α is a scalar. For an isotropic molecule, α is the scalar polarizability. The quantity $\bar{\alpha}$ of Eq. (12) then becomes

$$\bar{\alpha} = \alpha + \mu^2/3kT \tag{16}$$

and from Eqs. (3) and (11),

$$\chi^{(1)}(\omega) = N f_\omega \bar{\alpha}/\varepsilon_0 \tag{17}$$

Setting $\chi = \varepsilon - 1$ recovers the usual form of the Debye equation for a dipolar medium. The quantity

$$\mathscr{P} = N_{AV}(\varepsilon - 1)/(\varepsilon + 2) = \chi^{(1)}(\omega)/3f_\omega = N_{AV}\bar{\alpha}/\varepsilon_0 \tag{18}$$

(where N_{AV} is Avogadro's number) is called the molar polarization. At optical frequencies, where $\varepsilon = n_r^2$, it becomes the molar refraction $\mathscr{R}$, which is essentially a measure of the polarizability of the molecule. Values of $\mathscr{R}$ for groups of atoms commonly occurring in organic molecules have been tabulated and for many classes of compounds a high degree of additivity holds. In aromatic compounds, however, the additivity rules are not successful. Attempts to establish similar schemes for the hyperpolarizabilities have been described by Hauchecorne, Kerherve, and Mayer (1971) and by Finn and Ward (1974). Again, these have been successful for saturated compounds but fail for the type of conjugated π electron system that is treated in later sections. Of more importance in the theory of the nonlinear effects is the additivity of $\mathscr{P}$ and $\mathscr{R}$ over the different components of liquid mixtures and solutions, which enables solute properties to be isolated from dilute solution measurements.

Electric-field-induced SHG (EFISH) measurements designed to provide information on the β tensor can be analyzed in an analogous fashion (Levine and Bethea, 1975). A zero-frequency field $E_Z(0)$ is applied in the Z direction and the optical field $E_Z(\omega)$ is polarized along the same axis. The macroscopic equations for the frequency components of the polarization up to the second harmonic can be written in the form

$$P_Z(0) = \varepsilon_0 \chi^{(1)}(0,0) E_Z(0) \tag{19}$$

$$P_Z(\omega) = \varepsilon_0 \chi^{(1)}(-\omega,\omega) E_Z(\omega) \tag{20}$$

$$P_Z(2\omega) = \varepsilon_0 \chi^{(3)}(-2\omega,\omega,\omega,0) E_Z(\omega)^2 E_Z(0) \tag{21}$$

The even-order susceptibility $\chi^{(2)}$ of the liquid is 0 but there are third-order terms via $\chi^{(3)}(0,0,0,0)$, $\chi^{(3)}(0,0,\omega,-\omega)$ and $\chi^{(3)}(-\omega,\omega,\omega,-\omega)$ which contribute to $P_Z(0)$ and $P_Z(\omega)$, but which are omitted since they contribute only in higher orders to $P_Z(2\omega)$. At the molecular level,

$$P_Z(0) = N\langle w[\mu_Z + \alpha_{ZZ}(0)E_Z^L(0)]\rangle \tag{22}$$

$$P_Z(\omega) = N\langle w\alpha_{ZZ}(\omega)E_Z^L(\omega)\rangle \tag{23}$$

$$P_Z(2\omega) = N\langle w[\alpha_{ZZ}(2\omega)E_Z^L(2\omega) + \beta_{ZZZ}(-2\omega,\omega,\omega)E_Z^L(\omega)^2 + \gamma_{ZZZZ}(-2\omega,\omega,\omega,0)E_Z^L(\omega)^2 E_Z^L(0)]\rangle \tag{24}$$

The linear term in Eq. (24) results from the internal field of frequency, 2ω, produced by the nonlinear polarization at 2ω. Since there is no applied field at 2ω, it can be expressed in terms of $P_Z(2\omega)$ using Eq. (4), giving

$$(1 - N\bar{\alpha}(2\omega)g/\varepsilon_0)P_Z(2\omega) = \langle w[\beta_{ZZZ}(-2\omega, \omega, \omega)E_Z^L(\omega)^2 + \gamma_{ZZZZ}E_Z^L(\omega)^2E_Z^L(0)]\rangle \tag{25}$$

Using Eqs. (6) and (17), the factor multiplying $P_Z(2\omega)$ can be identified as $1/f_{2\omega}$, the reciprocal of the linear local field factor at 2ω. The linear factors at 0 and ω can be used to convert $E_Z^L(0)$ and $E_Z^L(\omega)$ to applied field values since Eqs. (22) and (23) are linear. Then, treating w as before,

$$P_Z(2\omega) = Nf_{2\omega}f_\omega^2f_0\langle\mu_Z\beta_{ZZZ}/kT + \gamma_{ZZZZ}\rangle E_Z(\omega)^2E_Z(0) \tag{26}$$

Again transforming to axes fixed in a molecule,

$$\langle\gamma_{ZZZZ}\rangle = \langle a_{iZ}a_{jZ}a_{kZ}a_{lZ}\rangle\gamma_{ijkl}$$

By taking the transformed coordinate indices in pairs and separating out the case where all four are the same,

$$\begin{aligned}\langle\gamma_{ZZZZ}\rangle &= \langle a_{iZ}^4\rangle\gamma_{iiii} + (1 - \delta_{ij})\langle a_{iZ}^2a_{jZ}^2\rangle(\gamma_{iijj} + \gamma_{ijij} + \gamma_{ijji}) \\ &= \tfrac{1}{5}\gamma_{iiii} + (1 - \delta_{ij})\tfrac{1}{15}(2\gamma_{iijj} + \gamma_{ijji}) \\ &= \tfrac{1}{15}(2\gamma_{iijj} + \gamma_{ijji})\end{aligned}$$

where the symmetry of $\gamma_{ijkl}(-2\omega, \omega, \omega, 0)$ with respect to interchange of the second and third indices has been used. The quantity

$$\gamma = \tfrac{1}{15}(2\gamma_{iijj} + \gamma_{ijji}) \tag{27}$$

is a scalar that is analogous to the quantity α in the first-order effect.

The term involving the second-order hyperpolarizability depends on the average,

$$\langle\mu_Z\beta_{ZZZ}\rangle = \langle a_{xZ}a_{jZ}a_{kZ}a_{lZ}\rangle\mu\beta_{jkl}$$

where the molecular x axis is taken in the direction of the permanent dipole. Again selecting nonvanishing combinations,

$$\begin{aligned}\langle\mu_Z\beta_{ZZZ}\rangle &= \langle a_{xZ}^4\rangle\mu\beta_{xxx} + \langle a_{xZ}^2a_{yZ}^2\rangle\mu(\beta_{xyy} + 2\beta_{yxy} + \beta_{xzz} + 2\beta_{zxz}) \\ &= \tfrac{1}{5}\mu\beta_x\end{aligned} \tag{28}$$

where β_x is the x component of the vector

$$\beta_i = \tfrac{1}{3}(\beta_{ijj} + 2\beta_{jij}) \tag{29}$$

The symmetry of $\beta_{ijk}(-2\omega, \omega, \omega)$ with respect to interchange of the last two indices has been used. The scalar and vector properties of γ and β_i follow from the fact that they are contractions of tensors.

If

$$\bar{\gamma} = \gamma + \tfrac{1}{5}\mu_i\beta_i/kT \tag{30}$$

then

$$P_Z(2\omega) = Nf_{2\omega}f_{\omega}^2 f_0 \bar{\gamma} E_Z(\omega)^2 E_Z(0) \tag{31}$$

$P_Z(2\omega)$ is the source, in Maxwell's equations, of the second harmonic field and can be experimentally deduced from measurements of the second-harmonic intensity. A macroscopic effective third-order susceptibility

$$\Gamma = Nf_{2\omega}f_{\omega}^2 f_0 \bar{\gamma} \tag{32}$$

can then be calculated. The f factors have to be determined from dielectric constants and refractive indices at the appropriate frequencies before $\bar{\gamma}$ can be calculated. If it is then assumed that γ is negligible (see, for example, Oudar *et al.*, 1977) and if the molecular dipole is known, then β_x, the component of the vector part of β_{ijk} along the molecular dipole, can be determined. [See also Levine and Bethea (1974) and Oudar and le Person (1975).]

As in the determination of molecular dipoles from solution measurements, procedures have been developed for extrapolating to infinite dilution to remove solvent–solute interactions. Values for linear parameters occurring in the field factors and the molecular dipole may be either taken as known constants in Eqs. (30) and (31) (Oudar *et al.*, 1977) or determined by a simultaneous extrapolation of all factors in the equation (Singer and Garito, 1981).

III. THE SUM OVER STATES METHOD

In this section a brief outline of the derivation of the formulas used in the application of the perturbation theory expansion over molecular states to the calculation of the hyperpolarizability tensor is given. The main part of the discussion relates to elementary perturbation theory, rather than quantum-field theory techniques, the aim being to provide a basis for the understanding of more specific points made in subsequent sections.

Adopting a semiclassical approach, in which the electromagnetic field is described classically and the molecules by quantum mechanics, the problem can be specified by a time-dependent Schrödinger equation

$$H\Psi = i\hbar\, \partial\Psi/\partial t \tag{33}$$

where the Hamiltonian

$$H = H_0 + H' = H_0 - e\mathbf{E} \cdot \mathbf{r} \tag{34}$$

is the sum of the unperturbed molecular Hamiltonian H_0 and a perturbation resulting from the interaction of the applied field with the molecule in the electric dipole approximation. The quantity $\mathbf{r}$ is the general dipole moment operator for the molecule (with the charges in units of e), but if only changes in the electronic states are to be considered, then

$$\mathbf{r} = \sum_{\alpha} \mathbf{r}_{\alpha} \tag{35}$$

where the sum is over all the electrons.

The complete set of unperturbed eigenfunctions,

$$\Psi_n(\mathbf{r}) = e^{-i\omega_n t}\psi_n = e^{-i\omega_n t}|n\rangle \tag{36}$$

forms a basis for the expansion of the perturbed function Ψ. The functions ψ_n or $|n\rangle$ are the time-independent eigenfunctions of H_0, and $\omega_n = E_n/\hbar$ is the corresponding eigenvalue in frequency units. Without loss of generality, the ground-state energy of the molecule is taken as zero and the time-dependent coefficients in the expansion

$$\Psi = \sum_n C_n(t)\Psi_n \tag{37}$$

are then obtained from the master equations

$$\dot{C}_n = -(i/\hbar)\sum_m H'_{nm} e^{i\omega_{nm}t} C_m \tag{38}$$

where

$$H'_{nm} = \langle n|H'|m\rangle$$

and

$$\omega_{nm} = \omega_n - \omega_m$$

If the coefficients C_n are written as a power series in the field, then

$$C_n(t) = \delta_{n0} + C_n^{(1)} + C_n^{(2)} + \cdots \tag{39}$$

where $C_n^{(1)}$ and $C_n^{(2)}$ are the linear and quadratic terms. The total polarization (dipole plus induced dipole) of the molecule can be expressed as

$$\begin{aligned} \mathbf{p} &= -e\langle\Psi|\mathbf{r}|\Psi\rangle \\ &= -e\left\langle \Psi_g + \sum_n (C_n^{(1)} + C_n^{(2)})\Psi_n \middle| \mathbf{r} \middle| \Psi_g + \sum_m (C_m^{(1)} + C_m^{(2)})\Psi_m \right\rangle \end{aligned} \tag{40}$$

If the part of $\mathbf{p}$ depending on quadratic terms in the field is written as $\mathbf{p}^{(2)}$, then

$$\mathbf{p}^{(2)} = -e\left[\sum_n\sum_m C_n^{(1)*}C_m^{(1)}\langle\Psi_n|\mathbf{r}|\Psi_m\rangle + \sum_n (C_n^{(2)*}\langle\Psi_n|\mathbf{r}|\Psi_g\rangle + C_n^{(2)}\langle\Psi_g|r|\Psi_n\rangle)\right] \tag{41}$$

The first-order coefficients $C_n^{(1)}$ contain the matrix elements

$$\langle g|H'|n\rangle = \mathbf{E}\cdot\langle g|\mathbf{r}|n\rangle e^{-i\omega_n t}$$

and the $C_n^{(2)}$ are sums over terms containing the combinations

$$\langle g|H'|n'\rangle\langle n'|H'|n\rangle = (\mathbf{E}\cdot\langle g|\mathbf{r}|n'\rangle)(\mathbf{E}\cdot\langle n'|\mathbf{r}|n\rangle)e^{-i\omega_n t}$$

Hence, when resolved into components, every term in Eq. (41) contains products of the form

$$\langle g|r_i|n'\rangle\langle n'|r_j|n\rangle\langle n|r_k|g\rangle \tag{42}$$

The order of appearance of the coordinates is interchanged in the three terms of Eq. (41). The coefficients also contain frequency factors obtained from the time integration of Eq. (38) in each order of approximation. These are algebraically complicated but can be evaluated using diagrammatic techniques.

Explicit tabulations of second- and third-order formulas are given by Ward (1965). As far as the second-order formulas are concerned, the modifications described by Orr and Ward (1971) are incorporated in the present treatment through the formal transformation to the electronic charge centroid system [see Eqs. (49)–(51)].

A general form of the perturbation for second-order effects can be written as the sum of two fields of different frequencies,

$$H' = -e\mathbf{r}\cdot[\mathbf{E}(\omega_1)\sin\omega_1 t + \mathbf{E}(\omega_2)\sin\omega_2 t] \tag{43}$$

The diagrammatic technique makes it possible to pick out the polarization term of a given frequency in the form

$$\mathbf{P}_i(\omega_1+\omega_2) = \beta_{ijk}(-\omega_1-\omega_2,\omega_1,\omega_2)E_j(\omega_1)E_k(\omega_2) \tag{44}$$

from which $\beta_{ijk}(-\omega_1-\omega_2,\omega_1,\omega_2)$ can be identified. Introducing the notation

$$f(a,b) = [(\omega_{n'}+\omega_a)(\omega_n+\omega_b)]^{-1}$$

and defining

$$\omega_3 = \omega_1 + \omega_2$$

so that, for example,

$$f(3,1) = [(\omega_{n'} + \omega_1 + \omega_2)(\omega_n + \omega_1)]^{-1}$$

$$f(\bar{3},\bar{1}) = [(\omega_{n'} - \omega_1 - \omega_2)(\omega_n - \omega_1)]^{-1}$$

and, following Ward, setting

$$\langle ijk \rangle = r_{gn'} r_{n'n} r_{ng} \tag{45}$$

where

$$r_{gn'} = \langle g | r_i | n' \rangle$$

etc. gives for the sum over states expression for three-wave mixing, from Table V of Ward (1965)

$$\begin{aligned}\beta_{ijk}(-\omega_1 - \omega_2, \omega_1, \omega_2) \\ = -(e^3/4\hbar^2)\sum_{n'}\sum_{n}\{\langle ikj\rangle[f(3,1) + f(\bar{3},\bar{1})] + \langle jki\rangle[f(\bar{1},\bar{3}) + f(1,3)] \\ + \langle ijk\rangle[f(3,2) + f(\bar{3},\bar{2})] + \langle kji\rangle[f(\bar{2},\bar{3}) + f(2,3)] \\ + \langle kij\rangle[f(\bar{2},1) + f(2,\bar{1})] + \langle jik\rangle[f(\bar{1},2) + f(1,\bar{2})]\}\end{aligned} \tag{46}$$

Equation (46) could be used to compute values of the β tensor for three-wave mixing experiments in which a wave of frequency $\omega_3 = \omega_1 + \omega_2$ is generated from input fields at ω_1 and ω_2. The tensor appropriate to the difference frequency is obtained simply by changing the sign of ω_2 throughout the expression.

The special case of second-harmonic generation where one input field of frequency ω acting quadratically produces output of 2ω is important as the most commonly used experimental technique. Before putting $\omega_1 = \omega_2$ in Eq. (46) it must be remembered that the perturbation that gave rise to Eq. (46), H' as in Eq. (43), also produces polarization at $2\omega_1$ and $2\omega_2$. These terms are, described by tensors already having the correct form for SHG. The cross-term in $\omega_1 + \omega_2$ in the expansion $[\mathbf{E}(\omega_1) + \mathbf{E}(\omega_2)]^2$ contains a factor of 2 compared with the squared terms, which has been absorbed into the tensor. $\beta(-2\omega, \omega, \omega)$ is therefore obtained from Eq. (46) by setting $\omega = \omega_1 = \omega_2$ and $\omega_3 = 2\omega$, and dividing the resulting formula by 2:

$$\begin{aligned}\beta_{ijk}(-2\omega, \omega, \omega) = -(e^3/8\hbar^2)\sum_{n'}\sum_{n}\{(\langle ikj\rangle + \langle ijk\rangle)[f(3,1) + f(\bar{3},\bar{1})] \\ + (\langle jki\rangle + \langle kji\rangle)[f(\bar{1},\bar{3}) + f(1,3)] \\ + (\langle kij\rangle + \langle jik\rangle)[f(\bar{1},1) + f(1,\bar{1})]\}\end{aligned} \tag{47}$$

Another special case of potential interest is $\beta_{ijk}(-\omega, \omega, 0)$, which is related to the form $\chi^{(2)}(-\omega, \omega, 0)$ occuring in the linear electro-optic or Pockels effect, where a static field induces a field-dependent change in the refractive index at frequency ω. (The effect is linear only in the applied static field.) While this effect is not directly observable except in noncentrosymmetric crystals, theoretical values of $\beta_{ijk}(-\omega, \omega, 0)$ are of interest since they contribute to the overall $\chi^{(2)}$. Again, the formula can be obtained from Eq. (46), but in this case it must be noted that a polarization $\omega_1 - \omega_2$ resulting from the H' of Eq. (43) will also contribute if $\omega_2 \to 0$. Also, to obtain a nonzero static field, the sine functions in Eq. (43) must be changed to cosines. A factor of -1 is thereby introduced. Setting $\omega_2 = 0$ and multiplying by a factor -2 then produces from Eq. (46)

$$\begin{aligned}\beta_{ijk}(-\omega, \omega, 0) = (e^3/2\hbar^2)\sum_{n'}\sum_{n}\{&(\langle ikj\rangle + \langle jki\rangle)[f(1,1) + f(\bar{1},\bar{1})] \\ &+ (\langle ijk\rangle + \langle jik\rangle)[f(1,0) + f(\bar{1},0)] \\ &+ (\langle kji\rangle + \langle kij\rangle)[f(0,1) + f(0,\bar{1})]\}\end{aligned} \tag{48}$$

Equations (46), (47), and (48) require further rearrangement to make them convenient for computation. If the origin of the coordinate system is taken as the charge centroid of the ground-state electron distribution, then

$$\langle g|\mathbf{r}|g\rangle = 0 \tag{49}$$

and

$$e\langle g|\mathbf{R}_{\mathrm{NUC}}|g\rangle = \boldsymbol{\mu}_g \tag{50}$$

Equation (49) implies that the ground state ($n = g$) may be omitted from the summations of Eq. (47). Only diagonal matrix elements are affected by the change of origin, and for these,

$$\begin{aligned}e\langle n|\mathbf{r}|n\rangle &= e[\langle n|\mathbf{r} - \mathbf{R}_{\mathrm{NUC}}|n\rangle + \langle n|\mathbf{R}_{\mathrm{NUC}}|n\rangle] \\ &= -(\boldsymbol{\mu}_n - \boldsymbol{\mu}_g) = -\Delta\boldsymbol{\mu}_n\end{aligned} \tag{51}$$

Hence in evaluating Eq. (47), $-e\mathbf{r}_{nn}$ is interpreted as the change in dipole moment between ground and excited state.

It is convenient for computation, and to identify more clearly the major contributing states, to separate the diagonal terms in Eq. (47). When $n = n'$, $f(a, b) = f(b, a)$ and $\langle ijk\rangle = \langle kji\rangle$, so that,

$$\begin{aligned}\beta_{ijk}^{\mathrm{DIAG}}(-2\omega, \omega, \omega) = -(e^3/8\hbar^2) \times \sum_{n}\{&2(\langle ikj\rangle + \langle ijk\rangle)[f(3,1) + f(\bar{3},\bar{1})] \\ &+ 4\langle jik\rangle f(\bar{1},1)\}\end{aligned} \tag{52}$$

leading to

$$\beta_{ijk}^{\mathrm{DIAG}}(-2\omega,\omega,\omega) = -(e^3/2\hbar^2) \times \sum_n \left[r_{gn}^i(r_{gn}^j r_{nn}^k + r_{gn}^k r_{nn}^j) \cdot \frac{(\omega_n^2 + 2\omega^2)}{(\omega_n^2 - 4\omega^2)(\omega_n^2 - \omega^2)} + r_{gn}^k r_{gn}^j r_{nn}^i \cdot \frac{1}{(\omega_n^2 - \omega^2)} \right] \tag{53}$$

For the case where one diagonal charge-transfer term along one molecular axis makes the major contribution, setting i, j, $k = x$ and restricting the summation to one term,

$$\beta_{xxx}^{\mathrm{DIAG}}(-2\omega,\omega,\omega) = \left(\frac{3e^3}{2\hbar^2}\right)(r_{gn}^x)^2 r_{nn}^x \cdot \frac{\omega_n^2}{(\omega_n^2 - 4\omega^2)(\omega_n^2 - \omega^2)} \tag{54}$$

In this equation $(r_{gn}^x)^2$ is essentially the oscillator strength of the transition, r_{nn}^x the change in dipole between ground and excited states, and ω_n the transition frequency. The resonant form of the term is clearly exhibited. This formula is the basis of the extensively used two-state approximation.

The above outline is based on the semiclassical approach in the case where only secular perturbations of the system are included. Even for an isolated atom, quantization of the field would give rise to small linewidths. For all other systems, stochastic perturbations with a random frequency spectrum arising from interactions with eigenstates of the system that have not been explicitly included and from intermolecular interactions will give rise to much greater linewidths. The technique for formally including such effects, in the context of nonlinear optics, via the ensemble-averaged density matrix, has been described by Bloembergen (1965). The result is that the effects can be represented by adding an imaginary part to each eigenvalue

$$\omega_n \to \omega_n - i\Gamma_n$$

and replacing ω_n by ω_n^* when it has arisen from $\langle\Psi_n|$. The formulas replacing Eqs. (46)–(48) are then somewhat more complicated, since the cancellation of certain terms in the derivation no longer occurs. In particular, a second component of $P(2\omega)$ phase-shifted by $\pi/2$ from the first appears.

The phase-shifted polarization component is not directly associated with absorption of energy, since its frequency is double that of the driving field, leading to a zero value for the cycle-averaged energy dissipation. In the region where $|\omega_n - 2\omega| \approx \Gamma_n$ there is a first-order absorption of both components of the wave generated at 2ω as it proceeds through the material.

In the following most of the work discussed relates to frequency ranges where the linewidth can be neglected, although considerable enhancement of β values by reduction of the denominator $(\omega_n - 2\omega)$ is achieved.

IV. MOLECULAR ORBITAL CALCULATIONS OF THE HYPERPOLARIZABILITY

Relatively few molecular orbital calculations of the molecular hyperpolarizability using the ground-state and excited-state perturbation approaches have been reported, with the exception of the work of Morrell and Albrecht (1979), Garito and collaborators (Lalama and Garito, 1979; Lalama *et al.*, 1981; Teng and Garito, 1983a, 1983b), and Docherty *et al.* (1985). All of these calculations, which have been restricted generally to variations of both the CNDO (Pople and Beveridge, 1970) and the CNDO/S methods (Del Bene and Jaffe, 1968) can be described as follows:

(1) the calculation of ground-state molecular orbitals by the appropriate SCF equations;
(2) the generation of more accurate spectroscopic states by configuration interaction between a specified number of ground and excited states; and
(3) the calculation of the hyperpolarizability tensor using the matrix elements of the dipole operator between unperturbed ground and excited states on the one hand and that between two excited states on the other.

The evaluation of (3) can be accomplished by means of molecular orbital Hartree–Fock theory using the Roothan method. Here the unperturbed ground-state molecular wave function ψ_g is expressed as a single antisymmetrized product of molecular orbitals ψ_i:

$$\psi_g = \det\{\psi_1\bar{\psi}_1 \cdots \psi_i\bar{\psi}_i\} \tag{55}$$

The elements of the determinant are usually written as linear combinations of a set of basis functions ϕ_k which are normally centered at the atoms and defined as

$$\psi_i = \sum_k C_{ki}\phi_k \tag{56}$$

The general problem of deriving the LCAO coefficients C_{ki} by the variational method was solved by Roothan to obtain the wave function and the one-electron energies ε_i,

$$\sum(F_{kl} - \varepsilon_i S_{kl})C_{li} = 0 \tag{57}$$

where the Fock matrix F_{kl} and overlap integrals S_{kl} are defined in the following way:

$$\begin{aligned} F_{kl} &= H_{kl} + \sum_{m,n} P_{mn}[(kl|mn) - \tfrac{1}{2}(kn|ml)] \\ S_{kl} &= \langle\phi_k\phi_l\rangle \end{aligned} \tag{58}$$

H_{kl} is the core Hamiltonian for an electron in the field of bare nuclei and $(kl|mn)$ is the two-electron integral, defined as

$$(kl|mn) = \langle \phi_k(1)\phi_l(1)|r_{12}^{-1}|\phi_m(2)\phi_n(2)\rangle$$

If there are two electrons per molecular orbital, the density matrix P_{kl} is given by

$$P_{kl} = 2\sum_{i=1}^{i_{\mathrm{occ}}} C_{ki}C_{li} \tag{59}$$

Finally, the ground-state total molecular Hamiltonian H is defined in terms of the core Hamiltonian and the electron repulsion terms:

$$\begin{aligned} H_i^{\mathrm{core}} &= -\tfrac{1}{2}\nabla_i^2 - \sum_A Z_{\mathrm{A}}/r_{\mathrm{A}i} \\ H &= H^{\mathrm{core}} + \sum_j \frac{1}{r_{ij}} \end{aligned} \tag{60}$$

The most difficult part of LCAO–SCF theory is usually the evaluation of a large number of two-electron integrals. At the CNDO/2 level, only the valence electrons are treated explicitly, the smaller two-electron integrals that involve overlap between different ϕ functions are neglected, and other integrals are treated in a simplified manner. With these approximations, the Fock matrix is given by

$$\begin{aligned} F_{kk} &= U_{kk} + (P_{\mathrm{AA}} - \tfrac{1}{2}P_{kk})\Gamma_{\mathrm{AA}} + (P_{\mathrm{BB}} - Z_{\mathrm{A}})\Gamma_{\mathrm{AB}} \\ F_{kl} &= \tfrac{1}{2}(\beta_{\mathrm{A}}^0 + \beta_{\mathrm{B}}^0)S_{kl} - \tfrac{1}{2}P_{kl}\Gamma_{\mathrm{AB}} \end{aligned} \tag{61}$$

where U_{kk} is the core Hamiltonian matrix element for an electron in orbital k, P_{AA} and P_{kk} are the atomic and orbital charge densities, respectively, P_{kl} is the bond order, Z_{A} is the core charge of atom A, Γ_{AA} and Γ_{AB} are the one- and two-center bielectronic repulsion integrals, β_{A}^0 and β_{B}^0, are empirically derived bonding parameters for atoms A and B, and S_{kl} is the overlap integral between Slater orbitals k at atom A and l at atom B.

The electronic transition energies ΔE are calculated from the ground-state orbital energies using both the occupied and virtual orbitals. Here the energy required to promote an electron from occupied orbital Ψ_i to virtual orbital Ψ_j without reversal of spin is given by

$$\underset{i\to j}{\Delta E} = (E_j - E_i) - J_{ij} + 2K_{ij} \tag{62}$$

where E_i and E_j are the energies of the initial and final orbitals, and J_{ij} and K_{ij} are the molecular coulomb and exchange integrals, which are defined at the

CNDO/S level as

$$J_{ij} = \sum_{m,n} C_{im} C_{jn} \Gamma_{mn}$$
$$K_{ij} = \sum_{m,n} C_{im} C_{jm} C_{in} C_{jn} \Gamma_{mn} \qquad (63)$$

respectively. The energies are further refined by a configuration interaction treatment where the excited-state wave functions Ψ_n are expressed in terms of a linear combination of singly excited configurations (Pariser, 1956), i.e.,

$$\Psi_n = \sum_{i,j} A_{n,i \to j} X_{i \to j} \qquad (64)$$

where $X_{i \to j}$ is an antisymmetrized product of one-electron molecular orbitals. The wave functions and energies of the excited states are derived by solving the appropriate equations:

$$\sum_{i,j,o,p} A_{n,i \to j} A_{n,o \to p} \langle X_{i \to j} | H - E | X_{o \to p} \rangle = 0 \qquad (65)$$

The number of configurations adopted is clearly of considerable importance in the calculation of accurate transition energies.

In Morrell and Albrecht's approach, the core Hamiltonian matrix elements are evaluated at the CNDO/2 level, i.e.,

$$U_{kk} = -\tfrac{1}{2}(I_k + A_k) - (Z_A - \tfrac{1}{2})\Gamma_{AA} \qquad (66)$$

where I_k and A_k are the orbital ionization potential and electron affinity respectively. The monocentric electron repulsion integrals (Γ_{AA}) are evaluated using the CNDO/S method by taking the difference between these terms:

$$\Gamma_{AA} = (I_k - A_k) \qquad (67)$$

The bonding parameters adopted for the off-diagonal terms of the core Hamiltonian are independent of orbital type, and the same value is used for *s*, *p*, and *d* orbitals. However, the σ and π orbitals are distinguished by the adoption of a spectroscopic constant κ, so that

$$H_{kl} = \tfrac{1}{2}(\beta_A^0 + \beta_B^0) S_{kl}$$
$$H_{kl} = \tfrac{1}{2}\kappa(\beta_A^0 + \beta_B^0) S_{kl} \qquad (68)$$

The bicentric electron repulsion integrals (Γ_{AB}) are evaluated from the monocentric terms using the Nishimoto–Mataga approximation, which makes the repulsion dependent on the internuclear distance r_{AB} (Nishimoto and Mataga, 1957):

$$\Gamma_{AB} = [r_{AB} + 2/(\Gamma_{AA} + \Gamma_{BB})]^{-1} \qquad (69)$$

In contrast, the core Hamiltonian matrix elements are evaluated in Garito's approach using a modification of the CNDO/1 method (Lipari and Duke, 1975), i.e.,

$$U_{kk} = -I_k - (Z_A - 1)\Gamma_{AA} \tag{70}$$

where I_k is the orbital ionization potential as before. However, the s and p orbitals are distinguished by the adoption of separate bonding parameters β_A^0 for carbon, nitrogen, and oxygen. Furthermore, the nitrogen atoms are assigned unique values depending on whether the atom is present in an amino group or a nitro group. The Slater orbital exponents (used at the CNDO/2 and CNDO/S levels) have been modified also and optimized specifically for aniline, nitrobenzene, and 4-nitroaniline. The CNDO/S spectroscopic constant has been retained along with the one- and two- center electron repulsion integral approximations.

The most recent approach (Docherty *et al.*, 1985) follows the CNDO/S treatment except that theoretical monocentric bielectronic repulsion integrals are used, but made a function of shell and orbital type in a similar way to the approach of François *et al.* (1977, 1979), so that

$$\Gamma_{AA} = C_{AA}\langle \phi_k^2(1) | r_{12}^{-1} | \phi_l^2(2) \rangle \tag{71}$$

$$C_{AA} = [C_k(n, l) C_l(n, l)]^{1/2} \tag{72}$$

where n is the principal quantum number (shell), l is the azimuthal quantum number (orbital), and C_k and C_l are shell coefficients for atomic orbitals centered on atom A. Both the theoretical monocentric repulsion integrals and the bicentric terms have been varied concomitantly using the shell coefficients (the latter are evaluated by the Nishimoto–Mataga expression) along with the spectroscopic constant (κ) to derive the best combination of ground-state dipole moments and excited-state transition energies for a limited range of organic compounds.

For the computation of β_{ijk}, specific formulas are required for the matrix elements $\mathbf{r}_{gn}$ and $\mathbf{r}_{n'n}$:

$$\mathbf{r}_{n'n} = \langle n' | \mathbf{r} | n \rangle = \sum_{i \to i', j \to j'} A_{n', i \to i'} A_{n, j \to j'} \langle X_{i \to i'} | \mathbf{r} | X_{j \to j'} \rangle \tag{73}$$

In an arbitrary coordinate system, from standard quantum chemistry texts (e.g., Szabo and Ostlund, 1982),

$$\langle X_{i \to i'} | \mathbf{r} | X_{j \to j'} \rangle = \langle g | \mathbf{r} | g \rangle \delta_{ij} \delta_{i'j'} + (i' | \mathbf{r}_\alpha | j') \delta_{ij} - (i | \mathbf{r}_\alpha | j) \delta_{i'j'} \tag{74a}$$

where the rounded brackets refer to one-electron integrals over specified molecular orbitals. Substituting in Eq. (73) and using the orthonormality of the A matrix gives

$$\mathbf{r}_{n'n} = \langle g|\mathbf{r}|g\rangle \delta_{n'n} + \sum_{i\to i', j\to j'} A_{n',i\to i'} A_{n,j\to j'} [(i'|\mathbf{r}_\alpha|j')\delta_{ij} - (i|\mathbf{r}_\alpha|j)\delta_{i'j'}] \quad (74b)$$

Transforming to the electronic charge centroid as above gives

$$\mathbf{r}_{n'n} = \sum_{i\to i', j\to j'} A_{n',i\to i'} A_{n,j\to j'} [(i'|\mathbf{r}_\alpha|j')\delta_{ij} - (i|\mathbf{r}_\alpha|j)\delta_{i'j'}] \quad (75)$$

The matrix elements $\mathbf{r}_{gn}$ contain a factor $\sqrt{2}$ because of the lack of degenerate singlet spin configurations in the ground state:

$$\mathbf{r}_{gn} = \sqrt{2} \sum_{i\to i'} A_{n,i\to i'} (i|\mathbf{r}_\alpha|i') \quad (76)$$

The reduction to integrals over the atomic basis set is then accomplished via Eq. (56).

V. THE β TENSOR FOR 4-NITROANILINE

A. Calculations

The first results were reported by Morrell and Albrecht, who used the nonplanar crystal structure of PNA for calculation with a basis set of 60 states both for the configuration interaction treatment and for the singlet excited states. The calculated value of β_{xxx}, which is the dominant component of the tensor (if the molecule is orientated with the nitrogen atoms of the substituents along the x axis), was calculated as -32.7×10^{-30} esu (at an applied frequency of 1.17 eV) out of a total value of -34.3×10^{-30} esu for β_x. While excited states 3, 6, and 9 are the primary contributors to the value, the third excited state, which is a charge-transfer state, contributes a value of -25.1×10^{-30} esu and thus accounts for more than 75% of the 60-configuration calculation. The convergence of the perturbation occurs fairly rapidly, and the first 12 states account for most of the final value. An analysis of the diagonal and off-diagonal elements of the transition moment operator using all 60 states shows that the former alone contributes 78% of the final value of β_x.

In contrast, Lalama and Garito (1979) calculated an essentially planar structure that was derived from a standard benzene ring coupled with substituent geometries averaged from the crystal structure (to simulate the structure in solution) and a total of nine states for the configuration interaction and in the sum over singlet excited states. The β_{xxx} component is shown to be the major contributor. They provide a detailed analysis of symmetries and electron distributions of the molecular states contributing to β_{xxx} and show that only those states that transform as the A_1 representation of the C_{2v} point group are allowed by symmetry to contribute to the value of β_{xxx}. Every term in the perturbation expansion for β_{xxx} involves one or two excited states coupled to the ground state, which is itself of A_1 symmetry, so that the excited states must also have A_1 symmetry. The calculated components β_{xyy}, β_{yyx}, β_{xzz}, and β_{zzx} contained in β_x are smaller than β_{xxx} because the dipole-moment difference between ground and excited state, $\mathbf{r}_{nm}$, has no y or z component as a consequence of the C_{2v} symmetry. Only off-diagonal dipole-moment matrix elements and transition moments contribute to the components of β_{ijk} involving y or z. An analysis of the A_1 excited states with the limited-configuration interaction shows that the lowest-energy excited state, which is composed primarily of the configuration $X_{26\to27}$, is responsible for almost the entire magnitude of β_{xxx}. The transition $X_{26\to27}$ is shown to be associated with considerable electronic charge transfer from the donor amino group of the molecule to the nitro group and so to result in the largest change in dipole moment (6.5 D) between ground and excited state. The large value of $\Delta\mu_n$ leads to the predominant contribution in the perturbation sum arising from the diagonal term associated with this transition [see Eqs. (47) and (54)].

The calculated transition energy of the second excited state (A_1 symmetry) after CI treatment (4.37 eV, 282.5 nm) correlates well with the corresponding experimental gas-phase value (4.35 eV, 283.8 nm), though the solution value in hexane is lower (3.85 eV, 320.7 nm). The calculated value of β_{xxx} is 8.635×10^{-30} esu and the overall value of β_x at the same applied field frequency (1.17 eV, 1.06 μm) is 8.73×10^{-30} esu. Teng and Garito (1983a, 1983b) have examined both the effects of dispersion and the influence of the solvent within the theoretical framework of the method of Lalama and Garito (1979). Their work is discussed below in connection with the assessment of the experimental results.

The most recent theoretical results, reported by Docherty *et al.* (1985), using a structure similar to that described by Garito and Lalama, with a basis set of 50 states for both the CI and in the excited singlet state summation, give a calculated β_x value of 12.18 at 1.17 eV. An analysis of the effect of the number of states on the β_x and β_{xxx} values (Fig. 1) shows that these remain stable to within 5% after the addition of 30 transitions, while the major part of their value is contributed by three particular states that occur within this range, as reported earlier by Morrell and Albrecht. A more detailed study of the results reveals that substantial matrix-element products, $r^i_{gn}r^j_{nm'}r^k_{n'g}$, continue to occur at intervals as n increases throughout the range of states included. The maximum contribution of individual terms then declines approximately as the square of the excitation energy, but the number of excited states in equal energy intervals increases in an approximately linear manner as the excitation energy increases above the absorption edge. The number of excited state products (n, n') increases at a rate that tends to compensate for the decline in

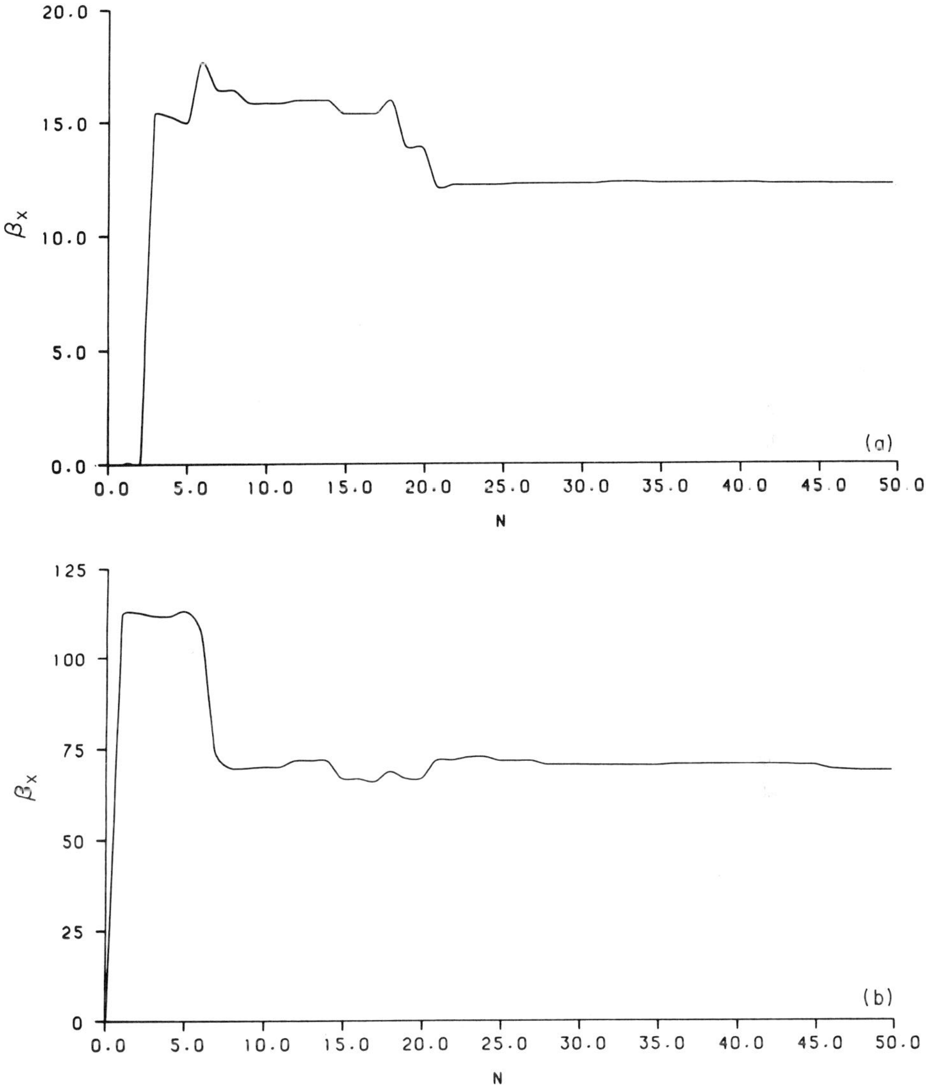

Fig. 1. Effect at 1.17 eV of the number of configurations and singlet states (N) on the calculated values of β_x at 1.17 eV for (a) 4-nitroaniline and (b) 4-amino-4′-nitro-*trans*-stilbene.

the magnitude of individual terms, leaving the irregular oscillations in the signs of the individually diminishing terms as a crucial factor in producing rapid convergence.

Again, as in the earlier work, the third excited state is the major contributor to the overall value, and an analysis of the change in total charge distribution from the ground state to the third excited state (Fig. 2) after CI shows a considerable movement of electrons from the donor amino group and the ortho ring carbons to the oxygen atoms of the nitro group and the meta ring carbons, which results in an enhancement of the calculated dipole moment from 7.95 to 15.55 D, compared with an experimental change from 5.80 to 14.4 D (Liptay, 1974).

B. Comparison of Calculated and Experimental Results

The experimental results for PNA obtained from EFISH measurements with corrections for internal field effects are shown in Table II. The calculated dispersion relations of Teng and Garito (1983a, 1983b), based on the method of Lalama and Garito (1979), and by the method of Docherty *et al.* (1985) are shown in Table III. After adjustment of the molecular eigenvalues to allow for

TABLE II

Experimental Values of β_x for 4-Nitroaniline

Reference	Solvent	λ (μm)	β_x ($\times 10^{-30}$ esu)
Levine and Bethea (1978)	*trans*-Stilbene	1.318	6.4
Teng and Garito (1983)	Dioxan	1.907	9.6
		1.370	11.8
		1.060	16.9
		0.909	25
		0.830	40
Levine and Bethea (1976)	Melt	1.318	21.1
	CH_3OH	1.06	20.0–24.5
Oudar and Chemla (1977)	CH_3OH	1.06	34.5
Dulcic and Sauteret (1978)	$(CH_3)_2SO$	1.89	46

TABLE III

Calculated Values of β_x for 4-Nitroaniline[a]

	λ (μm)					
Reference	∞	1.907	1.370	1.060	0.909	0.830
1	7.137	8.296	9.678	12.298	15.774	19.636
2		5.7	6.4	7.7	9.2	10.7
3		9.5	11.6	15.9	24	35

[a] $\beta_x \times 10^{-30}$ esu.

[b] References: (1) Calculated by the method of Docherty *et al.* (1985); (2) Teng and Garito (1983a), calculated by the method of Garito and Lalama (1979); (3) Teng and Garito (1983a), with adjustment for solvent effect.

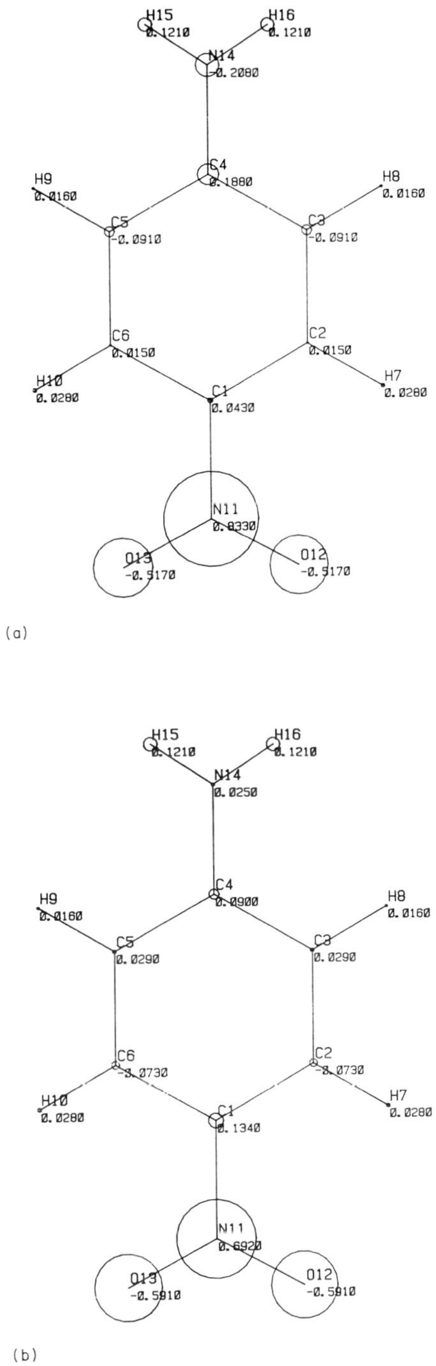

Fig. 2. Total charge distribution in 4-nitroaniline for (a) ground state and (b) third excited state. (The circles are proportional to the magnitude of the electronic charge.)

the solvatochromism in dioxan (see below), the revised dispersion relation of Teng and Garito produces values that coincide with their experimental results.

The experimental results of Teng and Garito are obtained by extrapolation to infinite dilution using the method if Singer and Garito (1981). Solute–solute interaction effects should therefore have been eliminated and, since dioxan is a nonpolar solvent, it is expected that the effects of molecular association between solute and solvent will be small.

Experimentally, the β_x values in dioxan increase smoothly as the fundamental frequency ω is increased and begin to diverge as 2ω approaches the first major optical transition that corresponds to the charge-transfer state. However, the frequency-dependent values calculated by Teng and Garito are smaller than the corresponding experimental values. This difference in dispersion behavior has been attributed to solvent-induced (solvatochromic) shifts, $\hbar\,\Delta\omega_{ng}$ of the transition energies that enter the perturbation-theory denominators, to longer wavelengths. Solvent-induced shifts of this type, which affect the transition energy, are thought to be caused principally by dipole-mediated interactions, which change the charge distributions in ground and excited states. In an endeavor to simulate the solvent effects, Teng and Garito (1983a) have modified the calculated transition energies of Garito and Lalama using the following relation [Amos and Burrows (1973)]:

$$\hbar\,\Delta\omega_{ng} = A\,\Delta\mu_{ng}(\mu_n + \mu_g) + B\,\Delta\mu_{ng}\mu_g$$

where $\hbar\,\Delta\omega_{ng}$ is the shift in the CNDO transition energy, μ_g and μ_n are the solute dipole moments for ground and excited states, respectively, and

$$A = (1/a^3)[(n^2 - 1)/(2n^2 + 1)]$$

$$B = (2/a^3)[(\varepsilon - 1)/(2\varepsilon + 1) - (n^2 - 1)/(2n^2 + 1)]$$

where n and ε are the refractive index and dielectric constant of the solvent, respectively, and a is the effective cavity radius of the molecular site. The first and second terms are the reaction field experienced by the permanent dipole moment of the solute due to the interaction with both the induced dipole moments of the surrounding molecules and the permanent dipole moments of the solvent and of other solute molecules at relatively high concentrations. As a result of these corrections, the calculated second transition energy (charge-transfer state) for PNA is moved from 4.37 (282.5 nm) to 3.46 eV (356.8 nm) and the correlation between the theoretical and experimental dispersion results in dioxan is greatly improved.

In comparing the absolute values of these measurements with theory, it should be noted that the reparametrized CNDO method of Lalama and Garito includes an element of fitting to the β_x values of aniline and nitrobenzene. Nevertheless, the excellent agreement between theoretical and experimental dispersion relations, when transitions energies are modified to agree with the solution spectrum, suggests very strongly that the contribution of the low-energy virtual $\pi \rightarrow \pi^x$ transitions accounts for a very large proportion of the observed value of β_x.

The calculations of Docherty *et al.* (1985) have been parameterized to approximate the spectra and dipole moments of a range of organic molecules in nonpolar solvents, a common procedure in quantum chemistry, but with no specific adjustments for PNA in dioxan or to observed hyperpolarizabilities. The measure of absolute agreement with the results of Teng and Garito again seems sufficient to be of quantitative significance and to confirm the identification of the electronic origins of the hyperpolarizability.

Morrell and Albrecht (1979) quote a calculated value of 34.3×10^{-30} esu at 1.06 μm. Their equation for the induced polarization includes a factor of $\frac{1}{2}$ in the quadratic term, and it seems probable that the appropriate value for comparison with the other results is 17.15×10^{-30} esu, again in good agreement with the results of Teng and Garito.

It is possible to come to the tentative conclusion that, allowing for the variations due to solvatochromic effects specific to particular solvents, the CNDO-based methods are at least in semiquantitative agreement with measurements on PNA in dilute solutions of nonpolar solvents. This conclusion evidently provides an indication as to how similar calculations on other systems should be assessed.

The apparent discrepancy between the values obtained in methanol by Levine and Bethea (1976) on the one hand and by Oudar and Chemla (1977) on the other is almost certainly attributable to the difference in the values of the reference (d_{11}) coefficient of quartz employed in the two cases ($d_{11} = 0.8 \times 10^{-9}$ and $d_{11} = 1.2 \times 10^{-9}$ esu, respectively). All other values in Table II, including those of Teng and Garito, are based on the latter value (Choy and Byer, 1976) and, if this is accepted, the results of Levine and Bethea (1976) should be multiplied by a factor of $\frac{3}{2}$. Their measured value for PNA in *trans*-stilbene then becomes 9.6×10^{-30} esu at 1.06 μm, similar to the value obtained by Teng and Garito in dioxan. It is clear from Table II that the values of β_x, when the molecule is in a polar environment (methanol, DMSO, molten), are substantially greater, often by a factor of 2 or more, at the corresponding frequencies. This effect is probably partly the result of solvatochromism and partly the result of molecular association, and both effects are the consequence of strong dipole–dipole interactions. Molecular association would make the size of the effective reorienting entity in the EFISH measurements larger than a single molecule.

The question of molecular association in relation to EFISH measurements has been addressed by Levine and Bethea (1976), although not specifically for PNA solutions. They attempt to account for variations in β_x at different concentrations by introducing a correction factor in the internal field calculations as prescribed by the theories of Kirkwood and Frohlich (see, for example, Hill *et al.*, 1969). The results are sometimes anomalous. For example, while pure nitrobenzene appears to require no correction for association, more dilute solutions of nitrobenzene in nonpolar solvents appear to be associated. It seems likely that in these systems the more drastic effects of the energy level shifts cannot be adequately accounted for by the Kirkwood–Frohlich theory. It is possible that the theory could be developed to include the effects of both solvatochromism and molecular association.

Levine and Bethea (1978) have also investigated the quantities

$$\beta_x(-2\omega, \omega, \omega) \quad \text{and} \quad \beta_x(-3\omega, 2\omega, \omega)$$

by combining information from SHG measurements and from measurements in which the original and doubled frequencies were mixed. The YAG line (1.06 μm, 1.17 eV) corresponds to ω and the other frequencies are therefore at 2.34 and 3.51 eV. The trebled frequency is above the first absorption band, and real and imaginary contributions to β_x must be included. To relate the values at different frequencies, a specific excited-state model is required. A two-state model ($n = g, n = 1$) was adopted and it was found that the nondispersive part of β_x gave a value of 16×10^{-30} esu, with a corresponding resonant contribution of 6.9×10^{-30} esu. Since only the latter is associated with the virtual transition of the two-state model, Levine and Bethea (1978) were unable to account for the large nonresonant contribution and attributed the result to the failure of the two-state model: more recent theoretical results tend to confirm this conclusion. While a major part of β_x arises from the dominant charge-transfer state, there are considerable contributions from other states that would be enhanced at the 3ω frequency and that invalidate an analysis based on the two-state model. In addition, computations of the sum over states method, which have been extended over the whole of the $\pi \rightarrow \pi^*$ spectrum and a sample of $\sigma \rightarrow \sigma^*$ transitions, have given no indication of any large cumulative nonresonant background.

VI. CALCULATIONS FOR OTHER MOLECULES

A number of other molecules, such as aniline, nitrobenzene, and 2-methyl-4-nitroaniline, have been calculated by Garito *et al.* to give values that are generally smaller and transition energies that are greater than the corresponding experimental results in solution. However, the calculated values for the 2-methyl derivative have been modified by the use of solvent parameters (which change the computed transition energies), and the resulting correlation between theory and experiment is considerably improved, as with the data for 4-nitroaniline. Theoretical calculations have also been carried out by Lalama *et al.* (1981) on 7,7-diamino-8,8-dicyanoquinodimethane (**I**) using a structure based on the crystal structure of 7,7,8,8-tetracyanoquinodimethane and the results compared with experimental data on the related molecule 2-(4-dicyanomethylenecyclohexa-2,5-dienylidiene)-imidiazoline (**II**).

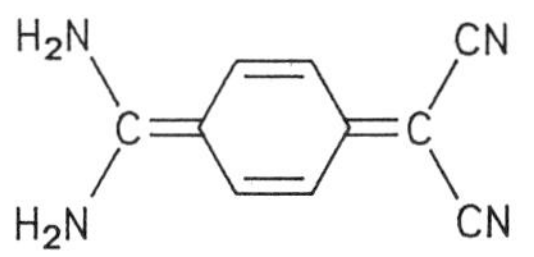

I

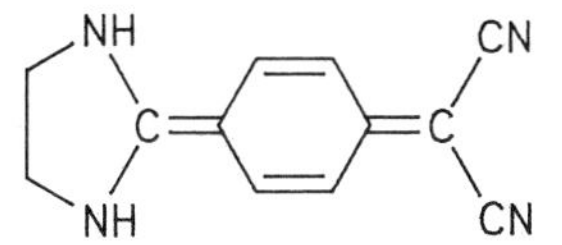

II

The ground-state charge distribution of **I** shows the characteristic single- and double-bond alternation of quinoid structures and the classical donor and acceptor properties of the amino and cyano substituent groups. Limited CI calculations using four states give both a singlet excitation energy and β_x value of -167×10^{-30} esu, which compares well with the experimental spectrum and β_x value of -240×10^{-30} esu. An increase from four states to 16 states surprisingly gives no further improvement to the final result of the calculation.

The negative sign arises from the change in dipole moment after excitation. Thus, β_x is dominated by the β_{xxx} component, which contains the product $\Delta\mu_n^x(r_{ng}^x)^2$ in the perturbation expansion. Again, as in other cases, a single transition is the major contributor to the β_{xxx} value, and the $\Delta\mu_n^x$ value for this state relative to the ground state is -10.84 D compared with a ground-state value of $+9.48$ D. Unlike the results for substituted benzenes, charge is transferred from the acceptor group to the donor group on excitation, resulting in a negative sign for the β_x tensor.

Work by Docherty *et al.* (1985) on a series of 17 compounds has generally given good correlations between the calculated values of β_x and those derived from a variety of experimental sources. In addition to studies on the smaller molecules such as 4-nitroaniline, larger systems such as 4-amino-4′-nitro-*trans*-stilbene (**III**) have been calculated using up to 100 states for the configuration interaction and singlet state expansion (Fig. 1).

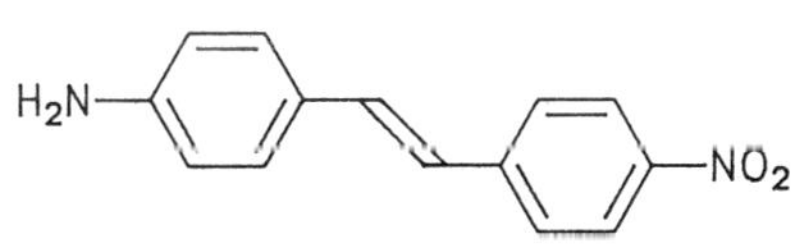

III

The results show that at least 50 states are required before convergence of the perturbation sum occurs, in apparent contrast to the results obtained for

TABLE IV

Calculated and Experimental Hyperpolarizabilities of Selected Organic Molecules[a]

Molecule	λ	β^{calc}	β^{expt}	Solvent	Reference[d]
Aniline	1.318		0.79	Liquid	1
	1.064	2.4260	1.1	Liquid	2
	1.064		2.46	Liquid	3
	[b]		1.17	Liquid	12
	1.064		0.89	Liquid	11
Nitrobenzene	1.318	0.9034	1.97	Liquid	1
	1.064	1.0757	2.2	Liquid	2
	1.064		4.6	Liquid	3
	[b]		2.27	Liquid	12
	1.064		2.04 ± 0.31	Liquid	13
	1.064		1.11 ± 0.17	Hexane	13
	1.064		1.39 ± 0.20	DMSO	13
	1.064		2.04	Liquid	11
4-Nitroaniline	1.064	12.2982	16.7 ± 0.5	Dioxane	5
	1.318	10.6701	21.1	Liquid	1
	1.064		34.5	Methanol	2
	1.890	8.9014	46.8 ± 0.9	DMSO	4
	1.064		34.5	Liquid	6
	1.318		6.4	Stilbene	9
	1.064		22.0	Methanol	10
2-Nitroaniline	1.318	4.7017	6.4	Liquid	1
	1.064	6.0984	10.2	Acetone	2
3-Nitroaniline	1.318	5.2210	4.2	Liquid	1
	1.064	6.2214	6.0	Acetone	2
	[b]		6.1	Acetone	12
2,4-Dinitroaniline	1.064	9.2858	21	Acetone	2
N,*N*-Dimethylaniline	[b]	6.9430	1.27	Liquid	12
2-Chloro-4-nitroaniline	1.064	11.7532	34	Methanol	10

2-Bromo-4-nitroaniline	1.064	12.4438	31	Methanol	10
2-Methyl-4-nitroaniline	1.064	13.7058	42	Methanol	10
	1.064		16.7 ± 0.5	Dioxane	5
N,*N*-Dimethyl-4-nitroaniline	1.890	13.9286	52 ± 0.8	DMSO	4
4-Cyanoaniline	1.890	6.5878	13.34	DMSO	4
4-Cyano-*N*,*N*-dimethylaniline	1.890	10.4644	14.24	DMSO	4
4-Aminoacetophenone	1.890	7.0716	2.40	DMSO	4
3-Aminophenol	[b]	2.4536	1.2	Acetone	12
Phenol	[b]	2.0134	0.17	Liquid	12
Benzonitrile	[b]	0.7323	0.48	Liquid	12
Fluorobenzene	1.064	0.8535	1.06	Liquid	3
	1.064		0.44	Liquid	11
	[b]		−0.70	Liquid	12
Chlorobenzene	1.064	0.4665	0.281	Liquid	11
	[b]		−0.33	Liquid	12
Bromobenzene	1.064	0.2051	0.042	Liquid	11
	[b]		−0.20	Liquid	12
Iodobenzene	1.064	0.4366	−0.279	Liquid	11
	[b]		0.46	Liquid	12
1,2-Diiodobenzene	1.064	0.0082	−0.291	Liquid	11
Toluene	1.064	0.3308	0.18	Liquid	11
4-Nitrotoluene	1.890	1.8822	8.24 ± 0.9	DMSO	4
4-Cyanotoluene	1.890	1.4147	2.93 ± 0.2	DMSO	4
4-Nitroanisole	1.890	6.798	15.9 ± 1.6	DMSO	4
4-Cyanoanisole	1.890	5.1200	4.8 ± 0.3	DMSO	4
1,3-Dinitrobenzene	[b]	1.0969	1.2	Benzene	12
1,3-Dihydroxybenzene	[b]	1.2118	0.2	Ethanol	12
2-Fluoronitrobenzene	1.064	1.3624	−1.75	Liquid	11
3-Fluoronitrobenzene	1.064	2.0467	−1.64	Liquid	11
4-Fluoronitrobenzene	1.064	3.1664	−2.14	Liquid	11
3-Bromonitrobenzene	[b]	1.1688	1.2	Benzene	12
3-Chloronitrobenzene	[b]	0.7991	1.6	Benzene	12

(*continues*)

TABLE IV (*Continued*)

Molecule	λ	β^{calc}	β^{expt}	Solvent	Reference[d]
4-Nitro-*trans*-stilbene	1.064	20.0593	29 ± 10	Benzene	6
4-Amino-*trans*-stilbene	1.064	31.5228	12 ± 7	Benzene	6
4-Dimethylamino-*trans*-stilbene	1.064	59.9308	29 ± 8	Benzene	6
4-Chloro-*trans*-stilbene	1.064	5.2390	-3.6 ± 2	Benzene	6
4-Chloro-4′-nitro-*trans*-stilbene	1.064	13.1528	39 ± 10	Chloroform	6
4-Chloro-4′-dimethylamino-*trans*-stilbene	1.064	87.1776	42 ± 10	Chloroform	6
4-Cyano-4′-dimethylamino-*trans*-stilbene	1.890	44.9865	134.07[c]	DMSO	8
4-Cyano-4′-methoxy-*trans*-stilbene	1.890	22.9604	19.70[c]	DMSO	8
4-Methyl-4′-nitro-*trans*-stilbene	1.890	15.6167	38.66[c]	DMSO	8
4-Amino-4′-nitro-*trans*-stilbene	1.064	69.1317	260 ± 35	Acetone	6
4-Dimethylamino-4′-nitro-*trans*-stilbene	1.064	117.1400	450 ± 90	Acetone	6
	1.318		39	Stilbene	9
	1.890		596.48[c]	DMSO	8
4-[(1-Methyl-4-(1*H*)-pyridinylidene)ethylidene]-2,5-cyclohexadien-1-one	1.890	47.1550	869.56[c]	DMSO	7
	1.318		-100	Methanol	15
			-210	Pyridene	15
4-Dimethylamino-β-nitrostyrene	1.064	126.9200	220 ± 40	Chloroform	6
Methyl-*N*-(2,4-dinitrophenyl) alaninate	1.064	16.6923	22	Acetone	2
trans-4-Azastilbene	1.890	4.6956	15.00[c]	Ethanol	7
4-Dimethylaminocinnamaldehyde	1.890	42.5674	68.02[c]	DMSO	8
4-Methoxycinnamaldehyde	1.890	20.8695	16.07[c]	DMSO	8

1-(4-Dimethylaminophenyl)-4-nitrobuta-1,3-diene	1.064	355.7150	630 ± 160	Acetone	6
1-(4-Cyanophenyl)-4-(4-dimethylaminophenyl)-buta-1,3-diene	1.890	69.6036	264.98[c]	DMSO	8
2-(4-Thianomethylenecyclohexa-2,5-dienylidene)-imidazolidene	1.064		−240 + 16	DMSO	14
3-Ethyl-2-[2-(4-oxo-2,5-cyclohexadienylidene)ethylidene]-2,3-dihydrobenzothiazole	1.890	39.68	54.45	DMSO	7
2-[6-(4-Carboxyphenylimino)hexa-2,4-dienylidene]-3-ethyl-2,3-dihydroxybenzothiazole	1.890	118.82	196.85	DMSO	7

[a] The experimental results, β^{expt} in units of 10^{-30} esu, are based on dc electric-field-induced second-harmonic generation in solution (except where stated otherwise); λ is the wavelength of the incident laser radiation in micrometers; liquid refers to neat materials or the molten state; and DMSO is dimethyl sulfoxide. The calculated results, β^{calc} in the same units, are taken from Docherty *et al.* (1985).

[b] Experimental measurements from static-field-induced SHG and tunable four-wave mixing. The calculated values in these cases have been obtained at 1.064 μm.

[c] Experimental results derived from the product $\mu\beta$ using dipole moments calculated by molecular orbital methods.

[d] References: (1) Levine (1976); (2) Oudar and Chemla (1977); (3) Zyss (1979a, 1979b, 1979c); (4) Dulcic and Sauteret (1978); (5) Teng and Garito (1983); (6) Oudar (1977); (7) Dulcic (1978, 1979); (8) Dulcic *et al.* (1981); (9) Levine and Bethea (1978); (10) Levine *et al.* (1979); (11) Levine and Bethea (1974, 1976); (12) Oudar *et al.* (1977); (13) Singer and Garito (1981); (14) Lalama *et al.* (1981); (15) Levine *et al.* (1978).

structure **II**. In all cases there is a substantial increase in the dipole moment from the unperturbed ground state to the dominant excited state, although as the molecular size increases then clearly other excited states begin to contribute significantly to the overall β_x value. A selection of experimental and calculated β_x values is shown in Table IV.

REFERENCES

Amos, A. T., and Burrows, B. L. (1973). *Adv. Quantum Chem.* **7,** 289.
Andrews, D. L. (1980). *J. Phys. B* **13,** 4091.
Bloembergen, N. (1965). "Nonlinear Optics." Benjamin, New York.
Bottcher, C. J. F. (1952). "Theory of Electric Polarisation." Elsevier, Amsterdam.
Choy, M., and Byer, R. L. (1976). *Phys. Rev. B* **14,** 1693.
Cyvin, S. J., Rauch, J. E., and Decius, J. C. (1965). *J. Chem. Phys.* **43,** 4083.
Del Bene, J., and Jaffe, H. H. (1968). *J. Chem. Phys.* **48,** 1807.
Docherty, V. J., Pugh, D., and Morley, J. O. (1985). *J. Chem. Soc., Faraday Trans. 2* **81,** 1179.
Dulcic, A. (1978). *Opt. Commun.* **25,** 402.
Dulcic, A. (1979). *Chem. Phys.* **37,** 57.
Dulcic, A., and Sauteret, C. (1978). *J. Chem. Phys.* **69,** 3454.
Dulcic, A., Flytzanis, C., Tang, C. L., Pepin, D., Fetizon, and Hoppilliard, Y. (1981). *J. Chem. Phys.* **74,** 1559.
Edwards, L., and Gouterman, M. (1970). *J. Mol. Spectrosc.* **33,** 292.
Finn, R. S., and Ward, J. F. (1973). *J. Chem. Phys.* **60,** 454.
François, P., Carles, P., and Rajzmann, M. (1977). *J. Chim. Phys.* **74,** 606.
François, P., Carles, P., and Rajzmann, M. (1979). *J. Chim. Phys.* **76,** 328.
Garito, A. F., Teng, C. C., Wong, K. Y., and Eamman 'Khamiri, O. (1984). *Mol. Cryst. Liq. Cryst.* **106,** 219.
Hauchecorne, G., Kerherve, F., and Mayer, G. (1971). *J. de Phys. (Les Ulis, Fr.)* **32,** 47.
Hill, N. E., Vaughan, W. E., Price, A. H., and Davies, M. (1969). "Dielectric Properties and Molecular Behaviour." Van Nostrand. Reinhold, Princeton, New Jersey.
Lalama, S. J., and Garito, A. F. (1979). *Phys. Rev. A* **20,** 1179.
Lalama, S. J., Singer, K. D., Garito, A. F., and Desai, K. N. (1981). *Appl. Phys. Lett.* **39,** 940.
Levine, B. F. (1976). *Chem. Phys. Lett.* **37,** 516.
Levine, B. F., and Bethea, C. G. (1974). *Appl. Phys. Lett.* **24,** 445.
Levine, B. F., and Bethea, C. G. (1975). *J. Chem. Phys.* **63,** 2666.
Levine, B. F., and Bethea, C. G. (1976). *J. Chem. Phys.* **65,** 2429.
Levine, B. F., and Bethea, C. G. (1978). *J. Chem. Phys.* **69,** 5240.
Levine, B. F., Bethea, C. G., Thurmond, R. T., and Berstein, J. L. (1979). *J. Appl. Phys.* **50(4),** 2523.
Levine, B. F., Bethea, C. G., Wasserman, E., and Leenders, L. (1978). *J. Chem. Phys.* **68,** 5042.
Lipari, N. O., and Duke, C. B. (1975). *J. Chem. Phys.* **63,** 1748.
Liptay, W. (1974). *In* "Excited States", Vol. 1 (E. C. Lim, ed.), p. 129. Academic Press, New York.
Lyons, L. E., Walsh, J. R., and White, J. W. (1960). *J. Chem. Soc.,* p. 167.
Morrell, J. A., and Albrecht, A. C. (1979). *Chem. Phys. Lett.* **64,** 46.
Nishimoto, K., and Mataga, N. (1957). *Z. Phys. Chem.* (*Wiesbaden*) **12,** 335.
Onsager, L. (1936). *J. Am. Chem. Soc.* **58,** 1486.
Orr, B. J., and Ward, J. F. (1971). *Mol. Phys.* **20,** 513.
Oudar, J. L. (1977). *J. Chem. Phys.* **67,** 446.

Oudar, J. L., and Chemla, D. S. (1977). *J. Chem. Phys.* **66,** 2664.
Oudar, J. L., and Le Person, H. (1975). *Opt. Commun.* **15,** 258.
Oudar, J. L., Chemla, D. S., and Batifol, E. (1977). *J. Chem. Phys.* **67,** 1626.
Pariser, R. (1956). *J. Chem. Phys.* **24,** 250.
Pople, J. A., and Beveridge, D. L. (1970). "Approximate Molecular Orbital Theory." Mcgraw-Hill, New York.
Sheppard, S. E., and Newsome, P. T. (1942). *J. Am. Chem. Soc.* **64,** 2937.
Singer, K. D., and Garito, A. F. (1981). *J. Chem. Phys.* **75,** 3572.
Szabo, A., and Ostlund, N. S. (1982). "Modern Quantum Chemistry." Macmillan, New York.
Teng, C. C., and Garito, A. F. (1983a). *Phys. Rev. Lett.,* **50,** 350.
Teng, C. C., and Garito, A. F., (1983b). *Phys. Rev. B* **28,** 6766.
Ward, J. (1965). *Rev. Mod. Phys.* **37,** 1.
Zyss, J. (1979a). *J. Chem. Phys.* **70,** 3333.
Zyss, J. (1979b). *J. Chem. Phys.* **70,** 3341.
Zyss, J. (1979c). *J. Chem. Phys.* **71,** 909.

Chapter II-3

Design and Synthesis of Organic Molecular Compounds for Efficient Second-Harmonic Generation

J. F. NICOUD

Ecole Supérieure de Physique et Chimie Industrielles
de la Ville de Paris
Laboratoire de Chimie et Electrochimie des
Matériaux Moléculaires
75231 Paris Cedex 05, France

R. J. TWIEG

IBM
Almaden Research Center
San Jose, California 95102-6099

I. INTRODUCTION

The laser, now 25 years young, has opened up enormous opportunities to study nonlinear optical phenomena. Within a year after the first successful laser, the feeble optical second-harmonic generation (SHG) from quartz was observed (Franken *et al.*, 1961), followed by the first observation of SHG in organic crystals of a few polynuclear aromatic hydrocarbons (Rentzepis and

Pao, 1964). Since this early work there has been a remarkable evolution in the basic understanding of the structure of the organic materials that are the critical media in which the nonlinear optical effects described in this volume occur. In the new age of optics, the organic crystals described here, and others yet to be discovered, will play a much more significant role alongside the more conventional inorganic substances. The monopoly of inorganic materials in optics is becoming as antiquated as the concepts that had defined organic chemistry prior to 1828, when Wöhler (1828) first synthesized the organic molecule urea (**1**) from the inorganic salt ammonium cyanate. It is particularly appropriate that urea, which then played such a pivital role in the definition of organic chemistry, is today one of the most important and thoroughly studied nonlinear optical (NLO) organic materials. After more than two decades of study of the theory of nonlinear optics, crowned by the 1981 Nobel Prize to Bloembergen, organic chemists assist today in the development of new NLO materials with a wide range of fascinating properties.

The collaboration and the persistence of those skilled in optics, physics, X-ray crystallography, crystal growth, and organic synthesis are the key components in the "molecular engineering" of optimal NLO materials (Chemla *et al.*, 1981). It is the goal of this chapter to describe this synergistic process from the viewpoint of the organic chemist who recognizes that certain deep-seated molecular properties manifest themselves in many different ways. In particular, the chemical reactivity and many spectroscopic features of molecules involve properties that are closely related or identical to those responsible for nonlinear behavior and optical hyperpolarizability.

The ensuing discussion concentrates on molecular crystals for SHG, mindful that the specific materials and principles involved are of general relevance to other organized media and optical phenomena. The ultimate practical organic NLO material must fulfill a plethora of requirements, some very general and some dictated by the particular application, as, for instance, in the case of NLO crystals for frequency doubling of diode laser light and laser fusion. Our discussion concentrates on the design of molecular crystals with a favorable balance of optimum doubling efficiency and transparency properties. However, this is not to underestimate a variety of equally important issues such as synthetic accessibility, cost, crystal growth capability, toxicity, environmental stability, optical damage properties, and a host of others.

The subsequent discussion involves not an arbitrary, but mandatory, separation of microscopic molecular properties and macroscopic crystal properties. The difficult challenges in simultaneously reconciling these molecular and crystal properties are considered, and finally a variety of specific examples of NLO molecules are described in some detail. Certain molecules for which single-crystal properties are available are described first, followed by those promising materials for which only preliminary powder tests are available. Due to space constraints, the discussion of specific materials is only representative and not comprehensive. A more complete compilation of organic NLO materials is found in Appendices I and II (Vol. II).

II. ORIGINS OF OPTICAL NONLINEARITY IN ORGANIC MOLECULAR CRYSTALS

A. Consideration of Organic Molecular Properties

1. Bonding in Molecules

In order to design an optimal NLO crystal, the origins of the nonlinear response to the optical field must be understood. This, in turn, requires knowledge of the constitution of the crystal, including the bonding properties of the atoms in the molecules and then the molecules in the crystal. It will be assumed that the molecules function independently of one another in the crystal and that only their net orientations are important. This assumption is usually reasonable, since the mode of bonding on these different scales is so different. In fact, the degree of molecular interaction is never zero but varies in mode and extent from material to material. For example, in the routine systematic structural modification within a series of compounds there is sometimes encountered an extraordinary change in solid-state absorption characteristics, usually attributed to intermolecular charge transfer or exitonic effects. These interactions are of themselves of great current interest and importance but are dealt with in detail in only a few instances since the subject is largely outside the scope of the present chapter (Gutman and Lyons, 1967; Mataga and Kubota, 1970; Meier, 1974; Pope and Swenberg, 1982).

The diverse properties of organic compounds are largely due to the unsurpassed ability of the carbon atom to form a variety of stable hybridized bonds with itself and many other elements. Energetically, a typical two-electron covalent σ C_{sp^3}—C_{sp^3} bond as in ethane is worth about 85 kcal/mol, a typical four-electron $\sigma + \pi$ C_{sp^2}—C_{sp^2} bond as in ethylene is worth about 150 kcal/mol, and σ and π bonds from carbon to other heteroatoms gravitate in this range (Gould, 1959; Roberts, 1965; March, 1977; Ladd, 1979). The average organic molecule may contain dozens of such σ and π bonds totalling thousands of kilocalories per mole. It is the π electrons that differentiate organic NLO materials from inorganic systems in which few such conjugated counterparts exist. Unlike the electrons in σ bonds, the electron density of a π bond, or series of π bonds, is much more mobile (Streitweiser, 1961; Dewar,

1969). This difference in electron mobility is easily explained in a number of ways, including an analysis of the relevant bond energies; note that the four-electron $\sigma + \pi$ bond is not twice as strong as the two-electron σ bond. In fact, it is about 20 kcal/mol less than what might be expected, or 5 kcal/mol per electron less, and as a result on the average the π electrons are not as tightly bound as are σ electrons. In a conjugated series of π bonds the total π-bond strength is further decreased, and this secondary conjugative perturbation is the "mesomeric energy." These and other fundamental issues of chemical bonding are discussed thoroughly by Pauling (1960) and Ingold (1969). The electron distribution can be skewed by substituents; the extent of redistribution is measured by the dipole moment, and the ease of redistribution by the (hyper)polarizability. The electron distribution can be perturbed by interaction with a radiation field such as the intense optical E-fields produced by a laser, and if the perturbation is asymmetric a quadratic nonlinearity results (Chemla *et al.*, 1981). Virtually all organic NLO molecules contain π bonds such as the single short polar π bond of the carbonyl as in urea or a more extended series of π bonds as in benzene and substituted benzene derivatives.

2. Molecular Dipole Moment and (Hyper)Polarizability

In organic chemistry, the understanding of dipole moments, polarizability, and chemical bonding and structure are historically strongly intertwined. A variety of linear and nonlinear electro-optic effects, such as the Pockels and Kerr effects, were known and studied long before the advent of the laser (Bogaard and Orr, 1975). Before modern chemical instrumentation, the determination of refractive index and molar refractivity for organic compounds was routine and many empirical relationships with structure were known (Smyth, 1955). Certain classes of organic compounds were found to have unusual molecular refractivities. For example, nearly a century ago Brühl (1886) noted that molecules containing conjugated double bonds possess an "optical exaltation" of molecular refraction in excess of the simple sum of individual bond refractivities.

As early as 1931 the origins of this enhanced "electron shift" that gives rise to high dipole moments and molar refractivities in some simple conjugated compounds were explained (Sutton, 1931). In a remarkable series of papers 20 years later this work was extended to deal with molecules that are strongly polarizable, such as the donor–acceptor-substituted stilbenes (Sutton *et al.*, 1951), which were later found to have very large optical hyperpolarizabilities (Oudar, 1977a). In Sutton's papers, the relationships of substituents, π conjugation, substituent-induced electron density redistribution (mesomerism), steric effects, positional effects, and interaction moments with molecular polarizability were clearly laid out. Subsequently, the structural origins of polarizability have been systematically studied in great detail, particularly by Le Fèvre and co-workers (Le Fèvre, 1965, 1970). A few particularly relevant papers of the hundreds from this group include the studies on diphenylpolyenes (Bramley and Le Fèvre, 1960), phenylpolyenals (Bramley and Le Fèvre, 1962), and polar aromatic derivatives including 4-nitroaniline (**2**) (Aroney *et al.*, 1968). The molecular properties responsible for polarizability are the same as those fundamentally important for molecular hyperpolarizability. This was recognized by Davydov *et al.* (1970, 1971, 1973) for the important role of mesomeric interaction and intramolecular charge transfer (CT) between conjugated donor and acceptor substituents as in **2** and by Bergman *et al.* (1972), who noted the relationship of linear polarizability and CT resonance structures in 5-nitrouracil (**3**). This important CT absorption is characteristic of many molecules that contain simple polar π bonds, such as a carbonyl group, or more extensive and conjugated systems, as in many aromatic molecules, which will be discussed (Nagakura, 1963).

Equation (1) gives the polarization p of a molecule as the sum of the ground-state polarization p_0 and the induced polarization by an external electric field E expressed as a power series (Flytzanis, 1975). Alternatively, one often finds this equation expressed in terms of the permanent molecular polarization or dipole moment. Here the coefficient α is the linear polarizability, related to the refractive index of the material, β is the quadratic hyperpolarizability, which must be optimized for SHG, and γ is the cubic hyperpolarizability, which is important for a variety of other NLO properties. The linear first term of the equation is adequate only until the induced moment is comparable in magnitude to the molecular dipole moment (Coulson *et al.*, 1952; Buckingham and Orr, 1967). Although this expression is concise and physically accurate, it really offers the organic chemist little insight into how to proceed in designing organic molecules with optimal properties:

$$p = p_0 + \alpha E + \beta EE + \gamma EEE + \cdots \tag{1}$$

From the complete sum frequency generation perturbation calculation, an approximation of β, which is the CT component β_{CT} given by Eq. (2), has been obtained (Oudar, 1977a). Here β_{CT} is a function of the laser fundamental photon energy $\hbar\omega$, the electron charge e and the mass m, the energy gap W, the oscillator strength f of the CT transition in the molecule, and the change in the dipole moment $\Delta\mu$ involved in that transition. Clearly, to maximize β it is favorable to seek molecules that have an electronic transition with a small energy gap near (but not including) 2ω accompanied by a large oscillator strength and dipole moment change. The spectroscopic energy gap W is related to the frequency (or wavelength λ) of absorptions found in the ultraviolet–visible (UV–VIS) spectra of the molecule and the oscillator strength is related to the extinction coefficient ε of the absorption. These are all properties familiar to the chemist and have well understood physical origins and for which many structural correlations exist, particularly in the field of dye chemistry (Fabian and Hartman, 1980; Gordon and Gregory, 1983).

$$\beta_{CT} = \frac{3e^2\hbar^2 Wf\,\Delta\mu}{2m[W^2 - (2\hbar\omega)^2][W^2 - (\hbar\omega)^2]} \tag{2}$$

$$\Delta\mu = \mu_e - \mu_g \tag{3}$$

In order to maximize β, it is necessary to maximize the absolute value of the induced dipole moment $\Delta\mu$ equal to the dipole moment change between the excited-state dipole moment μ_e and the ground-state dipole moment μ_g, as in Eq. (3). If the β_{CT} term predominates, then the sign of β is a function of the relative magnitudes of μ_e and μ_g, and generally $\mu_e > \mu_g$ and β is positive with few exceptions (Levine *et al.*, 1978; Lalama *et al.*, 1981). The ground-state dipole moments of a great variety of organic molecules have been studied for many years (Exner, 1975), and extensive compilations of data exist (McClellan, 1963, 1974). Excited-state dipole moments are less well understood because the experimental measurements to examine the short-lived excited states are much more difficult (Bakhshiev *et al.*, 1969; Minkin *et al.*, 1970; Koutek, 1984). This situation is unfortunate, since it is the absolute difference $|\Delta\mu|$ between the excited- and ground-state dipoles that is of most relevance to SHG. Nonetheless, the perusal of tables of ground-state dipole moments of organic compounds can provide useful insights to identify chemical functionality with the appropriate properties.

3. *Insights from Calculations and Theory*

The quantitative relationships describing the length of the π system and the modes and extent to which π electron density is distorted by substituents and their effect on molecular hyperpolarizability come from both calculations and theory. A large number of calculations that have been done to predict or rationalize the (hyper)polarizabilities of molecules are beyond the scope of this chapter, but their importance must not be underestimated. For example, the cyanine and merocyanine dyes are among the most polar organic molecules, and calculations predicted extremely large (hyper)polarizabilities (Schweig, 1967a,b). Subsequently, the large polarizabilities (Dähne and Nolte, 1972; Nolte and Dähne, 1976) and hyperpolarizabilities (Dulcic and Flytzanis, 1978; Dulcic, 1979; Levine *et al.*, 1978) were experimentally verified.

Almost simultaneous with Sutton's important work, Davies (1952) derived Eq. (4), which relates the polarizability α to the length L of the conjugated system. Later, corresponding equations for β [Eq. (5)] (Oudar and Chemla, 1975; Jain *et al.*, 1981) and for γ [Eq. (6)] (Rustagi and Ducuing, 1974) were derived and the strong nonlinear dependence of the hyperpolarizabilities on the length of the π system became apparent:

$$\alpha = 8e^2b^2(2m+1)^3/\pi^5\gamma \propto L^3 \tag{4}$$

$$\beta = 3\,\Delta\mu\gamma_0/\alpha \propto L^3 \tag{5}$$

$$\gamma = l^5/a_0^3e^2 \propto L^5 \tag{6}$$

Just as the length of the conjugated π electron system is an important factor to produce (hyper)polarizability, in addition the appropriate functionality at the ends of the π system can enhance the required asymmetric electronic distribution in either or both the ground-state and excited-state configurations. Functional groups are divided into two catagories, based on their ability to donate electrons into or accept electrons from the π system (Fig. 1). The common donor (D) groups are characterized by predominant p character (sp^3) bonding and often have available electron pairs on a p orbital [as in an amine —$\ddot{N}R_2$ such as aniline (**8**)]. The common acceptor (A) groups are characterized by more s character (sp^2 or sp) bonding, as in the nitro NO_2 in nitrobenzene (**7**), in which nitrogen is formally sp^2, or nitrile (—C≡N) in which the carbon is sp-hybridized, or as with the presence of vacant orbitals (as in certain boron compounds). The fully charged substituents such as O^- (phenoxide) and N_2^+ (diazonium ion) are among the most potent D and A groups but have found very limited use in NLO organic materials to date.

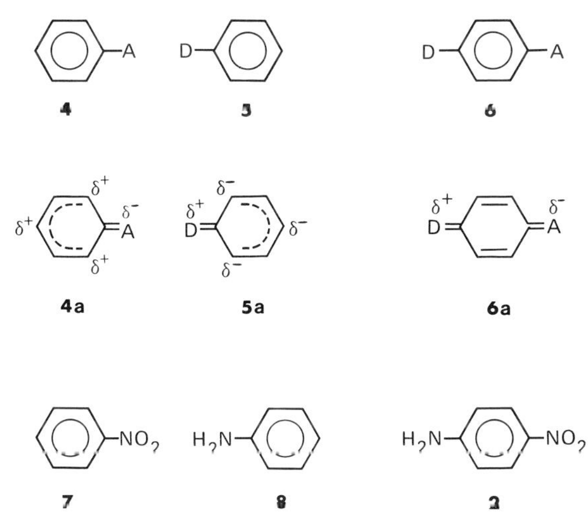

Fig. 1. Substituent-induced electronic perturbations. Substitution of benzene by an acceptor A group (**4**), resultant ground state resonance charge redistribution (**4a**), and typical example, nitrobenzene (**7**); substitution of benzene by an donor D group (**5**), resultant ground-state resonance charge redistribution (**5a**), and typical example, aniline (**8**); substitution of benzene by a donor–acceptor (D–A) pair (**6**), resultant ground-state mesomeric charge redistribution (**6a**), and typical example, nitroaniline (**2**).

Some typical D and A groups often found in NLO organic compounds are found in Table I, along with a variety of specific σ substituent values. There are many approaches to define and measure σ values, and each has its own particular origins and merits, which are beyond the scope of the present discussion, but, in general, for D groups $\sigma < 0$, for A groups $\sigma > 0$, and hydrogen has $\sigma = 0$ by definition.

Either a D or A group alone on the π system will perturb the electron density via resonance with the main perturbations in aromatic molecules at the ortho and para positions (**4, 4a, 5, 5a**). The most dramatic effect is found when both a D and an A group interact (**6, 6a**) in a mesomeric fashion (Fig. 1). For donor–acceptor (D-A) systems, particularly polarized aromatic systems, a variety of treatments have been successfully developed to explain the origins of the large β values (Levine and Bethea, 1975; Levine, 1976; Oudar and Chemla, 1975; 1977; Oudar, 1977a) and these are further discussed by Williams (1984). In a typical treatment, the β is seen to be the sum of two contributions due to an additivity term and a CT term as in Eq. (7). The β_{add} involves a substituent-induced charge asymmetry, and β_{CT} is the contribution due to through bond charge transfer or mesomerism. For the generic D–A substituted aromatic system (**6**), the relative contributions of β_{add} and β_{CT} are determined by the individual characteristics of the specific A and D groups, as in (**4a**) and (**5a**), respectively, as well as their interactive properties as in (**6a**). For example, in the specific case of *p*-nitroaniline (**2**), β_{add} is given by the vector sum of β

TABLE I

Typical Substituent Constants σ and Transition Constants l' and x

Substituent	σ_P [a]	$\sigma_{P(corr)}$ [b]	σ_R [a]	σ_I [a]	σ_R^0 [c]	l' [d]	x [d]
O^-	−0.52	—	—	—	−0.59	6.58	1.62
NMe_2	−0.17	−0.82	—	—	−0.53	—	—
NH_2	−0.17	−0.66	−0.76	0.10	−0.47	5.65	1.42
OCH_3	−0.11	−0.26	−0.51	0.17	−0.43	3.05	0.80
CH_3	−0.13	−0.14	−0.11	−0.05	−0.10	0.71	0.12
H	0.0	0.0	0.0	0.0	0.0	0.0	0.0
CN	0.67	0.69	0.10	0.58	0.09	4.49	−0.88
CF_3	—	0.53	0.09	0.41	0.11	—	—
CHO	0.45	—	—	—	0.24	9.05	−1.11
$COCH_3$	0.50	0.44	0.15	0.28	0.22	8.35	−1.13
NO_2	0.78	0.80	0.16	0.63	0.17	11.89	−1.50
N_2^+	1.80	—	—	—	0.30	—	—

[a] March (1968).
[b] Exner (1966).
[c] Katritsky and Topsom (1970).
[d] Stevenson (1965).

contributions from nitrobenzene (**7**) and aniline (**8**) and β_{CT} is the predominating contribution due to charge transfer between these substituents, as in **6a**:

$$\beta_{tot} = \beta_{add} + \beta_{CT} \tag{7}$$

4. *Insights from Chemical Intuition*

The various formulas related to Eq. (7) derived to explain the origin of β are very similar to the linear free energy relationships devised by physical organic chemists to describe the various quantitative relationships between structure and chemical reactivity or spectroscopic behavior (Hammett, 1970). In particular, the dual-substituent parameter (DSP) extensions such as Eq. (8) (Ehrenson *et al.*, 1973; Craik and Brownlee, 1983) of the Hammett equation are, like Eq. (7), two-level treatments. Here some physical property δ of a molecule is described by inductive σ_I and resonance σ_R terms for the substituents (Table I) and transmission factors ρ_I and ρ_R determined by the bonding between the substituents or some other specific attributes:

$$\delta = \rho_I\sigma_I + \rho_R\sigma_R \tag{8}$$

This equation has been applied in a variety of forms to numerous spectroscopic and physical phenomena of organic molecules, many of which have immediate relevance to the understanding of NLO molecules. Just a few particularly relevant cases are the σ^+ parameter useful for through-resonance interactions important for β_{CT} [still a controversial topic (Lipkowitz, 1982; Hiberty and Ohanessian, 1984)]; the carbon nuclear magnetic resonance (NMR) treatment of polarized styrenes (Bromilow and Brownlee, 1979), some of which have been found to have extremely large optical hyperpolarizabilities (Oudar, 1977a); the variation of substituent resonance as a function of electron demand (Happer and Wright, 1979); the separation of inductive and mesomeric effects in meta- and para-substituted benzene derivatives (Exner, 1966); the attempted correlations between substituents and dipole moment (Exner, 1975); and the correlation between X-ray aromatic bond angles and substituent parameters (Domenicano *et al.*, 1976). In addition, there are free-energy correlations between chemical reactivity and polarizability. The first is a correlation between nucleophilicity and the optical molar refractivity and the basicity of the nucleophile (Edwards, 1956), and the second is the correlation of the reactivity of halodinitrobenzenes as a function of their polarizability in the aromatic nucleophilic substitution reaction by which methyl (2,4-dinitrophenyl)aminopropanoate (MAP, **9**) (Oudar, 1977b; Oudar and Hierle, 1977) and many other organic NLO materials have been prepared (Bartoli *et al.*, 1977). Because the synthetic organic chemist knows both how substituents control chemical reactivity and the spectroscopic properties of molecules, any extensions required for a generic understanding of the origins of molecular hyperpolarizability are not great. It is no accident that the substituents most often found in the aromatic organic NLO materials have the most extreme and disparate σ values and that proper use of disparate pairs of substituents on a conjugated system can produce the required molecular charge asymmetry.

5. Electric-Field-Induced Second-Harmonic Generation

Fortunately, there are a variety of techniques by which the β value of a molecule can be determined without the intervention of a set of complications due to crystal symmetry properties, which will be discussed in the next section. The most popular technique, electric-field-induced second-harmonic generation (EFISH), has been gradually optimized experimentally and theoretically (Meredith, 1982; Appendix II, Volume II). Great care must be taken for both the dipole moment determination and EFISH measurements, and solvent effects must be considered (Levine *et al.*, 1978). Knowledge of the relative molecular β values determined by this technique establishes a "benchmark" for the synthetic chemist to attempt to achieve in the crystalline state for a class of materials. In this way the EFISH experiments can bridge the molecular and crystal aspects of the "molecular engineering" process. Further studies are required on a systematic set of organic molecules to attempt to establish a DSP-type correlation (δ related to hyperpolarizability) and to calibrate the accuracy of quantum-mechanical calculations. Although it is difficult to factor the vector aspects into the free-energy relationships, as was found in the case of dipole moments correlations (Exner, 1966, 1975), the undertaking should be very worthwhile in order to provide some semi-quantitative guidance.

B. Consideration of Organic Crystal Properties

1. Bonding in Crystals

The bonds that hold organic molecules together in crystals are different and significantly weaker than the bonds that join the atoms in these molecules. The total crystalline bonding energy is approximated by the sublimation energy, which is only on the order of 20 kcal/mol for a molecule like nitrobenzene (**7**). The packing of molecules in the crystal is the result of a variety of influences (Kitaigorodsky, 1973; Ladd, 1979). The energetics of the crystal bonding are approximated by Eq. (9); where U_{tot} is the total crystal binding energy, which is given by the sum of the van der Waals (U_n), multipolar (U_m) [itself the sum of dipolar (U_d) and quadrupolar (U_q) terms], and hydrogen bonding (U_H)

contributions:

$$U_{\text{tot}} \approx U_{\text{n}} + U_{\text{m}} + U_{\text{H}} \tag{9}$$

First, the molecules must "close-pack" and fill at least 60% of the available space, otherwise crystallization will not occur (close-packing of spheres utilizes 74% of the available volume). Due to this fact, the very shape of a molecule and its conformational mobility to adapt to a variety of packing situations may ultimately determine the space group to a great extent. Second, van der Waals forces operate and as molecules approach one another electron synchronization occurs and the molecules mutually align and attract until an energy minimum is reached, after which further compression leads to strong repulsion. Third, the molecular dipolar and quadrupolar forces may influence this energy minimum as the molecules will attempt to align themselves to minimize any multipolar repulsion. Fourth, and last, if the molecules contain hydrogen-bonding substituents, these interactions can be of significant influence on the energy minimum. The U_{n} term dominates because it varies quickly with structure parameters as r^6 whereas the multipolar and hydrogen-bonding interactions vary more slowly as r^2 (here r is the distance between molecules or parts of molecules). Overall, the intermolecular bonding components are of a variety of types and, in general, are more diffuse, less well understood, and significantly less prone to rational manipulation than covalent bonds. Although the crystals of rare gases and simple planar, rigid, and nonpolar polynuclear aromatics can be calculated quite accurately (Kitiagorodsky, 1973), the situation for typical NLO organics, which are often more conformationally mobile and polar, is beyond present-day expertise.

2. *The Optically Ideal Crystal Classes*

The search for the efficient organic nonlinear crystal is, in fact, the search for the "polar crystal" in which the macroscopic properties reflect the internal asymmetric molecular relationships (Curtin and Paul, 1981). Such a crystal may have external morphology with enantiomorphous (hemihedral) faces that have differing chemical reactivity and physical properties, such as the pyroelectric, piezoelectric, or ferroelectric effects, in addition to the optical properties we are discussing here. In addition to the ideal properties required to obtain high molecular hyperpolarizability, there is an optimal arrangement of those molecules in a crystal given by Eq. (10), which is the macroscopic equivalent of Eq. (1) (Bloembergen, 1965; Zernike and Midwinter, 1973). Here P is the polarization induced in a macroscopic medium under the influence of field E and the $\chi^{(n)}$ are the macroscopic coefficients that measure the projection of the microscopic molecular tensors on the unit cell:

$$P = P_0 + \chi^{(1)}E + \chi^{(2)}EE + \chi^{(3)}EEE + \cdots \tag{10}$$

Since the $\chi^{(n)}$ are tensor quantities, the unit-cell characteristics and the orientations of the molecules within the crystal lattice are of great importance if the molecular properties are to manifest themselves fully. Basic axioms predict 230 unique ways among which objects may become organized in three dimensional space [but see Shechtman *et al.*, (1984)]. The conventional 230 space groups are collected in 32 unique point groups and these in turn in seven unique lattice systems. For the molecules in an NLO SHG crystal, noncentrosymmetry is a prerequisite condition for nonvanishing $\chi^{(2)}$ (Nye, 1960); and this condition exists in 21 of the 32 three-dimensional point groups representing all of the seven systems. The appropriate optical tensor coefficients vanish in three of the 21 groups (422, 432, and 622) due to symmetry and Kleinman (1962) considerations. This leaves 18 useful point groups, which are biaxial (triclinic 1; monoclinic 2, *m*; orthorombic 222, *mm*2), uniaxial (trigonal 3, 32, 3*m*; tetragonal 4, 4*mm*, $\bar{4}$, $\bar{4}2m$; hexagonal 6, $\bar{6}$, 6*mm*, $\bar{6}m2$); and cubic (23, $\bar{4}3m$). The utility of these 18 surviving point groups is not equal if one invokes the need to phase match the crystal in some application. The point-group-dependent phase-matching conditions were derived by Midwinter and Warner (1965) for uniaxial systems and by Hobden (1967) for biaxial crystals. For straightforward device fabrication, it is desirable to have crystals that are noncritically phase-matchable (NCPM) since restrictions, such as acceptance angle and phase-matching bandwidth, are somewhat relaxed.

The important relationships between microscopic molecular hyperpolarizability and macroscopic crystal hyperpolarizability were alluded to by Shigorin and Shipulo (1975) and neatly formalized by Zyss and Oudar (1982) in a quantitative fashion. They found that even under the most ideal conditions (when the material crystallizes in point groups 1, 2, *m*, or *mm*2), only 38% of the molecular nonlinearity can ever contribute to the noncritical phase-matched macroscopic crystal nonlinear coefficient, and this maximum is obtained only when the molecular dipoles are oriented in a optimal direction θ_m relative to the optic axes of the crystal. (All the point groups will be subdivided by their maximal efficiencies in Table III.)

A comparison of the crystal packing of the three nitrobenzene derivatives 4-nitrodimethylaniline (NDMA, **11**) (Mak and Trotter, 1965) (Fig. 2), 3-acetamido-4-dimethylaminonitrobenzene (DAN, **12**) (Twieg and Jain, 1983) (Fig. 3), and *N*-(4-nitrophenyl)-(L)-prolinol (NPP, **10**) (Zyss *et al.*, 1984) (Fig. 4) is highly instructive. These three molecules are similar in their absorption and CT properties, all crystallize in space group $P2_1$ and yet their powder SHG efficiencies differ greatly (1, 115, and 150 × urea, respectively). NDMA is almost centrosymmetric and has $\theta = 89.89°$, DAN has $\theta = 70.84°$, and NPP has $\theta = 58.6°$, which is very close to the ideal (Zyss and Oudar, 1982) for this space group where $\theta_m = 54.74°$.

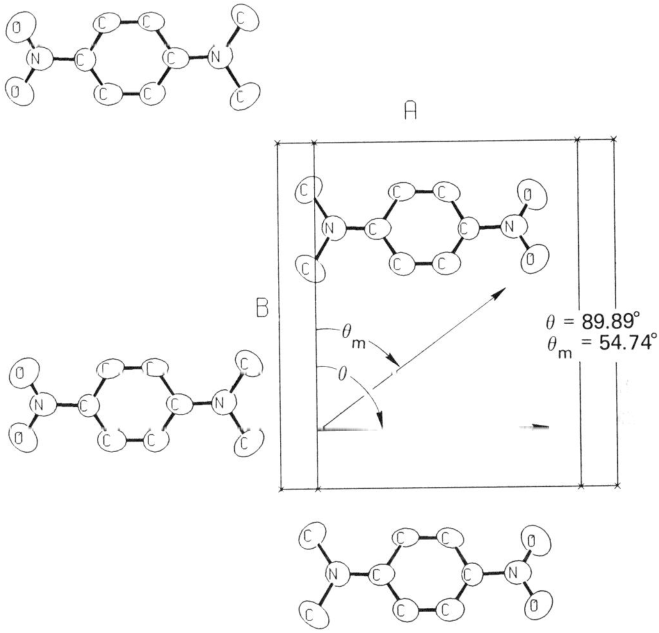

Fig. 2. ORTEP of 4-nitrodimethylaniline (NDMA, **11**). The molecules are almost centrosymmetrically disposed with molecular dipole angle θ nearly orthogonal to the B axis in the $P2_1$ space group.

3. Crystal Space Group Populations

The space groups that are ideal in an optical sense have just been described, and it is of interest to get an idea of how organic molecules distribute themselves among them. A number of studies of the distribution of organic compounds in the various space groups have been done, although it should be recognized that the "natural" distribution may never be obtainable since the nonobjective chemist or crystallographer always intervenes to select the compounds for structure determination. There are also cases in which too low a symmetry was determined for the crystal structure and the populations of more symmetric groups are underestimated. In spite of these shortcomings the analysis leads to some very valuable insights.

The available relevant data from studies of the distribution of organic compounds between centrosymmetric and noncentrosymmetric space groups and the most popular noncentrosymmetric space groups are found in Table II. The studies differ in the size of the basis set available at the time from an early study by Rez (1960) on piezoelectrics (all noncentrosymmetric point groups

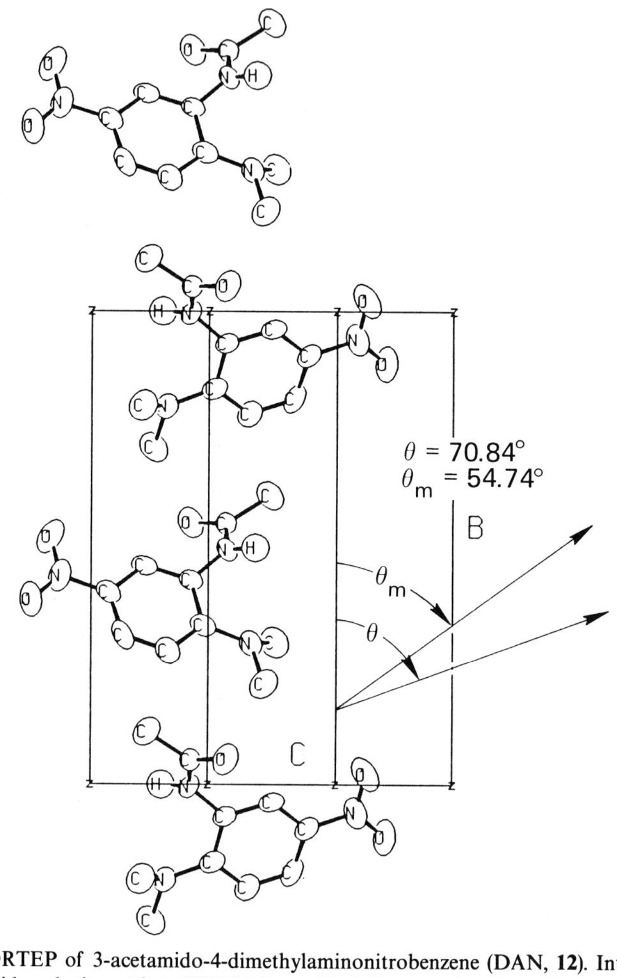

Fig. 3. ORTEP of 3-acetamido-4-dimethylaminonitrobenzene (DAN, **12**). Introduction of the 3-acetamido substituent into NDMA favorably influences the disposition of the molecular dipole angle θ relative to the B axis in the $P2_1$ space group.

except 432) through two later intermediate-size sets studied by Bel'skii and Zorkii (1971) and Curtin and Paul (1981) and finally two studies by Mighell *et al.* (1983) and Dirk and Twieg (unpublished results, 1985) on respective basis sets of over 29,000 organic compounds. The last two studies on contemporaneous NBS Files and Cambridge Files are the most comprehensive and, given their good agreement, are the best current source of data. It is of interest that the five most populated noncentrosymmetric space groups account for fully 90% of the total population of all the noncentrosymmetric

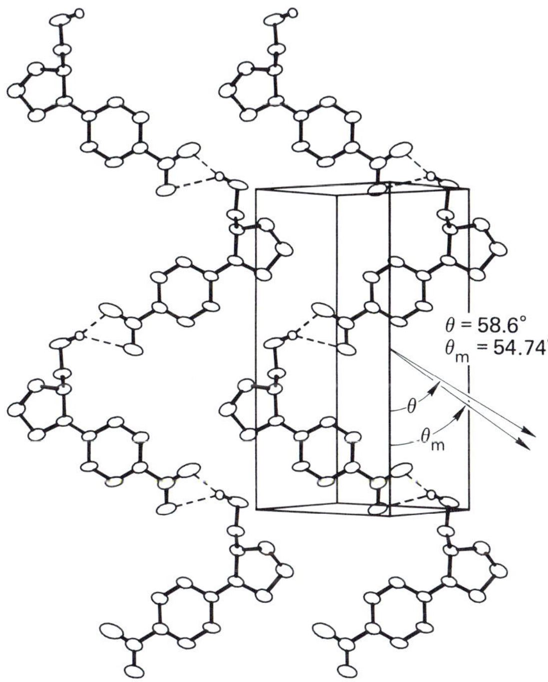

Fig. 4. ORTEP of *N*-(4-nitrophenyl)-(L)-prolinol (NPP, **10**). The molecules zig-zag about the (101) plane and are held head to tail by intermolecular hydrogen bonds (----) between the alcohol of one molecule and the nitro group of the next. The molecular dipole is disposed within 4° of the ideal alignment, $\theta_m = 54.7°$, for the $P2_1$ space group.

space groups. In addition to the data on populations, the Cambridge File contains a wealth of data (Allen *et al.*, 1983) useful for a preliminary screening and evaluation of these compounds prior to their actual preparation and NLO analysis.

The crystallographic data can be interpreted in a variety of ways, and an attempt to prioritize the point groups useful for crystals for SHG is presented in the "SHG crystal sieve" (Table III). Like a sieve, Table III attempts to separate out the most useful point groups first at the top, and the useless point groups fall through to the bottom. The sieving gradient is the optimal efficiency (38%, 25%, 19%, 0%) of the point group according to Zyss and Oudar (1982), and within each catagory the point groups are prioritized by population and a distinction is made to show which are the enantiomorphous space groups accessible by direct use of optically active enantiomeric compounds. An alternative and useful sieve might operate by distinguishing the populations as a first priority.

TABLE II

Comparison of Noncentrosymmetric Space Group Population Surveys of Organic Compounds

		Rez (1960)	Bel'skii and Zorkii (1971)	Curtin and Paul (1981)[a]	Mighell *et al.* (1981)	Dirk and Twieg (unpublished results, 1985)
Total number of of compounds studied		1,179	3,259	5,926	29,059	29,052
Percent noncentrosymmetric		42%[b]	>22%	~25%	29.6%	27.6%
Space groups		Percent of total noncentrosymmetric				
Number	Group					
19	$P2_12_12_1$	NA	59%	47%	39.0%	39.5%
4	$P2_1$	NA	28%	30%	22.7%	23.3%
33	*Pna* $Pna2_1$	NA	8%	6.5%	6.0%	5.8%
9	Cc Bb	NA	4%	NA	3.2%	3.2%
5	C2 B2	NA	NA	5.3%	3.2%	3.2%
18	$P2_12_12$	NA	NA	7%	2.2%	2.0%

[a] Data from Mighell and Ondik (1977); space group percent based on 1500 noncentrosymmetric compounds.
[b] Percent of compounds potentially piezoelectric.

The "SHG crystal sieve" clearly demonstrates a number of useful statistics about the compounds in the noncentrosymmetric groups. First, among all the noncentrosymmetric compounds, fully 90% are biaxial, only 9% are uniaxial, and a mere 1% are cubic. Second, all the 38% efficiency groups are biaxial, all the 25% efficiency groups are uniaxial, and the 19% efficiency groups are a mixture of the highly populated biaxial 222 group, a lesser number of compounds in uniaxial groups, and all of the cubic groups that have such a small population. By far the most outstanding point group is 2, which has the ultimate favorable attributes of being potentially 38% efficient, being enantiomorphous, and having a large population. It is certainly no accident that many of the most outstanding NLO organics among the large number of nitroaromatics studied fall into this group. The next best groups include *mm*2, with a good potential efficiency of 38% and a decent population without being enantiomorphous, and 222, which has a low maximal efficiency of 19% and no possibility of NCPM but has the attributes of the largest population and being enantiomorphous. It must be recognized that this prioritization is by no means intended to exclude any of the other noncentrosymmetric space groups, since each NLO material must be individually and ultimately judged on a variety of properties, of which the space group is just one.

Analysis of these surveys reveals some interesting differences between the crystallographic population distribution of organic and inorganic substances with important consequences on their relative utility for NLO applications. In general, organics tend to favor crystallization in low-symmetry space groups in an overall nonuniform distribution. On the other hand, inorganics favor high-symmetry groups, and the concentrations are overall more uniformly distributed (Mighell and Rodgers, 1980). The dichotomy in distribution uniformity between organics and inorganics is clearly evident in that the five most populated organic space groups contain fully 75% of the organic compounds, whereas the five most populated inorganic groups contain only some 31% of all the inorganic compounds. Among the 32 most populated organic space groups (which contain 95% of all structures reported), 14 are noncentrosymmetric, and most of these are in the 38% efficiency class. For the comparable 32 most populated inorganic space groups (which contain only 72% of all inorganic materials reported), only four are noncentrosymmetric, and these are in the 19% efficiency class. The low-symmetry point group priority for organics versus the high-symmetry point group priority for inorganics is clearly seen from the summary at the bottom of Table IV. Here, in the center column, are listed the point groups with the total number of space groups each contains, in order of symmetry. For example, the low-symmetry triclinic point group has two possible space groups and the high-symmetry cubic point group has 36 possible space groups. To the left and right of this center column, respectively, are listed the actual numbers of inorganic and

TABLE III
SHG Crystal Sieve[a,b]

Max. eff.	Biaxial 7738, 90.0%			Uniaxial 807, 9.3%				Totals in phase-matching % groups 8598
Point group population totals	Triclinic 305, 3.5%	Monoclinic 2643, 30.8%	Orthorombic 4790, 55.7%	Trigonal 217, 2.5%	Tetragonal 467, 5.4%	Hexagonal 123, 1.4%	Cubic 62, 0.7%	
38%	1 (1)305	2 (3–5)2241 m (6–9)402	*mm*2 (25–46)1125					2546E
Subtotal	305	2643	1125	—	—			4073, 47.4%
25%				3 (143–146)81 32 (149–155)63 3*m* (156–161)73		$\bar{6}$ (174)1 $\bar{6}m2$ (187–190)9		144E

Subtotal	—	—	—	217	—	10		227, 2.6%
19%			222 (16–24)3665		4 (75–80)79 422 (89–98)166 $4mm$ (99–110)26 $\bar{4}2m$ (111–122)130 $\bar{4}$ (81–82)66	6 (168–173)68 622 (177–182)20 $6mm$ (183–186)16	23 (195–199)18 432 (207–214)6 $\bar{4}3m$ (215–220)38	4022E
Subtotal	—	—	3665	—	467	104	62	4298, 50.0%
0%	$\bar{1}$	$2/m$	mmm	$\bar{3}m$ $\bar{3}$	$4/mmm$ $4/m$	6 $6/m$	m^3m m^3	

[a] Data from Mighell *et al.* (1983).

[b] Example:
3 — 3 point group
(143–146)81 — (143–146) space groups; 81 population
Enantiomorphous point groups at left of columns.
Total E(enantiomorphous) 6712, 78.0%.

TABLE IV

The 32 Most Populated Inorganic and Organic Space Groups

Inorganic[a]					Organic[b]				
Number	Group	Form	%	PM%[d]	Number	Group	Form	%	PM%[d]
225	$Fm3m$	Cubic	8.5	—	14	$P2_1/c$	Mono	35.9	—
62	$Pnma$	Ortho	6.8	—	2	$P\bar{1}$	Tri	13.7	—
227	$Fd3m$	Cubic	5.5	—	19	$P2_12_12_1$	Ortho	11.6	19
194	$P6_3/mmc$	Hex	5.3	—	4	$P2_1$	Mono	6.7	38
14	$P2_1/c$	Mono	4.9	—	15	$C2/c$	Mono	6.6	—
221	$Pm3m$	Cubic	4.8	—	61	$Pbca$	Ortho	4.3	—
166	$R\bar{3}m$	Trig	3.7	—	62	$Pnma$	Ortho	1.9	—
12	$C2/m$	Mono	2.5	—	33	$Pna2_1$	Ortho	1.8	38
15	$C2/c$	Mono	2.5	—	60	$Pbcn$	Ortho	1.2	—
63	$Cmcm$	Ortho	2.2	—	1	$P1$	Tri	1.0	38
148	$R\bar{3}$	Trig	1.9	—	9	Cc	Mono	0.9	38
164	$P\bar{3}m1$	Trig	1.9	—	5	$C2$	Mono	0.9	38
191	$P6/mmm$	Hex	1.8	—	29	$Pca2_1$	Ortho	0.8	38
139	$I4/mmm$	Tet	1.8	—	11	$P2_1/m$	Mono	0.8	—
2	$P\bar{1}$	Tri	1.7	—	12	$C2/m$	Mono	0.6	—
129	$P4/nmm$	Tet	1.3	—	18	$P2_12_12$	Ortho	0.6	19
141	$I4_1/amd$	Tet	1.3	—	13	$P2/c$	Mono	0.5	—
216	$F\bar{4}3m$	Cubic	1.2	19	148	$R\bar{3}$	Trig	0.4	—
205	$Pa3$	Cubic	1.2	—	43	$Fdd2$	Ortho	0.4	38
88	$I4_1/a$	Tet	1.1	—	7	Pc	Mono	0.3	38

230	*I23d*	Cubic	1.1	—	56	*Pccn*	Ortho	0.3	—
11	$P2_1/m$	Mono	1.0	—	92	$P4_12_12$	Tet	0.3	19
60	*Pbcn*	Ortho	1.0	—	88	$I4_1/a$	Tet	0.3	—
136	$P4_2/mnm$	Tet	0.9	—	64	*Cmca*	Ortho	0.3	—
167	$R\bar{3}c$	Tet	0.9	—	20	$C222_1$	Ortho	0.3	—
186	$P6_3mc$	Hex	0.9	19	176	$P6_3/m$	Hex	0.3	—
176	$P6_3/m$	Hex	0.7	—	114	$P\bar{4}2_1c$	Tet	0.2	19
140	*I4/mcm*	Tet	0.7	—	57	*Pbcm*	Ortho	0.2	—
61	*Pbca*	Ortho	0.7	—	82	$I\bar{4}$	Tet	0.2	[c]
193	$P6_3/mcm$	Hex	0.7	—	63	*Cmcm*	Ortho	0.2	—
19	$P2_12_12_1$	Ortho	0.6	19	36	$Cmc2_1$	Ortho	0.2	38
220	$I\bar{4}3d$	Cubic	0.6	19	86	$P4_2/n$	Tet	0.2	—
Totals			71.7					95.0	
				1	Tri (2)	2			
				4	Mono (13)	9			
				5	Ortho (59)	14			
				7	Tet (68)	5			
				3	Trig (25)	1			
				5	Hex (27)	1			
				7	Cubic (36)	0			

[a] Mighell and Rodgers (1980), Fig. 4.
[b] Mighell *et al.* (1983).
[c] Not NCPM.
[d] Maximum NCPM (Zyss and Oudar, 1982).

organic space groups appearing in the body of Table IV. There is a clear population trend in the inorganics to favor the high-symmetry groups and an equally clear but contrary trend in the organics to favor the low-symmetry groups.

Hence, it appears that organics have a much greater tendency to crystallize in space groups useful for SHG than their inorganic counterparts. This difference in abundance may reflect the very fundamentally different forces that influence the crystal packing of organics and inorganics. It undoubtedly also reflects the fact that many organic compounds that have been studied were naturally occurring, optically active, and therefore in enantiomorphous and noncentrosymmetric space groups. As will be seen shortly, this latter fact has been exploited in the molecular engineering of NLO organic compounds.

III. MOLECULAR ENGINEERING

A. Definition

Molecular engineering has been defined as "the planned synthesis of materials having prescribed bulk properties" (Le Fèvre, 1970). In the present context, the molecular engineering of nonlinear molecular crystals, this definition may be refined as the planned synthesis of materials providing the simultaneous optimization of molecular and crystal properties responsible for efficient SHG. The important molecular and crystal properties that provide efficient SHG have been described, and each in its own right is a challange to optimize. Unfortunately, as one modifies the molecular properties it becomes more difficult to simultaneously obtain the desirable absorption and crystal properties. The most important conflict arises between optimization of molecular hyperpolarizability and the transparency of the molecular crystal that results. The second conflict involves the optimization of molecular hyperpolarizability and the ability of such molecules to pack in the crystal in a useful fashion. The resolution of these trade-offs is presently the greatest challenge in molecular engineering.

B. Trade-Offs and Conflicts Involved in Optimization

1. The Efficiency–Transparency Trade-Off

The optimal molecular and crystal properties have now been discussed in some detail and the importance of charge transfer to molecular hyperpolarizability is clear. A trade-off that occurs in conjugated and polar molecules possessing the CT characteristics needed for large hyperpolarizability is an almost inevitable loss of transparency. There are a variety of equations that relate the frequency shift of the $^1A-^1L_a$ CT band due to interacting substituents (Platt, 1964; Cowles, 1986), including Eq. (11), which has been derived by Stevenson (1965). Here $\Delta\nu$ is the shift in λ_{max} in the substituted compound relative to benzene, K is an attenuation factor to compensate for the number of substituents, the first l' term on the right in m sums the substituent "monoshifts" that each substituent alone contributes, and the second term in x provides the mesomeric interactive contribution of any para-disposed substituents. An examination of the appropriate l' and x values (Table I) indicates that the best donors and acceptors, including amino and nitro groups, give the greatest shift in the CT band. Note that Eq. (11) separates the monofunctional and difunctional contributions to absorbance, just as Eq. (7) does for the contributions to hyperpolarizability:

$$\Delta\nu = K\left(\sum_m l'_m\right) + (x_0 - x_3)^2 \tag{11}$$

Let us now put what we know to work and attempt to design a molecule that will provide molecular crystals suitable for doubling a high-power YAG laser from the fundamental at 1.06 to 0.53 μm. In order to utilize preresonance effects, it is useful to take advantage of dispersion and maximize β by operating near the absorption cutoff λ_c. However, for practical purposes the crystal should not absorb at either fundamental or harmonic wavelengths, otherwise damage is more likely to occur. Let us then attempt to "tune" the chromophore of the molecule and hopefully retain a large part of the hyperpolarizability in the process. Since the 4-nitroaniline class of molecules is currently the best understood, it will be used as a model for our discussion, but the principles involved are relevant to many other molecules. In qualitative valence-bond (VB) theory the ground state of a molecule is assumed to resemble that of the most stable valence structure, whereas the first excited state resembles some less stable charge-separated structures. In the case of polar aromatics these two states **6** and **6a** are depicted in Fig. 5 (some prefer resonance form **6b** over **6a**). The interaction of the donor D with the acceptor A through the aromatic ring produces a ground-state wave function Ψ_g, which is the combination of the "no-bond" wave function Ψ_{DA} and a "charge-transfer" wavefunction $\Psi_{D^+A^-}$ in which the electron density has moved from the donor to acceptor as given in Eq. (12) for the ground state and in Eq. (13) for the excited state. The coefficients α, β, γ, and δ [not related to the (hyper)polarizability coefficients] partition the contributions of no-bond and charge-transfer wave functions:

$$\Psi_g = \alpha\Psi_{DA} + \beta\Psi_{D^+A^-} \tag{12}$$

$$\Psi_{CT} = \gamma\Psi_{DA} + \delta\Psi_{D^+A^-} \tag{13}$$

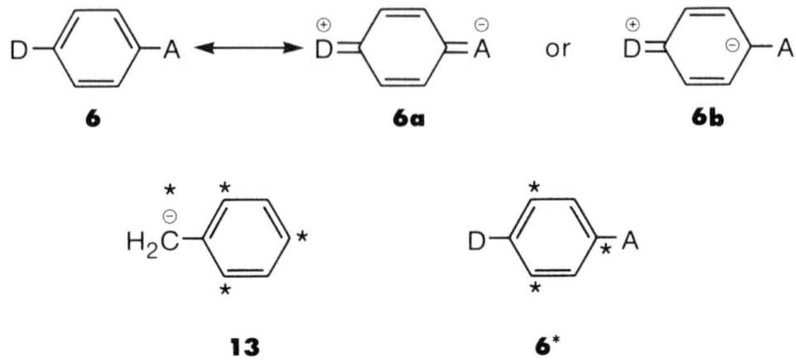

Fig. 5. Charge separation and Dewar's starring convention for D–A-substituted benzene. The D and A groups as in **6** produce a ground-state charge separation, which may be represented by either **6a** or **6b**. The starring system as in **6*** is based on that for the benzyl anion **13**.

This VB treatment has the following two consequences in designing molecules for nonlinear optical application. First, the absorption maximum λ_{max}, and to a lesser extent the absorption cut-off λ_c, of the intramolecular charge-transfer band are related to the difference in energy between the Ψ_g and Ψ_{CT} states. Second, in general, if $\alpha > \beta$ then $\gamma < \delta$, which leads to a corresponding difference in the dipole moments of these states such that $\mu_e > \mu_g$. The effect of substituents and heteroatoms on the energies of these states and the resultant chromophore of the molecule can be predicted by Dewar's rules, which have found wide utility in explaining the effects of such modifications on the chromophores of dyes (Dewar, 1950; Gordon and Gregory, 1983). Following Dewar's convention, the benzylic anion (**13**) serves as a model and the atoms in the conjugated backbone of an even alternant molecule such as the disubstituted benzene (**6***) are paired as "starred" or "unstarred" by their positions relative to a terminal donor group which is defined as a starred position (Fig. 5). The relevance of Dewar's rules is illustrated by the substituent and heteroatom substitution of nitroaniline and resultant changes in the absorption λ_m of the resultant molecules:

(1) The placement of an electron-withdrawing substituent A at a starred position or any electron-releasing D substituent at an unstarred position will produce a large hypsochromic (short-wavelength) shift.

(2) The placement of an electron-withdrawing A substituent at an unstarred position or an electron-releasing D substituent at a starred position will produce a smaller bathochromic (long-wavelength) shift.

(3) The replacement of carbon in the conjugated molecular backbone by a more electronegative atom has the same effect as an electron-withdrawing A substituent at that position. This replacement must not alter the total number of π electrons, and only a terminal carbon in the conjugated backbone can be replaced by any heteroatom other than nitrogen.

Dewar's rules fail to explain the intense bathochromic shift that occurs between aniline and *p*-nitroaniline, in which the electron-withdrawing nitro group is placed at a starred position. However, this "mutual conjugation" effect (Dewar, 1969) can be explained by VB theory in which the charge-separated excited state is stabilized. Using either **2** or 4-(3-hydroxypropylamino)nitrobenzene (**17**) as parent systems, examination of Table V reveals that when nitrogen atoms are introduced into a starred position of **6*** to produce the pyridine (**20**) or pyrimidine (**21**) analog, a hypsochromic shift of about 30 nm/N atom occurs. The same hypsochromic shift is obtained when an electron-withdrawing A group such as nitro in **14**, nitrile in **23**, or trifluoromethyl in **18** is introduced at the starred positions, but a bathochromic shift is obtained if the trifluoromethyl group is introduced into an unstarred position as in **19**. As expected, the introduction of a electron-releasing D group, such as an amine in **15** or ether in **22**, at the starred position gives a bathochromic shift, while introduction of the amino group at an unstarred position as in **16** gives a hypsochromic shift.

TABLE V

Influence of Substituents on Chromophore[a]

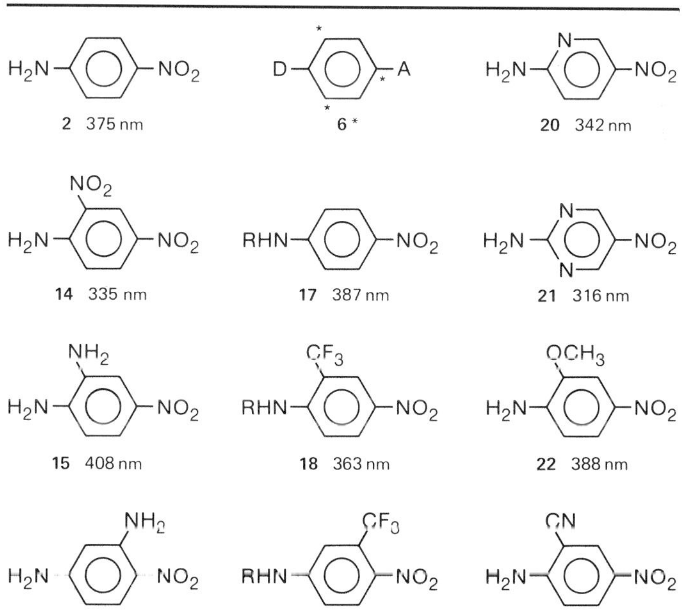

2 375 nm; **6***; **20** 342 nm

14 335 nm; **17** 387 nm; **21** 316 nm

15 408 nm; **18** 363 nm; **22** 388 nm

16 365 nm; **19** 413 nm; **23** 346 nm

[a] λ_{max}(EtOH), R = $CH_2CH_2CH_2OH$ (Nicoud, unpublished results, 1985).

The application of Dewar's rules to the molecular engineering of a more conjugated and polarizable system is also of interest. An ideal candidate for this purpose is the donor–acceptor-substituted stilbene 4-dimethylamino-4′-nitrostilbene (DANS, **24**), which has been quite extensively studied by the EFISH technique (Oudar, 1977a). The DANS analogs (**25**, **26**) in which a nitrogen replaces a carbon in the central stilbene double bond (benzylidene anilines), are known compounds (Skrabal *et al.*, 1975), but their hyperpolarizability properties remain to be determined. Examination of the absorption data indicate that Dewar's predictions are correct: the nitrogen at the starred position produces a hypsochromic shift while substitution at the unstarred position produces the corresponding bathochromic shift. Based on our confidence in Dewar's rules, it is reasonable to expect that a multiply heteroatom-substituted system such as the stilbene derivative **27** with five appropriately placed nitrogens will have a maximal hypsochromic shift (Fig. 6)

The absorption maximum (and to a lesser extent the absorption edge) can be manipulated at will by the propitious use of heteroatoms and substituents, but the direction and extent of any simultaneous effect on the hyperpolarizability of such modified molecules cannot be ignored. Although little quantitative data is presently available, it is clear that the molecular modification intended to produce a hypsochromic shift in absorption will exact its toll on the hyperpolarizability available from the molecule as well. What is not presently clear and what requires significant future experimental and theoretical examination is the extent of this trade-off. An initial effort to compare λ_{max} to $\mu_0\beta$ has been done by Dulcic and Sauteret (1978) on simple D-A-substituted aromatics, and further refinement and extension of this work is needed. The

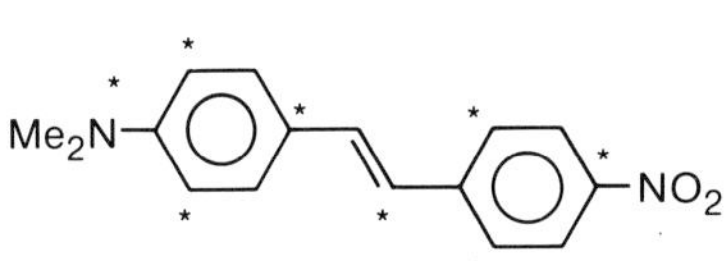

24 (420 nm, ϵ = 33, 00, cyclohexane)

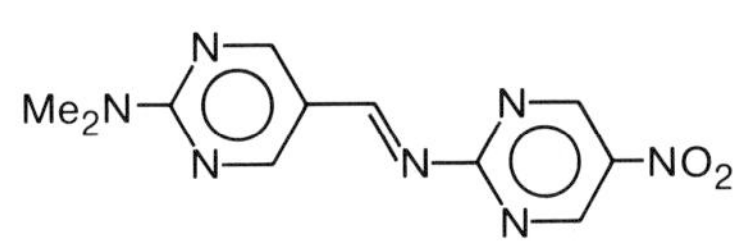

27

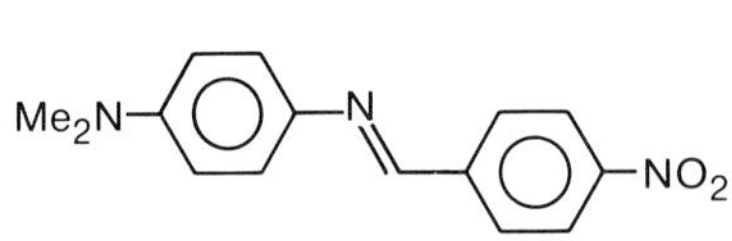

25 (438 nm, ϵ = 20,300, cyclohexane)

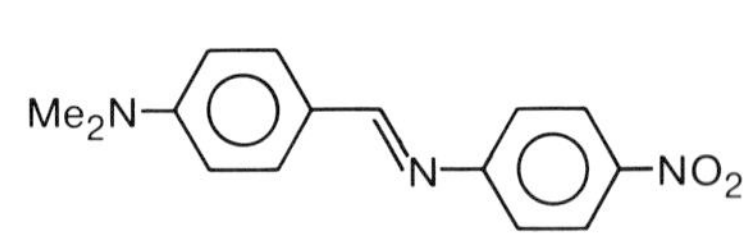

26 (387 nm, ϵ = 29,100, cyclohexane)

Fig. 6. 4-Dimethylamino-4′-nitrostilbene (DANS, **24**) molecule, with starring convention and effects of aza substitution on chromophore. Absorption data from Güsten and Salzwedel (1967) and Skrabal *et al.* (1975).

extensive and complicated synthesis involved in the molecular engineering of the DANS molecule to azasubstituted analogs such as **27** or **34** is a fruitless exercise if one finally ends up with a highly sophisticated molecule with the same absorption and hyperpolarizability of a mononuclear nitroaniline derivative that is much easier to prepare!

Presently the only consistant set of information comes from calculations. Papadopoulos and Waite (1985) found that the substitution of nitrogen for carbon in benzene leads to a diminuition of both α and γ for the parent heterocycles (pyridine, pyridizine, etc.), and the same trend is certainly true for β. This effect is easily rationalized on the basis of the lower atomic polarizability of nitrogen and/or its greater electronegativity. A series of PPP-SCF-Cl calculations on a set of donor acceptor substituted systems (**2**, **20**, **21**, **28**–**34**) is found in Table VI (Dirk, Twieg, and Wagnière, unpublished results, 1985). The clear prediction is that the hyperpolarizability is affected not only by the number of substituents but by their location is accord with the application of Dewar's rules. Further refined calculations based on crystallographic coordinates and the EFISH data on this set of molecules (and similar sets) need to be pursued.

2. Dipole and Multipole Repulsion

The structural changes that lead to large β also very often produce an increase in ground-state dipole moment that accompanies the bathochromic shift in the chromophores just discussed. It is clear that as the molecular ground-state dipoles become larger, the electrostatic interaction between adjacent molecules increases and the net molecular dipole alignment required to achieve the maximum crystal anisotropy becomes more energetically unfavorable. What is not yet adequately understood is a quantitative sense of the magnitude of the impact this has on the crystal structures of typical organic NLO materials. Kitaigorodsky (1973) and Kitaigorodsky and Mirskaya (1965) developed a calculation to evaluate the contribution of the dipolar energy U_{d} to the total lattice energy U_{tot} and found that the intermolecular dipolar contribution to the lattice energy U_{d} is given by Eq. (14), in which the point dipoles with moments μ are vectors r_i with directional components s_i. Hence, different space groups will have differing intrinsic abilities to tolerate the dipole moment alignments that are required in NLO polar crystals:

$$U_{\mathrm{d}} = U'\frac{\mu^2}{a^3} = \frac{\mu^2}{2}\sum_i f(s_i, r_i) \qquad (14)$$

Although maximum and minimum values of U' (related to U_{d} by packing coefficients determined by the size and shape of the molecule) are available for

TABLE VI

Wavelength and Hyperpolarizability Trade-Offs in Heterocyclic Analogs of Donor–Acceptor Benzene and Stilbene

Number	Compound	λ_{max} Exp.	λ_{max} Calc.	μ_0 Exp.	μ_0 Calc.	β Exp.	β Calc.
2	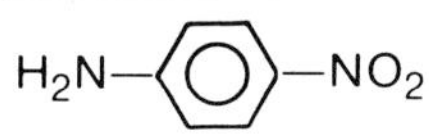	325	325	6.8	8.2	34.5	35.3
20	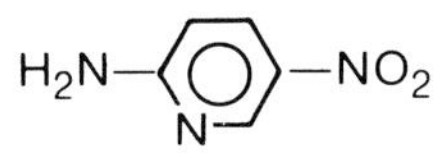	312	310	6.3	8.2	—	33.9
28	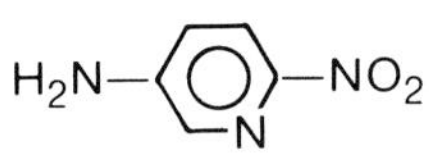	—	355	—	8.7	—	46.7
21	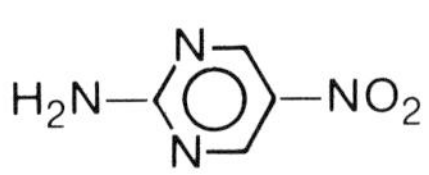	—	283	—	7.2	—	25.6
29	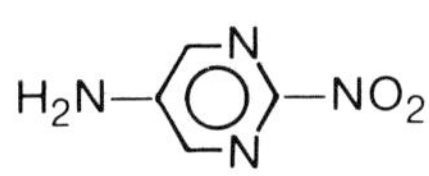	—	354	—	8.2	—	47.6
30	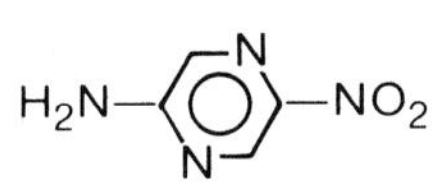	—	349	—	9.5	—	34.8
31	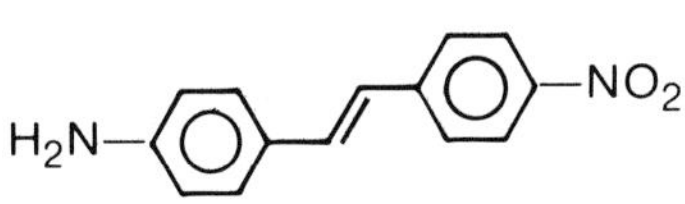	—	384	—	10.0	—	248.0
32	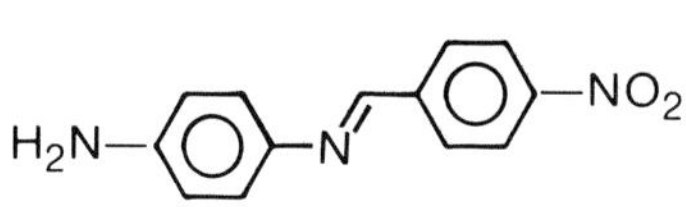	—	410	—	6.0	—	371.0
33	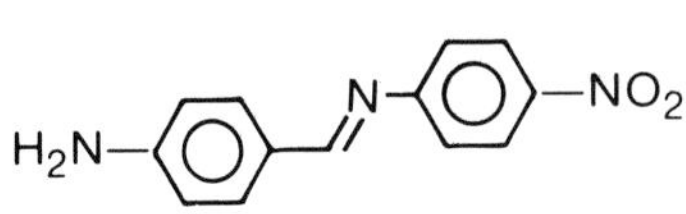	—	337	—	13.3	—	77.0
34	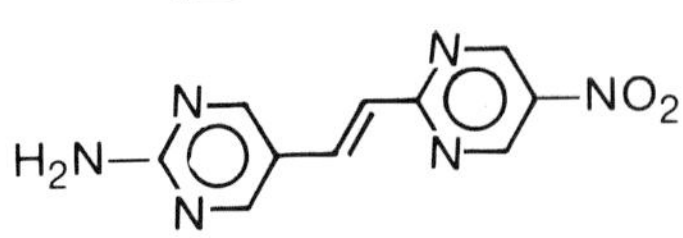	—	335	—	9.4	—	106.0

a few space groups, it appears that there have been few efforts to apply Kitaigorodsky's techniques to any useful quantity of specific molecules. Zorkii *et al.* (1971) have treated a few molecules such as acenaphthequinone and succinic anhydride in detail and conclude that in these cases electrostatic interactions (related to but not directly proportional to μ) play an essential role in the determination of their ultimate crystal class. All that can be said at this time, to paraphrase Kitaigorodsky, is that superposition of the U_d energy surface on the U_n energy surface will produce only slight displacement of minima or an exchange of minima and that for high moments the most stable structure is not obliged to correspond to the lowest minimum in U_n. The crystallographic data that has now accumulated on the NLO organic compounds is an excellent source that should be exercised to better understand these molecular multipolar interactions and to simultaneously learn more about the specific NLO crystals themselves.

C. The Practice of Molecular Engineering

1. The Utility of Optical Activity and Enantiomers

It is very fortunate that molecular chirality provides the chemist with a means to guarantee that crystallization of a pure enantiomer will occur in one of the 11 enantiomorphous point groups that are a subset of the 21 noncentrosymmetric point groups (Henry and Lonsdale, 1969; Schudt and Weitz, 1971). The theoretical aspects and practical applications of chirality and related optical activity in organic molecules has been developed very thoroughly since the pioneering work of Pasteur (Jacques *et al.*, 1981; O'Loane, 1980). A molecule with one or more inverse symmetry elements (center, plane, improper axis) is achiral, but a molecule with none of these symmetry elements is chiral and can exist in two mirror-image enantiomorphous forms. Depending on the system of nomenclature, the pure enantiomers are designated as *d* or *l*, + or − (relating to the sign of optical rotation), or *R* or *S*, which specify the absolute configuration of each chiral center; the chiral center in such a molecule is often designated with an asterisk (not to be confused with Dewar's starred atoms). Nature provides a great number of optically active substances that can be utilized directly (amino acids and carbohydrates) or in a synthetically modified form (amino alcohols) (Kuhn and Martin, 1962; Szazbo and Lee, 1980; Morrison, 1983; Scott, 1984; Solladié and Zimmermann, 1984). These pure enantiomers, in turn, may be utilized for the optical resolution of a wide variety of other racemic substances by the process of resolution, which involves the separation of diastereomeric mixtures, taking advantage of some difference in physical property (usually solubility) (Jacques *et al.*, 1981; Wilen, 1972; Newman, 1978, 1981, 1984).

Achiral molecules generally crystallize in a centric crystal class but occasionally crystallize in a noncentrosymmetric class [e.g., urea (**1**), benzil (**35**), *m*-nitroaniline (mNA, **36**), 2-methyl-4-nitroaniline (MNA, **37**), 3-methyl-4-nitropyridine *N*-oxide (POM, **38**), and DAN (**12**)]. A pure enantiomer must crystallize in an enantiomorphous space group [e.g., MAP (**9**), 2-(α-methylbenzylamino)-5-nitropyridine (MBANP, **39**), NPP (**10**)], but a mixture of equal quantities of both enantiomers, a racemate, almost invariably crystallizes in point groups with inverse symmetry elements but may in rare cases crystallize in an enantiomorphous space group. Racemates fall into three separate classes, which must be distinguished in order to correctly interpret the SHG powder efficiency test. The first class is a conglomerate that contains a mechanical mixture of crystals of the pure enantiomers; the second is a racemic compound that contains both enantiomers in some ordered configuration; and the third is a pseudo-racemate in which the enantiomers coexist in an unordered fashion in the crystal.

The utility of chirality for NLO organic materials was recognized very early (Rieckhoff and Peticolas, 1965) for amino acids and sugars and was extended to a variety of other readily available and naturally occurring substances, including, among others, steroids (cholesterol) and terpenoids (menthol) (Kurtz and Perry, 1968). These common molecules, of course, do not contain the necessary conjugation and polar functionality required for large (hyper)polarizability. The important relevant advances for NLO materials came with the recognition that molecules containing the strongly polarizable groups could be rendered optically active by attachment of a "chiral auxiliary." This was done for a number of optically active thiourea derivatives (Davydov *et al.*, 1970) and for the 2,4-dinitroaniline derivative MAP (**9**), in which the natural, optically active amino acid (S)-(+)-alanine was employed as its methyl ester (Oudar, 1977a,b). The great value of this technique has been amply demonstrated in numerous other molecules, of which many to be shown in Table XVI are exemplary. To date, only the chirality of the tetrahedral carbon has been utilized in this fashion; the five other general chirality types (asymmetric heteroatom, asymmetric octahedral site, restricted rotation of dissymmetric planes, helical dissymmetry, and dissymetric knots and braids) remain to be exploited (March, 1977; O'Loane, 1980). Future application of these other chirality classes are certain to provide useful NLO materials with novel structures.

The fact that a molecule is optically active does not alone guarantee that it will be a useful material for SHG, since the necessary polarizable functionality must still be present in the appropriate alignment (see the disparate results for the nitrobenzene derivatives in Table XV). The chiral auxiliary dilutes the polarizable charge transfer portion of the molecule in the crystal volume, and the synthetic situation can be complicated by enantiomeric purity issues. In spite of these complications, the utilization of optical activity has been the single most successful molecular engineering technique.

2. Hydrogen Bonding

According to simple valence theory, a hydrogen atom should be capable of forming only one chemical bond. However, in certain cases an electron-deficient hydrogen attains a higher valency when it interacts with another region of higher electron density (Kollman and Allen, 1972). The hydrogen bond bridges two atoms that have higher electronegativity (such as O, N) than hydrogen, and this can have an enormous influence on a variety of physical and spectroscopic molecular properties.

The hydrogen bonds most often found in NLO organics involve an interaction between a hydrogen bound to sp^3 nitrogen or oxygen (amine, alcohol) and an oxygen atom with more s-bond character (nitro, carbonyl), although more symmetrical interactions also occur. The energies of these bonds are 3–6 kcal/mol (12 kcal/mol for a carboxylic acid dimer) and can make a significant contribution to the total bonding energy in a crystal and a particularly valuable contribution if U_H can assist to "neutralize" a destabilizing U_m interaction (Kitaigorodsky, 1973). Hydrogen bonds may be intramolecular or intermolecular, and the interactions may be two-center, three-center, or even "cooperative," in which case chains or networks may arise (Taylor and Kennard, 1984).

An example of an intramolecular hydrogen bond is found in the molecule 2-(α-methylbenzylamino)-3,5-dinitropyridine (MBADNP) (**40**) (Twieg *et al.*, 1982b), in which the NH forms a six-member ring with the oxygen atom of the adjacent nitro group (Fig. 7). This internal hydrogen bond is common in *o*-nitroaniline and its derivatives and in other *o*-aminonitroaromatics such as the NLO material MAP (**9**) (Knossow *et al.*, 1976). The ORTEP of DAN (**12**) clearly shows the series of hydrogen bonds that hold the adjacent molecules together and maintain the required, yet energetically unfavorable, dipole alignment (Fig. 8). Here the hydrogen bonds involve a hydrogen on an acetamido nitrogen with the acetamido carbonyl in an adjacent molecule in an infinite chain in the crystal lattice. Urea (**1**) is an extreme case of a hydrogen-bonding network (Fig. 9). The urea molecule has only eight atoms, yet each molecule has eight hydrogen bonds (four to each carbonyl and one from each hydrogen).

The profound influences of hydrogen bonding on crystal structure are clear, but the ramifications of the presence of these bonds on the hyperpolarizabilities of a molecule or set of molecules are not presently well understood. The effects will be different depending on whether the hydrogen bond is

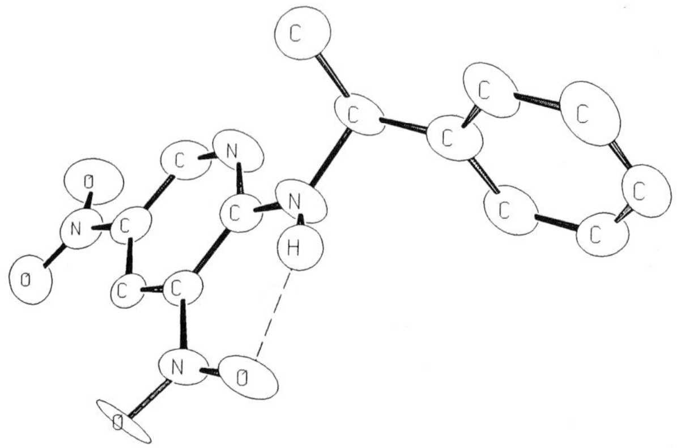

Fig. 7. ORTEP of 2-*N*-(α-methylbenzyamino)-5-nitropyridine (MBADNP, **40**). The intramolecular hydrogen bond (----) is between the adjacent amine and nitro groups.

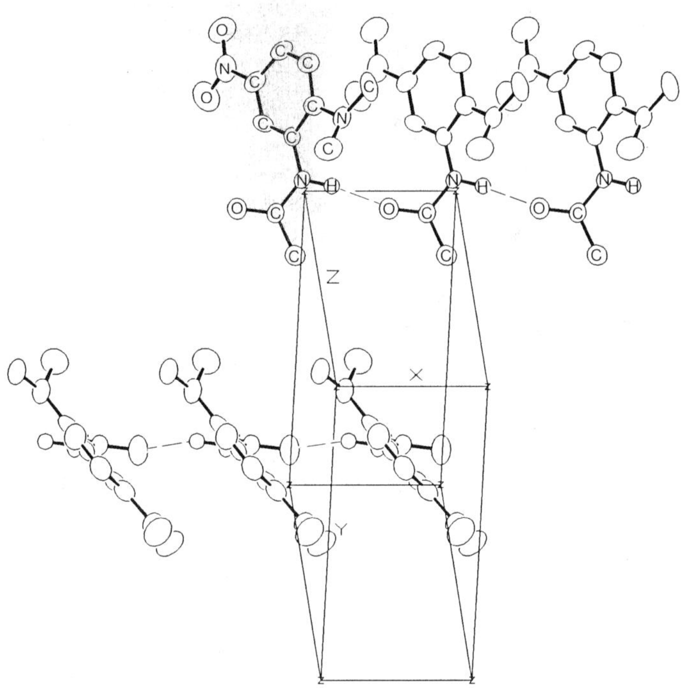

Fig. 8. ORTEP of 3-acetamido-4-dimethylaminonitrobenzene (DAN, **12**). The intermolecular hydrogen bonds (----) are between the acetamido nitrogen in one molecule and the acetamido oxygen of an adjacent molecule, forming an infinite chain in the crystal lattice.

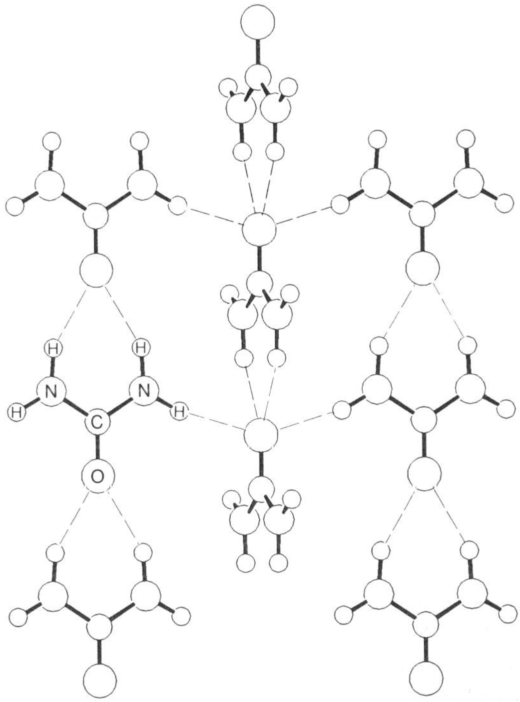

Fig. 9. ORTEP of urea (**1**). An intermolecular hydrogen bond (----) network exists between the carbonyl of one molecule and the nitrogens of three adjacent molecules.

intramolecular or intermolecular, and in the latter case on the specific extent and mode of interaction. The hydrogen bond undoubtedly influences the local electronic situation, as manifested by perturbed infrared and ultraviolet (UV) absorption properties, while the overall larger-scale electronic effect is unclear. In urea the hydrogen bonding is felt to be beneficial and to enhance the hyperpolarizability (Zyss and Berthier, 1982), but it is presently dangerous to generalize what the effects of hydrogen bonding are except on a case-to-case basis.

3. *Molecular Substituents and Steric Effects*

A large part of the extraordinary diversity of organic molecules involves the response of their physical properties to substituent modification. Among the modifications discussed thus far are the use of optically active substituents to control crystal point group, the use of hydrogen-bonding substituents to influence molecular dipole alignment, and the use of substituents to electronically modify the molecular chromophore. Some more subtle features about substituents that are often overlooked are a host of effects that involve their space-filling or steric properties, and these also can have a large impact on the chromophore, dipole moment, and ultimately the (hyper)polarizability of the molecule. The most important of these involves steric inhibition of mesomerism, which will influence all these properties (Table VII). One example is the reduction of the dipole moment from 6.93 to 4.11 in going from 4-nitrodimethylaniline (**11**) to 4-nitrodimethylaminodurene (**41**), demonstrating that optimal mesomerism (and hence optimal β) requires coplanar disposition of D and A groups (Ingham and Hampson, 1939). The importance of such steric effects was also pointed out by Sutton *et al.* (1951) for linear polarizability. Another example involves the change in the chromophore of 4-nitropyridine *N*-oxide (**42**) as methyl groups are introduced and influence the interaction of the nitro group. POM (**38**) is the derivative with one methyl group, and the derivative with two ortho methyl groups present (**43**) has only a very weak CT band, with hyperpolarizability greatly diminished. Hence, in a situation in which the maximum hyperpolarizability is important the mesomeric partners must be free to interact as efficiently as possible

TABLE VII

Stereoelectronic Effects

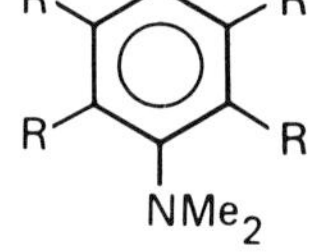

Number	R	μ_g [a]
11	H	6.87
41	Me	4.11

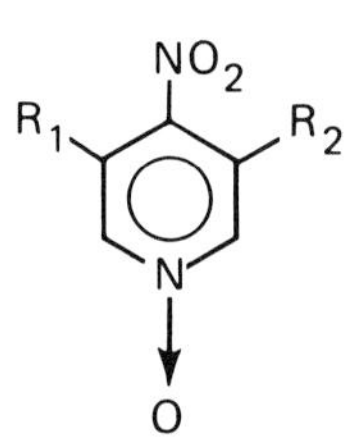

Number	R_1	R_2	CT band [b] λ_m (nm)	CT band [b] ε	θ [c]
42	H	H	346	15,900	0°
38	H	Me	344	15,300	11°
43	Me	Me	None		55°

[a] Ingham and Hampson (1939).
[b] Yamakawa *et al.* (1974).
[c] Estimated nitropyridine torsion angle.

(Table VII). On the other hand, it is possible to tune the cut-off to some extent, with a concomitant change in hyperpolarizability, by the introduction of substituents (Twieg and Jain, 1983).

As will be seen, the introduction of substituents can also grossly change the crystal packing of molecules and when properly understood can be used to advantage. In attempting to correlate molecular structure with crystal class, one seeks threads of continuity in order to synthesize SHG-active analogs (Table VIII). For example, the MNA family the methyl group can be replaced with halogens to give the chloro (**44**) and bromo (**45**) analogs, which occupy a similar volume, and noncentrosymmetric crystal structure is retained (Koreneva *et al.*, 1975). The same is true for substitutions of these halogens in

TABLE VIII

Isostructural Continuity

	Number	R	Efficiency ($\times U$)[a]	Reference
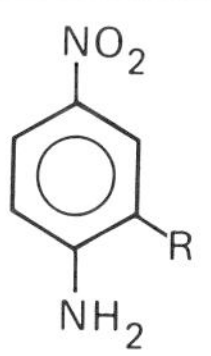	**37**	Me	20–50	Koreneva *et al.* (1975)
	44	Cl	20–40	Shigorin and Shipulo (1975)
	45	Br	4	Shigorin and Shipulo (1975)
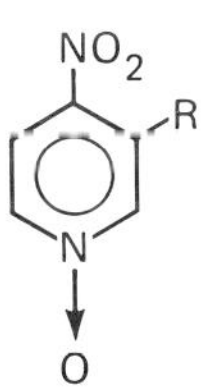	**38**	Me	13 (High)	Zyss *et al.* (1981)
	46	Cl	(Medium)	Zyss *et al.* (1981)
	47	Br	(Low)	Zyss *et al.* (1981)
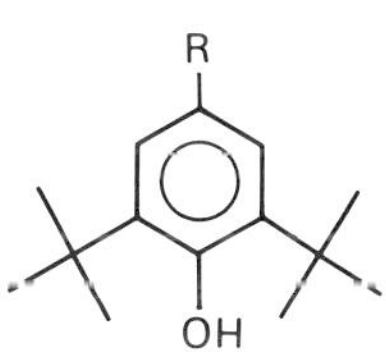	**48**	Me	0.1	Dirk and Tweig (unpublished results, 1985)
	49	CHO	8	Davydov *et al.* (1977)
	50	NO_2	10	Dirk and Twieg (unpublished results, 1985)
	**51**	—	2	Davydov *et al.* (1977)

[a] Numerical powder efficiencies from Twieg and Dobrowolski (unpublished results, 1985).

TABLE IX

Typical Synthetic Procedures

(a)

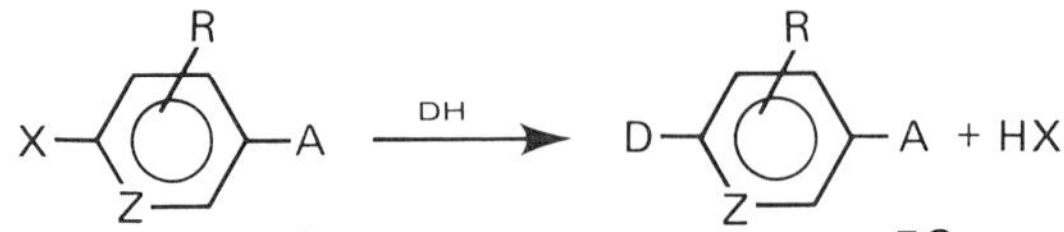

52 **53**

A = NO_2, CN, SO_2R', COR', CHO
X = F, Cl, Br
D = NR'R'', OR', SR'
Z = CH, N
R, R', R'' = wide variety

(b)

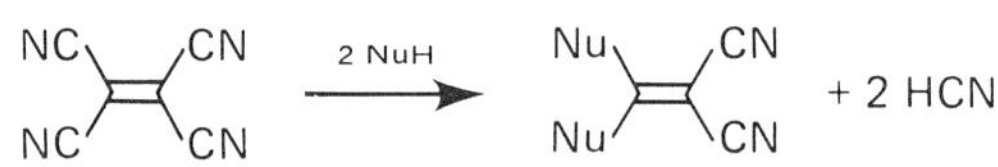

54 **55**

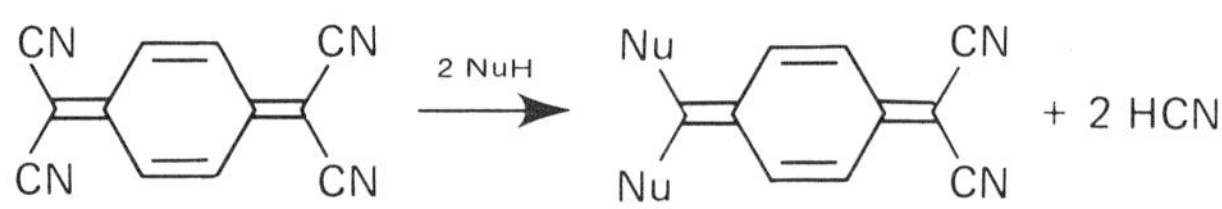

56 **57**

NuH = ROH, RR'NH, RSH, bifunctional systems

(c)

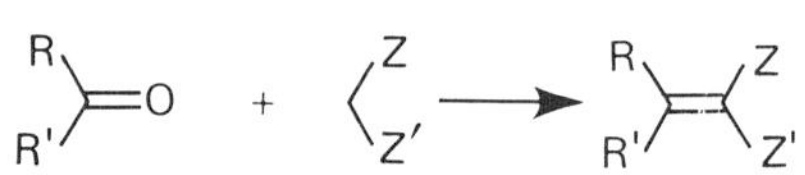

58 **59** **60**

R, R' = Ar, H, alkyl, etc.
Z, Z' = common A groups, etc.

the POM family (**46, 47**) (Zyss *et al.*, 1981). In another set of molecules it was noted (Dirk and Twieg, unpublished results, 1985) that derivatives of 2,6-di-*t*-butylphenol such as the 4-methyl derivative (**48**) (Maze-Baudet, 1973) appeared to crystallize with a large incidence of noncentrosymmetry. It was found that in certain cases introduction of the appropriate A group such as CHO or NO_2 at the para position of 2,6-di-*t*-butylphenol gave molecules **49** and **50**, which are quite strongly nonlinear (6–10 × urea).

As a final comment on molecular engineering, it is important to point out the value of an unprejudiced and objective approach. The molecule 4-

nitrobenzonitrile (**51**) breaks all the rules, and with a strict dogmatic approach one would reject the substance out of hand. This molecule crystallizes in space group $P2_1$ (Higashi and Osaki, 1977) with $\theta = 35.3°$, $b_{yyy} = 54.4\%$, and $b_{yxx} = 70.7\%$. The molecular alignment is quite good and, although electronically the molecule is not ideally substituted (there are two A groups disposed at para positions on the aromatic ring), it has a powder efficiency (2 × urea), which is larger than many other well-intended and well-engineered D-A molecules.

4. Typical Synthetic Procedures

Since organic NLO molecules are often very polar, a special few of the many reactions in the repertoire of synthetic organic chemistry have proved to be particularly useful. The most important of these is the SNAr aromatic nucleophilic substitution reaction (Table IX, procedure a), which has many variations, a few of which may be generically represented for conversion of simple halobenzene (Z = CH) and pyridine (Z = N) systems (**52**) to their respective D-A-substituted products (**53**) (March, 1985; Miller, 1968; Stock, 1968). Here D and A are donor and acceptor groups in the usual sense, but in addition in terms of chemical reactivity the A groups enhance the activity of X to leave and D has some nucleophilic properties to displace X. As a generality, the trends of reactivity for the D and A groups parallel their respective σ values (Table I). For acceptor groups, the larger and more positive the σ value, the faster is the SNAr reaction; for donors there is only a tenuous relationship between the nucleophilicity of DH and the σ value of D. The reactivity of X is again dependent on the specific reaction, but usually F > Br ≈ Cl and for Z most often N > CH. These reactions are often run in a dipolar aprotic solvent [dimethyl sulfoxide (DMSO), *N*-methylpyrrolidinone (NMP), dimethylformamide (DMF), etc.] and additional base (excess of DH if D = simple NRR′ or Et_3N, K_2CO_3, etc.) to neutralize by-product HX. The aromatic nucleophilic displacement reactions are very often efficient, straightforward to run, and in many cases ideally suited for the preparation of the D-A-substituted aromatics.

A related set of reactions (Table IX, procedure b) involves conversion of tetracyanoethylene (TCNE) (**54**) and 7,7,8,8-tetracyanoquinodimethane (TCNQ) (**56**) to their respective "push–pull" derivatives (**55**, **57**) (Ericsson *et al.*, 1970; Gompper *et al.*, 1968). Here the nitrile (—C≡N) groups serve the dual function of activation and leaving groups in reaction with a variety of nucleophiles Nu:.

The Knoevenagel condensation (Jones, 1967; March, 1985) is a particularly useful reaction to prepare the olefinic double bond in many push-pull systems (Table IX, procedure c). Here R and R′ in the electrophilic component (**58**) are variable, but very often R is a polar aromatic group and R′ is hydrogen, as in the case of a group of particularly useful D-substituted aldehydes prepared by the SnAr reaction (Bader *et al.*, 1966). There is also a great deal of variation possible in the activated methylene compound ZCH_2Z' (**59**) in which Z and/or Z′ are often A groups and render the methylene group acidic. Typical compounds represented by structure **60** prepared by this route are DANS (**24**), merocyanine, and cyanines, including **61** and a host of other highly polarized systems. It is even possible in a few cases to perform concurrent Knoevenagel and SnAr reactions (Brunskill *et al.*, 1978).

The few specific reactions discussed here have found great utility for the preparation of NLO organic molecules, particularly those prepared from commercially available starting materials. In cases where desired starting materials are not commercially available, the myriad other synthetic reactions come into play.

IV. EXTENSIVELY STUDIED ORGANIC MATERIALS

A. Early Studies

In the early 1960s very little was known about the appropriate structures and physical properties of organic molecules that might have NLO properties, and little was known about the appropriate methodology to search for these molecules. So the first compounds were examined on the basis of their ready availability, and in some cases something may have been known about the optical properties of the material or a related physical effect such as the piezoelectric, pyroelectric, or ferroelectric effect. In spite of this rather random approach (which admittedly certainly continues to this date!), many of the molecules that were identified set precedents and guided the way for future developments. Among these early compounds are a few (Table X) that were quite extensively studied as single crystals whose structures do not fit particularly well in the other catagories discussed later. These compounds include hexamethylenetetramine (**62**) (Heilmeier *et al.*, 1964), hippuric acid (**63**) (Orlov, 1966), and benzil (**35**) (Gott, 1971), for all of which a variety of optical properties were already known (Winchell, 1954), and 4-methyl-7-diethylaminocoumarin (**64**) (Bass *et al.*, 1969). All these molecules are alike in that they are achiral and yet crystallize in noncentrosymmetric point groups, and each has some unique and instructive features as well.

Hexamethylenetetramine is a polycyclic heterocycle that crystallizes in the unusual space group $I\bar{4}3m$, and it was already a known linear electro-optical material. Hippuric acid (*N*-benzoylglycine) has been studied as single crystals,

TABLE X
Early Studies

Number		Space group	$\chi^{(2)}$, d_{ij} [$\mathrm{pmV^{-1}}$], Efficiency, reference
62	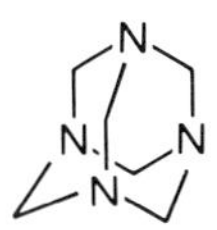	$I\bar{4}3m$	10 × KDP, $d = 6.3$, Heilmeier *et al.* (1964)
63	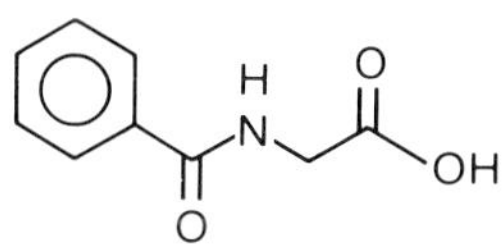	$P2_12_12_1$	6 × KDP, Orlov (1966)
35	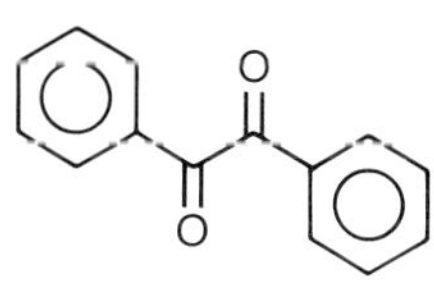	$P32$	9 × KDP, Jerphagnon (1971)
64	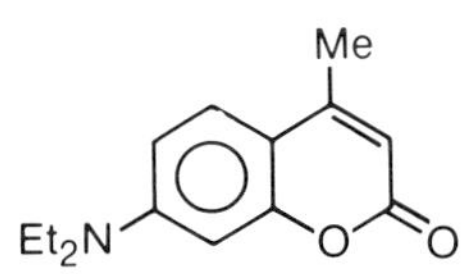	$P2_1$	10 × KDP, $d_{12} = 6.7$, Southgate and Hall (1971)

and the hyperpolarizability in this molecule probably arises from the amide functionality. Benzil is a particularly interesting molecule, since one might expect that in the crystal it would exist with a plane or center of symmetry through the central —CO—CO— bond. This, however, it not the case; and the molecule is twisted around this bond and crystallizes in a helical state, and the resultant crystal is noncentrosymmetric. The SHG properties of 4-methyl-7-diethylaminocoumarin were undoubtedly discovered due to its proximity to laser labs (it is the laser dye Coumarin 1). It is an example of one of the first donor–acceptor charge-transfer molecules discovered, although this fact was not appreciated at the time. In coumarin 1 the diethylamino donor at the 7 position is in direct conjugation with the coumarin carbonyl group through the benzene ring and the double bond. All these molecules are now of largely historic interest, since they have been superceded by superior materials.

The early studies were important because the physical techniques to screen nonlinear molecules were developed simultaneously. In particular, Kurtz and Perry (1968) and Dougherty and Kurtz (1976) refined the powder technique and in the course of their work devised a useful classification for nonlinear materials based on efficiency and phase matchability. It became clear that the most efficient molecules were polar and π-bonded, whereas the molecules with saturated bonds (cholesterol, menthol, and glucose) were intrinsically less efficient. The SHG powder technique is also of some use for structure analysis in that a simple test for centrosymmetry is possible (Shigorin and Shipulo, 1973; Coda and Pandarese, 1976).

B. Urea and Derivatives

The SHG properties of urea were first reported by Kurtz and Perry (1968) and subsequently studied in great detail theoretically (Zyss and Berthier 1982) for determination of its single-crystal optical properties (Betzler *et al.*, 1978; Cassidy *et al.*, 1979) and for device application (Chapter III-6). Urea is presently almost unique, given its large birefringence and nonlinear coefficients relative to its excellent transparency (200 nm) and excellent optical damage threshold, which render it valuable for frequency conversion into the ultraviolet. Although a significant amount of effort has gone into the development of crystal growth techniques for urea, this still seems to be the shortcoming of this molecule since it has yet to become commercially available. A number of other urea derivatives have been studied, such as *sym*-dimethylurea (**65**) (Halbout *et al.*, 1980) and a host of others (Jain *et al.*, 1981), but none of them appear to have been exploited (Table XI). Not only is urea valuable in its own right, but the urea (or thiourea) functionality is often found in molecules which are noncentrosymmetric and hence SHG-active.

C. *m*-Disubstituted Benzene Derivatives

The *m*-disubstituted benzene derivatives appear to have a propensity to crystallize in noncentrosymmetric space groups relative to their ortho and para analogs (Curtin and Paul, 1981), and early studies of NLO materials

TABLE XI
Urea and Derivatives

Number		Space group	$\chi^{(2)}$, d_{ij} [pmV^{-1}], Efficiency, reference
1	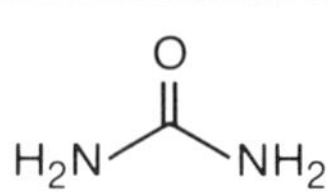	$P\bar{4}2_1m$	3 × KDP, $d_{36} = 2.3$, Betzler *et al.* (1978) Halbout *et al.* (1979)
65	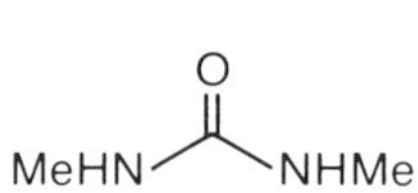	*mm*2	0.33 × U, Halbout *et al.* (1980)

concentrated heavily on this group. The *m*-disubstituted aromatics that have been studied in great detail often do not have the desirable combination of donor and acceptor groups, and the *m*-substitution pattern automatically rules out the favorable mesomeric interaction between the substituents. The many studies of this class include the compounds *m*-aminophenol (**66**), *m*-dinitrobenzene (**67**), *m*-dihydroxybenzene (resorcinol) (**68**), and finally and most important *m*-nitroaniline [mNA (**36**), not to be confused with MNA (**37**)] (Southgate and Hall, 1972) are most often encountered. Another compound related to this class, although formally trisubstituted, is 2,4-diaminotoluene (*m*TDA) (**69**) (Shigorin *et al*., 1976). Good-quality crystals of this diamine have been grown with transparency between 0.33 and 1.50 μm and a *d* value some three times that of KDP (Table XII). The only compound in this series which

TABLE XII

m-Substituted Benzenes

Number		Space group	$\chi^{(2)}$, d_{ij} [pmV^{-1}], Efficiency, reference
66	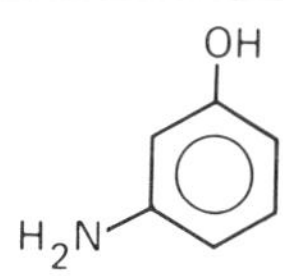	*Pca*	~3 KDP, Southgate and Hall (1972)
67	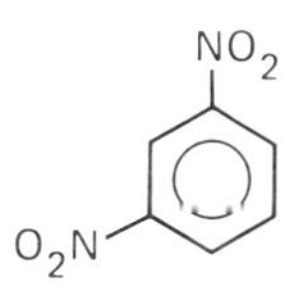	*Pbn*	1–5 KDP, Southgate and Hall (1972)
68	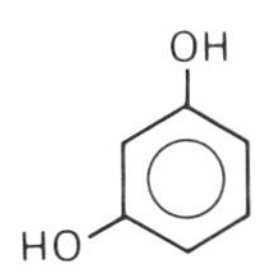	*Pna*	~2 KDP, Southgate and Hall (1972)
36	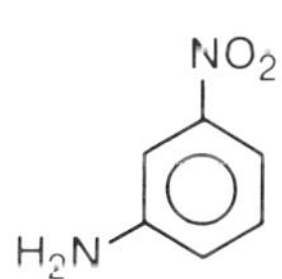	$Pbc2_1$	51 × KDP, $d_{14} = 34$, Southgate and Hall (1971), Kato (1980)
69	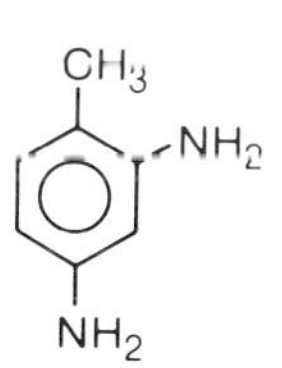	$Pna2_1$	2.9 × KDP, $d = 1.8$, Shigorin *et al.* (1976), Filipenko *et al.* (1980)

has been developed for practical application is mNA, which has been found by Kato (1980) to give a 65% conversion efficiency in a 3-mm crystal. Here again, interest in most of these compounds has been superceded by others that were discovered later.

D. Aromatic Nitro Compounds

As theoretical concepts developed for the NLO properties of organic compounds, more and more materials with the appropriate intramolecular charge-transfer properties have appeared. Since the nitro group is one of the strongest electron-withdrawing A groups, it is found in many of the well-studied materials (Table XIII). The first thoroughly studied aromatic nitro compound was 5-nitrouracil (5-NU, **3**), which was discovered by Jerphagnon (1971) and studied in detail by Bergman *et al.* (1972). The origins of the charge transfer in this molecule are not really clear, since the molecule cannot be accurately represented by a single tautomeric form. Nonetheless, if one examines the various resonance forms, the nitro group clearly functions as an acceptor, and the hydroxyl (carbonyl) between the ring nitrogens may be the primary donor. Measurements of single crystals of 5-NU, transparent between 0.5 and 1.5 μm, indicate that it is phase-matchable with a coefficient comparable to lithium niobate. Here again, no further studies of 5-NU appear to have been done, since subsequent compounds with comparable or superior properties were later discovered.

Methyl (2,4-dinitrophenyl)aminopropanoate (MAP, **9**) was described by Oudar and Hierle (1977) and is one of the most thoroughly studied organic NLO materials. MAP is a derivative of 2,4-dinitroaniline in which the simple amine donor has been replaced by the methyl ester of the optically active amino acid alanine (—NHCH*CH_3COOCH_3), which guarantees a noncentrosymmetric crystal structure. This molecular engineering technique has subsequently been employed in many other molecules. MAP is phase-matchable over its entire transparency range between 0.5 and 2.0 μm and has a figure of merit of d^2/n^3 15 times that of $LiNbO_3$. Excellent-quality crystals have been grown, and a 30% SHG conversion efficiency can be obtained from only a 1-mm-thick crystal, which has a damage threshold in excess of 1 GW/cm^2. More than any other molecule, MAP influenced the design of NLO organics that were to follow.

Two years after MAP was described, single-crystal studies of the nitroaniline derivative 2-methyl-4-nitroaniline (MNA, **37**) appeared that described the SHG (Levine *et al.*, 1979) and the linear electro-optic effect (Lipscomb *et al.*, 1981a,b) properties. It appears that earlier powder SHG studies on MNA discussed by Koreneva *et al.* (1975) and referenced by Vizgert *et al.*

TABLE XIII
Aromatic Nitro Compounds

Number			Space group	$\chi^{(2)}$, d_{ij} [pmV^{-1}], Efficiency, reference
3	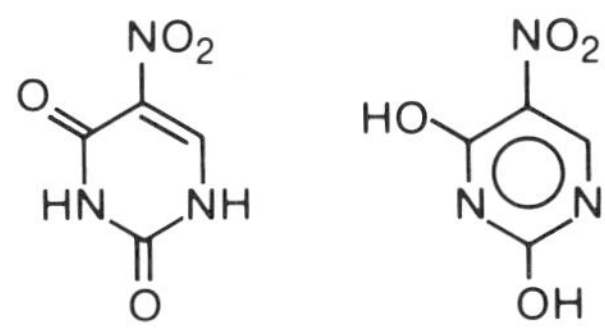	5NU	$P2_12_12_1$	4 × KDP, $d_{36} = 7.1$, Bergman *et al.* (1972)
9	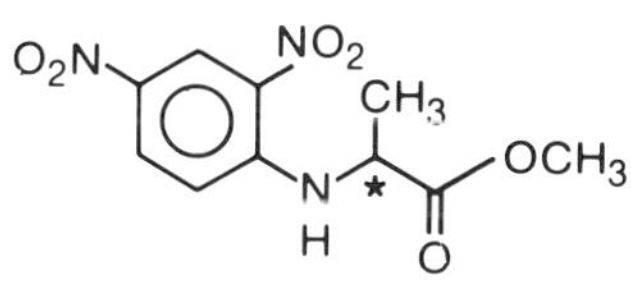	MAP	$P2_1$	$d_{22} = 18.4$, Oudar (1977b)
37	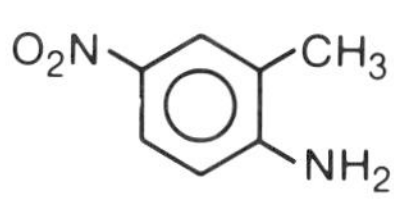	MNA	*Cc*	$d_{11} = 250$, $d_{12} = 38$, Levine *et al.* (1979), Lipscomb *et al.* (1981a,b)
38	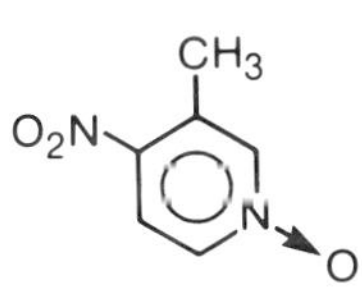	POM	$P2_12_12_1$	$d = 10$, Zyss *et al.* (1981)
70	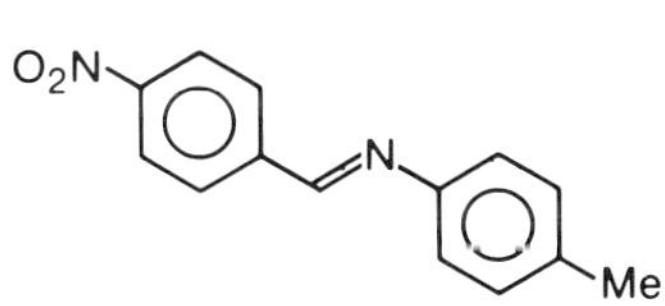	NMBA	*Pb*	$d = 16.7$, Filipenko *et al.* (1977)
10	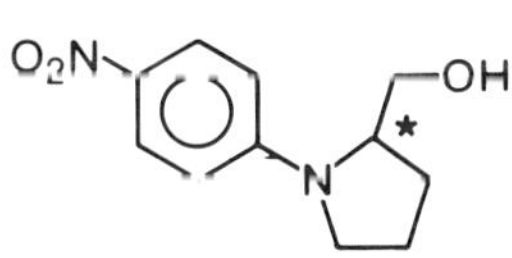	NPP	$P2_1$	$I^{2\omega} \approx 10$ × MAP, Zyss *et al.* (1984) $d_{21} = 84$, Ledoux *et al.* (1986)

(1982) may have underestimated the nonlinearity of MNA. The d_{12} coefficient is the largest that is phase-matchable and gives a merit figure of d^2/n^3 fully 45 times larger than lithium niobate. The d_{11} coefficient, which is not phase-matchable, has an enormous merit figure some 2000 times that of lithium niobate. Although the MNA crystal is optimized for the electro-optic effect rather than for SHG, recent applications of MNA for SHG in waveguides (Stegeman and Liao, 1983; Sasaki *et al.*, 1984) have appeared.

Another nitroaromatic that has been thoroughly studied is the 3-methyl-4-nitropyridine *N*-oxide (POM, **38**) (Zyss *et al.*, 1981). This molecule is unique in that it has a small ground-state dipole moment (from cancellation of the formal contributions of nitrobenzene and pyridine *N*-oxide), which is believed to enhance the material's ability to crystallize in a noncentrosymmetric fashion. POM is highly transparent relative to other nitroaromatics and has an NLO efficiency some 13.5 times that of KDP. Crystals of POM are among the easiest to grow and are of excellent optical quality (Chapter III-4), which makes it a prime candidate for device applications. The further enhancement of SHG efficiency from the POM family of compounds seems reaonable, since calculations indicate that the merit figure of 3.3 for POM crystals could conceivably reach 12.2 and also since POM crystallizes in the low 19% efficiency point group 222, which is not noncritically phase-matchable. Further effort in the molecular engineering in the POM family is strongly warranted.

Another nitro compound with a novel structure is the benzylidene aniline derivative 4-nitro-4′-methylbenzylidene aniline (NMBA, **70**) (Filipenko *et al.*, 1977). The quadratic optical susceptibilities have only been estimated from X-ray structural data and are intriguing in that the coefficients are almost as large as those for MAP and the crystals are more transparent than MAP crystals. Although NMBA is a highly conjugated molecule, its chromophore is hypsochromically shifted, perhaps due to rotation about the imine bonds. It is unfortunate that no further development has occurred for this particular material and for the benzylidene aniline class as a whole.

The ultimate pay-offs that come from persistance in molecular engineering are embodied in the nitroaniline derivative *N*-(4-nitrophenyl)-(S)-prolinol (NPP, **10**) (Zyss *et al.*, 1984). Here the donor portion of the molecule is derived from a chemically reduced form of the natural amino acid (S)-(−)-proline and serves as the chiral auxiliary. The pyrrolidine group is known to be a very potent donor and is only slightly twisted out of an ideal coplanar position. The most remarkable feature of the NPP crystal is that the angle between the symmetry axis and the dipole moment transition axis of 58.6° is very close to the optimal 54.74° required in the $P2_1$ class (Fig. 4). Powder tests on NPP indicate that it is fully an order of magnitude more efficient than MAP and in addition has an absorption cut-off only slightly bathochromically shifted from

MAP. Single-crystal thin films of NPP with areas on the order of 10 mm^2 can be grown between silica plates, and transverse SHG measurements have been performed (Ledoux, *et al.*, 1986). The d_{21} coefficient of 84 pmV^{-1} so obtained is the highest phase-matchable coefficient ever obtained with a transparency range of 0.5 to 2.0 μm.

E. Organic Salts

Although the organic salts are not formally molecular compounds in a structural sense, they bridge the covalent organics and fully ionic inorganic materials. A number of compilations of crystals studied for SHG exist; among the most comprehensive are those due to Bechmann and Kurtz (1969), Kurtz *et al.* (1979), and Jerphagnon *et al.* (1984). The vast majority of entries in these compilations are fully inorganic, but a few organic salts and even a few covalent organic compounds are included. The organic part of the salts may be the anion or cation, and either may be optically active. Among the organic anions, the simple carboxylic acids predominate (formate, propionate, oxalate, etc.), although more complicated (phthalate) and optically active (tartrate) anions have been employed. The organic cations are more rare, but the amino acids glutamic acid and arginine, in which the basic functionalities are protonated, are examples.

A few more recent examples of organic salts are of particular interest (Table XIV). Racemic potassium malate (**71**) has efficiencies similar to KDP (Schuler *et al.*, 1982). Many formate salts have been examined, among them the group III formates such as yttrium formate (**72**), which has been found to have a *d* coefficient twice potassium dihydrogen phosphate (KDP) and is transparent between 0.25 and 1.4 μm (Furmanova *et al.*, 1984). The novel L-arginine phosphate monohydrate (LAP, **73**) is transparent between 0.19 and 0.60 μm and is some 3.5 times as efficient as KDP (Xu *et al.*, 1983).

The neutral merocyanine and cationic cyanine dyes are well known to be highly polar, and as such they are prime candidates for molecular engineering. Although the chemistry of the dye molecules is quite well understood because of their great technological importance as spectral sensitizers, the great challenge for SHG application has involved how to manipulate macroscopic properties to obtain noncentrosymmetric crystal habit. The only successful technique reported to date involves the metathesis of 4-dimethylamino-*N*-methyl-4′-stilbazolium iodide to introduce new anions (Meredith *et al.*, 1985). The most efficient derivative, **61**, contains the pseudo-tetrahedral methosulfate anion, and in the powder form this dye is fully 30 times as efficient as mNA. This large efficiency is associated with its inevitable trade-off of loss of transparency [λ_{max} 473 nm (MeOH), λ_c 588 nm]. The crystallographic study

TABLE XIV
Organic Salts

Number			Crystal group	$\chi^{(2)}$, d_{ij} [pmV^{-1}], Efficiency, reference
71	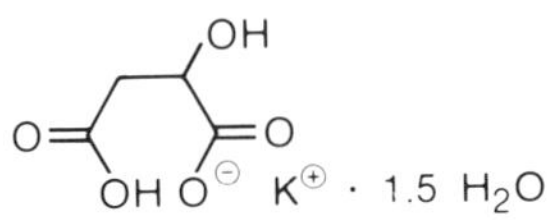	KM	*m*	1.4 × KDP, $d_{31} = 0.88$, Schuler *et al.* (1982)
72	Y^{+3} $(HCOO^{\ominus})_3$	YF3	*R*3*m*	*d* = 0.84, Furmanova *et al.* (1984)
73	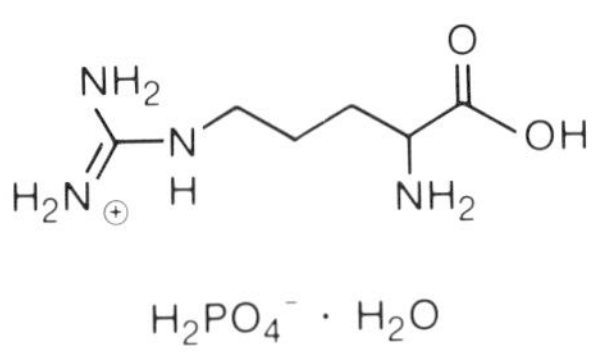	LAP	*P*2$_1$	3.5 × KDP, Xu *et al.* (1983)
61	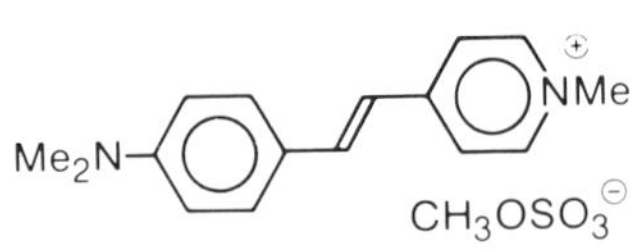		*Cmc*2$_1$	$I^{2\omega}$ = 30 mNA (1.91 μm), Meredith *et al.* (1985)

of this compound indicates it is well optimized for both electro-optics and phase-matched SHG.

V. LESS THOROUGHLY STUDIED ORGANIC MATERIALS

In this section we will examine representative materials that look promising on the basis of powder measurements but in most cases for which no detailed single-crystal optical data is available. As in the case of the well-studied materials, the nitroaromatics are by far the most popular. In the discussion sections, the nitrobenzene and nitropyridines are treated separately, but the tabulated data will be separated into chiral nitrobenzenes and pyridines (Table XVI) and achiral nitrobenzenes and pyridines (Table XVII) in an attempt to better compare and contrast their properties. In addition to the nitrobenzenes and pyridines, other heterocycle derivatives are discussed, with, finally, a rather wide variety of other structural types.

A. Nitrobenzenes

The nitrobenzene class of materials is the single most studied group, since the largest variety of precursors are readily commercially or synthetically accessible and in a pragmatic sense the products are often highly SHG-active. There are numerous nitrobenzene derivatives that have been examined by the powder technique for which single-crystal studies remain to be done; many of these compounds are closely related structurally to materials that have already been discussed.

The combination of substituent and chiral auxiliary has a profound effect on the efficiency of these materials, and the subtleties involved are particularly evident if the donor amine, in this case optically active (S)-(+)-prolinol, is kept constant and the nitrobenzene is modified with additional substituents (**10**, **74–86**) (Table XV). Not only are the SHG efficiencies greatly different due to changes in crystal packing, but in a few cases the compounds are not even crystalline at room temperature. Among these prolinol derivatives, the new compound *N*-(3-hydroxy-4-nitrophenyl)-(S)-prolinol (HNPP)(**83**) has an

TABLE XV

Optically Active Nitroaromatic Prolinol Derivatives

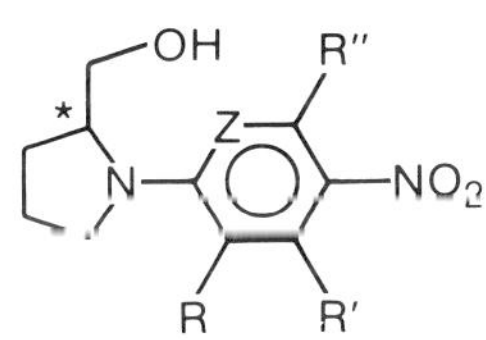

Number	R	R′	R″	Z	mp (°C)	Eff. (× *U*)
10	H	H	H	CH	116	150
74	H	H	H	N	90	140
75	F	F	F	CF	—	2
76	Me	H	H	CH	NC	—
77	H	Me	H	CH	106	3
78	CF_3	H	H	CH	NC	—
79	H	CF_3	H	CH	98	Weak
80	CN	H	H	CH	121	22
81	CH_2OH	H	H	CH	—	Weak
82	COOH	H	H	CH	—	4
83	H	OH	H	CH	—	140
84	NO_2	H	H	CH	NC	—
85	NO_2	H	F	CH	—	3
86	NO_2	H	Cl	CH	—	9

[a] NC, not crystalline.

additional hydroxyl group and has a very high powder SHG efficiency comparable to NPP (Nicoud, unpublished results, 1985). There is a similar family of compounds related to MAP in which the amino acid (Davydov *et al.*, 1977), amino acid ester (Oudar, 1977b; Zyss, 1982), and other nuclear substituents (Twieg and Jain, 1983) have been modified. Many of the particularly efficient new chiral nitrobenzenes (Table XVI) have chiral auxiliaries containing an additional functional group such as an alcohol (**91**–**95**) or an amine (**107**, **108**) which may play a role in hydrogen bonding, and in certain cases the secondary functionality is at a chiral carbon, as in the ephedrine family (**96**, **98**) (Nicoud, unpublished results, 1985).

TABLE XVI

Chiral Nitrobenzene and Nitropyridine Derivatives

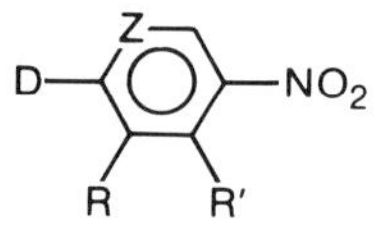

Number	D	R	R′	Z	Efficiency (× *U*)
87	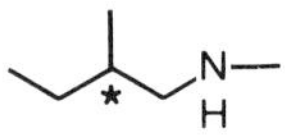	H	H	CH	22
88		H	H	N	20
89	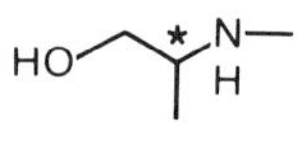	H	H	CH	22
90		H	H	N	92 (NPA)
91		Me	H	CH	10
92		CN	H	CH	10
93	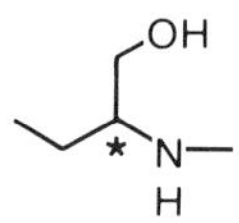	H	H	CH	12
10	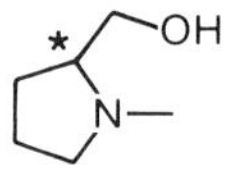	H	H	CH	150 (NPP)
94		CN	H	CH	22
83		H	OH	CH	140 (HNPP)
74		H	H	N	140 (PNP)
95	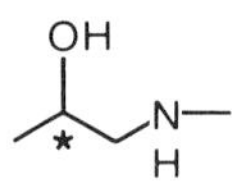	H	H	CH	15
96	Ph—CH(OH)—CH(Me)—N(Me)— (+) pseudoephedrine	H	H	CH	22
97		H	H	N	4

TABLE XVI (*Continued*)

Number	D	R	R′	Z	Efficiency (× *U*)
98	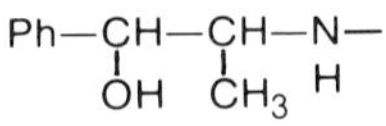 (−) norpseudoephedrine	H	H	CH	8
99	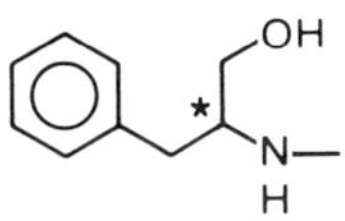	H	H	N	130 (NPPA)
100	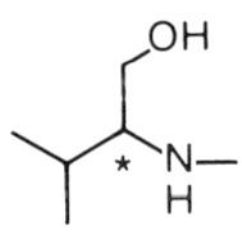	H	H	N	8
101	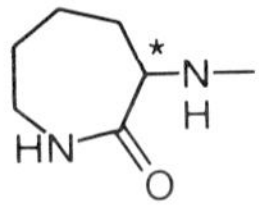	H	H	N	8
102	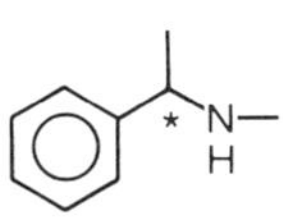	Me	H	CH	6
40		NO_2	H	N	7
39		H	H	N	25 (MBANP)
103		Me	H	N	7
104	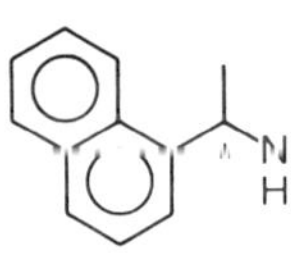	H	H	N	2
105		Me	H	N	25
106		NO_2	H	N	16
107	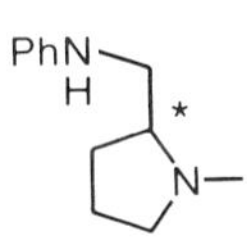	H	H	CH	8
108	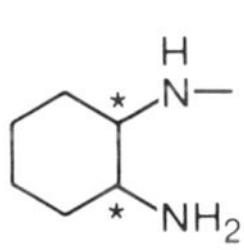	H	H	CH	10
109	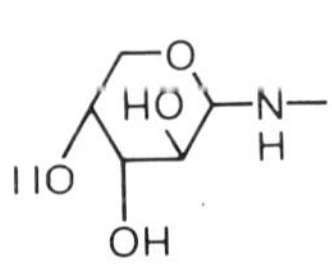	H	H	CH	10

Just as amino acids and their derivatives can serve as chiral auxiliaries as in MAP, so can the carbohydrates. The amine *N*-glycoside derivatives of nitroaniline and a variety of simple sugars have been prepared, and among them the compound from nitroaniline and arabinose, **109**, has reasonable SHG efficiency (10 × urea) and like the parent carbohydrate has the virtue of good crystallinity. In addition, the carbohydrate derivatives of a variety of more polarizable and absorbing molecules such as 4-amino-4′-nitrostilbene (**31**) have been prepared, but their powder efficiencies await measurement (Twieg and Cheng, 1985).

Many achiral nitrobenzene derivatives that crystallize noncentrosymmetrically are now known and have good powder SHG efficiencies. For example, there are many close relatives of MNA in which the methyl group has been replaced by other substituents, such as halogen as in **44**. In addition, the amino group has been replaced by a methylamino group in a certain few cases, as in 2-methyl-4-nitro-*N*-methylaniline (MNMA) and 2-chloro-4-nitroaniline (CNMA) (**110**, **111**) (Twieg and Jain, 1983). Many of these compounds have additional functional groups (hydroxy, carboxylic acid, amide, nitrile, etc.) to influence crystal packing by hydrogen-bonding or polar effects. Among them, several new aminopropanol derivatives, including *N*-(4-nitrophenyl)-3-amino-1-propanol (APNP, **17**), and their trifluoromethyl analogs (**18**, **19**) and the butanoic acid derivative *N*-(4-nitrophenyl)-4-aminobutanoic acid (BANP, **120**) are very efficient (Nicoud, unpublished results, 1985). Of particular interest are the derivatives containing a nitrile-modified amine donor, such as *N*-(4-nitrophenyl)-N-methylaminoacetonitrile (NPAN, **119**) (Nicoud, unpublished results, 1985). This very simple achiral compound has a powder efficiency comparable to the best chiral nitrobenzene NPP, and X-ray study (Coquillay, 1986) of this promising new material reveals that the polar cyano group [3.47 D; Minkin *et al.* (1970)] plays an important role in the packing of this Fdd2 space group crystal. Conceptually, the utilization of a substituent with large ground-state dipole and relatively small optical hyperpolarizability to influence crystal packing is a very attractive molecular engineering technique that demands further attention.

A series of disubstituted nitrobenzenes formally derived from 3,4-diaminonitrobenzene is of interest. The parent compound 3,4-diaminonitrobenzene is deep red in the crystalline state, but if the 3-amine is acetylated the color reverts to that exhibited by the parent 4-nitroaniline. Typical of this class is 3-acetamido-4-dimethylaminonitrobenzene (DAN, **12**), which was discussed earlier in terms of its optimized crystal structure and hydrogen bonding. DAN is a member of a small family of compounds (**123**–**125**) that are strongly nonlinear and vary in the identities of the 4-amino substituent and the mode of 3-amino derivatization (Twieg and Jain, 1983). It is interesting that the related dinitro compound 2-acetamido-4,5-dinitroaniline is also optically nonlinear (Vizgert *et al.*, 1982).

B. Nitropyridines

The nitropyridines have been developed along with the nitrobenzenes and are generally more transparent by about 30 nm. Among the best early examples in this series are the α-methylbenzylamino derivative MBANP (**39**) (powder test 25 × urea, $P2_1$, $\theta = 33.2°$, $b_{yyy} = 58.5\%$, and $b_{yxx} = 65.3\%$ for a one-dimensional treatment that ignores the pyridine N contribution) and the picoline analog with the optically active napthylethylamine donor (**105**) (powder test 25 × urea), for which crystallograpic details are not available (Twieg *et al.*, 1982a,b). Single-crystal NLO measurements have been done on MBANP, but details have not yet appeared (Nayar, 1984). The pyridine analog of NPP, *N*-(5-nitro-2-pyridyl)-(S)-prolinol (PNP, **74**), has similar extraordinary properties (powder test 140 × urea, $P2_1$, $\theta = 59.6°$, $b_{yyy} = 12.9\%$, $b_{yxx} = 97.8\%$, one-dimensional treatment) (Twieg and Dirk, 1986). This coincidence of excellent nonlinear behavior between NPP and PNP is fortuitous, since correspondence in nonlinear SHG behavior between the respective pairs of amine derivatives of nitrobenzene and nitropyridine is very rare (Tables XVI and XVII). This fact is demonstrative of the enormous changes in crystal packing that can result from what appears to be superficially minor change from a carbon to a nitrogen atom. Other

TABLE XVII

Achiral Nitrobenzene and Nitropyridine Derivatives

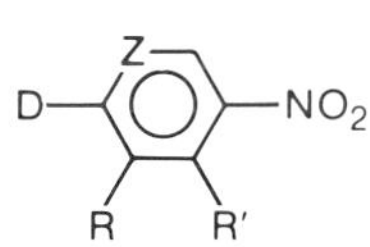

Number	D	R	R′	Z	Eff. (× *U*)
37	H_2N—	Me	H	CH	22 (MNA)
44		Cl	H	CH	40 (CNA)
110	MeN—H	Me	H	CH	80 (MNMA)
111		Cl	H	CH	20 (CNMA)
112	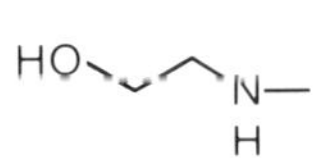	COOH	H	CH	15
113		H	H	N	3
17	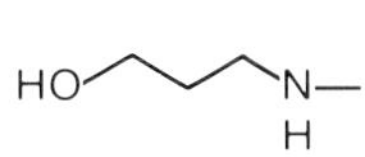	H	H	CH	80 (APNP)
114		Me	H	CH	8
18		CF_3	H	CH	30
19		H	CF_3	CH	80

(*continues*)

TABLE XVII (*Continued*)

Number	D	R	R′	Z	Eff. ($\times U$)
115	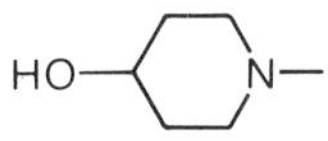	H	H	CH	1
116		H	CF_3	CH	18
117		CN	H	CH	8
118	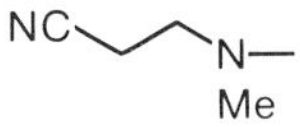	H	H	CH	85 (NPPN)
119	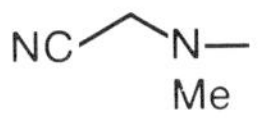	H	H	CH	140 (NPAN)
120	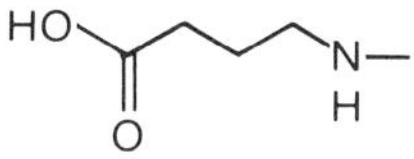	H	H	CH	115 (BANP)
121	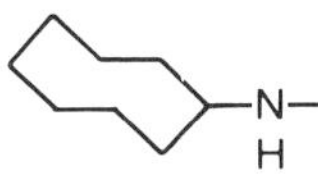	H	H	CH	7
122		H	H	N	50 (COANP)
12	Me_2N—	NHCOMe	H	CH	115 (DAN)
123		$NHCOCF_3$	H	CH	70
124	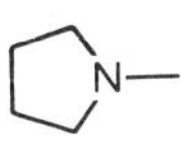	NHCOMe	H	CH	80 (PAN)
125	H_2N—	NHCOMe	H	CH	20
126	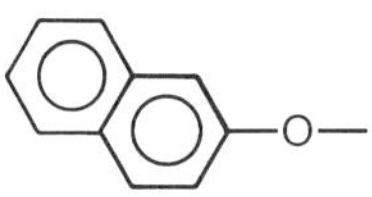	H	H	N	8
127	Cl—	H	H	N	1
128		NO_2	H	N	5
129	MeO—	H	H	N	2

chiral aminoalcohols derived from naturally occurring amino acids have more recently been found to function effectively as chiral auxilliaries for the nitropyridines. Thus, the nitropyridine derivatives from alaninol, *N*-(5-nitro-2-pyridyl)-(S)-alaninol (NPA, **90**), and phenylalaninol, *N*-(5-nitro-2-pyridyl)-(S)-phenylalaninol (NPPA, **99**), both possess very large powder efficiencies in the range of NPP and PNP (Nicoud, unpublished results, 1985). The full crystal structure data on these important new NLO molecules is not yet available.

All the really efficient NLO nitropyridines discussed so far are pure enantiomers, and the incidence of efficient materials among the achiral nitropyridines studied to date is quite rare. (This behavior is in contrast to that of the nitrobenzenes, for which numerous efficient achiral compounds are now known.) A recent discovery in this achiral class is 2-*N*-cyclooctylamino-5-nitropyridine (COANP, **122** (powder test 50 × urea), which came from a systematic study of cycloalkylamine derivatives of nitroaromatics (Twieg and Dobrowolski, unpublished results, 1985). Unfortunately, the crystal structure data is not yet available for this compound, as it is of interest to learn how the cyclooctylamine portion of the molecule influences the noncentrosymmetric crystal packing. Other smaller cycloaliphatic amine donors (cyclopropylamine through cycloheptylamine) give derivatives with no SHG activity, and it is of obvious interest to examine nitropyridine derivatives from the less synthetically accessible medium-ring-size cycloaliphatic amines.

A variety of other donor atoms including halogen and oxygen have been studied in the achiral nitropyridine series with some success. Particularly interesting is the molecule 2-(β-napthyloxy)-5-nitropyridine (**126**), which exists as a colorless SHG-inactive form and a bright-yellow SHG-active form. This behavior is typical of the complications often encountered due to a multiplicity of crystal forms that often exist for a single compound. Other very simple derivatives include 2-chloro-5-nitropyridine (**127**), 2-chloro-3,5-dinitropyridine (**128**). and 2-methoxy-5-nitropyridine (**129**), which possess modest nonlinear activity with enhanced transparency relative to their amine donor counterparts.

C. Other Nitroheterocycles

The number of opportunities for new nonlinear compounds among the heterocycles is enormous and, except for pyridine derivatives, the area has not been adequately exploited (Table XVIII). Among the few reported cases are the optically active α-methylbenzylamino derivatives of nitropyrimidine (**130**) and nitrothiazole (**131**) (Twieg and Jain, 1983). The nitrofuran derivative nitrofurazone (**132**) (Davydov *et al.*, 1970) is a more conjugated system in which the semicarbazide nitrogen functions as a donor, similar to the donor imine functionality NMBA. The original powder measurements of this compound appear to have underestimated its SHG efficiency (Twieg and Dobrowolski, unpublished results, 1985). An intriging compound is 1-hydroxy-2-methyl-5-nitrobenzimidazole (**133**), which is reported to be colorless and yet has $I^{2\omega} = 20 \times$ KDP, which would be an excellent efficiency–transparency trade-off (Vizgert *et al.*, 1982).

TABLE XVIII
Other Nitroheterocycles

Number		Efficiency	Reference
130	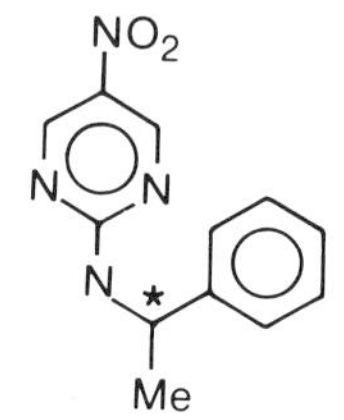	4*U*	Twieg and Jain (1983)
131	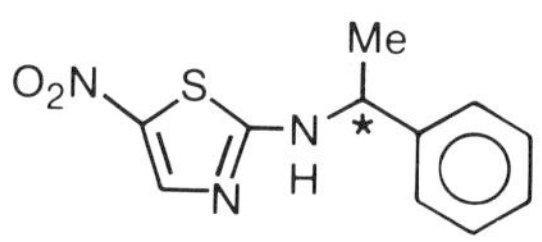	8*U*	Twieg and Jain (1983)
132	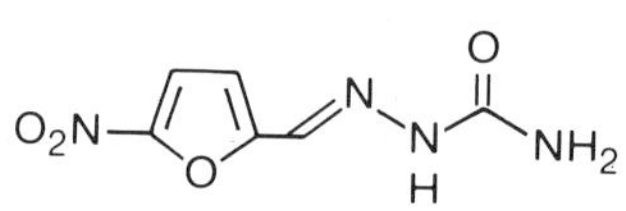	~$LiNbO_3$ ~30*U*	Davydov *et al.* (1970) Dirk and Twieg (unpublished results, 1985)
133	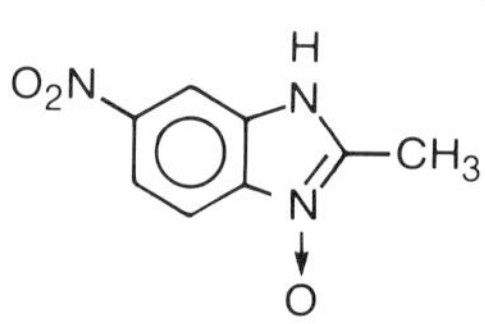	20 × KDP	Vizgert *et al.* (1982)

D. Donor–Acceptor Aromatics Besides Nitroaromatics

The reader must not be left with the impression that only aromatic nitro compounds are useful for SHG. Numerous other combinations of donor and acceptor groups on an aromatic ring have been examined (Table XIX), and although none has been found that rivals the efficiency of the D-A combination of amine and nitro group, the other groups often give materials that are significantly more transparent. Other donors that have been examined include oxygen and sulfur, and the acceptors are nitrile, carbonyl (aldehyde, ketone, ester, acid, amide), and sulfone or sulfonamide. Typical examples include the chiral benzophenone (**134**), the urea benzolog 4-aminobenzamide

TABLE XIX

D-A Aromatics Besides Nitroaromatics

Number		Eff.	Space Group	Reference
134	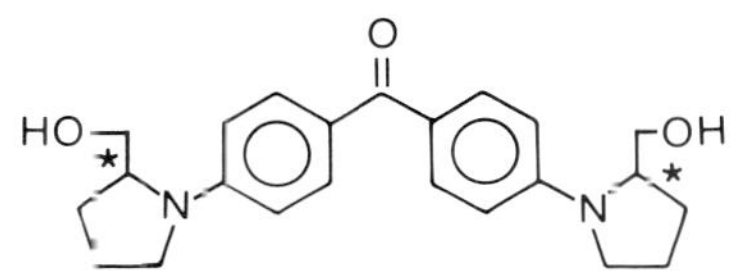	5*U*	—	Nicoud (unpublished results, 1985)
135	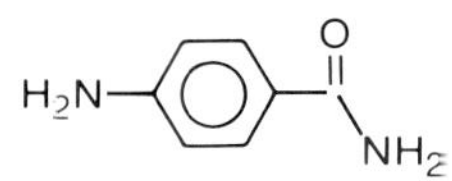	2–4*U*	$P2_1$	Nicoud (unpublished results, 1985) Dirk and Twieg (unpublished results, 1985) Alleaume (1967)
136	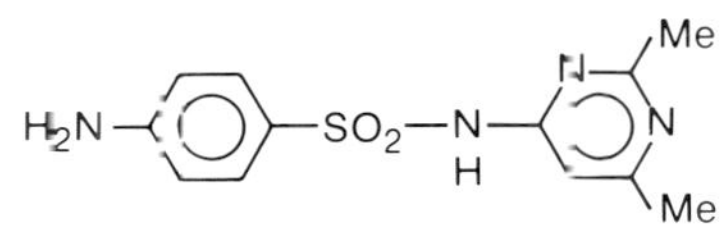	2*U*	$Pca2_1$	Dirk and Twieg (unpublished results, 1985) Reinhardt *et al.* (1980)
137	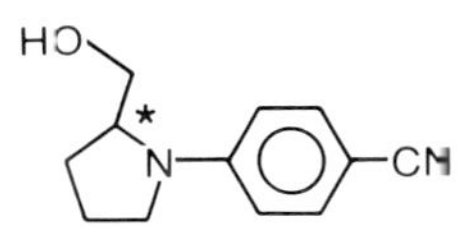	2*U*	—	Twieg and Dobrowolski (unpublished results, 1985)
138	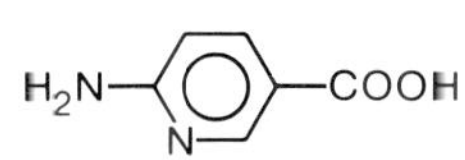	2*U*	—	Dirk and Twieg (unpublished results, 1985)

(**135**), which has a noncentric crystal structure (Alléaume, 1967; Alléaume and Decap, 1965) and powder efficiency a few times that of urea (Dirk and Twieg, unpublished results, 1985; Nicoud, unpublished results, 1985), as well as other aromatic derivatives including the sulfonamide sulfosomidine (**136**), the nitrile analog of NPP (**137**), and 6-aminonicotinic acid (**138**) (Dirk and Twieg, unpublished results, 1985).

E. Polarized Olefins

It is surprising that so little work has been done to exploit polarized olefins as nonlinear optical materials, in spite of the fact that the chemistry of these systems has been quite thoroughly examined (Sandström, 1983) and the crystal structures of many of these materials have been obtained by Adhikesavalu and Venkatesan (1983) and others. These so called "push-pull" molecules are of an enormous variety and may be generically represented as **139** (Fig. 10) to describe the presence of an ethylene-bearing acceptor substituents A and/or B at one end and donor substituents D and/or E at the other end to produce a ground-state charge separation as in **139a**.

Granted this structural flexibility, the resultant properties of these molecules are varied (Table XX) and are most conveniently divided by their chromophore into colorless, yellow, and more highly absorbing catagories. Among the colorless compounds is the ethylene glycol adduct of TCNE (**140**) (Nicoud, unpublished results, 1985), the product of its subsequent reaction with ammonia (**141**) (Moret *et al.*, 1983), and a number of compounds including 2-amino-1,1,3-tricyanopropene (**142**) (Klewe, 1972), and the TCNE adducts of *N,N'*-dimethyl-1,3-diaminopropane (**143**) (Adhikesavalu and Venkatesan, 1982) and the mixed system (**144**) with one nitrogen and one sulfur donor (Adhikesavalu and Venkatesan, 1983). All of these simple compounds have powder efficiencies comparable to urea. Among the yellow compounds are the dicyanoheterofulvene (**145**), the indanedione derivative (**146**), and a variety of derivatives of optically active amino alcohols with TCNQ such as (**147**) (Nicoud, unpublished results, 1985; Moret *et al.*, 1983). The synthesis of these optically active TCNQ derivatives was motivated by the report of the large hyperpolarizabilities determined in solution by the EFISH technique on achiral diamine derivatives (Lalama *et al.*, 1981). All attempts to date to realize these large solution hyperpolarizabilities in a noncentrosymmetric powder or crystal have been disappointing, although

139

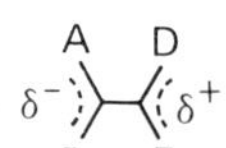

139a

Fig. 10. The generic push–pull olefin (**139**). The donor groups (D, E) and acceptor groups (A, B) produce a ground-state charge separation as in **139a**.

the exactly analogous optically active diamine derivatives of TCNQ (which are significantly more absorbing) remain to be prepared and examined in a thorough fashion.

Among the more absorbing red compounds are perhaps the greatest opportunities to exploit the push-pull systems. The indanedione derivative (**148**) is highly nonlinear, and here we can include the new benzylidene aniline systems in which the polarized olefin is replaced by a polarized imine (**149**, **150**) (Nicoud, unpublished results, 1985). It should be noted that formally even the polar stilbenes and cyanines already described elsewhere can be considered as polarized olefins if the entire appended aromatic rings are considered as the D and A substituents.

In general, given the strong polarization known to exist in these push-pull molecules, the maximum optical hyperpolarizability seems not to have been achieved yet. Much insight needs to be gained from refined EFISH measurements and crystal structure analysis on molecules already in hand to qualify the push-pull molecules as a useful class of NLO materials that warrant further significant synthetic effort.

F. Carbohydrates and Amino Acids

The carbohydrates, amino acids, and amino acid derivatives have been demonstrated to function as highly useful chiral auxiliaries for a variety of nitroaromatrics and other D-A molecules with large hyperpolarizabilities. In addition to this function, these molecules are in their own right also of potential value for harmonic generation into the UV (Table XXI). A very thorough study of the powder efficiencies of a variety of biological systems has been reported (Delfino, 1979), in addition to some earlier studies on amino acids. Rock candy (sucrose) (**151**) is very transparent (cut-off 192 nm) and has been thoroughly studied (Halbout and Tang, 1982), but various other carbohydrates including lactose (**152**) have only been measured as powders (Rosker and Tang, 1984). Simple derivatives of the carbohydrates and amino acids have not been adequately examined, and many substances potentially useful for harmonic generation into the UV remain to be found. For example, among the derivatives of amino acids, *N*-acetyl cysteine (**153**) is highly crystalline, damage-resistant, and has a powder efficiency about one-third that of urea (Dirk and Twieg, unpublished results, 1985).

G. Compounds with Unusual Structure

For harmonic generation in the UV there are a large variety of novel compounds (Table XXII) that have been studied by the powder method and for which further efforts at molecules engineering should be attempted. In the class of organic salts, the carbon disulfide adduct pyrrolidinium pyrrolidine

TABLE XX
Polarized Olefins

Number		Efficiency	Space group	Reference
140	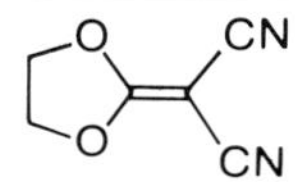	2*U*	—	Nicoud (unpublished results, 1985)
141	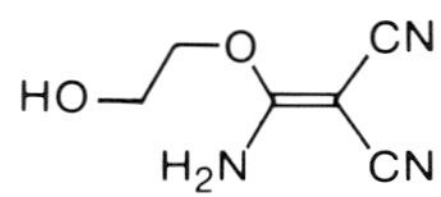	1*U*	—	Moret *et al.* (1983)
142	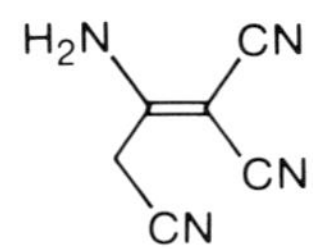	1*U*	$P2_1$	Twieg and Dobrowolski (unpublished results, 1985) Klewe (1972)
143	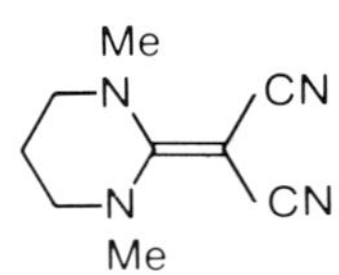	2*U*	$P2_1cn$	Nicoud (unpublished results, 1985) Adhikesavalu and Venkatesan (1982)
144	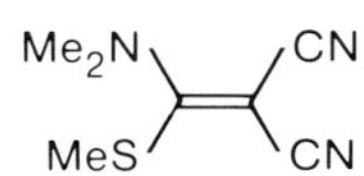	2*U*	$Pna2_1$	Nicoud (unpublished results, 1985) Adhikesavalu *et al.* (1983)
145	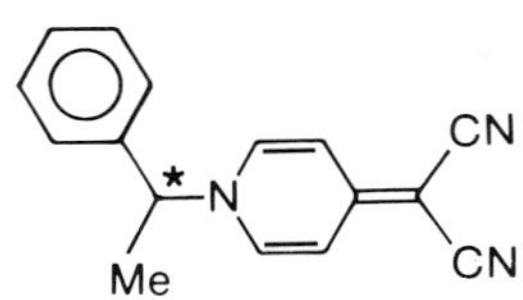	8*U*	—	Nicoud (unpublished results, 1985)

146	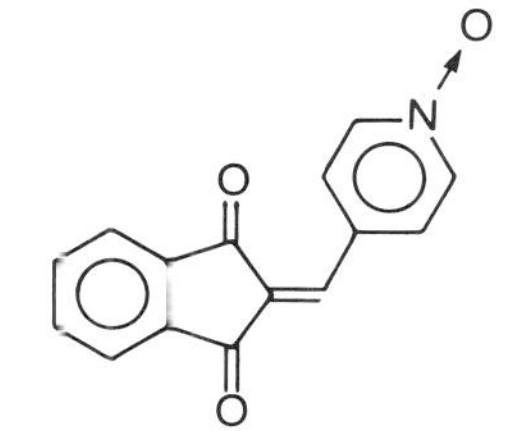	1*U*	—	Nicoud (unpublished results, 1985)
147	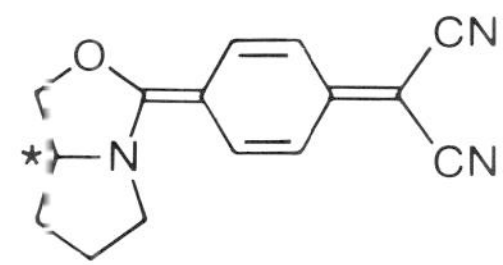	1*U* 5*U*	—	Nicoud (unpublished results, 1985) Moret *et al.* (1983)
148	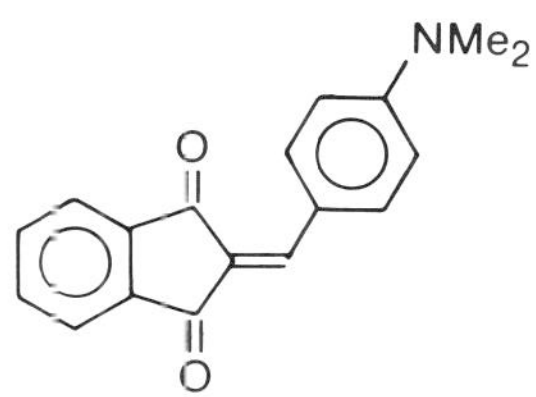	High	*Pna*2_1	Nicoud (unpublished results, 1985) Magomeda *et al.* (1980)
149	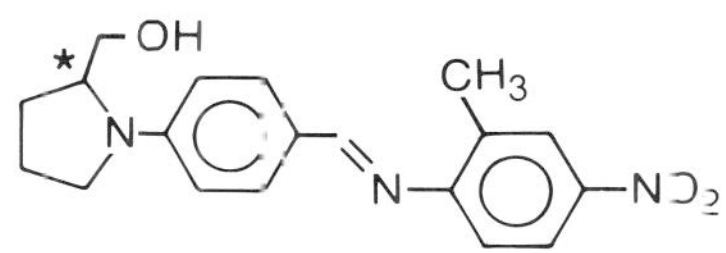	20*U*	—	Nicoud (unpublished results, 1985)
150	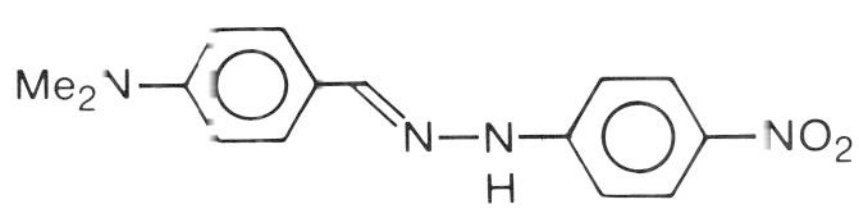 1/4 C_6H_6	High	*P*2_12_12	Nicoud (unpublished results, 1985) Brown *et al.* (1980)

TABLE XXI

Carbohydrates, Amino Acids

Number		Efficiency	Space group	Reference
151	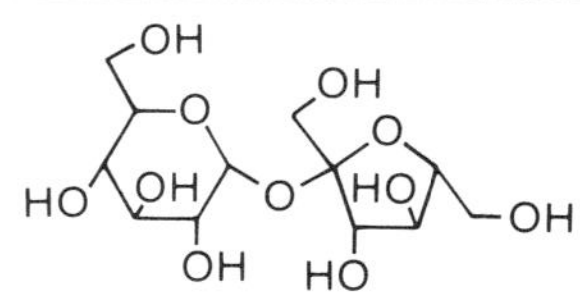 Sucrose	0.2 × KDP	$P2_1$	Rieckhoff and Peticolas (1965) Halbout and Tang (1982) Brown and Levy (1973)
152	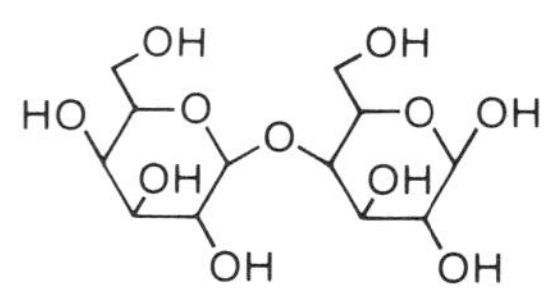 Lactose	13 × Sucrose	$P2_1$	Rosker and Tang (1984) Hirotsu and Shimeda (1974)
153	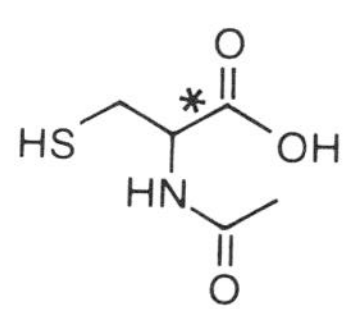	0.3*U*	—	Dirk and Twieg (unpublished results, 1985)

TABLE XXII

Compounds with Unusual Structure

Number		Efficiency	Space group	Reference
154	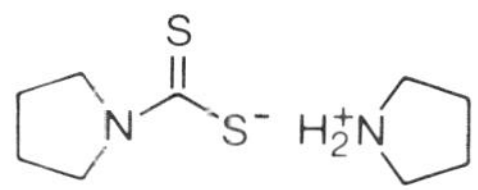	1*U*	$P2_1$	Dirk and Twieg (unpublished results, 1985) Wahlberg (1979)
155	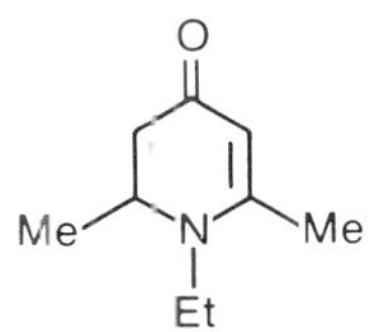	1*U*	—	Nicoud (unpublished results, 1985)
156	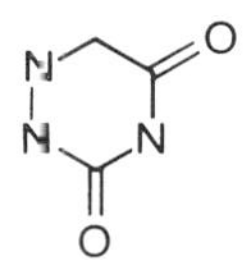	2.5*U*	$P2_12_12_1$	Nicoud (unpublished results, 1985) Singh and Hodgson (1974)
157	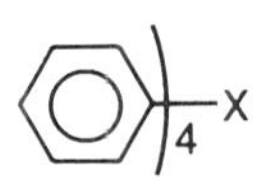	<*U*	$P\bar{4}2_1c$	Schlotter and Twieg (1984) Kitaigorodsky (1973)

(continues)

TABLE XXII (*Continued*)

Number		Efficiency	Space group	Reference
158	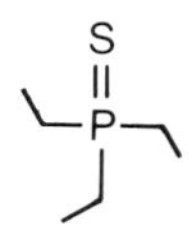	1*U*	$P6_3mc$	Dirk and Twieg (unpublished results, 1985) Van Meerssche and Leonard (1959)
159	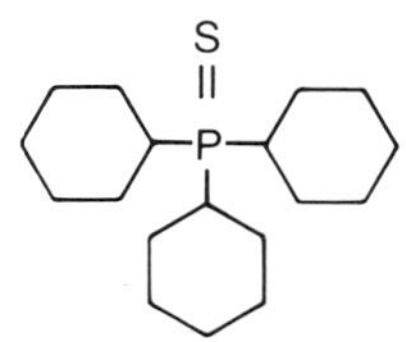	Weak	$Pn2_1a$	Dirk and Twieg (unpublished results, 1985) Ann Kerr *et al.* (1977)
160	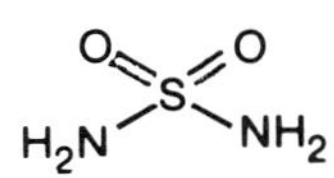	0.3*U*	*Fdd*2	Dirk and Twieg (unpublished results, 1985) Trueblood and Mayer (1956)

dithiocarbamate (**154**) is known to have an acentric crystal structure (Wahlberg, 1979), and powder tests on this substance indicate that it has comparable SHG efficiency and good damage stability like urea (Dirk and Twieg, unpublished results, 1985). The *N*-alkyl-4-pyridones are an attractive group of materials with a relatively large μ_g near 6 D and a CT band with λ_{max} near 260 nm. Among them, *N*-ethyl-2,6-dimethyl-4-pyridone (**155**) has an efficiency comparable to urea (Nicoud, unpublished results, 1985). The novel molecule 6-azauracil (**156**), for which an acentric crystal structure is known (Singh and Hodgson, 1974), is formally a derivative of urea and has a powder SHG efficiency about twice that of urea (Nicoud, unpublished results, 1985). The isomorphous series Ph_4X (X = C, Si, Ge, Sn, Pb) (**157**) has $\bar{4}$ molecular symmetry (Kitaigorodsky, 1973) in the crystal due to hindered rotation of the aromatic rings, and the resultant space groups are noncentric $P\bar{4}2_1c$. All these compounds are SHG-active, but it has proved impossible to obtain NLO measurements because even high-quality single crystals shatter under the influence of the YAG fundamental at 1.06 μm (Schlotter and Twieg, 1984). This pronounced optical instability is of interest in its own rights.

The crystal structure of triethylphosphine sulfide (TEPS, **158**) is known to be in an acentric hexagonal space group (Van Meerssche and Leonard, 1959). TEPS has a powder efficiency near that of urea (λ_{max} at 215 nm) and has its molecular dipole aligned right along b_{zzz} (although here phase matching is not possible). The isomorphous compounds in this group, such as triethylphosphine selenide and triethylarsine sulfide, are of obvious interest for examination (Dirk and Twieg, unpublished results, 1985). It is of interest that in the structure determination of tricyclohexylphosphine sulfide (**159**), SHG was utilized to assist determination of the space group (Ann Kerr *et al.*, 1977). The oxides of sulfur, the oxides and sulfides of phosphorus, and related compounds are certainly germane for further study as a general class of NLO materials. As an additional example, the urea analog (—CO— replaced by —SO_2—) sulfamide (**160**) has a powder efficiency about one-third that of urea, with comparable excellent transparency and damage resistance (Dirk and Twieg, unpublished results, 1985). The stickler will note that this molecule is not organic (since it contains no carbon), but the appropriate eventual molecular engineering with carbon-containing substituents will render the general class organic.

VI. CONCLUSION

The present work unambiguously demonstrates that remarkable progress has been made in the basic understanding and in the molecular engineering of single-crystal nonlinear optical materials. However, the level of progress has been uneven as a function of the absorption properties of these NLO materials. In the region of absorption cut-off of about 400–500 nm characterizing the nitroaromatic derivatives, near-ideal molecules with high merit figures are now known, but in the UV cut-off region of $\lambda < 400$ nm and in the longer-wavelength region of $\lambda > 500$ nm, significant improvement is anticipated.

In addition to the single-crystal organic materials, alternative organized media require more attention (and invention!). These media include crystal inclusion complexes, such as *p*-nitroaniline (pNA) in cyclodextrin (Tomaru *et al.*, 1984a), cocrystallites such as pNA and *p*-nitrophenol mixtures (Vizgert *et al.*, 1982), "quasi-crystals" from spiropyrans (Meredith *et al.*, 1983b), and thin-film organic crystals and polycrystallites (DeMicheli *et al.*, 1983; Hewig and Jain, 1983). The utility of Langmuir–Blodgett monolayers for SHG in reflection mode has been demonstrated (Aktsipetrov *et al.*, 1983), and efforts have begun on the more technically difficult Z configuration or alternating layers in X or Y configuration desirable for transmission SHG (Daniel and Smith, 1984; Blinov *et al.*, 1984; Girling, 1985). Significant progress has also been made in the application of Langmuir–Blodgett films to $\chi^{(3)}$ phenomena (Carter *et al.*, 1983; Kajzar *et al.*, 1983), since here the symmetry requirements are relaxed. Liquid crystals are a very attractive media for development, and a variety of NLO phenomena involving liquid crystals have already been examined (Arakelyan *et al.*, 1980). Much more attention must be paid to the applications of polymers as nonlinear media to provide an anisotropic environment by mechanical deformation, as a host for alignment of polarizable guest molecules and especially the preparation of specialized polymers with the polarizable functionality covalently bound as side groups or in the main chain. Liquid crystalline polymers are particularly attractive for such applications (Meredith *et al.*, 1983a). A number of these areas are dealt with in more detail in the other chapters of this volume.

Sustained progress demands diligence and cooperation between chemists and physicists, particularly as these new materials find device applications. Already crystals of urea have been utilized for optical parametric oscillation (Donaldson and Tang, 1984), and parametric interaction has been obtained from POM (Zyss *et al.*, 1985). Experience with growing NLO organic crystals in microenvironments such as benzil in the cores of optical fibers (Nayar, 1983) or 1,3-dinitrobenzene (Ballentyne and Al-Shukri, 1984) and mNA in channel waveguides (Tomaru *et al.*, 1984b) is accumulating, and technological applications are close at hand. The multitude of organic media to examine and the large variety of potential applications for them are more than adequate motivation for the organic chemist to venture into the lab and continue the search for these fascinating molecules.

ACKNOWLEDGMENTS

J. F. Nicoud wishes to thank Prof. H. Kagan of the Laboratoire de Synthèse Asymètrique at the Université d'Orsay, where many of the new compounds described in this chapter were prepared. The valuable collaboration of J. Prandi, M. F. Harris, and M. Coquillay is acknowledged, as well as the numerous fruitful discussions with J. Badan, R. Hierle, I. Ledoux, J. L. Oudar, A. Périgaud, P. Vidakovic, and J. Zyss of the CNET Bagneux. A portion of this work has been financially supported by an "ATP Optoélectronique" grant from the Centre National de la Recherche Scientifique (CNRS).

Robert Twieg wishes to acknowledge the contributions of his colleagues at IBM and visiting scientists who have been involved in the nonlinear organic materials program, especially Dr. Kanti Jain (now at IBM Yorktown Heights Research Laboratory) and Dr. Alain Azema (Université de Nice, Laboratoire d'Electro-optique). Particular thanks are due to more recent collaborators, including Prof. Georges Wagnière (Universität Zürich, Physikalisch-Chemisches Institut) and especially Dr. Carl Dirk and Dr. Diane Doborowolski for their invaluable contributions, including those involved in the preparation of this chapter.

REFERENCES

Adhikesavalu, D., and Venkatesan, K. (1982). *Acta Crystallogr., Sect. B* **B38,** 855.

Adhikesavalu, D., and Venkatesan, K. (1983). *Acta Crystallogr., Sect. C* **C39,** 589.

Adhikesavalu, D., Kamath, N. U., and Venkatesan, K. (1983). *Proc.—Indian Acad. Sci., Sect. A* Part 1 **92**(4&5), 449.

Aktsipetrov, O. A., Akhmediev, N. N., Mishina, E. D., and Novak, V. R. (1983). *Sov. Phys.—JETP Lett., (Engl. Transl.)* **37,** 207.

Alléaume, M. (1967). Thesis, University of Bordeaux.

Alléaume, M., and Decap, J. (1965). *C. R. Hebd. Seances Acad. Sci.* **260,** 5790.

Allen, F. H., Kennard, O., and Taylor, R. (1983). *Acc. Chem. Res.* **16,** 146.

Ann Kerr, K., Boorman, P. M., Misener, B. S., and van Roode, J. G. H. (1977). *Can J. Chem.* **55,** 3081.

Arakelyan, S. M., Lyakhov, G. A., and Chilingaryan, Yu. S. (1980). *Sov. Phys.—Usp. (Engl. Transl.)* **23**(5), 245.

Aroney, M. J., Calderbank, K. E., Le Fèvre, R. J. W., and Pierens, R. K., (1968). *J. Chem. Soc. B* p. 561.

Bader, H., Hansen, A. R., and McCarty, F. J. (1966). *J. Org. Chem.* **31,** 2319.

Bakhshiev, N. G., Knyazhanskii, M. I., Osipov, O. A., Minkin, V. I., and Saidov, G. V. (1969). *Usp. Khim* **38,** 1644.

Ballentyne D. W. G., and Al-Shukri, S. M. (1984). *J. Crystal Growth* **68,** 651.

Bartoli, G., Todesco, P. E., and Forentino, M. (1977). *J. Am. Chem. Soc.* **99,** 6874.

Bass, M., Bua, D., Mozzi, R., and Monchamp, R. R. (1969). *Appl. Phys. Lett.* **15**(12), 393.

Bechmann, R., and Kurtz, S. K. (1969). *In* "Landolt-Börnstein Tables," Vol. III, Part 2, p. 167. Springer-Verlag, Berlin and New York.

Bel'skii, V. K., and Zorkii, P. M. (1971). *Sov. Phys.—Crystallogr. (Engl. Transl.)* **15**(4), 607.

Bergman, J. G., Crane, G. R., Levine, B. F., and Bethea, C. G. (1972). *Appl. Phys. Lett.* **20**(1), 21.

Betzler, K., Hesse, H., and Loose, P. (1978). *J. Mol. Struct.* **47,** 393.

Blinov, L. M., Dubinin, N. V., Milchev, L. V., and Yudin, S. G. (1984). *Thin Solid Films* **120,** 161.

Bloembergen, N. (1965). "Nonlinear Optics." Benjamin, New York.

Bogaard, M. P., and Orr, B. J. (1975). *In* "Molecular Structure and Properties" (A. D. Buckingham, ed.), Phys. Chem. Ser. Two, Vol. 2, Chapter 5. Butterworth, London.
Bramley, R., and Le Fèvre, R. J. W. (1960). *J. Chem. Soc.* p. 1820.
Bramley, R., and Le Fèvre, R. J. W. (1962). *J. Chem. Soc.* p. 56.
Bromilow, J., and Brownlee, R. T. C. (1979). *J. Org. Chem.* **44**(8), 1261.
Brown, G. M., and Levy, H. A. (1973). *Acta Crystallogr., Sect. B* **B29,** 790.
Brown, J. N., Kutchan, T. H., and Rist, P. E. (1980). *Cryst. Struct. Commun.* **9,** 17.
Brühl, J. W. (1886). *Justus Liebig's Ann. Chem.* **235,** 1.
Brunskill, J. S. A., De, A., and Vas, G. M. F. (1978). *Synth. Commun.* **8**(1), 1.
Buckingham, A. D., and Orr, B. J. (1967). *Q. Rev., Chem. Soc.* **21,** 195.
Carter, G. M., Chen, Y. J., and Tripathy, S. K. (1983). *ACS Symp. Ser.* **233,** 213.
Cassidy, C., Halbout, J. M., Donaldson, W., and Tang, C. L., (1979). *Opt. Commun.* **29**(2), 243.
Chemla, D., Oudar, J. L., and Zyss, J. (1981). *Echo Rech.* **103,** 3.
Coda, A., and Pandarese, F. (1976). *J. Appl. Cryst.* **9,** 193.
Coulson, C. A., Maccoll, A., and Sutton, L. E. (1952). *Trans. Faraday Soc.* **48,** 106.
Coquillay, M. (1986). Personal communication.
Cowles, E. J. (1986). *J. Org. Chem.* **51**(8), 1336.
Craik, D. J., and Brownlee, R. T. C. (1983). *Prog. Phys. Org. Chem.* **14,** 1.
Curtin, D. Y., and Paul, I. C. (1981). *Chem. Rev.* **81**(6), 525.
Dähne, S., and Nolte, K-D. (1972). *J. Chem. Soc., Chem. Commun.* p. 1056.
Daniel, M. F., and Smith, G. W. (1984) *Mol. Cryst. Liq. Cryst.* **102,** 193.
Davies, P. L. (1952). *Trans. Faraday Soc.* **48,** 789.
Davydov, B. L., Derkacheva, L. D., Dunina, V. V., Zhabotinskii, M. E., Zolin, V. F., Koreneva, L. G., and Samokhina, M. A. (1970). *Sov. Phys.—JETP Lett. (Engl. Transl.)* **12**(1), 16.
Davydov, B. L., Derkacheva, L. D., Dunina, V. V., Shabotinskii, M. E., Zolin, V. F., Koreneva, L. G., and Samokhina, M. A. (1971). *Opt. Spectrosc. (Engl. Transl.)* **30,** 274.
Davydov, B. L., Dunina, V. V., Zolin, V. F., and Koreneva, L. G. (1973). *Opt. Spectrosc. (Engl. Transl.)* **34**(2), 150.
Davydov, B. L., Kotovschchikov, S. G., and Nefedov, V. A. (1977). *Sov. J. Quantum. Electron. (Engl. Transl.)* **7**(1), 129.
Delfino, M. (1979). *Mol. Cryst. Liq. Cryst.* **52,** 271.
DeMicheli, M., Zyss, J., and Azema, A. (1983). *Proc. SPIE Int. Soc. Opt. Eng.* **401,** 216.
Dewar, M. J. S. (1950). *J. Chem. Soc. p.* 2329.
Dewar, M. J. S. (1969). "The Molecular Orbital Theory of Organic Chemistry." McGraw-Hill, New York.
Domenicano, A., Mazzeo, P., and Vacaigo, A. (1976) *Tetrahedron Lett.* p. 1029.
Donaldson, W., and Tang, C. L. (1984). *Appl. Phys. Lett.* **44**(1), 25.
Dougherty, J. P., and Kurtz, S. K. (1976). *J. Appl. Cryst.* **9,** 145.
Dulcic, A. (1979). *Chem. Phys.* **37,** 57.
Dulcic, A., and Flytzanis, C. (1978). *Opt. Commun.* **25**(3), 402.
Dulcic, A., and Sauteret, C. (1978). *J. Chem. Phys.* **69**(8), 3453.
Edwards, J. O. (1956). *J. Am. Chem. Soc.* **78,** 1819.
Ehrenson, S., Brownlee, R. T. C., and Taft, R. W., (1973). *Prog. Phys. Org. Chem.* **10,** 1.
Ericsson, E., Sandström, J., Wennerbeck, I. (1970) *Acta Chem. Scand.* **24,** 3102.
Exner, O. (1966). *Collect. Czech. Chem. Commun.* **31,** 65.
Exner, O. (1975). "Dipole Moments in Organic Chemistry." Thieme, Stuttgart.
Fabian, J., and Hartman, H. (1980). "Light Absorption of Organic Colorants." Springer-Verlag, Berlin and New York.
Filipenko, O. S., Shigorin, V. D., Ponomarev, V. I., Atovmyan, L. O., Safina, Z. S., and Tarnopo'skii, B. L. (1977). *Sov. Phys.—Crystallogr. (Engl. Transl.)* **22**(3), 305.

Filipenko, O. S., Ponomarev, V. I., and Atovmyan, L. O. (1980). *Sov. Phys.—Crystallogr. (Engl. Transl.)* **25**(5), 549.

Flytzanis, C. (1975). *In* "Quantum Electronics: A Treatise" (H. Rabin and C. L. Tang, eds.), Vol. 1, Part A, p. 19. Academic Press, New York.

Franken, P. A., Hill, A. E., Peters, C. W., and Weinrech, G. (1961). *Phys. Rev. Lett.* **7**(4), 118.

Furmanova, N. G., Razmanova, Z. P., Soboleva, L. V., Maslyanitsyn, I. A., Siegert, H., Shigorin, V. D., and Shipulo, G. P. (1984). *Sov. Phys—Crystallogr. (Engl. Transl.)* **29**(3), 285.

Girling, I. R., Kolinsky, P. V., Cade, N. A., Earls, J. D., and Peterson, I. R. (1985). *Opt. Commun.* **55,** 289.

Gompper, R., Wagner, H. U., and Kutter, E. (1968). *Chem. Ber.* **101,** 4144.

Gordon, P. F., and Gregory, P. (1983). "Organic Chemistry in Colour." Springer-Verlag, Berlin and New York.

Gott, J. R. (1971). *J. Phys. B* **4,** 116.

Gould, E. S. (1959). "Mechanism and Structure in Organic Chemistry." Holt, New York.

Gusten, H., and Salzwedel, M. (1967). *Tetrahedron* **23,** 173.

Gutman, F., and Lyons, L. E. (1967). "Organic Semiconductors," Chapter 5. Wiley, New York.

Halbout, J.-M., and Tang, C. L., (1982). *IEEE J. Quantum Electron.* **QE-18**(3), 410.

Halbout, J.-M., Blit, S., Donaldson W., and Tang, C. L. (1979). *IEEE J. Quantum Electron.* **QE-15**(10), 1176.

Halbout, J.-M., Sarhangi, A., and Tang, C. L. (1980). *Appl. Phys. Lett.* **37**(10), 864.

Hammett, L. P. (1970). "Physical Organic Chemistry: Reaction Rates, Equilibria and Mechanisms," 2nd ed. McGraw-Hill, New York.

Happer, D. A. R., and Wright, G. J. (1979). *J. Chem. Soc. Perkin. Trans. 2*, p. 694.

Heilmeier, G. H., Ockman, N., Braunstein, R., and Kramer, D. A. (1964). *Appl. Phys. Lett.* **5**(11), 229.

Henry, N. F. M., and Lonsdale, K., eds. (1969). "International Tables for X-Ray Crystallography," Vol. 1. Kynoch Press, Birmingham.

Hewig, G. H., and Jain, K. (1983). *Opt. Commun.* **47**(5), 347.

Hiberty, P. C., and Ohanessian, G. (1984) *J. Am. Chem. Soc.* **106,** 6963.

Higashi, T., and Osaki, K. (1977). *Acta Crystallogr., Sect. B* **B33,** 2337.

Hirotsu, K., and Shimada, A. (1974). *Bull Chem. Soc. Jpn.* **47,** 1872.

Hobden, M. V. (1967). *J. Appl. Phys.* **38**(11), 4365.

Ingham, C. E., and Hampson, G. C. (1939). *J. Chem. Soc.* p. 981.

Ingold, C. K. (1969). "Structure and Mechanism in Organic Chemistry," 2nd ed. Cornell Univ. Press, Ithaca, New York.

Jacques, J., Collet, A., and Wilen, S. (1981). "Enantiomers, Racemates and Resolutions." Wiley (Interscience), New York.

Jain, K., Crowley, J. I., Hewig, G. H., Cheng, Y. Y., and Twieg, R. J. (1981). *Opt. Laser Technol.* 297.

Jerphagnon, J. (1971). *IEEE J. Quantum Electron.* **QE-7,** 42.

Jerphagnon, J., Kurtz, S. K., and Oudar, J. L. (1984). *In* "Landolt–Bornstein," (K.-H. Hellwege, ed.), Vol. 18, Springer–Verlag, Berlin and New York.

Jones, G. (1967) *Org. React.* **15,** 204.

Kajzar, F., Messier, J., Zyss, J., and Ledoux, I. (1983). *Opt. Commun.* **45**(2), 133.

Kato, K. (1980) *IEEE J. Quantum Electron.* **QE-16**(12), 1288.

Katritzky, A. R., and Topsom, R. D. (1970). *Angew. Chem., Int. Ed. Engl.* **9**(2), 87.

Kitaigorodsky, A. I. (1973). "Molecular Crystals and Molecules" Academic Press, New York.

Kitaigorodsky, A. I., and Mirskaya, K. V. (1965). *Sov. Phys.—Crystallogr. (Engl. Transl.)* **9**(5), 539.

Kleinman, D. A. (1962). *Phys. Rev.* **128,** 1761.

Klewe, B. (1972). *Acta Chem. Scand.* **26,** 317.

Knossow, M., Mauguen, Y., and de Rango, C. (1976). *Cryst. Struct. Commun.* **5,** 719.
Kollman, P. A., and Allen, L. C. (1972). *Chem. Rev.* **72**(3), 283.
Koreneva, L. G., Zolin, V. F., and Davydov, B. L. (1975). "Molecular Crystals in Nonlinear Optics." Nauka, Moscow.
Koutek, B. (1984). *Collect. Czech. Chem. Commun.* **49,** 1680.
Kuhn, W., and Martin, H. (1962). *In* "Landolt-Börnstein Tables" (K.-H. Hellwege, ed.), Vol. II, Part 8, p. 676. Springer-Verlag, Berlin and New York.
Kurtz, S. K., and Perry, T. T. (1968). *J. Appl. Phys.* **39**(8), 3798.
Kurtz, S. K., Jerphagnon, J., and Choy, M. M. (1979). *In* "Landolt-Börnstein Tables" (K.-H. Hellwege, ed.)., Vol. III, Part 11, p. 671. Springer-Verlag, Berlin and New York.
Ladd, M. F. C. (1979). "Structure and Bonding in Solid State Chemistry." Wiley, New York.
Lalama, S. J., Singer, K. D., Garito, A. F. and Desai, K. N. (1981). *Appl. Phys. Lett.* **39**(12), 940.
Ledoux, I., Josse, D., Vidaković, P., Zyss, J. (1986). *Opt. Engineering* **25**(2), 202.
Le Févre, R. J. W. (1965) *Adv. Phys. Org. Chem.* **3,** 1.
Le Fèvre, R. J. W. (1970). *Rev. Pure Appl. Chem.* **20,** 67.
Levine, B. F. (1976). *Chem. Phys. Lett.* **37,** 516.
Levine, B. F., and Bethea, C. G. (1975). *J. Chem. Phys.* **63,** 2666.
Levine, B. F., Bethea, C. G., Wasserman, E., and Leenders, L. (1978). *J. Chem. Phys.* **68**(11), 5042.
Levine, B. F., Bethea, C. G., Thurmond, C. D., Lynch, R. T., and Bernstein J. L. (1979). *J. Appl. Phys.* **50**(4), 2523.
Lipkowitz, K. B. (1982). *J. Am. Chem. Soc.* **104,** 2647.
Lipscomb, G. F., Garito, A. F., and Narang, R. S. (1981a). *Appl. Phys. Lett.* **38**(9), 663.
Lipscomb, G. F., Garito, A. F., and Narang, R. S. (1981b). *J. Chem. Phys.* **75**(3), 1509.
McClellan, A. L. (1963). "Tables of Experimental Dipole Moments," Vol. I. Freeman, San Francisco, California.
McClellan, A. L. (1974). "Tables of Experimental Dipole Moments," Vol. 2. Rahara Enterprises, El Cerrito, California.
Magomeda, N., Zvonkava, Z., Neigauz, M., and Novakovskaya, L. (1980). *Sov. Phys.—Crystallogr. (Engl. Transl.)* **9,** 17.
Mak, T. C. W., and Trotter, J. (1965). *Acta Crystallogr.* **18,** 68.
March, J. (1968). "Advanced Organic Chemistry: Reactions, Mechanism, Structure," McGraw-Hill, New York.
March, J. (1985). "Advanced Organic Chemistry: Reactions, Mechanism, Structure," 3rd ed. McGraw-Hill, New York.
Mataga, N., and Kubota, T. (1970). "Molecular Interactions and Electronic Spectra." Dekker, New York.
Maze-Baudet, M. (1973). *Acta Crystallogr., Sect. B* **B29,** 602.
Meier, H. (1974). "Organic Semiconductors," Chapter 5. Wiley, New York.
Meredith, G. R. (1982). *Rev. Sci. Instrum.* **53,** 48.
Meredith, G. R., Vandusen, J. G., and Williams, D. J. (1983a). *ACS Symp. Ser.* **233,** 109.
Meredith, G. R., Williams, D. J., Fishman, S. N., Goldburt, E. S., and Krongauz, V. A. (1983b). *ACS Symp. Ser.* **233,** 135.
Meredith, G. R., Weagley, R. J., Williams, D. J., and Ziolo, R. F. (1985). Personal communication.
Midwinter, J. E., and Warner, J. (1965). *Br. J. Appl. Phys.* **16,** 1135.
Mighell, A. D., and Ondik, H. M. (1977). *J. Phys. Chem. Ref. Data* **6,** 675.
Mighell, A. D., and Rodgers, J. R. (1980). *Acta Crystallogr., Sect. A* **A36,** 321.
Mighell, A. D., Himes, V. L., and Rodgers, J. R. (1983). *Acta Crystallogr., Sect. A* **A39,** 737.
Miller, J. (1968). "Aromatic Nucleophilic Substitution." Elsevier, Amsterdam.
Minkin, V. I., Osipov, O. A., and Zhdanov, Yu. A. (1970). "Dipole Moments in Organic Chemistry" (Engl. Transl.). Plenum, New York.
Moret, E., Twieg, R. J., and Jain, K. (1983). *IBM Disclosure Bull.* **26**(1), 422.

Morrison, J. D. (1983). *In* "Asymmetric Synthesis," Vol. 1, Ch. 1. Academic Press, Orlando, Florida.
Nagakura, S. (1963). *Pure Appl. Chem.* **7,** 79.
Nayar, B. K. (1983). *ACS Symp. Ser.* **233,** 153.
Nayar, B. K. (1984). *CLEO '84, Conf. Abstr.* p. 259.
Newman, P. (1978). "Optical Resolution Procedures for Chemical Compounds," Vol. I. Optical Resolution Information Center, Riverdale, New York.
Newman, P. (1981). "Optical Resolution Procedures for Chemical Compounds," Vol. II. Optical Resolution Information Center, Riverdale, New York.
Newman, P. (1984). "Optical Resolution Procedures for Chemical Compounds," Vol. III. Optical Resolution Information Center, Riverdale, New York.
Nolte, K.-D., and Dähne, S. (1976). *J. Prakt. Chem.* **318**(4), 643.
Nye, J. F. (1960). "Physical Properties of Crystals." Oxford Univ. Press (Clarendon), London and New York.
O'Loane, J. K. (1980). *Chem. Rev.* **80,** 41.
Orlov, R. Y. (1966). *Sov. Phys.—Crystallogr. (Engl. Transl.)* **11**(3), 410.
Oudar, J. L. (1977a). *J. Chem. Phys.* **67,** 446.
Oudar, J. L. (1977b). Ph.D. Thesis, p. 158. University of Pierre and Marie Curie, Paris.
Oudar, J. L., and Chemla, D. S. (1975). *Opt. Commun.* **13,** 164.
Oudar, J. L., and Chemla, D. S. (1977). *J. Chem. Phys.* **66,** 2664.
Oudar, J. L., and Hierle, R. (1977). *J. Appl. Phys.* **48,** 2699.
Papadopoulos, M. G., and Waite, J. (1985). *J. Chem. Phys.* **82**(3), 1435.
Pauling, L. (1960). "Nature of the Chemical Bond." Cornell Univ. Press, Ithaca, New York.
Platt, J. R. (1964). "Systematics of the Electronic Spectra of Conjugated Molecules." Wiley, New York.
Pope, M., and Swenberg, C. (1982). "Electronic Processes in Organic Crystals," Chapter 1. Oxford Univ. Press (Clarendon), London and New York.
Reinhardt, R., Tiwari, R., and Singh, T. (1980). *Curr. Sci.* **49,** 586.
Rentzepis, P. M., and Pao, Y-H. (1964). *Appl. Phys. Lett.* **5**(8), 156.
Rez, I. S., (1960). *Sov. Phys.—Crystallogr. (Engl. Transl.)* **5,** 54.
Rieckhoff, K., and Peticolas, W. F. (1965). *Science* **147,** 611.
Roberts, J. D., and Caserio, M. C. (1965). "Basic Principles of Organic Chemistry." Benjamin, New York.
Rosker, M. J., and Tang, C. L. (1984). *IEEE J. Quantum Electron.* **QE-20**(4), 334.
Rustagi, K. C., and Ducuing, J. (1974). *Opt. Commun.* **10,** 258.
Sandström, J. (1983). *Top. Stereochem.* **14,** 83.
Sasaki, K., Kinoshita, T., and Karasawa, N. (1984). *Appl. Phys. Lett.* **45**(4), 333.
Schlotter, N. E., and Twieg, R. J. (1984). *IBM Disclosure Bull.* **27**(3)1, 1538.
Schudt, E., and Weitz, G. (1971). *In* "Landolt-Börnstein Tables" (K.-H. Hellwege, ed.), Group III, Parts 5a and 5b. Springer–Verlag, Berlin and New York.
Schuler, L., Betzler, K., Hesse, H., and Kapphan, S. (1982). *Opt. Commun.* **43**(2), 157.
Schweig, A. (1967a). *Chem. Phys. Lett.* **1,** 163.
Schweig, A. (1967b). *Chem. Phys. Lett.* **1,** 195.
Scott, J. W. (1984). *In* "Asymmetric Synthesis" Vol. 4, (J. D. Morrison, ed.), Chapter 1. Academic Press, Orlando, Florida.
Shechtman, D., Blech, I., Gratias, D., and Cahn, J. W. (1984). *Phys. Rev. Lett.* **53**(20), 1951.
Shigorin, V. D., and Shipulo, G. P. (1973). *Sov. J. Quantum Elec.* **2**(4), 390.
Shigorin, V. D., and Shipulo, G. P. (1975). *Sov. Phys.—Crystallogr. (Engl. Transl.)* **19**(5), 622.
Shigorin, V. D., Shipulo, G. P., Grazhulene, S. S., Musikhin, L. A., and Shekhtman, V. Sh. (1976). *Sov. J. Quantum Electron. (Engl. Transl.)* **5**(11), 1393.
Singh, P., and Hodgson, D. J. (1974). *Acta Crystallogr. Sect. B* **B30,** 1430.

Skrabal, P., Steiger, J., and Zollinger, H. (1975). *Helv. Chim. Acta* **58**(3), 800.
Smyth, C. P. (1955). "Dielectric Behavior and Structure." McGraw-Hill, New York (University Microfilms Facsimile).
Solladié, G., and Zimmermann, R. G. (1984). *Angew. Chem., Int. Ed. Engl.* **23,** 348.
Southgate, P. D., and Hall, D. S. (1971). *Appl. Phys. Lett.* **18**(10), 456.
Southgate, P. D., and Hall, D. S. (1972). *J. Appl. Phys.* **43**(6), 2765.
Stegeman, G. I., and Liao, C. (1983). *Appl. Opt.* **22**(17), 2518.
Stevenson, P. E. (1965). *J. Mol. Spectrosc.* **15,** 220.
Stock, L. (1968). "Aromatic Substitution Reactions." Prentice-Hall, Englewood Cliffs, New Jersey.
Streitweiser, A. (1961). "Molecular Orbital Theory for Organic Chemists." Wiley, New York.
Sutton, L. E. (1931). *Proc. R. Soc. London* **133,** 668.
Sutton, L. E., Kumar, L., and Everad, K. B. (1951). *J. Chem. Soc.* pp. 2807, 2816, 2817, 2818, 2821, 2826.
Szazbo, W. A., and Lee, H. T. (1980). *Aldrichim. Acta* **13**(1), 13.
Taylor, R., and Kennard, O. (1984). *Acc. Chem. Res.* **17,** 320.
Tomaru, S., Zembutsu, S., Kawachi, M., and Kobayashi, H. (1984a). *J. Chem. Soc., Chem. Commun.* p. 1207.
Tomaru, S., Kawachi, M., and Kobayashi, M. (1984b). *Opt. Commun.* **50**(3), 154.
Trueblood, K. N., and Mayer, S. W. (1956). *Acta Crystallogr.* **9,** 628.
Twieg, R. J., Azema, A., Jain, K., and Cheng, Y. Y. (1982a). *Chem. Phys. Lett.* **92**(2), 208.
Twieg, R. J., and Cheng, Y. Y. (1985). *IBM Tech. Disclosure Bull.* **27**(12), 7017.
Twieg, R. J., and Dirk, C. (1986). *J. Chem. Phys.*, accepted.
Twieg, R. J., and Jain, K. (1983). *ACS Symp. Ser.* **233,** 57.
Twieg, R. J., Jain, K., Cheng, Y. Y., Crowley, J. I., and Azema, A. (1982b). *Polym. Prepr., Am. Chem. Soc. Div. Polym. Chem.* **23**(2), 147.
Van Meerssche, M., and Leonard, A. (1959). *Bull. Soc. Chim. Belg.* **68,** 683.
Vizgert, R. V., Davydov, B. L., Kotovshchikov, S. G., and Starodubtseva, M. P. (1982). *Sov. J. Quantum Electron. (Engl. Transl.)* **12**(2), 214.
Wahlberg, A. (1979). *Acta Crystallogr. Sect. B* **B35,** 485.
Wilen, S. H. (1972). "Tables of Resolving Agents and Optical Resolutions." Univ. of Notre Dame Press, Notre Dame, Indiana.
Williams, D. J. (1984). *Angew. Chem. Int. Ed. Engl.* **23,** 690.
Winchell, A. N. (1954). "The Optical Properties of Organic Compounds." Academic Press, New York.
Wöhler, F. (1828). *Ann. Phys. (Leipzlg)* [2] **12,** 253.
Xu, D., Jiang, M., and Tan, Z. (1983). *Acta Chim. Sin.* **2,** 230.
Yamakawa, M., Kubota, T., Ezumi, K., and Mizuno, Y. (1974). *Spectrochim. Acta* **30,** 2103.
Zernike, F., and Midwinter, J. E. (1973). "Applied Nonlinear Optics." Wiley (Interscience), New York.
Zorkii, P. M., Lazareva, S. G., and Zefirov, Yu. V. (1971). *Sov. Phys.—Crystallogr. (Engl. Transl.)* **15**(4), 602.
Zyss, J. (1982). *J. Non-Cryst. Solids* **47**(2), 211.
Zyss, J., and Berthier, G. (1982). *J. Chem. Phys.* **77**(7), 3635.
Zyss, J., and Oudar, J. L. (1982). *Phys. Rev. A* **26**(4), 2028.
Zyss, J., Chemla, D. S., and Nicoud, J. F. (1981). *J. Chem. Phys.* **74**(9), 4800.
Zyss, J., Nicoud, J. F., and Coquillay, M. (1984). *J. Chem. Phys.* **81**(9), 4160.
Zyss, J., Ledoux, I., Hierle, R., Raj, R., and Oudar, J.-L. (1985). *IEEE J. Quant. Elec.*, **QE-21**(8), 1286.

Chapter II-4

Growth and Characterization of Molecular Crystals

J. BADAN, R. HIERLE, A. PERIGAUD, and P. VIDAKOVIC

Centre National d'Etudes Telecommunication
92220 Bagneux, France

I. INTRODUCTION

The fast-growing development of optical fiber communication systems has stimulated the search for new highly nonlinear materials capable of fast and efficient processing of optical signals. Nonlinear organic crystals have proven to be interesting candidates for a number of applications: second-harmonic generation, frequency mixing, electro-optic modulation, optical parametric oscillation, optical bistability, etc. (Badan *et al.*, 1983). Furthermore, nonlinear materials of higher efficiency, either in crystal, thin film, or any other suitable shape, can provide such functions on a reduced scale and without a delaying electron–photon conversion process. Molecular features responsible for the enhancement of three-photon effects were initially identified after scanning a large number of organic compounds (Davidov *et al.*, 1970; Jerphagnon, 1971; Chemla *et al.*, 1975) using the second-harmonic generation (SHG) powder test (Kurtz, 1968). It has helped show the basic features of nonlinear molecules, namely, (1) the presence of highly conjugated and polarizable electronic systems, either linear, cyclic, or a combination of both, and (2) the occurrence of intramolecular charge transfer, which could be modulated by a suitable choice of acceptor and donor substituents. Moreover, one can in a similar way adjust the transparency of the system (usually between 0.3 and 2 μm). It is not sufficient to ensure a molecular nonlinear response: the crystal must also be noncentrosymmetric. From the fabrication and exploitation point of view, some other criteria must be added, like (1) easy and inexpensive synthesis, (2) certain physicochemical properties, such as good radiation and thermal and chemical resistance, which will influence the crystal growth method to be used, and (3) a tendency to crystallize easily. Moreover, the as-grown crystals should exhibit satisfactory mechanical properties, to follow for further device processing operations among which cutting and polishing are essential.

Certain classes of organic molecules have been identified for their interest in nonlinear optics (NLO), such as (1) disubstituted aromatic molecules: exemplified by *meta*-dinitrobenzene (*m*-DNB), *meta*-bromonitrobenzene (*m*-BNB), and nitroaniline derivatives, *meta*-nitroaniline (*m*-NA); *para*-nitroaniline (*p*-NA); methyl-(2,4-dinitrophenyl)-amino-2-propanoate (MAP); *N*-(4-nitrophenyl)-(L)-prolinol (NPP); and nitro-4-phenyl-*N*-(methyl-cyanomethyl)-amine (NPAN); (2) pyridine family: methyl-3-nitro-4-pyridine 1-oxide (POM); (3) stilbene family: *R*-(+)-*N*-methylidene-*gem*-(carbethoxyphenyl) amide of 4-*N*-dimethylaminobenzylidene cyanacetic acid (AMA); and (4) *para*-dimethylamino-β-nitrostyrene (MANS).

In this work, we first present the physicochemical properties of the above-mentioned molecules which prove relevant toward crystal growth. Synthesis and purification methods are briefly outlined in Section III. Section IV is devoted to the methods of crystal growth employed in our laboratory (i.e., growth from the melt and from solution). We discuss in Section V, different means of crystalline structure and quality characterization, such as X-ray (Laüe and Lambot techniques and Lang topography) and optical methods (polarized-light microscopy and conoscopic fringes). Finally, we discuss the elaboration of organic nonlinear waveguides, exemplified by organic crystal cored fibers (OCCF) and organic channel waveguides (OCWG).

II. PHYSICOCHEMICAL PROPERTIES OF NONLINEAR ORGANIC MATERIALS

Crystal growth is governed by a number of different physicochemical properties. The melting behavior will offer basic qualitative information on the material purity. A crude chemical synthesis product must be carefully purified before further use in growth. As an example, MAP without purification would remain for several days in the liquid state, even when strongly supercooled. This behavior precludes crystal growth of unpurified MAP from an supercooled melt. After appropriate treatments, it crystallizes spontaneously during cooling. Another property influencing the crystal growth of organic materials is their thermal and chemical stability. In order to be used for crystal growth from the supercooled melt, organic materials must show negligible thermal degradation and small vapor pressure. It is also well known (Pollara and von Korff, 1960; Böhm and Rüstov, 1969) that most organic materials are very sensitive to ultraviolet (UV) and visible light. To prevent photochemical decomposition, the crystal growth apparatus must be carefully protected from the light, either by darkening the laboratory or by individual apparatus protection schemes.

Crystal growth from solutions always involves solvent/crystallizing substance interactions, which show up in the formation of complexes of solvent with solute molecules as well as by solvation of crystal faces. All these effects depend strongly on the solvent polarity and can modify crystalline structure (by solvent inclusion, for example) or even stop the crystal growth. The solvent also influences the manner in which the solute is transported to the growing face. Bearing in mind NLO applications of interest in this chapter, platelet- and needlelike crystals, easily grown from solutions, are ruled out for practical applications. Generally, the crystals grow in supersaturated solutions, where the solute concentration exceeds the equilibrium value at a given temperature. The supersaturation domain is divided in two: the metastable zone, where crystal growth may be controlled, and the labile zone, where uncontrolled spontaneous crystallization occurs. Crystal growth can be controlled by the slow temperature decrease of a metastable supersaturated solution to yield crystals with well-developed faces.

Different physicochemical properties of organic materials we have studied are gathered in Table I.

III. MATERIAL PREPARATION FOR CRYSTAL GROWTH

A. Synthesis

The oxidation of 3-methylpyridine with hydrogen peroxide in glacial acetic acid leads to the liquefied 3-methylpyridine 1-oxide. This compound is diluted in cold sulfuric acid; then fuming nitric acid is added, while the temperature is slowly raised to 95–100°C over 30 min. From the crude product, POM is

TABLE I

Physicochemical Properties of Some Nonlinear Organic Materials[a]

Property	*p*-NA	*m*-NA	*m*-DNB	*m*-CNB[b]	*m*-BNB
Molecular weight	138	138	168	157	202
Melting point (°C)	148	114	90	46	56
Boiling point (°C)	33.1	307	291	235	256.5
Heat of fusion					
(kcal/mol)					
α_{20}^{λ} (deg)					
$\lambda = 6328$ Å					
$\lambda = 5890$ Å					
$\lambda = 5460$ Å					
Crystal class	Orthorhombic	Orthorhombic	Orthorhombic	Orthorhombic	Orthorhombic
Space group	$P2_{1/n}$	$Pca2_1$	$Pbn2_1$	$Pbn2_1$	$Pbn2_1$
Lattice parameters					
Z	4	4	4	4	4
a	12.336	19.23	13.257	6.0	5.92
b (Å)	6.07	6.48	14.048	21.40	21.52
c	8.592	5.06	3.806	5.35	5.34
β (deg)	91.45				
Calculated density ρ_c					
at 20°C	1.437	1.430	1.5656	1.582	1.969
at (T°C)		1.175	1.3644	1.343	1.651
		(160)	(90)	(46)	(59.5)

Property	MAP	NPP	NPAN[c]	POM	AMA	MANS
Molecular weight	269	222	191	154	378	192
Melting point (°C)	80.9	115.9	114	136.2	136.3	179
Boiling point (°C)	Decomp.[d]	Decomp.	Decomp.	Decomp.	Decomp.	Decomp.
Heat of fusion (kcal/mol)	3.71	6.6		5.75	7.66	
α_{20}^{λ}						
$\lambda = 6328$ Å					$+13.7498$	
$\lambda = 5890$ Å	-10.7	-92.1			$+25.6475$	
$\lambda = 5460$ Å	-25.3				$+50.2211$	
Crystal class	Monoclinic	Monoclinic	Orthorhombic	Orthorhombic	Monoclinic	Orthorhombic
Space group	$P2_1$	P_2	$Fdd2$	$P2_12_12_1$	$P2_1$	$Pcab$
Lattice parameters						
Z	2	2	16	4	2	8
a	6.829	5.257	4.33	21.359	20.963	7.30
b	11.121	7.178	26.2	6.111	5.244	10.15
c (Å)	8.116	14.893	34.2	5.134	9.392	25.10
β (deg)	95.59	109.20			99.47	
Calculated density ρ_c at 20°C	1.46	1.39	1.31	1.53	1.23	1.37

[a] Names of the compounds are detailed in Section I.
[b] *m*-Chloronitrobenzene.
[c] Detailed crystallographic analysis (Vidakovic *et al.*, 1986).
[d] Decomposes.

extracted with boiling chloroform. [For more details, see Zyss *et al.* (1981), and references therein.]

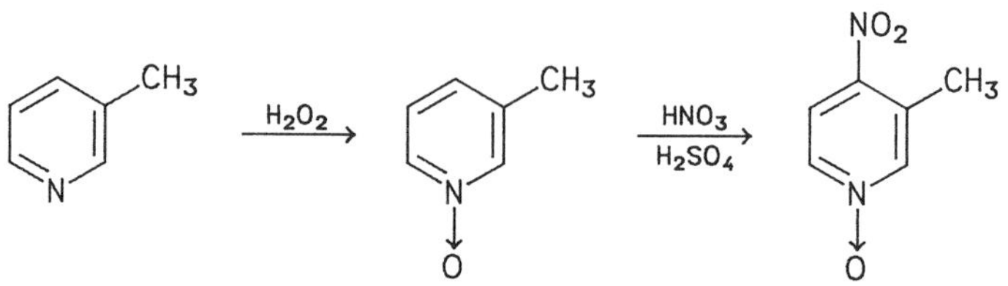

The reaction of 1-fluoro-4-nitrobenzene with L-prolinol in dimethyl sulfoxide (DMSO) leads to NPP. The crude material is recrystallized with toluene after treatment with activated charcoal (Zyss *et al.*, 1984).

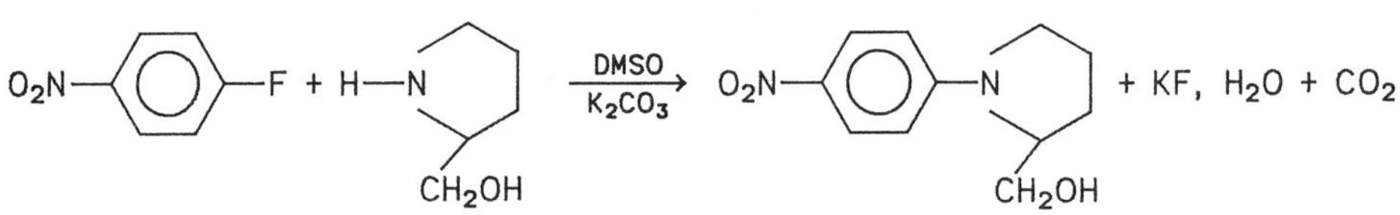

AMA is the condensation result of an acid chloride with the chlorhydrate of ethyl-*R*-(+)-α-phenylaminoacetate according to the reaction process shown in Fig. 1. Its complexity requires a considerable effort to obtain the large quantity of product necessary for crystal growth. Follow-up of the reaction

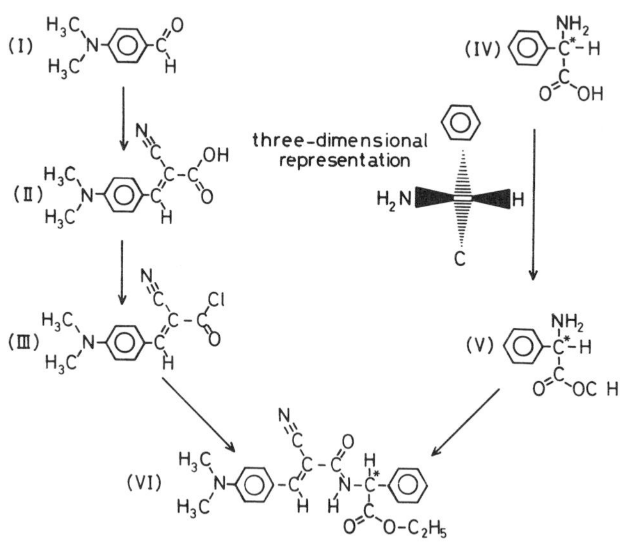

Fig. 1. Reactional scheme of the AMA synthesis.

kinetics by gas chromatography (GC) and nuclear magnetic resonance (NMR) has permitted the identification of optimal conditions for preparation of the crude product.

NPAN is obtained in a similar way (see Chapter II-3) following Sanger's reaction.

MAP is synthesized from L-(+)-α-alanine. 2,4-Dinitrophenyl-substituted acid is first prepared following Sanger's method (Oudar and Hierle, 1977). Two paths may lead to MAP: esterification by methanol in sulfuric medium, or reaction with an etheric solution of diazomethane in tetrahydrofuran.

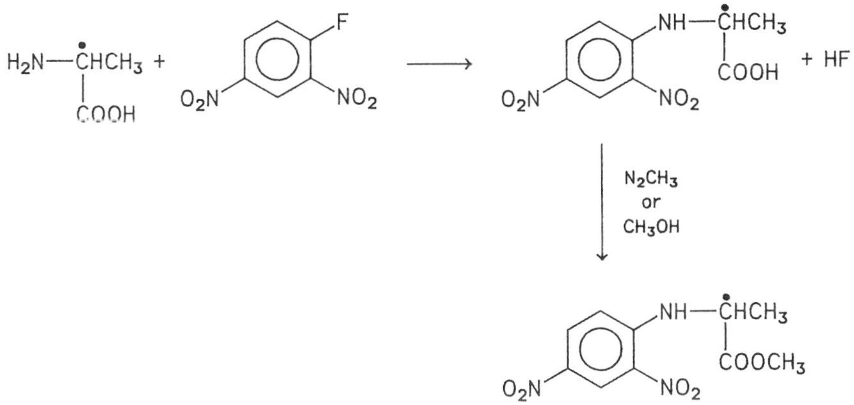

B. Purification

Generally, the first purification stage consists of one or several recrystallization steps of the organic material in an appropriate solvent. The material must display both a large solubility at its ebullition point and a reduced one at room temperature. Activated charcoal is added to adsorb and precipitate impurities before filtration at higher temperature. For some products—like POM—this process is sufficient. In most cases, the purification must go on with different physical methods, like distillation, sublimation, and sometimes evaporation, followed by zone melting. Table II sums up different means employed to purify the molecular materials studied in this chapter.

The most frequently used method for ultrapurification is zone refining, which has been discussed at large by Karl (1980). However, for MAP and NPP, this technique has proved completely inadequate owing to their thermal decomposition. For instance, products resulting from MAP degradation were identified by mass spectrometry to be 2,4-dinitrophenyl amino acid (DNPA)

TABLE II

Purification Methods for Studied Organic Materials

Methods	*m*-NA	*m*-DNB	*m*-BNB	MAP	NPP	NPAN	POM	AMA
Recrystallization	+	+	+	+	+	+	+	+
Distillation			+					
Sublimation			+					
Evaporation				+	+			
Zone refining				Decomp.[a]	Decomp.			
Preparative liquid chromatography				+				

[a] Decomposes.

and 2,4-dinitrophenylethylamine (DNPE). Table III shows the evolution of these impurities in relation to the number of passes (compared with MAP previously purified by recrystallization followed by evaporation) as revealed by high-performance liquid chromatography (HPLC).

Chromatographic methods (HPLC or GC) have high separation power. For preparative purposes, these techniques have the main disadvantage of being capable to separate only small quantities of purified product per cycle.

C. Characterization

Material synthesis requires some classical chemical analysis, either elemental (microanalysis) or functional (NMR). Melting area measurements with differential thermal analysis (DTA) or differential scanning calorimetry (DSC), chromatography (GC and/or HPLC), and spark emission spectrometry

TABLE III

HPLC Analysis of MAP Degradation Products Resulting from Zone Refining

	Number of passages				
Product[a]	0	30	35	40	45
DNPA	9	31	46	95	126
DNPE	0	15	36	63	88
Total impurities (ppm)	47	110	147	229	376

[a] DNPA, 2,4-dinitrophenyl amino acid; DNPE, 2,4-dinitrophenylethylamine. Values expressed in ppm.

supply information as to the product purity. At each stage of the purification and after the growth, these techniques are used to determine the impurity evolution. Some examples are listed in Table IV, while Fig. 2 shows the chemical purity of an NPP sample obtained by GC [injection in a cold capillary column (Badan *et al.*, 1986)] as well as by DSC.

IV. CRYSTAL GROWTH

A. Growth from the Melt

To be grown from the melt, organic materials must be thermally stable. Different crystal growth techniques such as the well-known Bridgman–Stockbarger (BS), Kyropoulos, and Czochralsky techniques have been

TABLE IV

Residual Impurities in MAP, NPP, NPAN, POM, and AMA at Different Stages of Purification

	Metallic impurities (ppm)[a]						
Name	Si	Mg	Fe	Al	Cu	Ca	Organic impurities (ppm)[b]
MAP							
crude	55	32	12	25	20	165	3×10^3
purified	0.4	22	—	—	7.5	—	47
monocrystal	2	0.5	—	—	—	—	60
NPP							
crude	10.2	430	—	—	—	85	10^4
purified	—	24.6	1.5	—	—	2.3	10^2
monocrystal	—	—	—				—
POM							
crude	15	21	—	10	2	23	4×10^3
purified	2	2	—	—	—	—	10
monocrystal	—	12	—	—	—	5	
AMA, purified	10	14.8	0.8	—	—	25.5	
NPAN							
crude							10^4
purified							10

[a] By spark emission spectroscopy.
[b] By HPLC/GC.

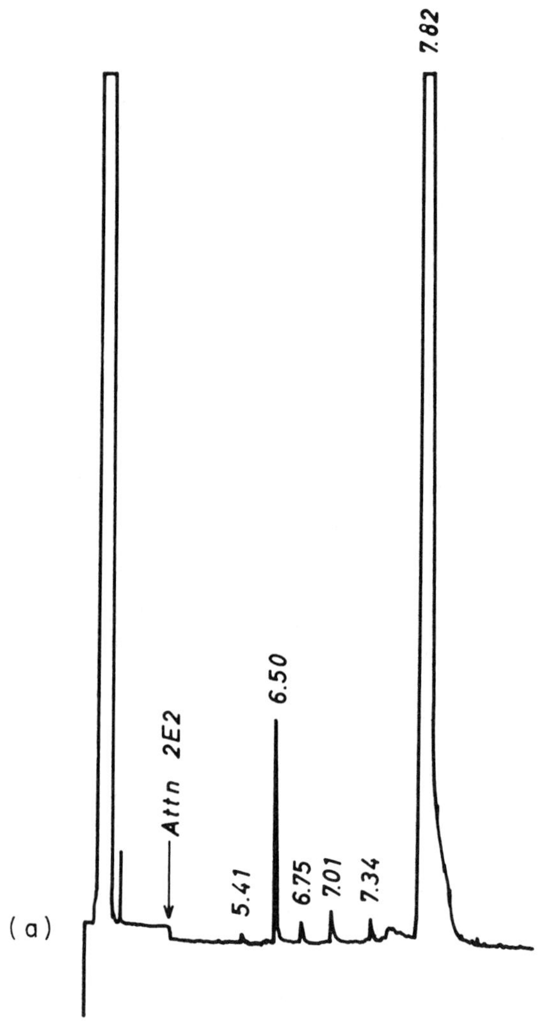

Fig. 2. Chemical purity of a NPP sample, shown by (a) gas chromatography in a capillary column and (b) differential scanning calorimetry on a 3.18-mg sample (*continues*).

developed. In the following, we will only evoke the basic principles of these techniques to be applied in the specific case of nonlinear (NL) organic crystals.

1. Bridgman–Stockbarger (BS) Method

a. Description. Developed by Bridgman (1925) and further modified by Stockbarger (1936, 1949), this method consists in the slow motion of an

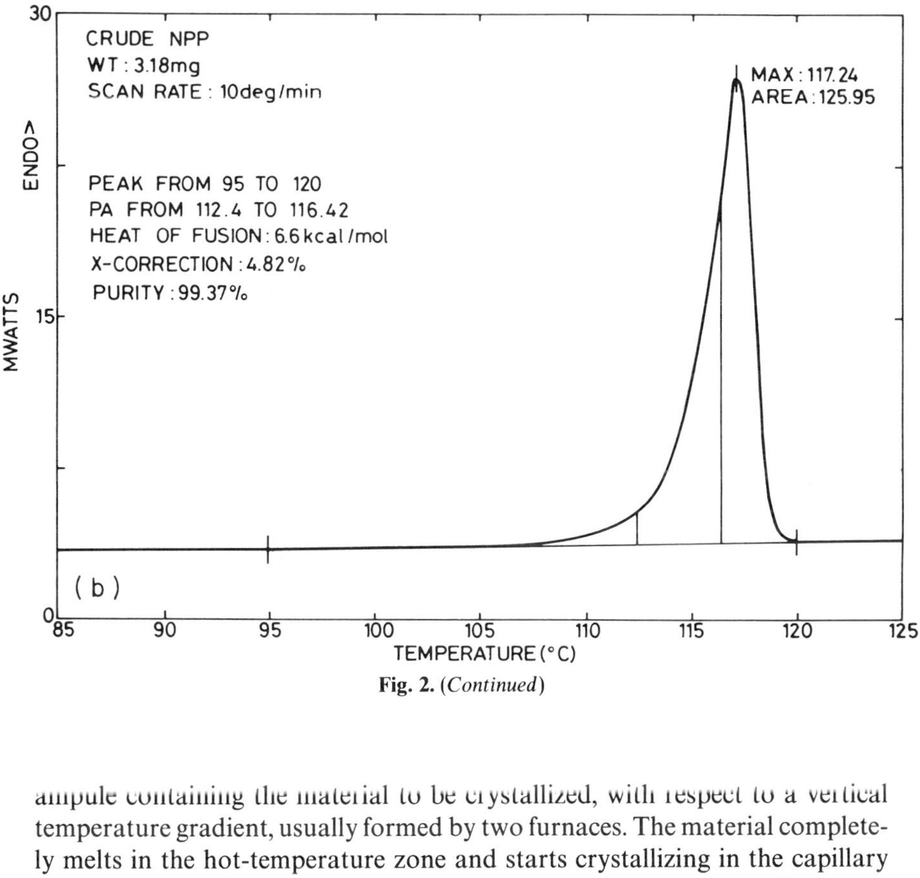

Fig. 2. (*Continued*)

ampule containing the material to be crystallized, with respect to a vertical temperature gradient, usually formed by two furnaces. The material completely melts in the hot-temperature zone and starts crystallizing in the capillary end of the ampoule as it enters the low-temperature zone. Owing to confinement in the capillary, one single crystalline seed is selected and develops into the crystal.

The temperature gradient must be as abrupt as possible, to prevent undesired supercooling. The BS method offers a unique possibility to place the interface in the hot zone by suitable adjustment of the two zone temperatures. This determines the form of the interface, flat or convex toward the liquid, permitting the outgrowing of dislocations, and ensuring crystalline quality.

b. Results. Single crystals of *m*-nitroaniline (*m*-NA) and *m*-dinitrobenzene (*m*-DNB) have been grown by the BS technique. The apparatus we have used here has been previously described by Guillaud *et al.* (1970). It consists of a vertical glass tube with a double-walled lower part, permitting the circulation of the thermostating liquid (low-temperature furnace). The high-temperature furnace is formed by coiling the wire around the upper part of the glass tube.

The lower-temperature part of the tube is then filled with mercury just to the level of the junction of the two furnaces. The resulting gradient was as high as 19°C cm^{-1} for both investigated materials.

Although the melting points of *m*-NA and *m*-DNB are 114°C and 90°C, respectively, the high-temperature zone was maintained at 135°C, which permitted a good degassing of the melt. Temperatures lower than 135°C were inefficient, leaving bubbles trapped in the crystal (Fig. 3).

Single crystals of 10–12 mm in diameter for *m*-NA and 20 mm in diameter for *m*-DNB have been successfully produced. *m*-NA single crystals have a tendance to cleave along the (010) plane containing the pulling axis. On the other hand, no cleavage of *m*-DNB crystals has been observed. However, small-diameter *m*-DNB crystals exposed to a small constant force acting perpendicularly to the crystal growth axis are deformed permanently, without rupture, showing the great plasticity of this material (Carenco *et al.*, 1977). This property is not favorable for crystal cutting and polishing.

2. Modified Bridgman–Stockbarger Method (BSM)

In the case of MAP and NPP, solution growth techniques are not relevant due to undesired solvent/compound associations occurring during the growth. Vapor-phase growth cannot be retained because of the too low vapor pressure of these organic compounds, entailing prohibitive growth time. The BS yields inhomogeneous organic crystals in relation to the progressive degradation of the material during growth. To bypass this phenomenon or to reduce its importance, we maintain the smallest possible quantity of product in the melt during a short period.

Figure 4 shows a schematic view of the apparatus. Its conception is totally modular. Two zones, one of them 2 cm long, are separated by a disk with an O-ring joint. In this technique, the circulating thermostated fluid is directly in contact with the product container in order to improve thermal exchange. Moreover, the glass ampule has very thin walls (0.3–0.5 mm). An internal plastic jacket enables linear dilatation between glass and material and hinders the adherence.

The melt temperature is chosen 3 or 4°C above the melting point while the crystallization temperature is about 10°C lower. The ampule pullin rate is very slow, ranging from 1 to 6 mm per day. Under such conditions, several single crystals of 1 cm in diameter and 7 cm long (Fig. 5) were obtained.

3. The Kyropoulos Method

a. Principle. The crystal grows from a seed (cooled by an external mean) immersed in the melt (Kyropoulos, 1926, 1930). The simplest way to obtain

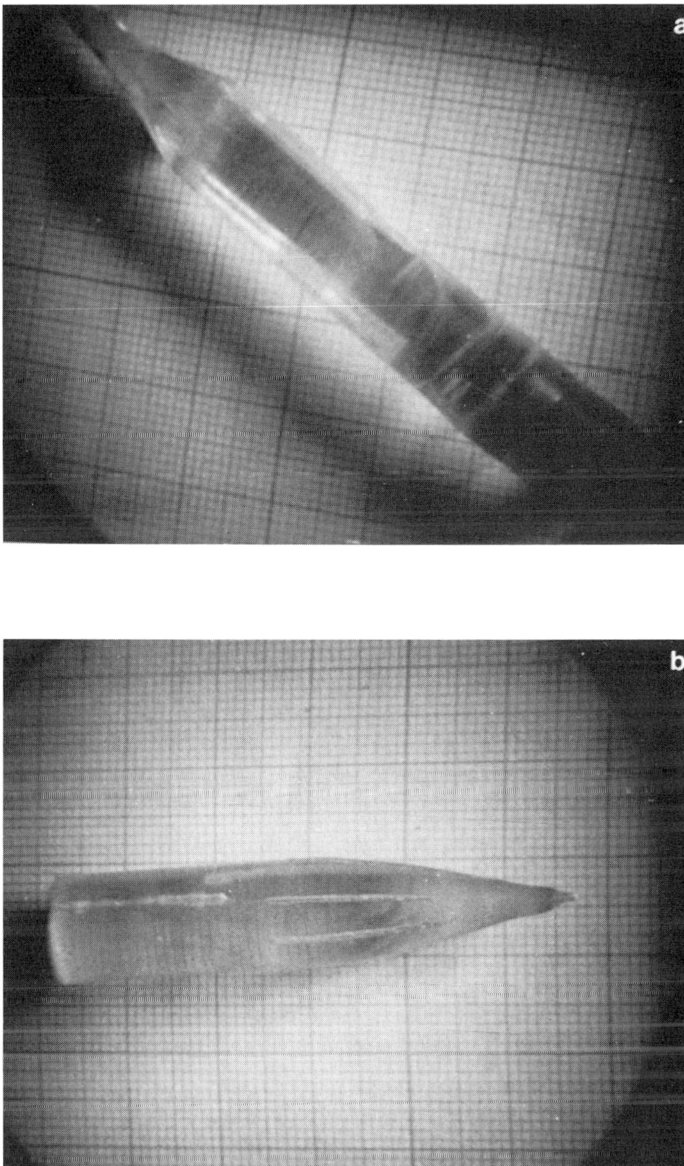

Fig. 3. *m* DNB single crystal obtained by the classical BS method. (a) High-temperature furnace maintained at 135°C. (b) High-temperature furnace maintained under 135°C.

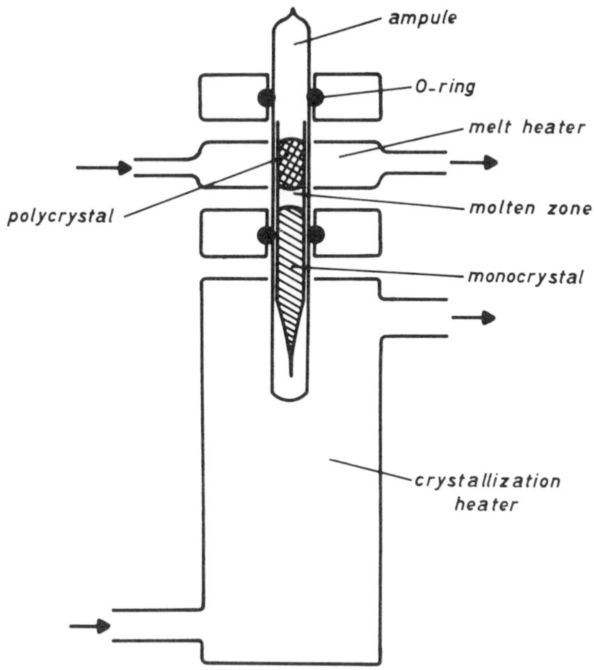

Fig. 4. Schematic representation of a modified Bridgman–Stockbarger apparatus.

Fig. 5. Monocrystal of MAP obtained by the modified Bridgman–Stockbarger method.

large organic crystals is to supercool their melt by 0.2–1°C and to immerse a crystal seed while maintaining the supercooling of the melt.

b. Results. This technique has been used to obtain single crystals of *m*-BNB (Fig. 6). This compound melts at 56°C, has a high vapor pressure, and crystallizes poorly from solution (benzene, toluene, acetic acid). The Kyropoulos method leads to single crystals with faces that can be used directly in nonlinear optical experiments. We briefly describe our original seed preparation. They are obtained by sublimation of *m*-BNB in a closed chamber heated near the melting point (50°C). A day later, thin needles 3–4 cm long appear. Monocrystalline needles selected by optical examination under polarized light are fixed by fusion at one end of a supporting tungsten wire. When dipped into a supercooled bath (0.2–0.3°C), the seed practically grows along two directions both perpendicular to the *c* crystallographic axis. Figure 5 shows single crystals grown by this technique. The large faces correspond to ⟨010⟩ planes and the small ones to ⟨102⟩ planes. Such a crystal is obtained in 3 hr. The optical quality of the samples is compatible for NLO purposes.

Mixed crystals of *m*-chloro- and *m*-bromonitrobenzene are grown in similar conditions. The cell parameters of the two products are very close, leaving the resulting crystal habit nearly unchanged. In the liquid state, they are entirely miscible: 25/75, 50/50, and 75/25 solutions are used to grow mixed

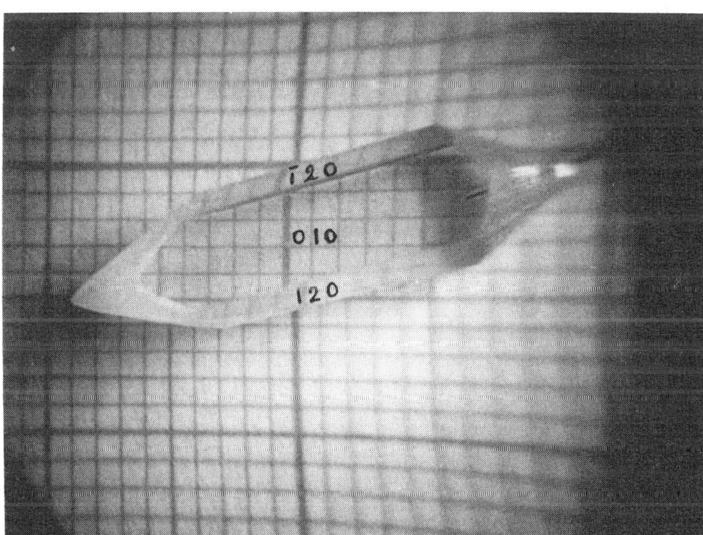

Fig. 6. *m*-BNB monocrystal obtained by the Kyropoulos method.

crystals of each composition. Figure 7 shows one of them grown from a 50/50 solution. Note the great modification of this crystal as compared to that shown in Fig. 6.

B. Crystal Growth from Solution

Solution growth for the production of POM crystals (Hierle *et al.*, 1984) is adopted for the potential capacity of this method to yield nearly constraint-free single crystals with more ease than growth from the melt. Crystals grow in supersaturated solutions containing a solute concentration exceeding the equilibrium value on the solubility curve and corresponding to the so-called metastable domain. Solutions are conveniently described in terms of solute concentration and temperature.

Let us recall that the concentration/temperature map is divided into unsaturated and supersaturated zones (see Fig. 8) by a line representing the solution–solute equilibrium. The supersaturated zone is itself divided in two: the metastable zone, and the so-called labile zone, where growth is spontaneous and uncontrollable (Petrov *et al.*, 1969). The width of the metastable zone depends on the nature of the solute, its concentration, the presence of impurities, temperature, and mechanical constraints. Some authors (Tarjan

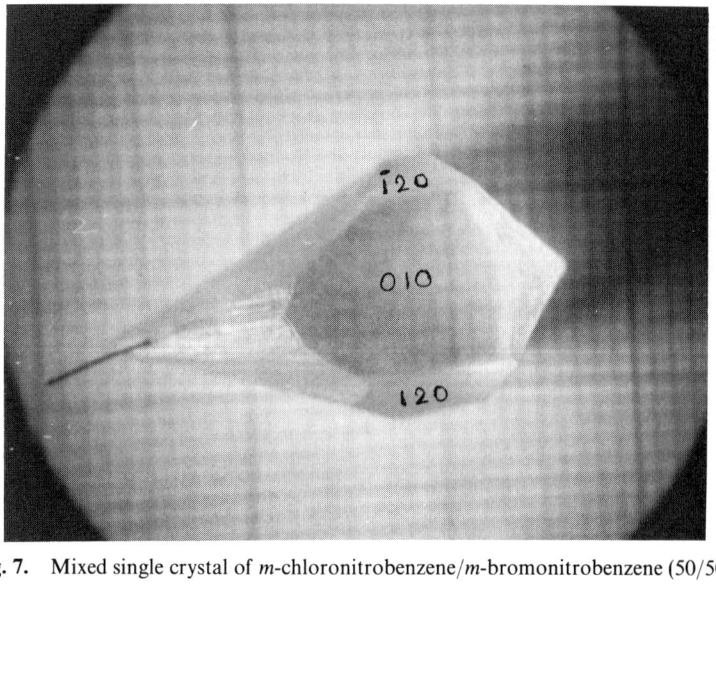

Fig. 7. Mixed single crystal of *m*-chloronitrobenzene/*m*-bromonitrobenzene (50/50).

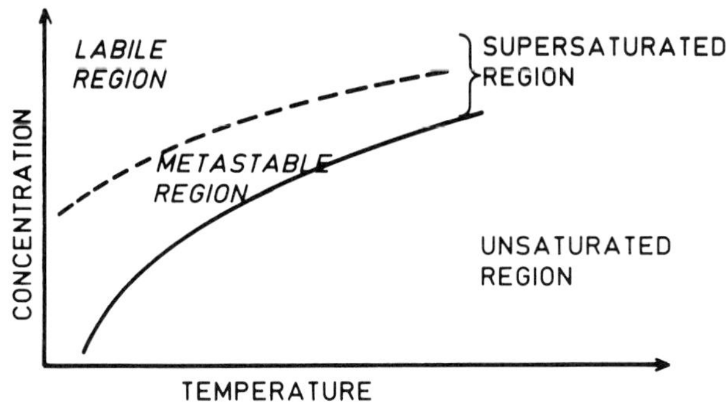

Fig. 8. Crystalline growth in solution. From Petrov *et al.* (1969), with permission.

and Matrai, 1972) have established a method defining this area under certain conditions. After a test in our laboratory on previous crystal growth (Perigaud and Marais, 1976), this method has proved capable of coping only partially with complex situations. It has nevertheless shown the troublesome narrowness of this area. We have proposed another optical method for growth control and monitoring based on refractometry measurements (Sigelle *et al.*, 1983).

The solution growth apparatus (Fig. 9) consists of a sealed glass cylinder of 1.5 liter capacity with two openings, the middle one with an inlet for a seed-supporting stirrer, the second one for the introduction of compounds.

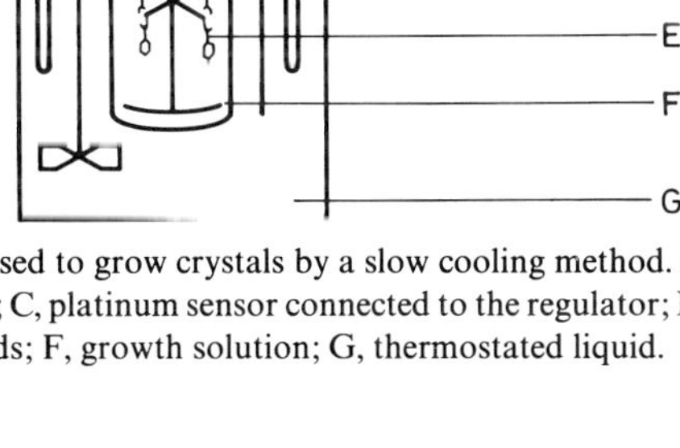

Fig. 9. Apparatus used to grow crystals by a slow cooling method. A, Motor; B, accelerated seed rotation technique; C, platinum sensor connected to the regulator; D, heating resistor driven by the regulator; E, seeds; F, growth solution; G, thermostated liquid.

This apparatus is immersed in a 20-liter water-filled thermostat, also equiped with a stirrer. The temperature of this water bath is controlled by a SERELEC electronic regulator with a platinum thermocouple sensor (100 Ω at 0°C), providing a regulation rate better than 0.01°C (including the drift resulting from time delays). The regulator is driven by an instruction emitter (CYBEREXACT E.C. 2485). Temperature decrease is defined by successive 0.01°C leaps at variable time intervals. Intervals shorter as the growth proceeds, so as to correspond to a step-by-step curve with a nearly parabolic envelope, a typical rate being 0.08°C per day.

Three solvents have been used for dissolution of POM: acetonitrile (Ac), methyl acetate (AM), and 1,2-dichloroethane (DCE).

As for most compounds, POM solubility increases linearly with temperature within the considered temperature–concentration boundary values. The following equations are derived from measurements:

Acetonitrile:	$y = 2.02t + 10.69$	(A curve, Fig. 10)
Methyl acetate:	$y = 1.67t - 2.31$	(B curve, Fig. 10)
1,2-Dichloroethane:	$y = 1.82t + 11.64$	(C curve, Fig. 10)

with t corresponding to the temperature in degrees Celsius and y the solubility expressed in grams per liter solution.

The growth solvents have been selected according to various criteria:

(1) POM appears to be soluble in either weakly dipolar solvents such as methyl acetate (1.85 D) and 1,2-dichloroethane (1.75 D) or dipolar ones such as acetonitrile [3.2 D, from Mellan (1970)].

(2) The solvent vapor pressure must be as low as possible within the considered temperature range (typically 35°C → 28°C). This criterion is satisfactorily met as shown by these values:

Acetonitrile:	150 mm Hg at 35°C,	73 mm Hg at 20°C
Methyl acetate:	284 mm Hg at 35°C,	173 mm Hg at 20°C
1,2-Dichloroethane:	117 mm Hg at 35°C,	99 mm Hg at 20°C

(3) The solvents should not be hygroscopic: water may at length interact with the seed surface to the point of inhibiting the growth process. Besides, any significant water inclusion would show up in the near-infrared spectrum, and subsequently increase the absorption in this range in contradiction with applications in view. Therefore, all solvents undergo a thermal treatment in presence of a dehydrating substance before distillation.

Five to six days are required to dissolve POM in 1.25 liter of solvent and to find the exact equilibrium temperature by adding small quantities of powder until its dissolution stops (Nyvlt, 1977).

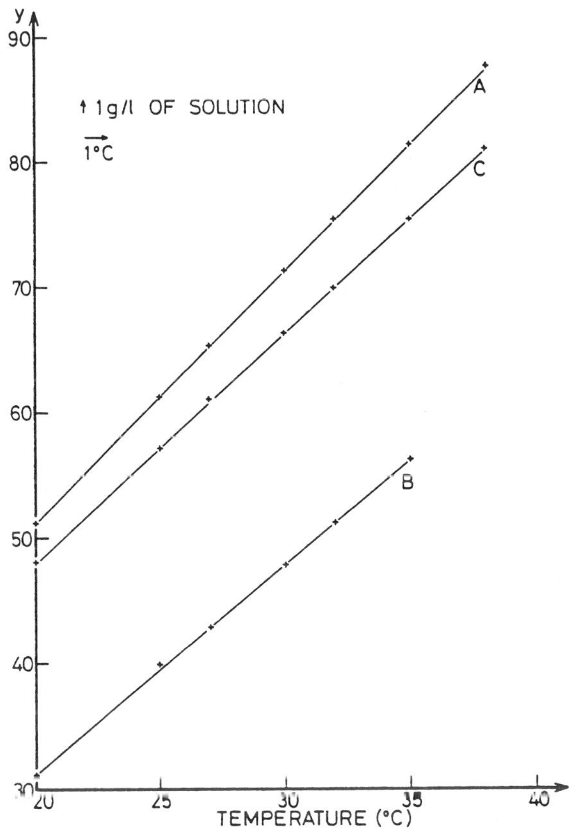

Fig. 10. Solubility curves of POM as a function of temperature *T* in three solvents: (A) acetonitrile, (B) methyl acetate, and (C) 1,2-dichloroethane.

The first seeds are obtained from the evaporation of a saturated POM solution in the same solvent as previously used. At a later production stage, parallelepipedic samples are obtained from single crystals by cutting parallel to the *a* and *c* (0.8 × 0.8 mm^2) crystallographic axes and cleaving perpendicular to axis *b*.

The seeds are then selected after X-ray Laüe examination, which is in our opinion the only valuable test to reveal their crystalline quality. Figure 11 shows nice crystals grown from seeds of good bulk crystalline quality but with surfaces polluted by the polishing material. Four seeds are attached to the seed carrier by platinum threads of 0.1 mm diameters. The seeds undergo alternative accelerated and decelerated rotations, followed by a rest period (A.S.R.T., accelerated seeds rotation technique). The duration of a cycle is

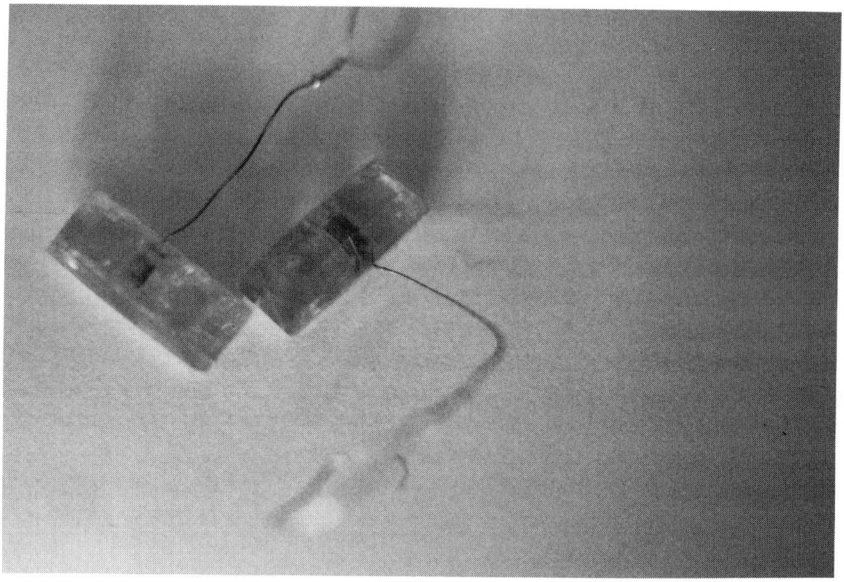

Fig. 11. POM crystal with polluted surfaces.

2 min 20 sec, and the maximum speed to be reached is 22 rotations/min. This original technique for growing organic crystals allows the seeds to be fed correctly on each face. The entire solution is well stirred while seeds are slowly rotating. Consequently, we have never observed any perturbation due to acceleration and deceleration vibrations. The crystals (Fig. 12) that are being discussed were grown following conditions stated in Table V.

It should be noted that the solvent evaporation problem was successfully solved, as shown by the absence of material deposition on the walls of the crystallisation bath (Table V). For each solvent, kinetic growth measurements have been carried out on more than 10 manipulations (40 crystals).

A few preliminary remarks should be made:

(1) For the three different solvents, the initial temperatures are not identical but come close (2–3°C deviation).

(2) The overall temperature decrease is variable and depends on the growth progress.

(3) The detailed structure of the sequence of constant-temperature plateau and leaps differs from one solvent to another, but the total duration of the process remains fairly constant. Discontinuous temperature variations may lead to related concentration discontinuities that cannot be accounted for within our time observation scale (around 30 min).

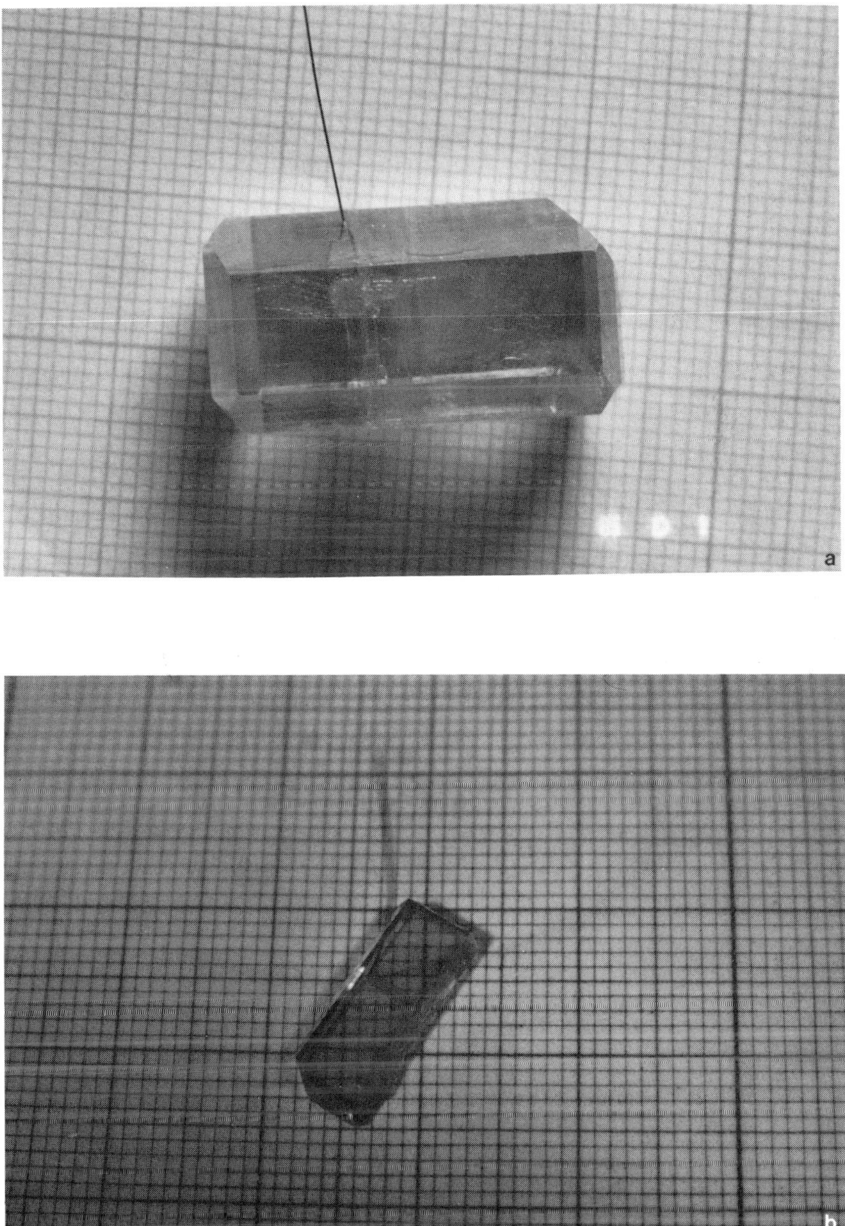

Fig. 12. POM crystals grown in three solvents: (a) acetonitrile, (b) methyl acetate, (c) 1,2-dichloroethane (*continues*).

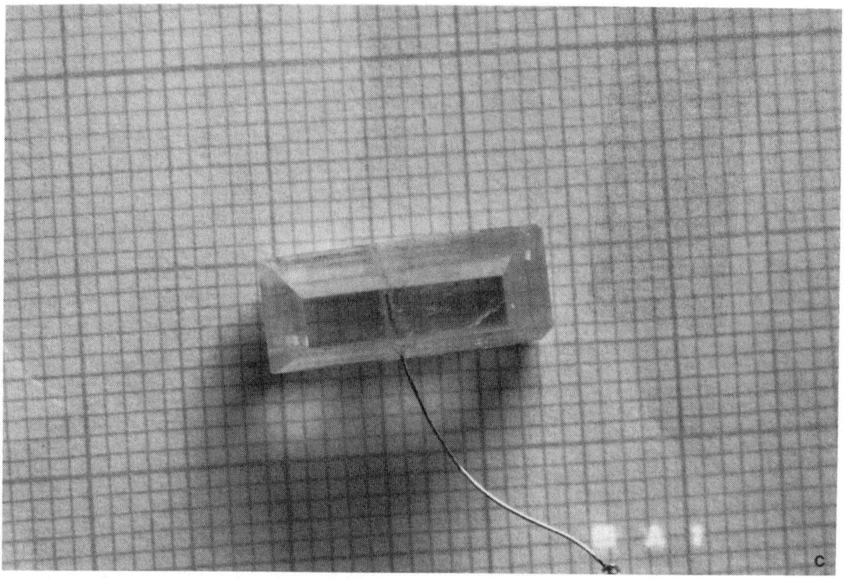

Fig. 12. (*Continued*)

Under such conditions, it is difficult to determine the precise growth kinetics parameters along the crystallographic axes. A "solvent effect" on growth seems to be significant. The seed is suspended by a thread and attached to it by gluing, fusing, or pinching: this, together with the relative orientation of the seed with respect to the thread, may cause specific crystalline defects that affect the growth kinetics. Growth speed along axis *c* is preferential (Table VI). This corresponds to the smallest dimension of the crystal unit cell. Growth speeds along the *a* and *b* axes are comparable, which cannot be simply accounted for in view of the very anisotropic cell dimensions within the (001) plane. In addition, according to Table IX of Shiro *et al.* (1977), the molecular least-square planes are relatively isotropically inclined over the crystalline axis (see Fig. 22c), showing no straightforward relationship with the growth anisotropy. In acetonitrile, growth speeds are comparatively close and not much altered by the seed suspension direction.

In the case of AMA, the technique used was similar to that of POM. Solubility curves for three mixtures with following composition:

I. Ethanol–acetonitrile (57/43% v/v)
II. Isopropanol–methylethyl acetone (50/50)
III. Xylene–acetone (70/30)

TABLE V

Temperatures, Concentration, and Duration of POM and AMA Growth from Solution in Different Solvents

Material and solvent	Temperature, initial (T_i) and final (T_F) (*C)	Concentration, initial (C_i) and final (C_F) (g/liter solution)	Duration (hr)
POM (Fig. 14)			
Acetonitrile	$T_i = 35.22$ $T_F = 31.25$	$C_i = 81.6$ $C_F = 72.67$	385
Methyl acetate	$T_i = 33.44$ $T_F = 28.68$	$C_i = 42.26$ $C_F = 40.60$	337
1,2-Dichloroethane	$T_i = 33.69$ $T_F = 31.40$	$C_i = 75.40$ $C_F = 73.78$	285
AMA (Fig. 14)			
I: Ethanol (57), acetonitrile (43)	$T_i = 19.1$ $T_F = 16.4$	$C_i = 41.3$ $C_F = 35.7$	816
II: Isopropanol (50), methyl ethyl ketone (50)	$T_i = 23.8$ $T_F = 20.5$	$C_i = 29.1$ $C_F = 22.3$	792
III: Xylene (70), acetone (30)	$T_i = 29.5$ $T_F = 28.7$	$C_i = 91.6$ $C_F = 89.5$	216

TABLE VI

POM Growth Kinetics along the Three Crystallographic Axes in Different Solvents: Acetonitrile (Ac), Methyl Acetate (AM), and 1,2-Dichloroethane (DCE)

	Axis *a*			Axis *b*			Axis *c*		
Crystallographic axes	Ac (1)[a]	AM (2)	DCE (3)	Ac (1)	AM (2)	DCE (3)	Ac (1)	AM (2)	DCE (3)
$a = 21.359$ Å	2[b]	7	5	5	6	6	3	7	5
$b = 6.111$ Å	5	4	6	7	6	5	3	6	6
$c = 5.132$ Å	7	20	10	7	15	15	5	8	15

[a] Conditions: (1) initial temperature 35°C, total decrease 2.86°C, duration 316 hr; (2) initial temperature 35°C, total decrease 4.36°C, duration 363 hr; (3) initial temperature 34°C, total decrease 3.90°C, duration 343 hr.

[b] Growth factor.

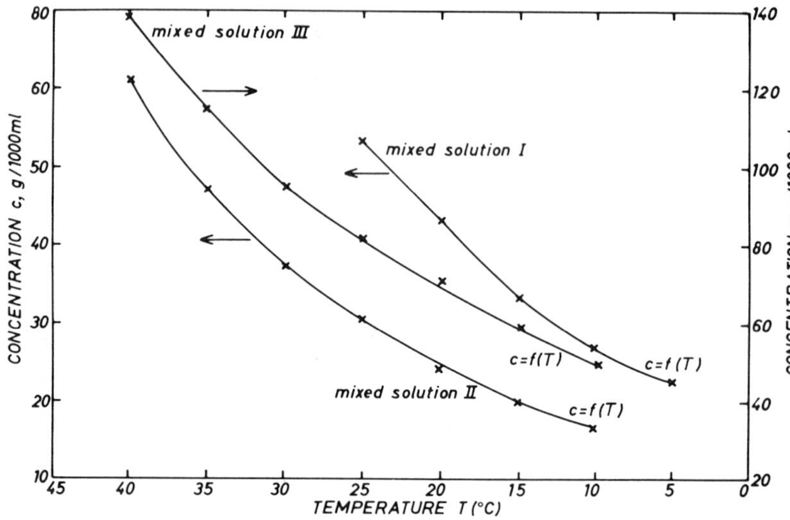

Fig. 13. Solubility curves of AMA as a function of temperature T in different ternary solutions: (I) ethanol–acetonitrile (57/43); (II) isopropanol–methyl ethyl ketone (50/50); and (III) xylene–acetone (70/30).

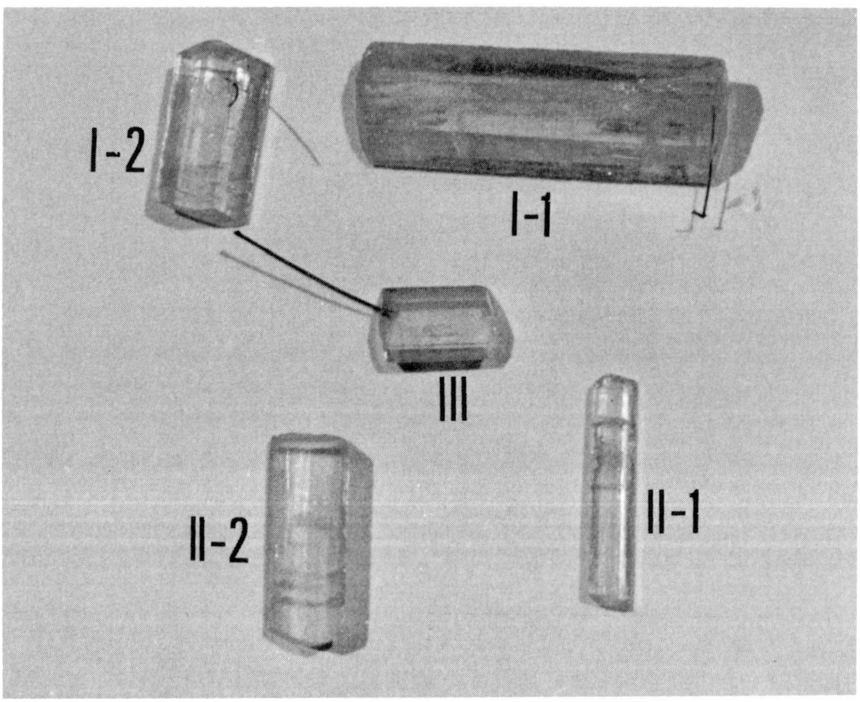

Fig. 14. Several AMA monocrystals grown in solutions (I), (II), and (III), as identified in Fig. 13.

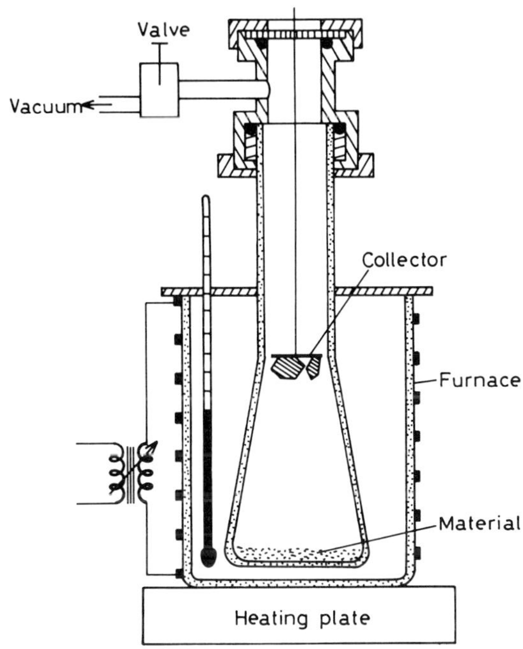

Fig. 15. Schematic view of the apparatus for the growth of large thin organic platelets by sublimation.

are given in Fig. 13, while growth conditions are stated in Table V. Figure 14 shows several AMA single crystals obtained from these different mixtures.

C. Growth by Sublimation

This method (Glockner, 1974; Karl, 1980) is highly suitable for growth of extremely thin (0.1–100 μm) platelike organic crystals, which are interesting for fundamental spectroscopy or waveguided optics. The apparatus (Fig. 15) consists of a metal sheet exposed to the highly saturated organic vapor (Bernard, 1984). The collector is colder (by about 30°C) than the organic material, which is maintained at a temperature just below the melting point. Crystal growth is very rapid. The time necessary to grow MANS and *p*-NA (Fig. 16) of about 10 mm × 10 mm × 80 μm was near 30 min, operating under reduced pressure (10^{-4} mm Hg).

Let us mention that MANS is centrosymmetric, and consequently inefficient for SHG. However, this material has a high third-order nonlinear coefficient and interesting fluorescence properties. For this purpose, it has also been grown from acetonitrile solution (Perigaud and Marais, 1976).

Fig. 16. (a) *p*-Dimethylamino-β-nitrostyrene (MANS) and (b) *p*-nitroaniline (*p*-NA) platelets.

V. CHARACTERIZATION

To control their crystalline and optical quality, the crystals should be cut and polished. For this purpose, the hardness of organic materials being lower than that of inorganic, the use of a wet wire saw is recommended. To obtain a good surface state, chemical polishing is desirable.

Simple observation of the crystals reveals macroscopic defects, such as color change, fractures, inclusions, veils, precipitates, cleavage, etc. Polarized-light observation evidences monocrystallinity of the samples. For MAP and NPP, the ingot form makes it difficult to locate the binary axis orientation by X-ray studies. Examination of conoscopic fringes helps orient the crystals. X Rays allow the determination of crystal morphology (Laüe diagrams), certify monocrystallinity (Lambot–Barraut negatives), and contribute to topographic studies of defects (dislocations, precipitates, striation bands, stacking faults, etc.).

In our case, the ultimate characterization is the measurement of NLO properties. It yields information such as light scattering on defects, residual absorption, parametric conversion, optical damage threshold, and the half-height peak of the phase-matched SHG.

A. Polarization Microscopy

Figure 17, (a) and (b), shows two MAP samples obtained from two different crystal growth runs and observed between crossed-polarizers. In Fig. 17(a) a mosaic structure accompanied by a small monocrystalline domain is shown. Direct observation reveals that the monocrystal, developed in the capillary section, does not entirely fill the cross section. As the ampule straigtens up, the two forms coexist. During growth, the polycrystal prevails over the single crystalline domain. Figure 17(b) also presents two zones, but both are monocrystalline. The same phenomenon occurs, giving a single crystal after a few centimeters. The Laüe diagram [Fig. 18(a)] shows the perfect monocrystallinity of the sample, but it does not permit us to find the binary axis. Conoscopic fringes (Born and Wolf, 1965) give the optical axis orientation, which is used to determine the binary axis. NPP samples obtained by cleavage and viewed between crossed polarizers are shown in Fig. 19.

B. Examination by X-Ray Methods

Laüe diagrams of some of our materials are presented in Fig. 20. Except for crystals obtained by BS methods, one can induce from their inspection crystallographic quality, symmetry axes, and crystal morphology.

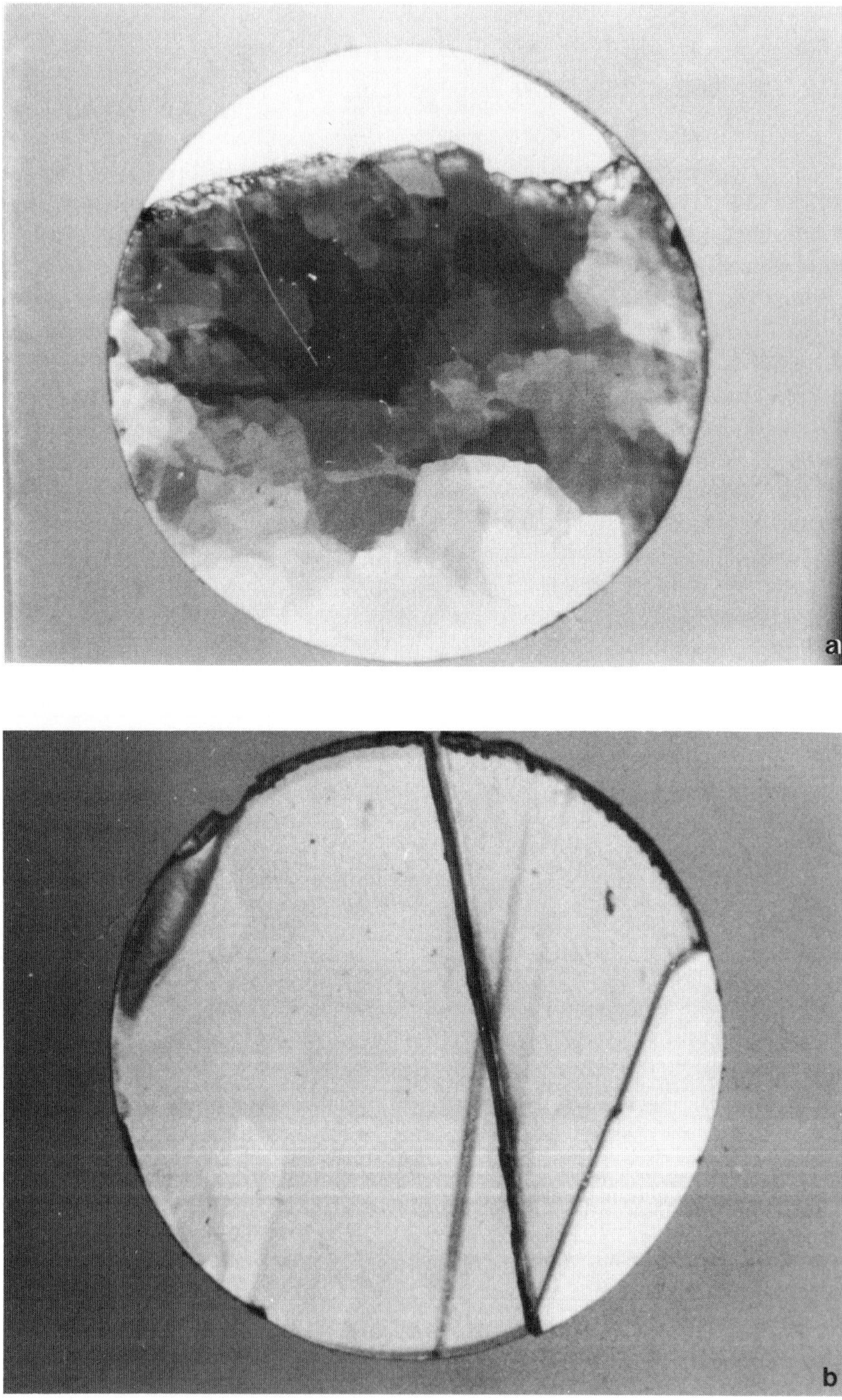

Fig. 17. Platelets of MAP cut from two different ingots, observed between crossed polarizers: (a) mosaic structure and (b) bicrystal.

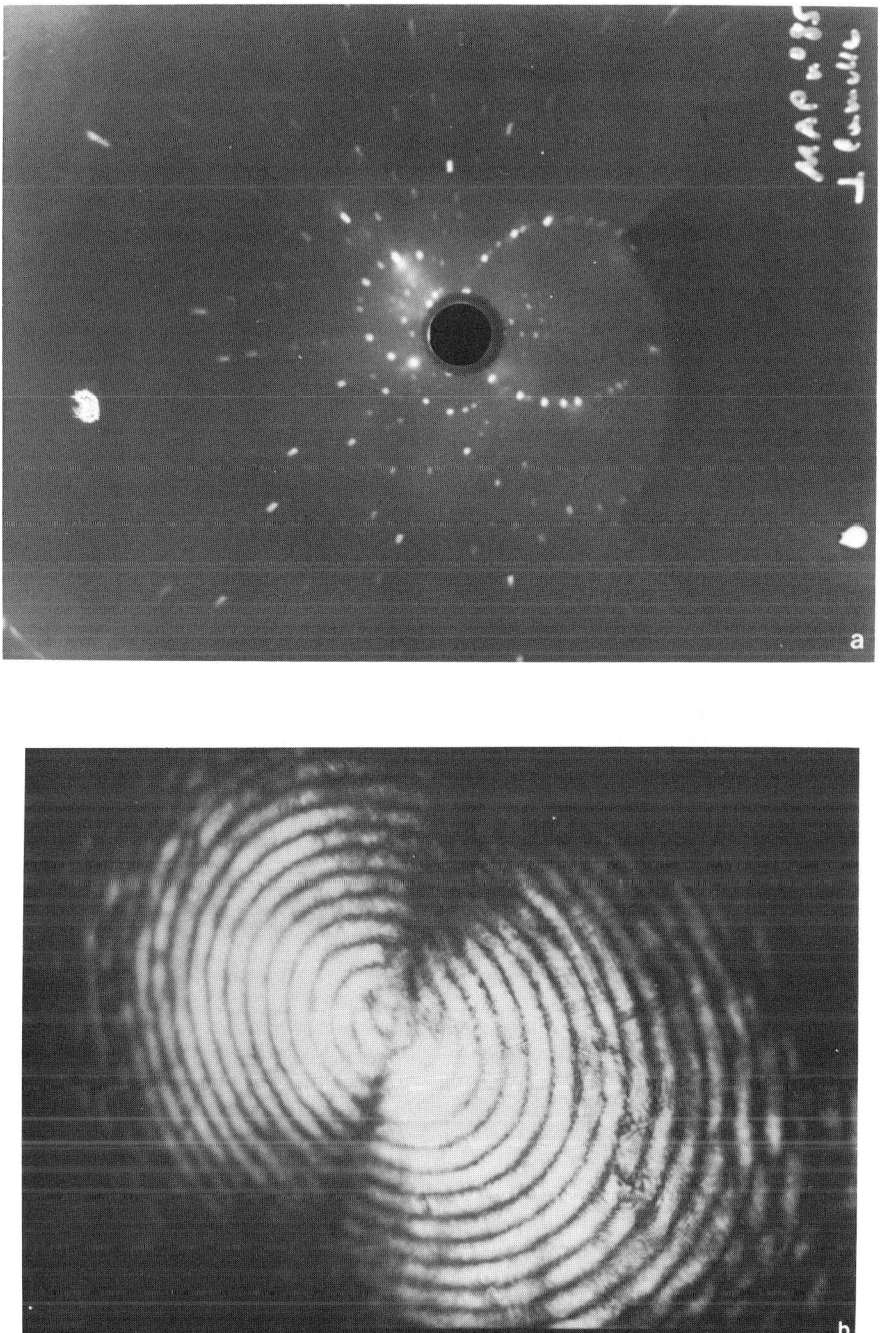

Fig. 18. (a) Laüe diagram of MAP platelet. The enlargement of the diffraction spots results from the sample thickness of 1.2 mm. (b) Conoscopic fringes revealing the optical axis orientation.

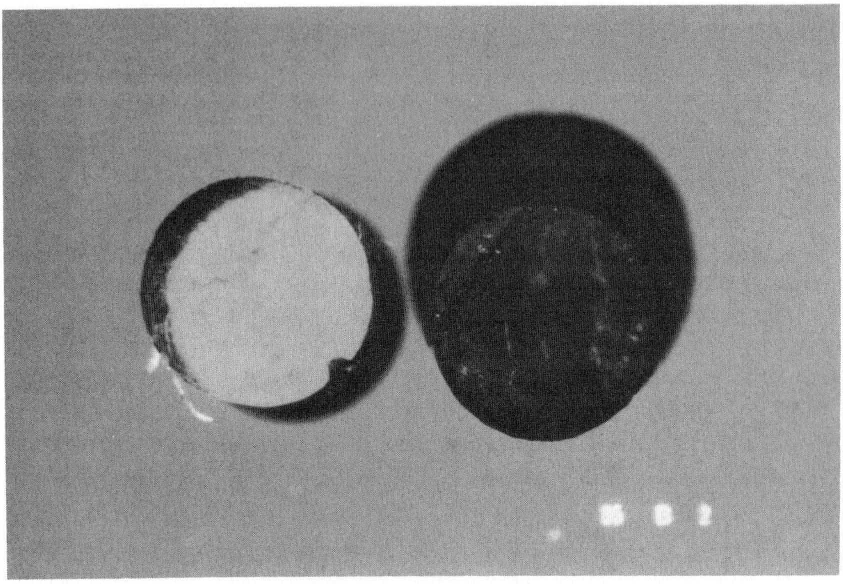

Fig. 19. Cleaved NPP crystals, 8 mm thick, viewed between crossed polarizers.

Fig. 20. Laüe diagram of some single crystals studied here: (a) NPAN (for more details, see Section VI); (b) POM; (c) AMA (continues).

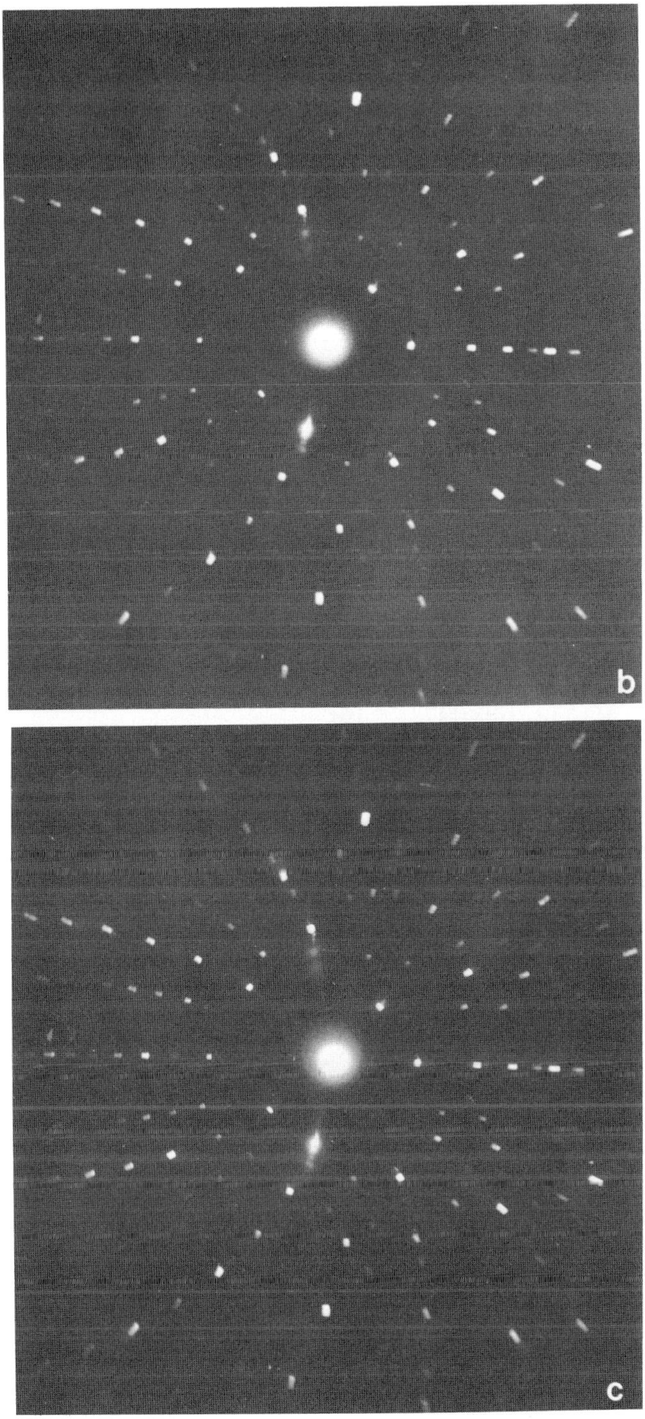

Fig. 20. (*Continued*)

When grown from solution, POM and AMA crystals are of good quality except for regions neighboring the seed, as evidenced by Laüe diffraction prints exposed through a 5-mm-thick sample [Fig. 20, (b) and (c)]. Face indexing of solution-grown crystals is also possible and is exemplified in Figs. 6, 7, and 21.

Lambot exposures of 10-mm-thick POM crystals along the *c* axis using the 020 Kα reflection show a 20-minute disorientation for the best samples and compare favorably to a similar exposure over a silicon sample using the 400 CuKα reflection, especially if one considers the greater penetration depth in the organic compound [Fig. 22(b)]. In addition, a Lang topography test of a crystal has not shown any dislocation in areas sufficiently apart from the original seed [Fig. 22(a)]. On the whole, these photographs reveal the remarkable crystal quality achieved.

Notice that AMA crystals cleave easily along (100) crystallographic planes (Maugen, 1981), like mica, which can be easily understood from Fig. 23. This

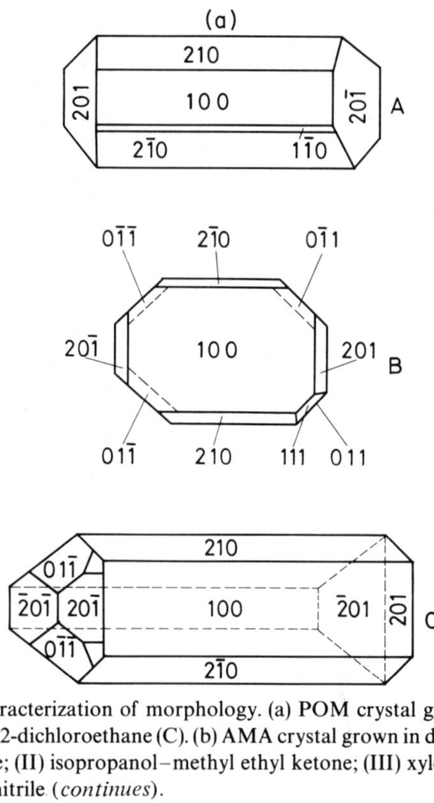

Fig. 21. X-Ray characterization of morphology. (a) POM crystal grown in acetonitrile (A), methyl acetate (B), or 1,2-dichloroethane (C). (b) AMA crystal grown in different mixed solutions: (I) ethanol–acetonitrile; (II) isopropanol–methyl ethyl ketone; (III) xylene–acetone. (c) MANS crystal grown in acetonitrile (*continues*).

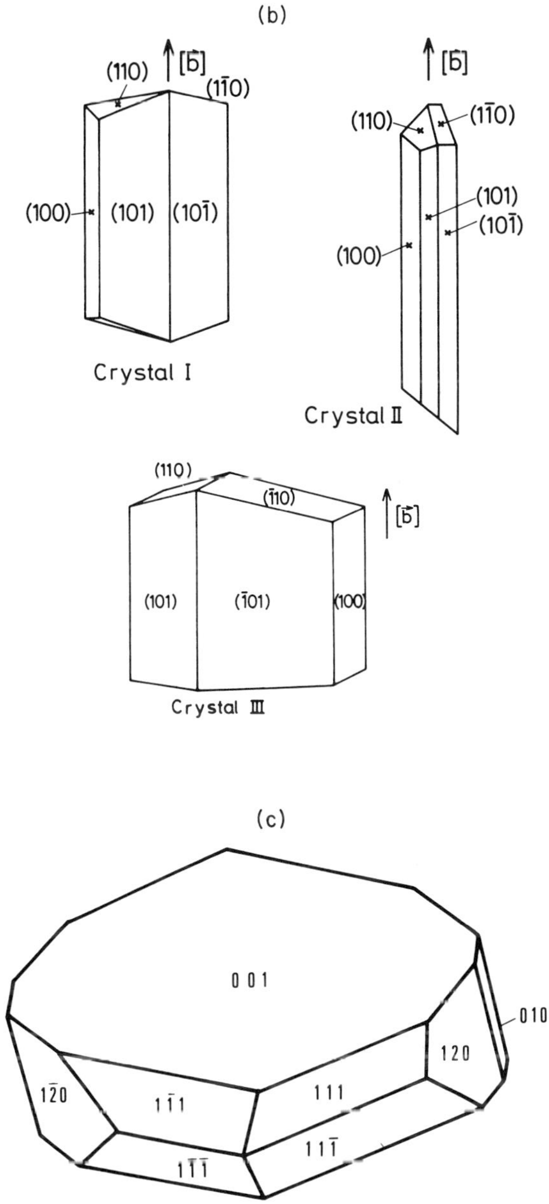

Fig. 21. (*Continued*)

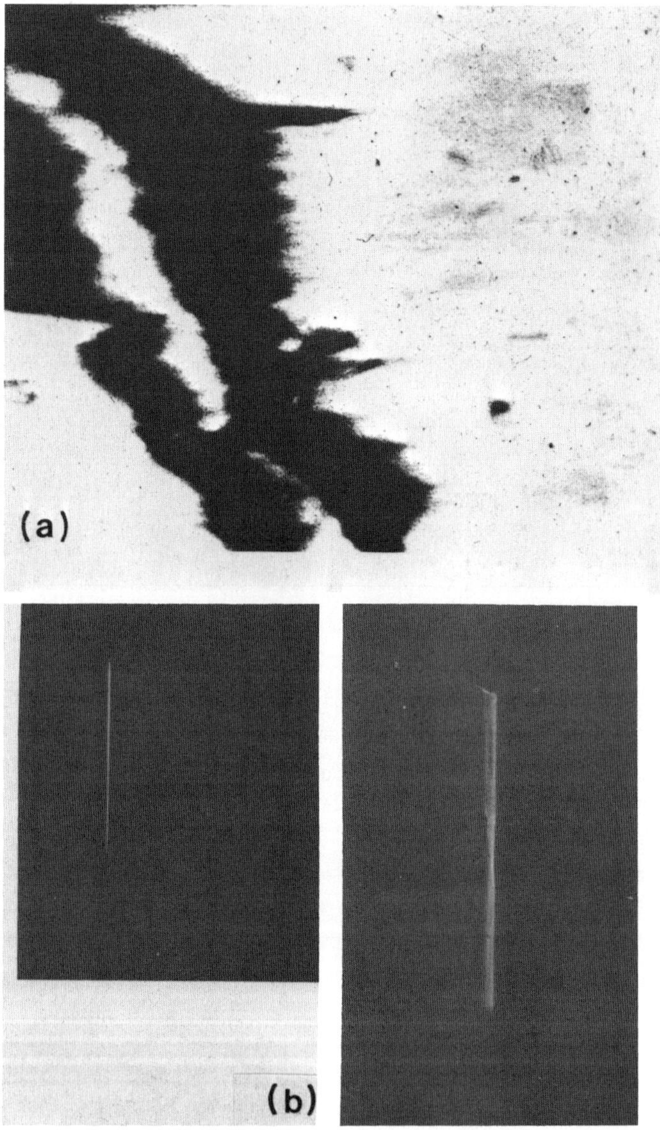

Fig. 22. Other X-ray characterization of POM crystals. (a) A photograph of Lang topography test of a polished crystal cut (010) with 010 $M_0 K\alpha$ reflection. (b) Lambot negative (right) compared to that of a similar sample of Si (left) for which the penetration depth is smaller. (c) Crystallographic structure of POM. From Shiro *et al.* (1977), with permission (*continues*).

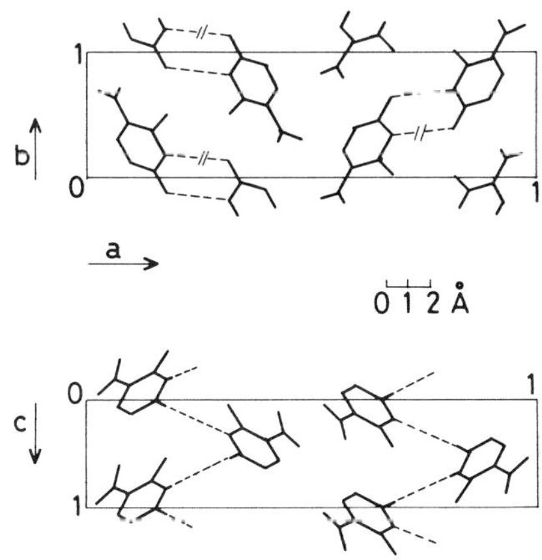

Fig. 22. (*Continued*)

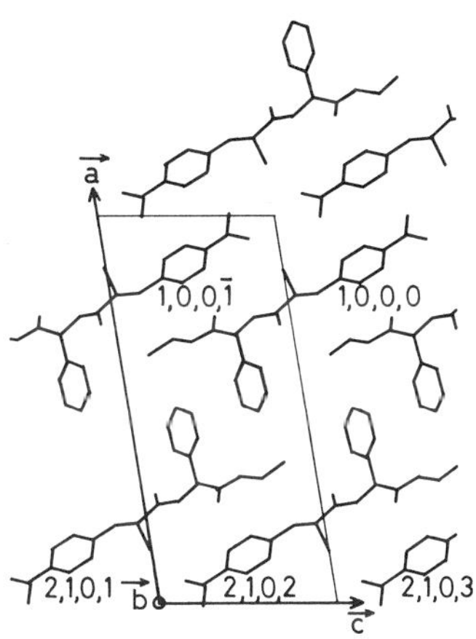

Fig. 23. Crystallographic structure of AMA.

property precludes the cutting and polishing of samples, as well as their utilization for NLO.

C. Other Methods

Besides X-ray characterization methods, we have also investigated the fusion behavior of POM and AMA single crystals (Fig. 24), as well as that of MAP (Perigaud *et al.*, to be published), to further investigate their chemical purity (Hierle *et al.*, 1984).

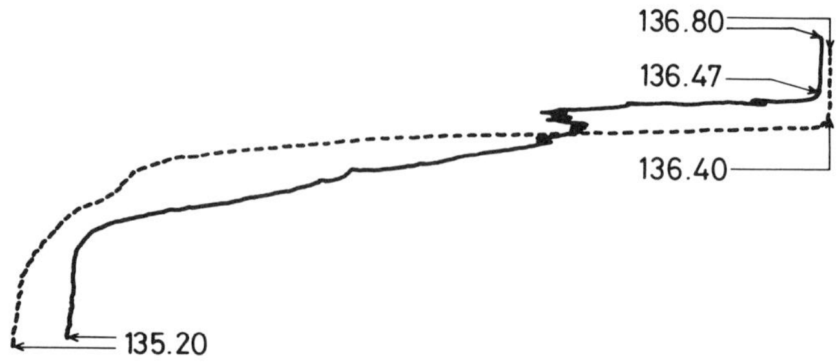

Fig. 24. Thermal behavior of POM near fusion as compared to a sample of AMA.

For POM in acetonitrile, UV and infrared (IR) transmission spectrophotometry, respectively, show maximal absorption at 340 nm and transparency from 0.5 to 2 μm (POM and MAP), except for first and second overtones of C—H vibrations around 1.7 and 1.2 μm, respectively (Fig. 25). More critical measurements using the ultimate nonlinear optical properties of the crystals have enabled the determination of the second harmonic generation (Zyss *et al.*, 1981; Zyss, 1982) and electro-optic coefficients (Sigelle and Hierle, 1981; Sigelle *et al.*, 1982).

In order to observe phase-matched second-harmonic generation at normal incidence and fundamental wavelength 1.06 μm, a 1900-μm-thick POM platelet was sliced and polished out of a (100) as-grown sample (Fig. 26). The phase-matching curve can be plotted when the sample is rotated around the *a* axis. As expected, the maximum intensity is reached at normal incidence. The half-height peak width is inversely proportional to the effective interaction

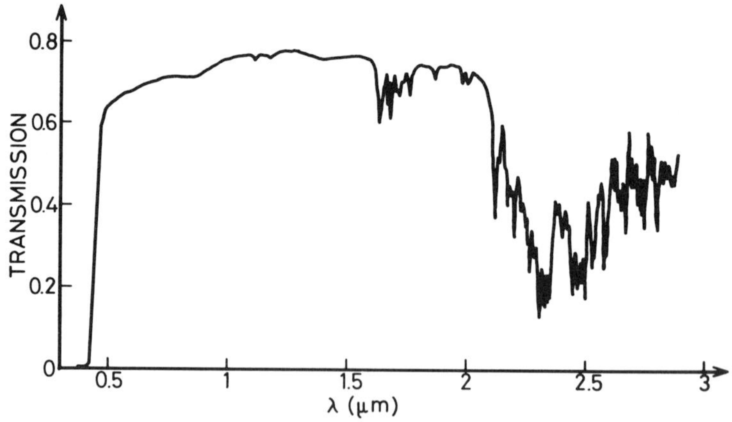

Fig. 25. Transparency of a 400-μm-thick plate of POM as a function of wavelength.

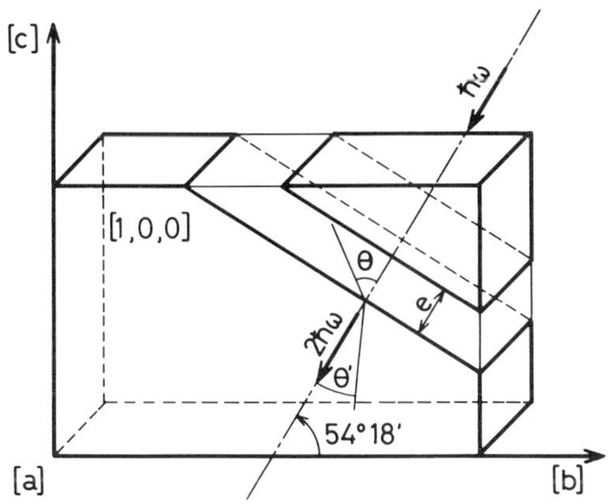

Fig. 26. Plate of POM 1900 μm thick cut from a (100) naturally oriented sample, for normal incidence of the 1.06-μm phase-matching direction.

length. Experimental results lead to $l_{eff} = 1870\ \mu$m, which, when related to the sample's actual width, $l = 1900\ \mu$m, shows a fairly constant crystalline orientation of the sample.

The optical damage threshold has been measured by Zyss *et al.* (1981) using the fundamental and doubled output of a YAG:Nd^{3+} pulsed laser (20 nsec pulse duration, 10 Hz repetition rate) and given an estimate value of more than 2 GW/cm^2 at 1.06 μm (fundamental) and 50 MW/cm^2 at 0.53 μm (doubled output). These relatively high damage thresholds are also accounted for by the single-crystal quality. Various configurations of three-wave parametric interactions are reported at the picosecond time scale in POM single crystals (Zyss *et al.*, 1985). Optical characterization of MAP and NPP is reported in Chapter III-1.

VI. ELABORATION OF ORGANIC NONLINEAR WAVEGUIDES

New organic nonlinear materials have been mainly used so far in high-power laser experiments. However, the remarkable progress made in development of low-power semiconductor lasers emphasizes a need for fully optimized nonlinear waveguide structures. Indeed, it is well known that NLO processes are efficient only if they are phase matched. In the bulk, this is usually achieved by compensation for the crystal birefringence by the index dispersion. Unfortunately, many materials with high NL coefficients, such as compound semiconductors, are inefficient due to their low birefringence and higher dispersion. This problem can be overcome by using organic waveguiding (WG) structures, which permit full exploitation of their large NL coefficients.

The advantages of WG are enumerated as follows:

(1) The natural material dispersion is replaced by modal dispersion, which can be tailored by a proper choice of WG and of substrate and superstrate material properties as well as of their dimensions.

(2) The NL interactions can take place either in the WG or in its environments.

(3) The low material dispersion and high bulk birefringence are not necessary conditions to achieve the phase matching.

(4) Large power densities can be easily confined in waveguides over long propagation paths, while in the bulk limitations to a few confocal length are caused by diffraction effects. The long propagation path combined with very high optical damage thresholds of organic materials (10^8 W cm^{-2}) make them very attractive. On the other hand, the higher NL coefficients of some organic materials make possible the use of relatively short interaction length (~ 10 mm) and ease the fabrication of WG as well as their applications in integrated optics.

In Table VII we report all NL organic materials investigated so far, which have been shaped in forms suitable for waveguiding: organic crystal cored fibers (OCCF) (Stevenson and Dyott, 1974; Stevenson, 1977; Babai *et al.*, 1977; Babai and White, 1980; Ballentyne and Al-Shukri, 1980, 1984; Nayar, 1983; Vidakovic *et al.*, 1984), organic channel WG (OCWG) (Tomaru *et al.*, 1984; Vidakovic *et al.*, 1984), and organic planar WG (OPWG) (Hewig and Jain, 1983; Sasaki *et al.*, 1984). Passive organic WG, although very important and instructive, is not included in this table (see Tien, 1971; Pitt, 1982).

The most thoroughly investigated NL organic WG structure is OCCF. The main reasons are relatively easy fabrication of uniform inner-wall-diameter capillaries as well as a shape compatible with glass optical fibers, a major element in optical communication systems. The modal WG dispersion, for fixed index ratio of cladding and core (or layer and substrate for planar WG), depends sensitively on the core diameter (or WG thickness). We can appreciate more quantitatively the advantages of WG over bulk NL interactions if we compare the respective second-harmonic conversion rates. In the case of WG, its magnitude is given (Azéma *et al.*, 1979) by

$$\rho(L) \equiv \frac{P(2\omega, L)}{P(\omega, 0)} \approx N d_{\text{eff}}^2 P(\omega, 0)\, f\left(\frac{\Delta\beta L}{2}\right) I_{\text{R}} \tag{1}$$

where N is a combination of the effective indexes of refraction at the wavelength of the pump beam $-\omega$ and that of the second harmonic -2ω; d_{eff} is the effective NL coefficient; L is the interaction length; $f(x) \equiv \sin^2 x/x^2$ is a

TABLE VII

Nonlinear Organic Materials Used for Fabrication of Different Waveguide Forms: Organic Crystal Cored Fibers (OCCF), Organic Channel Waveguides (OCWG), Planar WG, and Slab-Type WG

Name	Formula	WG dimensions (μm)	WG form	Reference
m-Nitroaniline (*m*-NA)	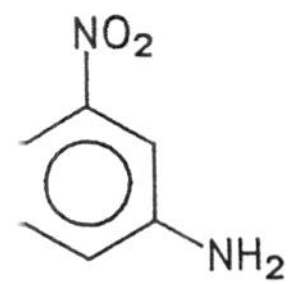	18–25 18–25 35 ~100	OCCF OCCF OCCF OCWG	Stevenson and Dyott (1974) Stevenson (1977) Babai and White (1980) Tomaru *et al.* (1984)
m-Dinitrobenzene (*m*-DNB)	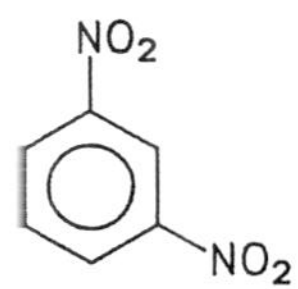	35–50 50 2–10 2–10	OCCF OCCF OCCF OCCF	Babai *et al.* (1977) Babai *et al.* (1980) Ballentyne and Al-Shukri (1980, 1984) Nayar (1982, 1983)
2-Bromo-4-nitroaniline (BNA)	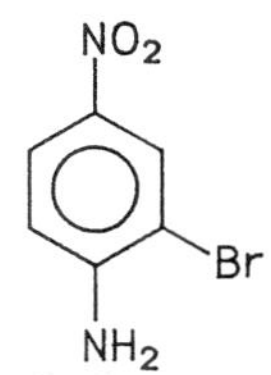	35–50 35–50	OCCF OCCF	Babai *et al.* (1977) Babai and White (1980)
m-Dihydroxybenzene (*m*-DHB)	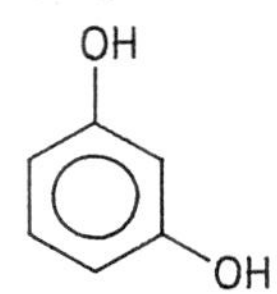	35	OCCF	Babai and White (1980)

(*continues*)

TABLE VII (*Continued*)

Name	Formula	WG dimensions (μm)	WG form	Reference
Formylnitrophenylhydrazine (FNPH)	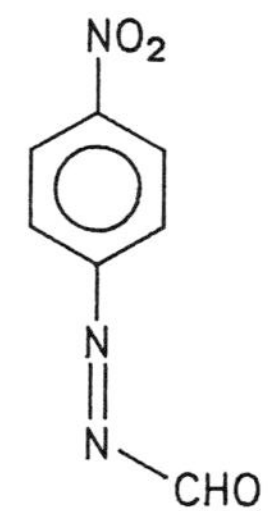	35–50	OCCF	Babai *et al.* (1977)
Benzil	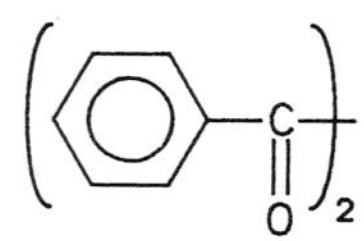	35	OCCF	Babai and White (1980)
		2–10	OCCF	Nayar (1982, 1983)
N-(4-Nitrophenyl)-(L)-prolinol	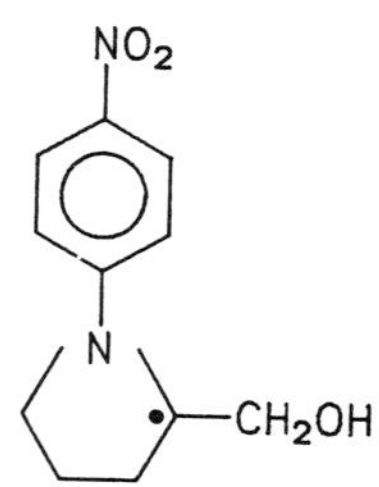	139	OCCF	Vidakovic *et al.* (1984)
(NPP)		~100	OCWG	Vidakovic *et al.* (1984)

Nitro-4-phenyl-*N*-(methylcyanomethyl)-amine (NPAN)	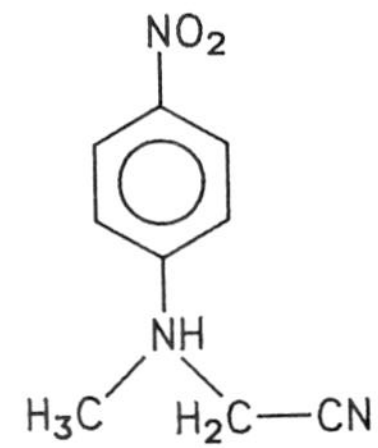	139	OCCF	This chapter
p-Chlorophenylurea (PCPU)	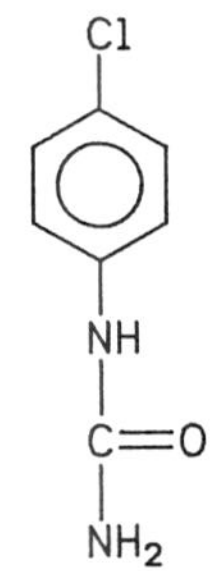	0.90	Planar WG	Hewig and Jain (1983)
2-Methyl-4-nitroaniline (MNA)	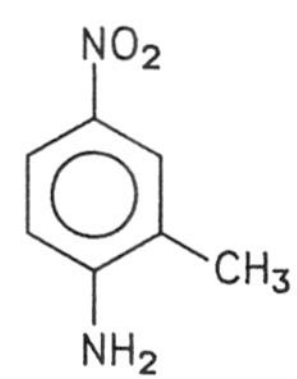	1.68 (max. SHG)	Tapered slab-type WG	Sasaki *et al.* (1984)

function reflecting the mismatch between a guided mode at ω [$\beta_l(\omega)$ being its propagation constant] and at 2ω [$\beta_m(2\omega)$]. The overlap between nonlinear polarization and harmonic modal field distributions over the guide cross section A is

$$I_{\mathrm{R}} = \int_A P_{\mathrm{NL}} E_m^{2\omega}\, dA \tag{2}$$

In practice, it is often difficult to optimize this factor (Ito and Inabe, 1978; Chen *et al.*, 1974). In the case of WG such as OCCF and OCWG, I_{R} is proportional to the inverse of an effective cross section r_{eff};

$$I_{\mathrm{R}} \approx r_{\mathrm{eff}}^{-2} \tag{3}$$

where r_{eff} is of the order of the wavelength λ. Taking into account Eqs. (3) and (2), in the case of perfect phase matching and supposing no pump depletion, Eq. (1) gives

$$\rho(L) \approx P(\omega, 0) L^2 \lambda^{-2} d_{\mathrm{eff}}^2 \tag{4}$$

The corresponding estimation of the second-harmonic rate $\rho_{\mathrm{B}}(L)$ in the case of bulk crystal (Yariv and Yeh, 1984) is given by

$$\rho_{\mathrm{B}}(L) \approx d_{\mathrm{eff}}^2 L^2 \frac{P(\omega, 0)}{\pi W_0^2} f\left(\frac{\Delta k L}{2}\right) \tag{5}$$

where $\Delta k = k_{2\omega} - 2k_\omega$ is the mismatch between fundamental and harmonic wavevectors and W_0 is the radius of the fundamental gaussian beams at its waist. In a more realistic case, taking into account the nature of a gaussian beam, the optimal conversion rate can be written, if L is of the order of few confocal parameters ($L \approx nW_0^2/\lambda$), as

$$\rho_{\mathrm{B}} \approx P(\omega, 0) L d^2 \lambda^{-1} \qquad \text{(phase matching)} \tag{6}$$

From Eqs. (4) and (6) one obtains the ratio $\rho(L)/\rho_{\mathrm{B}}(L)$,

$$\rho(L)/\rho_{\mathrm{B}}(L) \approx L/\lambda \tag{7}$$

Substituting $L = 1$ cm and $\lambda = 1\ \mu$m into Eq. (7), one obtains a very high ratio of 10^4, which is probably unrealistic due to the simplifications made in the above analysis, such as the absence of saturation. However, a ratio of 10^2 can be reasonably expected.

A first report on crystal growth of *m*-NA OCCF has been published by Stevenson and Dyott (1974). Although *m*-NA has high bulk second-harmonic (SH) and electro-optic (EO) coefficients, its crystalline structure is oriented with respect to the light propagation axis in such a way that the major part of its nonlinearity is made useless by the transverse nature of the electric field. Nevertheless, *m*-NA OCCF could serve as an efficient polarizer. Stevenson

(1977) has published a very exhaustive report on crystal growth of *m*-NA OCCF, giving a detailed description of the most frequently encountered defects. Some other organic materials, such as *m*-DNB, BNB, formylnitrophenylhydrazine (FNPH), and benzil, have been also shaped into OCCF (Babai *et al.*, 1977). It has been demonstrated, for instance, that *m*-DNB gives good-quality OCCF, contrary to the results obtained for bulk crystal growth.

Some improvements of classical BS and the use of an inverted BS methods have provided *m*-DNB and benzil OCCF of very good optical quality up to 5 cm in length (Babai and White, 1980; Ballentyne and Al-Shukri, 1980). Recently, Nayar (1982, 1983) has demonstrated successful crystal growth of benzil in 4-μm capillaries. He has also demonstrated for the first time SHG in OCCF. The HE_{11} fundamental mode at $\lambda = 1.06$ μm (Nd^{3+} : YAG laser as pumping source) has been doubled and phase matched with the continuum radiation field, giving a conversion efficiency proportional to the interaction length, instead of the squared length [see Eq. (1)] characteristic of two phase-matched waveguided modes.

In our laboratory, we have undertaken the growth of OCCF of a number of NL organic materials such as NPP (Zyss *et al.*, 1984) and NPAN (Vidakovic *et al.*, to be published).

In the case of NPP OCCF, the cleavage planes $\langle 101 \rangle$, containing the charge-transfer molecular axes, make an angle of $\sim 70°$ with the propagation axis (Fig. 27). That orientation permits us to take advantage of the optimized nonlinear coefficient d_{YXX}. The phase-matched SHG of an Nd^{3+}: YAG laser has been observed over short interaction lengths (~ 2 mm) limited by cleavage planes. Another material, NPAN, grown in capillaries has an NL efficiency of the same order as that of NPP and the additional advantage of no cleavage in OCCF. Detailed NL optical investigation of NPAN OCCF will be published in a separate paper.

A relatively new method for the crystal growth of organic channel waveguides (OCWG) has been proposed by Tomaru *et al.* (1984) and by Vidakovic *et al.* (1984). It is based on a zone-melting process generated by CO_2 laser scanning along the channel.

Hewig and Jain (1983) have used *para*-chlorophenyl urea (PCPU) to form a planar organic WG (~ 0.9 μm) on a single substrate via classical high-vacuum evaporation. They have used a single rutile prism for in- and out-coupling of a Raman-shifted Nd^{3+} : YAG-pumped dye laser. This method could prove very useful for assessment of crystalline quality and guiding properties of planar organic WG.

Sasaki *et al.* (1984) prepared flakelike methylnitroaniline (MNA) crystals by vapor-phase growth. This material is highly efficient (Levine *et al.*, 1979), but use of the d_{11} optimized coefficient is precluded in the bulk by the impossibility to ensure phase matching for parallel interacting polarizations. However, it

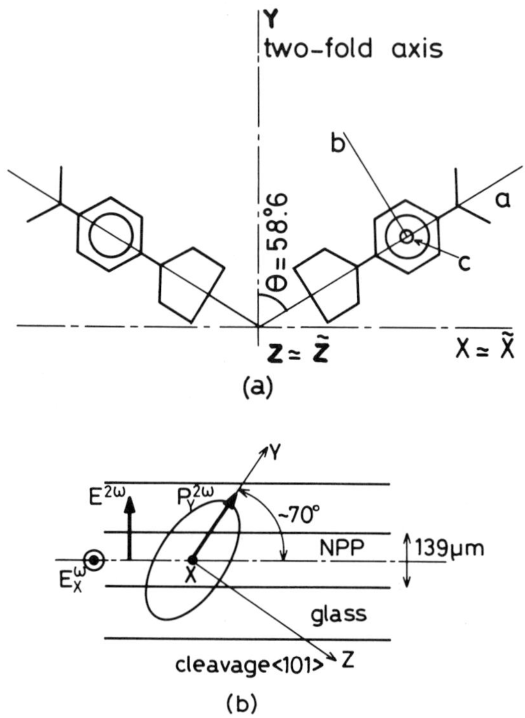

Fig. 27. (a) Simplified crystallographic NPP structure, showing planar charge-transfer aromatic molecules stacked in parallel planes. (b) The principal dielectric axis reference frame is represented with respect to the fiber axis: the molecular charge-transfer axis has a transverse orientation, allowing for NL interactions.

has been proposed (Levine *et al.*, 1979; Stegeman and Liao, 1983) to use MNA for frequency doubling of an Nd^{3+} : YAG laser (1.06 μm) in a properly tailored thin-film geometry.

In the reported experiment (Sasaki *et al.*, 1984), a thin crystal of MNA ($\sim$5 mm $\times$ 10 mm $\times$ 2 μm) was used as a high-index topping layer over a tapered slab-type glass WG. To obtain phase-matched SHG, one must adjust WG thickness by lateral movement of the tapered WG. Two prisms ensure in- and out-coupling of the light.

A. Fabrication of Capillaries and Channel WG

The two materials used here for the fabrication of capillaries and channel WG are silica glass and epoxy resin. Silica glass capillaries were obtained by standard motorized pulling technique with bore diameter and wall thickness

139 μm and 130 μm, respectively. Lindemann capillaries were used as purchased. Two-component thermally resistant epoxy resin (Rhône Poulenc RTV 101) has been used to probe the influence of mechanical properties of cladding on the core growth. The capillary was formed by using a metallic wire (~200 μm), around which the epoxy was polymerized.

The grooves of 100 μm for channel WG were formed in a silica glass substrate by abrasive action with a metallic wire. The quality and homogeneity of such grooves were sufficient for preliminary experiments on crystal growth. Another method for their fabrication, based on CO_2 laser processing, has also been proposed (Kawachi *et al.*, 1983; Tomaru *et al.*, 1984).

B. Growth of Organic Crystal Cored Fibres (OCCF) and Organic Channel Waveguides (OCWG)

NPP and NPAN were firstly melted and degassed under reduced pressure (~10^{-1} mm Hg). The capillaries were then filled with molten material, using capillary force action. For fibers of an average core diameter of 139 mm, the heights of the molten columns of both materials are 6–8 cm.

Filled capillaries are then carefully solidified by slow cooling to the room temperature. This procedure is important to prevent from randomly spaced spontaneous nucleation, which, in addition to the large volume shrinkage on solidification (10–15%), usually leads to an interruption of the core. Moreover, another undesired effect of this intermediate step is the trapping of residual gas dissolved in the melt, which is the main source of defects (boundary voids, bubbles, etc.) created during the controlled crystal growth (Section VI,D). To reduce these defects, the best method is to eliminate completely this first recrystallization step.

The capillaries with polycrystalline cores are placed in a specially designed holder to support eight fibers, then introduced into the apparatus for controlled crystal growth (Fig. 4). Thermal gradients range from 5 to 7°C cm^{-1} and pulling rates from 1 to 6 mm hr^{-1}. Special attention has been paid to ensure that the solid/liquid interface is in the hot zone.

A new method of OCWG growth has been used by Tomaru *et al.* (1984). We have simultaneously undertaken preliminary experiences on NPP, using a CO_2 laser for zone melting (Vidakovic *et al.*, 1984). Previously prepared channels have been filled by molten NPP and recrystallized by slow cooling. Relatively rapid scanning (~1 mm hr^{-1}) of a CO_2 laser along the grooves (Fig. 28) was the second fabrication step. The source used here is a CO_2 WG laser, QUANTEL WG 03S. Nominal continuous-wave (CW) power is 3 W, and the beam diameter at output mirror is 1.3 mm at $1/e^2$ power distribution. Its divergence is diffraction-limited.

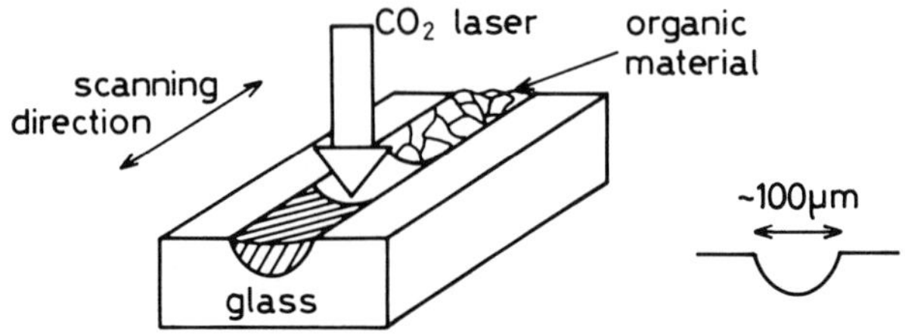

Fig. 28. Schematic representation of OCWG fabrication by CO_2 laser scanning.

Under the laser action, the organic material is locally melted. Depending on the scanning rate as well as on laser power, the behavior and the size of the subsequently created molten zone are different. For reduced scanning rate and laser power, we obtained a poor crystallinity of OCWG. For a given scanning rate of ~ 1 mm min^{-1}, increasing laser power (500 MW) has provoked a rapid decomposition of NPP, which has been identified by its color changing. Moreover, the NPP remains in a supercooled state, which is a characteristic of impurity-rich organic materials. Large oriented single-crystalline domains have been obtained using a scanning rate of ~ 1 mm min^{-1} and laser power ranging from 300 to 500 mW (Fig. 29). Further work on this method is in progress.

C. Results

Many organic materials synthesized for NLO have been grown by our modified BS technique. Here we discuss two specific materials of comparable NLO efficiencies: NPP and NPAN (see Table VII).

We recall first that our aim is to fabricate the waveguide structure composed of an organic crystal core and a glass cladding. To ensure good waveguiding properties, the crystal core must be free of any major defects. Although NPP grows satisfactorily in the capillaries, yielding single-crystal cores up to 4 cm in length, its major defect is the occurrence of cleavage planes during the cooling step (Fig. 30). This problem is common to many NL organic materials, such as *m*-NA, NPP, and MAP, and may exclude their use for fabrication of OCCF. The problem originates from strong adherence of organic materials on the walls as well as from inelastic crystal–wall interactions. These effects may induce stresses and fractures of the crystal during its growth. To verify this hypothesis, we have used epoxy resin capillaries, for which there is practically neither adherence nor induced stresses. Indeed, observations between crossed polarizers have confirmed our suppositions (Fig. 31). Unfortunately, a very rough crystal–epoxy interface and strong diffusion of NPP into the polymer matrix preclude their use in NL guided optics. These problems limit the

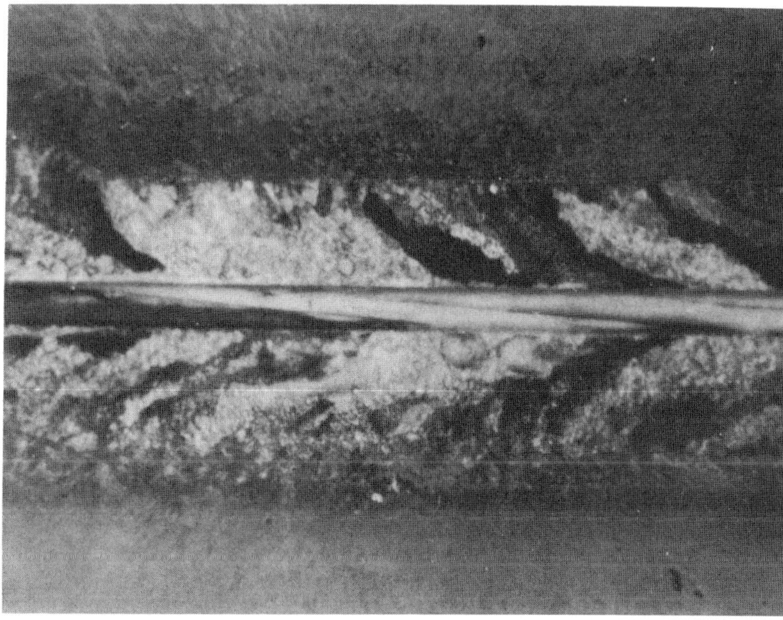

Fig. 29. NPP OCWG obtained by CO_2 laser scanning, viewed between crossed polarizers. Scanning rate, 1 mm min^{-1}; laser power, 500 mW. Note a change of crystal domain orientation, induced by off-axis laser scanning (at the left of photo).

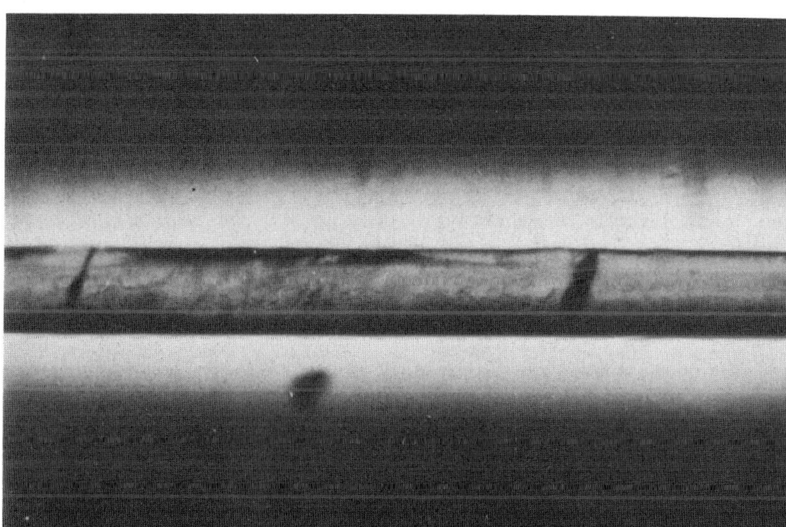

Fig. 30. NPP OCCF obtained by a modified BS method. Cleavage plane (101) makes an angle of ~70° with the fiber axis. Capillary diameter is 200 μm.

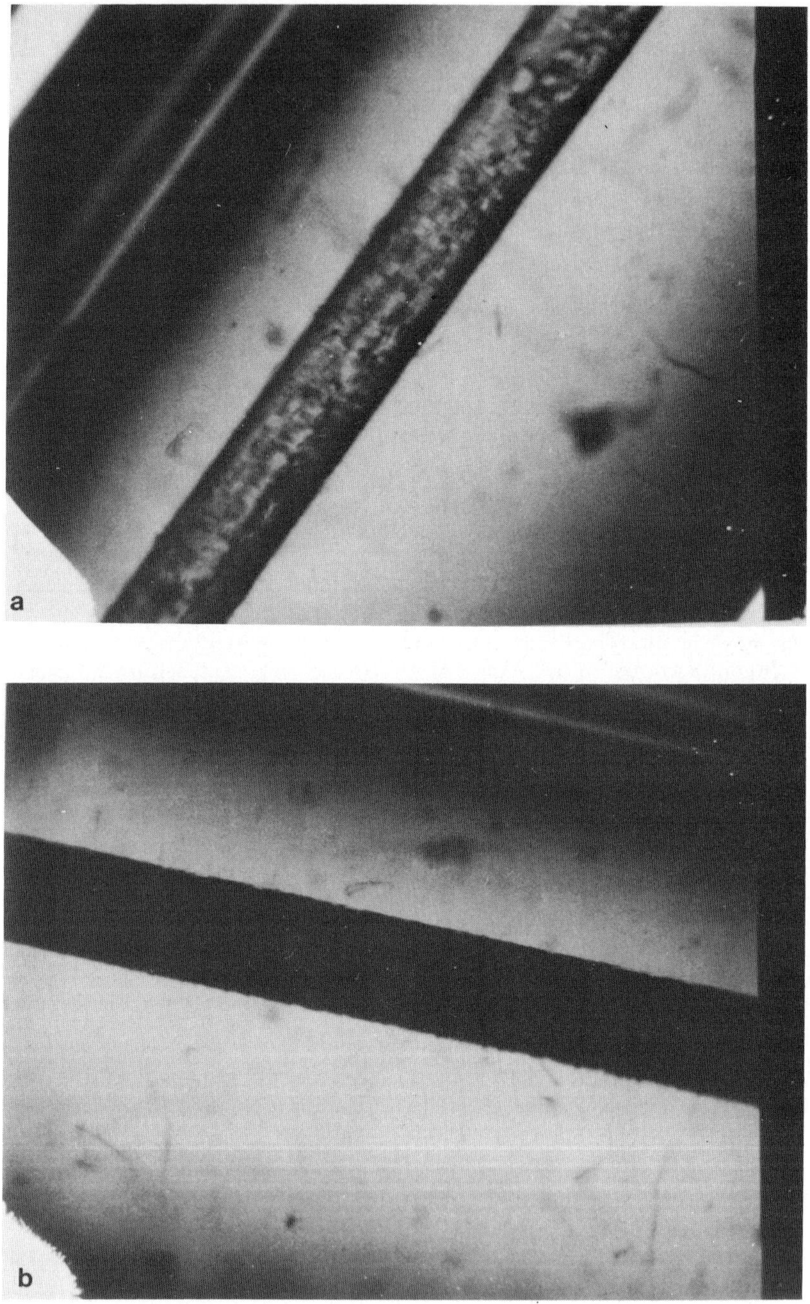

Fig. 31. NPP OCCF (epoxy resin capillary) obtained by a modified BS method, viewed between (a) parallel polarizers or (b) crossed polarizers. Note the absence of cleavage, as well as a rough epoxy/crystal interface.

number of NL organic materials useful for OCCF. However, a suitable surface treatment of capillary walls should relax the constraints imposed by adherence. Furthermore, crystal growth in open containers, such as channels or planar WG, could reduce the stresses due to complete confinement of the crystal. Consequently, organic materials belonging to the first category (NPP, MAP, etc.) may be good candidates for NL organic WG.

On the other hand, no cleavage of NPAN (belonging to the second class of organic materials) has been observed over a 3-cm-long single-crystal core (Fig. 32). The only really persistent defects in this case are of boundary voids, which are discussed in Section VI,D (see Fig. 34).

1. X-Ray and Microscopic Characterization of OCCF and OCWG

These two methods were our principal guidance toward the optimization of crystal growth conditions. The crystalline quality, form, and nature of observed defects have been found to be insensitive to minor changes made to the pulling rate or zone temperatures, as long as the solid–liquid interface remains in the hot zone.

It should be emphasized that X-ray analysis of OCCF is not a straightforward technique. Indeed, as we are particularly interested in crystal core orientation with respect to fiber axis, which is relevant for NL waveguided interaction, OCCF must be examined as grown. More precisely, an X ray incident on an OCCF traverses two capillary walls and an organic core. The intensity of scattered X rays is proportional to the cladding-to-core volume ratio. On the other hand, the X-ray absorption depends on the heavy-metal content of capillary glass. Same effects can be considered negligible for organic crystal core. Aforementioned effects could be reduced by a suitable choice of wall thickness (Babai and White, 1980). Combining a small cladding-to-core ratio with the use of special glasses with low heavy-metal content, such as Lindemann glass, we get a further reduction of the absorption by an order of magnitude. In Table VIII are given the values of linear absorption coefficients of three common glasses—Pyrex 7740, fused quartz, and Lindemann—at three X-radiation wavelengths: Ag$K\bar{\alpha}$, Mo$K\bar{\alpha}$, and Cu$K\bar{\alpha}$ (De Titta, 1985).

For our X-ray investigation, we have used Lindemann capillaries, with internal-wall diameter and wall thickness of 0.3 and 0.1 mm, respectively. Some of them with NPAN cores have served for the complete X-ray analysis of unknown NPAN crystalline structure (Vidakovic *et al.*, 1986).

Both X-ray and microscopic observation of NPP (Vidakovic *et al.*, 1984) and NPAN (this chapter) OCCF have confirmed that a constant crystal orientation with respect to fiber axis was preserved over lengths up to 3 or 4 cm (Fig. 20). The comparison of angles made by cleavage planes and capillary axis, for both silica and Lindemann glasses, has shown no influence

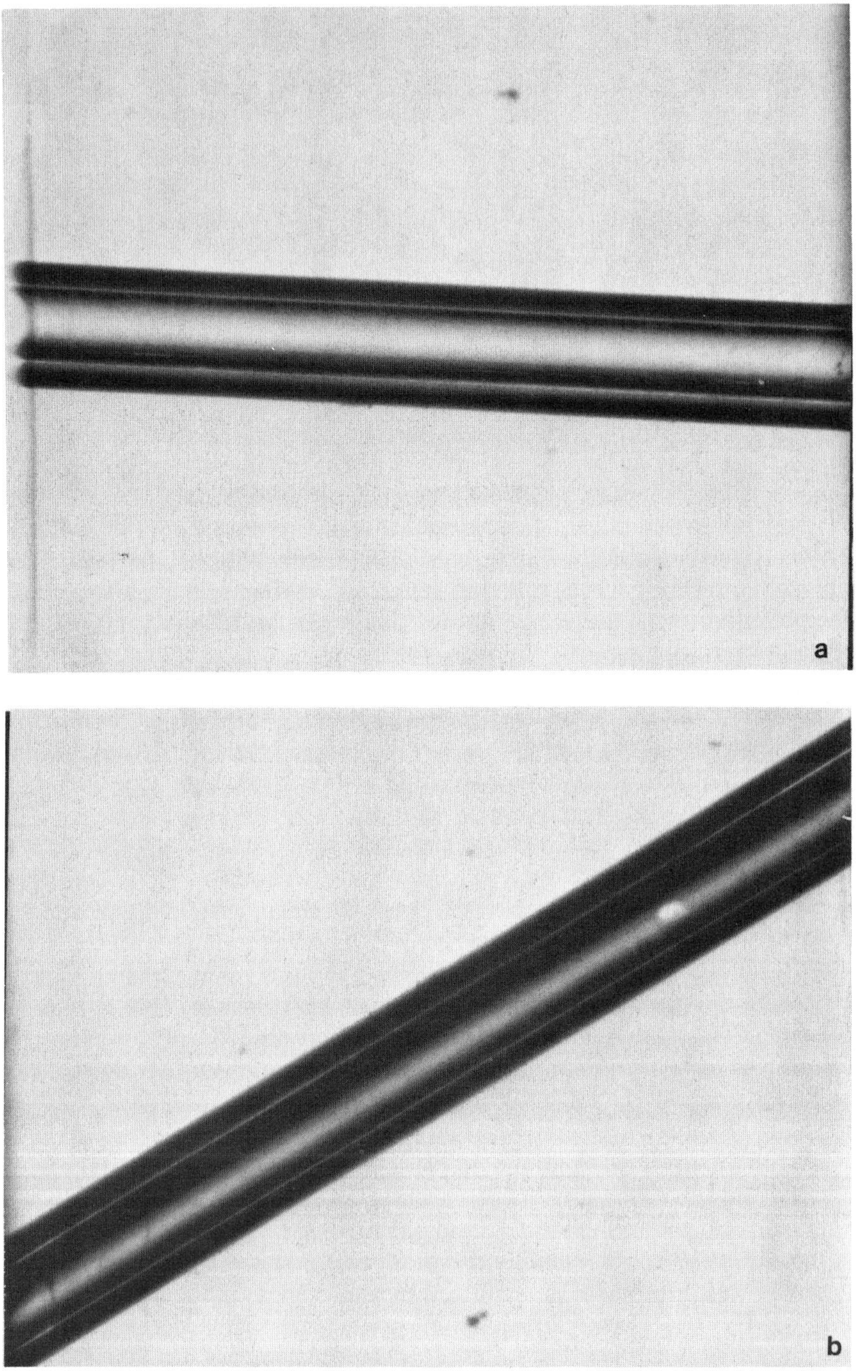

Fig. 32. NAPN/silica glass OCCF obtained by a modified BS method, viewed between (a) parallel polarizers or (b) almost cross polarizers. A bright spot (at the right) is due to a thin polycrystalline layer deposited at the crystal/glass interface, without perturbing crystal core autoalignment

TABLE VIII

Linear Absorption Coefficients of Common Glasses at Three Wavelengths of X Radiation: $AgK\bar{\alpha}$, $MoK\bar{\alpha}$, and $CuK\bar{\alpha}$

Glass	Density ($g\ cm^{-3}$)	Linear absorption coefficient (mm^{-1})		
		$AgK\bar{\alpha}$	$MoK\bar{\alpha}$	$CuK\bar{\alpha}$
Pyrex 7740	2.23	0.375	0.733	7.600
Fused quartz	2.20	0.700	1.360	13.298
Lindemann	1.77	0.094	0.161	1.484

of the different nature of the physicochemical interactions involved. If any influence exists, it seems be negligible compared to the prevailing driving force of heat transfer on crystal growth processes.

D. Discussion

The crystal growth of organic materials in capillaries can be considered a general method of assessment of very useful information on new materials, regardless of their possible applications. The advantages of this method may be singled out as follows:

(1) A small quantity of organic material, as may result very often from a complicated chemical synthesis, is sufficient to obtain good-quality single crystals.

(2) When other methods of crystal growth fail, it turns out that single crystals may still be obtained by growth in capillaries, with good quality and nonnegligible size. Indeed, in the case of capillaries, usual conditions imposed on heat transfer throughout the crystal and container walls are somewhat relaxed. It is well known that heat transfer throughout a growing crystal is hindered as its linear dimensions increase. This is particularly pronounced for organic materials, being usually very poor conductors. For instance, if we compare the thermal conductivity of Pyrex glass ($k_s = 1.13\ W\ m^{-1}\ K$) to that of salol ($k_s = 0.251\ W\ m^{-1}\ K$), we find that the former is about 4.5 times higher than the latter. Thus, a suitable choice of the capillary diameter and wall thickness favors the thermal exchange.

(3) The method is rapid. Common rates of crystallization of organic materials in capillaries range from 1 to 10 $mm\ hr^{-1}$.

Using this method, one finds two kinds of problems. First, the thermal decomposition of organic materials, although common to all methods of crystal growth from the melt, is less stringent in this case, due to high crystal growth rate. Internal stresses and fractures, due to the presence of container walls, are the common flaws of the BS method. However, these problems can be partially or completely resolved by a suitable choice of ampoule material, which should considerably decrease the adhesion on walls, together with undesired elastic interactions with them.

Although the second class of problems appears also in bulk crystal growth, it is particularly amplified in OCCF. The bubbles and gas inclusions are more easily trapped in the crystal core and at the capillary walls than in the bulk. Moreover, dimensions of these are of the same order as that of capillary core and consequently cannot be considered to be small perturbations. In OCCF or OCWG, the light will be strongly scattered by defects and guided modes converted into radiation modes, resulting in a significant reduction of their NL efficiencies.

In NPP and NPAN OCCF, we have observed such defects as loss of crystal core alignment, bubbles (Fig. 33), boundary voids (Fig. 34), and highly stressed regions, already discussed previously (Stevenson, 1977; Babai and White, 1980). The loss of crystal core alignment found in NPAN fibers is often related to the existence of a small quantity of polycrystalline material localized at the walls of the capillary. It is probably correlated with bubbles

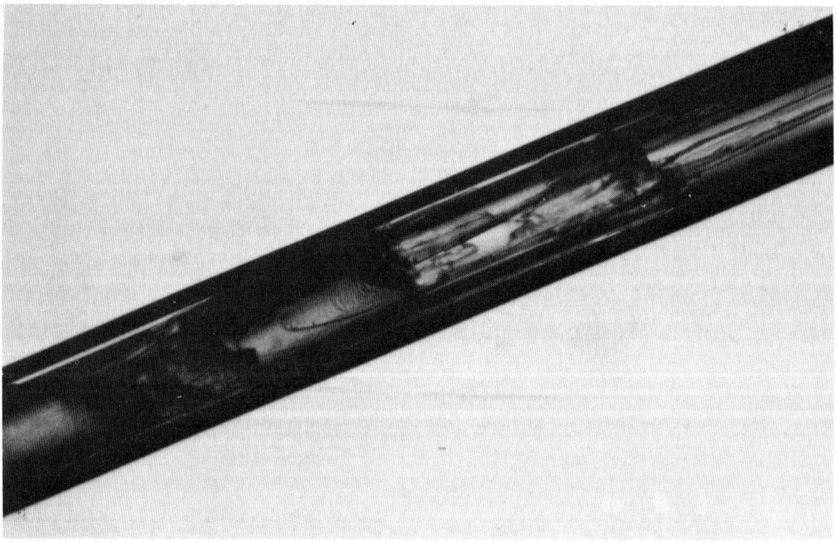

Fig. 33. NPAN/silica glass OCCF obtained by a modified BS method, viewed between crossed polarizers. A bubble has completely interrupted the NPAN crystal core. Pleochroism of such obtained segments is due to their stressed state.

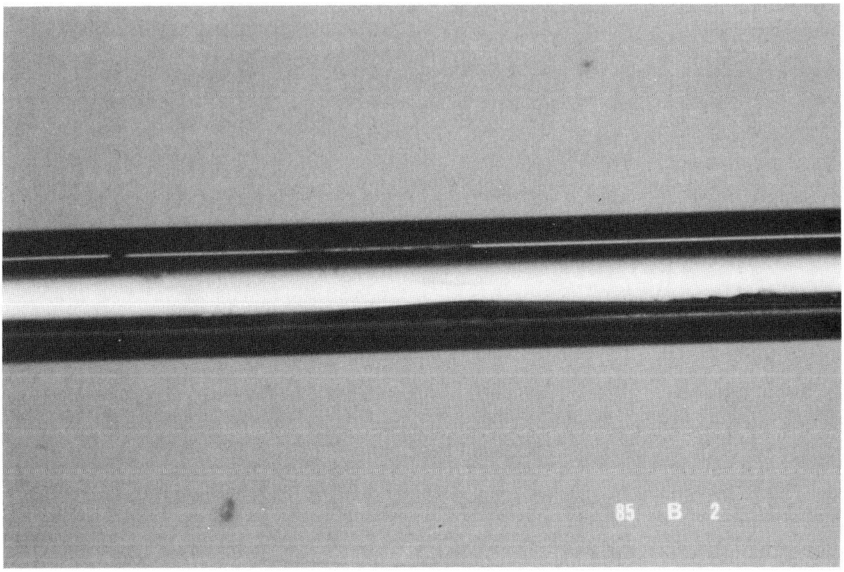

Fig. 34. NPAN/silica OCCF obtained by our modified BS method. A boundary void can be seen at the bottom.

filled with NPAN vapor that crystallize spontaneously in the cold zone of the BS apparatus. The much higher vapor pressure of NPAN as compared to that of NPP could explain the absence of this defect in NPP OCCF.

Complete breaking of the crystal core, leaving no trace on the walls and without loss of core autoalignment, which has been characterized by Stevenson (1977) as a "perplexing defect," is very rare (1%) in our fibers. One possible explanation is that a thin film of a few angstroms at the capillary walls is sufficient to transmit the information of crystal growth direction. Depending on plasticity, adhesion on walls, and volume shrinkage of organic material, this film interrupts, leaving no trace on the walls. The crystal core near the interruption is highly stressed, as was observed by Stevenson (1977).

The regular periodic interruption of the crystal core resulting in the formation of segments a few hundred angstroms long separated by cylindrical voids, has been observed in 4-μm *m*-DNB OCCF (Ballentyne and Al-Shukri, 1980, 1984).

Ballentyne and Al-Shukri (1980) have analyzed the classical and inverted BS method (where the two temperature zones are inverted) as applied to crystal growth in capillaries. They have shown (see Appendix) for this method the existence of a critical relation between the height of the liquid column (H) and the radius of the capillary (r), contrary to the classical method [Fig. 35 (a) and

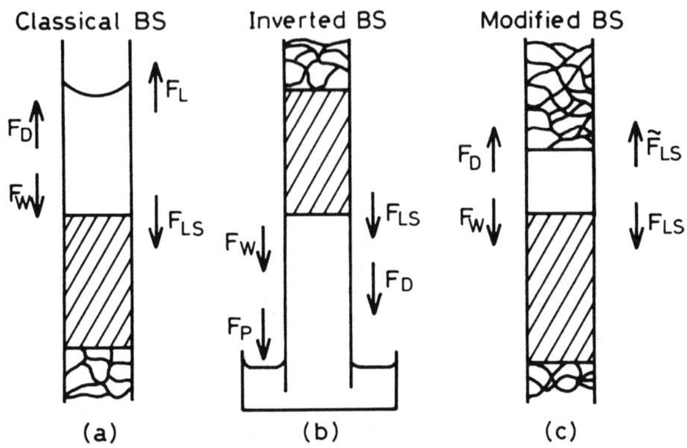

Fig. 35. Simplified scheme of different methods for OCCF growth: (a) classical BS, (b) inverted BS, and (c) modified BS. Principal forces acting on crystal growth are also represented.

(b)]. *m*-DNB OCCF grown by inverted BS method were free of voids (Ballentyne and Al-Shukri, 1980).

Our method, which is neither the classical nor the inverted BS, has not been, to our knowledge, previously analyzed in the case of crystal growth in capillaries. All numerical examples given in the present analysis are based on values measured for *m*-DNB from Ballentyne and Al-Shukri (1984), such as viscosity (η), surface tension (σ_L), liquid density (ρ_L), and crystal density (ρ_S).

In modified BS (MBS) method, far from the ends of the fiber, the molten column of height H [Fig. 35 (c)] is closed between two interfaces: single crystal/melt and polycrystal/melt. At the end of the crystal growth run, the later interface is replaced by a melt/air interface, such as in classical BS (see Appendix). The height of the molten column H is assumed to remain constant during growth. Its weight is given by

$$F_W = \pi r^2 \rho_L g H \tag{8}$$

This force contributes to supplying the growing crystal with molten material.

During growth, most of the organic material shrinks, due to the difference in phase densities. This shrinkage implies a melt flow toward the interface, creating a dragging force F_D in the opposite direction. For small crystal growth rates, this flow can be considered as streamline, so that F_D is given by

$$F_D = 8\pi\eta H v^* \tag{9}$$

where v^* is the melt velocity related to the mass transfer towards the crystals. From the continuity condition, v^* is given by

$$v^* = (\rho_S/\rho_L)v \tag{10}$$

where v is the crystal growth velocity. At the single-crystal/melt interface, the interfacial tension σ_{SL} creates the force F_{SL}:

$$F_{SL} = 2r\sigma_{SL} \tag{11}$$

The term σ_{SL} can be calculated, using Eq. (12) as proposed by Mayer and Kuhn (1961), if σ_L, ρ_L, and ρ_S are known:

$$\sigma_{SL} = \left(\frac{\rho_S}{\rho_L} - 1\right)\sigma_L \tag{12}$$

Similar expressions could be written for polycrystal/melt interface, taking into account the remarks given below. The polycrystalline part is composed of microcrystals with different crystal faces in contact with the melt. Furthermore, it is known that they are not always completely wetted by their own melt. Moreover, during the uncontrolled solidification, the grain boundaries and junctions are sites where a gas and other impurities are preferentially captured. Thus, $\tilde{\sigma}_{SL}$ and $\tilde{F}_{SL}$ are only averaged quantities and not necessary equal to σ_{SL} and F_{SL}, respectively. It should be also emphasized that the impurities segregated on the advancing front of crystallization could influence σ_{SL} and consequently F_{SL} in such a way that Eqs. (8)–(12) should be replaced by other expressions, taking these effects explicitly into account.

Thus, the column will be broken if

$$\tilde{F}_{SL} + F_D \geq F_{SL} + F_W \tag{13}$$

We will first suppose a simple case when $\tilde{F}_{SL} = F_{SL}$. Then, taking into account Eqs. (8)–(12), we obtain from Eq. (6)

$$v \geqslant \frac{\rho_L^2 g}{8\rho_S \eta} r^2 \tag{14}$$

This inequality gives the higher limit of crystal growth rate, beyond which the growth becomes unstable. It is different from the critical condition for core breaking, obtained in the case of classical BS (see Appendix). From Eq. (14), it is obvious that this condition becomes far less stringent for increased capillary diameter. For *m*-DNB (Ballentyne and Al-Shukri, 1984), Eq. (14) gives

$$v \geqslant 5.98 \times 10^5\, r^2 \qquad (\text{m sec}^{-1}) \tag{15}$$

If one put $r = 1\ \mu$m into Eq. (15), it gives $v = 0.22$ cm hr^{-1}, which is in the range of common crystal growth rates used for the fabrication of OCCF (Babai *et al.*, 1977; Babai and White, 1980).

For a general case, $\tilde{F}_{SL} = F_{SL}$, Eq. (13) becomes

$$v \geqslant \frac{1}{8\pi\eta}\left(\frac{\rho_L}{\rho_S}\right)\left[\pi\rho_L g r^2 - \frac{F_{SL}}{H}\left(\frac{\tilde{F}_{LS}}{F_{LS}} - 1\right)\right] \tag{16}$$

The crystal will grow, if

$$F_W \geqslant F_{SL}\delta \tag{17}$$

where

$$\delta = (\tilde{F}_{SL} - F_{SL})/F_{SL} \tag{18}$$

Equation (17) simply means that the weight of the liquid column must compensate for the difference between the interfacial tensions at the two interfaces, to allow for crystal growth.

Substituting Eqs. (8), (11), and (12) into Eq. (17), one obtains

$$\delta \leqslant \left(\frac{\pi g}{2}\right)\frac{\rho_L^2}{(\rho_S - \rho_L)\sigma_L}(rH) \tag{19}$$

For $r = 1$ μm and $H = 2$ cm in the case of *m*-DNB, the inequality of Eq. (19) gives

$$\delta \leqslant 7.7\% \tag{20}$$

From this very crude analysis, it follows that $\tilde{F}_{SL}$ may be greater than F_{SL}, without perturbing normal crystal growth. Moreover, the variations of σ_{SL} induced by impurity, temperature, or pressure fluctuations are not critical if Eq. (19) is satisfied. Taking again $r = 1$ μm and $H = 2$ cm, we may estimate the critical value of crystal growth rate from Eq. (16) by varying δ. For $\delta = -7\%$ and $+7\%$ one obtains $v \geq 4.4$ mm hr^{-1} and $v \geq 0.2$ mm hr^{-1}, respectively. These relatively limited differences in interfacial tensions ($|\delta| < 0.1$) favor lower crystal growth rates, which risk increasing undesired thermal decomposition for most organic materials. For fabrication of OCCF with 1 μm crystal core, the inverted BS method is more convenient, because there are no critical parameters imposed on crystal growth conditions (see Appendix).

The work on crystal growth of single-mode OCCF ($r \approx 2$ μm) is in progress and will be published in a separate paper (Vidakovic *et al.*, 1986).

VII. CONCLUSIONS

We have described for a number of organic compounds the basic growth principles and techniques used in our laboratory. Although NLO applications are very demanding in terms of crystalline quality, we have succeded in producing very efficient crystals such as MAP, NPP, NPAN, POM, and AMA. Certainly, other types of molecules or material processing techniques can be used. New techniques of fabrication of organic crystal waveguides are being developed, opening other possibilities for fully optimizing NLO devices that cannot be achieved with bulk crystals.

All nonlinear organic materials studied here are first grown by slow solvent evaporation, yielding small crystals. These are used to determine crystallographic structures as well as to investigate solvent interaction. Crystal growth from solution, when it is possible, leads to large good-quality crystals with well-developed faces, which are more convenient for further characterization. This technique is well exemplified by MANS, POM, and AMA. In these cases, excellent temperature regulation to less than 0.005°C, as well as smooth and steady crystal stirring over a period of the order of 1 month, are of the utmost importance.

In other cases, we have used the BS method or the MBS method depending on the degradation rate of materials. When this level is low, as is the case for *m*-DNB, *p*-NA, and *m*-BNB, the former method gives very good crystals. MAP and NPP, being of low thermal stability, have been grown by MBS, permitting the reduction of the time during with the compound is staying in the melt. A polymeric container reduces adhesion, and elastic interaction with ampule walls tends to induce defects and fractures.

To prevent damaging environment influences as well as to reinforce fragile organic crystals (AMA), we have employed epoxy resin claddings. Furthermore, such cladding allows easier cutting, polishing, and manipulation of polymer-covered crystals.

We have also demonstrated successful growth of NPP, NPAN, OCCF, and OCWG, presenting a promising route toward integrated or optical fiber NL devices.

Research toward new processing methods for nonlinear organic materials (other than single-crystals), such as oriented organic thin films, inclusion of organics in polymer, organic glasses, and others, is a fast-growing field that should lead to interesting advances in the near future.

ACKNOWLEDGMENTS

It is a pleasure to acknowledge fruitful and numerous discussions with J. Bernard (Université de Bordeaux I), D. S. Chemla (AT&T Bell Labs), M. Coquillay (Centre Pharmaceutique de Chatenay-Malabry), I. Ledoux, J. F. Nicoud (E.S.P.C.I., Paris), J. L. Oudar, and J. Zyss. The authors are obliged to J. F. Bayon (CNET, Lannion), who provided calibrated fibers, and to J. Primot, for X-ray analysis. They appreciate the constant and friendly support of R. Bertrand, G. Caumont, R. Chanat, P. Fremaux, and R. Mellet, and helpful assistance of F. Berdague, D. Bertin, P. Hisope, and A. Piquet for final presentation of the text.

APPENDIX

We follow here the analysis of the conventional and inverted BS methods applied on crystal growth in capillaries, previously exposed by Ballentyne and

Al-Shukri (1984). For conventional BS method [Fig. 35(a)], which also applies on our case at the end of a crystal growth run, the condition of breaking of liquid column is given by

$$F_L + F_D \geqslant F_W + F_{LS} \tag{A1}$$

F_L is the force due to the surface tension σ_L acting on the interface of air and melt:

$$F_L = 2\pi r \sigma_L \cos\theta \tag{A2}$$

where θ is the contact angle between the capillary wall and the melt.

In the case of *m*-DNB, we found

$$F_L \ll F_{LS} \ll F_W \ll F_D \tag{A3}$$

supposing $v = 1 \text{ mm hr}^{-1}$.

Neglecting F_D in Eq. (A1) and taking into account Eqs. (8)–(13) and (A2), the critical condition can be put in the following form:

$$Hr \leqslant \frac{2\sigma_L}{\pi g \rho_L}\left[\pi\cos\theta - \left(\frac{\rho_S}{\rho_L} - 1\right)\right] \tag{A4}$$

From Eq. (A4), it follows that for a given capillary diameter, the critical parameter is the height of the molten column (H). For instance, taking $r = 4\ \mu$m, H must be greater than 1 to obtain *m*-DNB OCCF that is not broken.

In the case of the inverted BS method [Fig. 35(b)], the condition of Eq. (A1) will be replaced by

$$F_W + F_D \geqslant F_P + F_{LS} \tag{A5}$$

where F_P is the force due to an external pressure P acting toward the surface of the crystal:

$$F_P = \pi r^2 \mathrm{P} \tag{A6}$$

For $P \approx 10^5$ Pa (atmospheric pressure), very simple analysis of Eq. (A5) shows that there is no critical parameter governing the crystal growth by the inverted BS method. Thus, if one compares the three methods—classical, modified, and inverted BS—the latter should be preferentially used for the fabrication of very small core diameter OCCF (1 μm).

REFERENCES

Azéma, A., Botineau, J., Gires, F., and Saissy, A. (1979). *SPIE Proc.* **213,** 26.
Babai, F. H., and White, E. A. D. (1980). *J. Cryst. Growth* **49,** 245.
Babai, F. H., Dyott, R. B., and White, E. A. D. (1977). *J. Mater. Sci.* **12,** 869.

Badan, J., Hierle, R., Perigaud, A., and Zyss, J. (1983). *ACS Symp. Ser.* **233,** 81–107.
Badan, J. *et al.* To be published.
Ballentyne, D. W. G., and Al-Shukri, S. M. (1980). *J. Cryst. Growth* **48,** 491.
Ballentyne, D. W. G., and Al-Shukri, S. M. (1984). *J. Cryst. Growth* **68,** 651.
Bernard, J. (1984). Ph.D. diss., Université de Bordeaux I. France.
Böhm, S. Rüstov, B. (1969). *Stud. Biophys.* **13,** (3) 169–176.
Born, M., and Wolf, E. (1965). In "Principles of Optics." Pergamon, Oxford.
Bridgman, P. W. (1925). *Proc. Am. Acad. Arts Sci.* **60,** 303–383.
Carenco, A., Jerphagnon, J., and Perigaud, A. (1977), *J. Chem. Phys.* **66,** 3806–3812.
Chemla, D. S., Oudar, J. L., and Jerphagnon, J. (1975). *Phys. Rev. B* **12,** 4534–4541.
Chen, B. U., Tang, C. L., and Telle, J. M. (1974). *Appl. Phys. Lett.* **25,** 495.
Davydov, B. L., Derkacheva, L. D., Dunina, V. V., Zhabolinski, M. E., Zolin, V. F., Koreneva, L. G., and Samokhina, M. A. (1970). *JETP Lett.* **12,** 16.
De Titta, G. T. (1985). *J. Appl. Crystallogr.* **18,** 75.
Glockner, E. (1974). Ph.D. Diss., Univierersity of Stuttgart.
Guillaud, G., Le Helley, M., and Mesnard, G. (1970). *Bull. Soc. Fr. Mineral. Cristallogr.* **93,** 131–132.
Hewig, G. H., and Jain, K. (1983). *Opt. Commun.* **47,** 347.
Hierle, R., Badan, J., and Zyss, J. (1984). *J. Cryst. Growth* **69,** 545–554.
Ito, H., and Inaba, H. (1978). *Opt. Lett.* **6,** 139.
Jerphagnon, J. (1971). *IEEE J. Quant. Electron.* **QE-7,** 42.
Karl, N. (1980). *In* "Crystals" (H. C. Freyhardt, ed.), Vol. 4, p. 63. Springer-Verlag, Berlin and New York.
Kawachi, M., Yasu, M., and Edahiro, T. (1983). *Electron. Lett.* **19,** 583.
Kurtz, S. K., and Perry, T. T. (1968). *J. Appl. Phys.* **39,** 3798.
Kyropoulos, S. (1926). *Z. Anorg. Chem.* **154,** 308.
Kyropoulos, S. (1930). *Z. Phys.* **63,** 55.
Levine, B. F., Bethea, C. G., Thurmond, C. D., Lynch, R. T., and Bernstein, J. L. (1979). *J. Appl. Phys.* **50,** 2523.
Mayer, H., and Kuhn, W. (1961). *Z. Phys. Chem.* **30,** 289.
Mellan, I. (1970). *In* "Industrial Solvents Handbook," pp. 424–433. Noyes Data Corp., Park Ridge, New Jersey.
Nayar, B. K. (1982). *Dig. Top. Meet. Integrated Guided Wave Opt., 6th,* Th A2.
Nayar, B. K. (1983). *ACS Symp. Ser.* **233,** 153–166.
Nyvlt, J. (1977). "Solid Liquid Phase Equilibria," p. 86. Am. Elsevier, New York.
Oudar, J. L., and Hierle, R. (1977). *J. Appl. Phys.* **48,** 2699–2703.
Perigaud, A., and Marais, M. (1976). *Mater. Res. Bull.* **11,** 23–27.
Perigaud, A. *et al.* To be published.
Petrov, T. G., Treivus, E. B., and Kasatkin, A. P. (1969). "Growing Crystals from Solution," pp. 8–9. Consultants Bureau, New York.
Pitt. C. W. (1982). *J. Non-Cryst. Solids* **47,** 159.
Pollara, B., and von Korff, R. W. (1960). *Biochim. Biophys. Acta* **39,** 364–376.
Sasaki, K. Kinoshita, T., and Karasawa, N. (1984). *Appl. Phys. Lett.* **45,** 333.
Shiro, M. Yamakawa, M., and Kubota, T. (1977). *Acta Crystallogr. Sect. B* **B33,** 1549–1552.
Sigelle, M., and Hierle, R. (1981). *J. Appl. Phys.* **52,** 4199–4204.
Sigelle, M., Zyss, J., and Hierle, R. (1982). *J. Non-Cryst. Solids* **47,** 287–290.
Sigelle, M., Flicstein, J. Hierle, R., and Badan, J. (1983). *J. Cryst. Growth* **61,** 229–234.
Stegeman, G. I., and Liao, C. (1983). *Appl. Opt.* **22,** 2518.
Stevenson, J. L. (1977). *J. Cryst. Growth* **37,** 116.
Stevenson, J. L., and Dyott, R.B. (1974). *Electron Lett.* **10,** 449.

Stockbarger, D. C. (1936). *Rev. Sci. Instrum.* **7,** 133–136.
Stockbarger, D. C. (1949). *Trans. Faraday Soc.* **5,** 294–306.
Tarjan, J., and Matrai, M. (1972). "Laboratory Manual on Crystal Growth," p. 19. Akadémiai Kiadó, Budapest.
Tien, P. K. (1971). *Appl. Opt.* **10,** 2395.
Tomaru, S., Kawachi, M., and Kobayashi, M. (1984). *Opt. Commun.* **50,** 154.
Vidakovic, P., Badan, J., Hierle, R., and Zyss, J. (1984). *Post Deadline Commun., IQEC 84,* PD-C5-1.
Vidakovic, P. *et al.* (1986). To be published.
Yariv, A., and Yeh, P. (1984). "Optical Waves in Crystals." Wiley, New York.
Zyss, J. (1982). *J. Non-Cryst. Solids* **47,** 211–226, and references therein.
Zyss, J., Chemla, D. S., and Nicoud, J. F. (1981). *J. Chem. Phys.* **74,** 4800–4811.
Zyss, J., Nicoud, J. F., and Coquillay, M. (1984). *J. Chem. Phys.* **81,** 4160–4167.
Zyss, J., Ledoux, I., Hierle, R., Raj, R., and Oudar, J. L. (1985). *IEEE J. Quant. Electron.* **QE-21,** 1286–1295.

Chapter II-5

Growth and Characterization of Organic Thin Films (Langmuir–Blodgett Films)

ANDRÉ BARRAUD and MICHEL VANDEVYVER

Centre d'Etudes Nucleaires de Saclay
91191 Gif-sur-Yvette Cedex, France

I. LANGMUIR–BLODGETT FILM FABRICATION

In the history of Langmuir–Blodgett (LB) films, three dates should be kept in mind:

(1) Transfer by Irvin Langmuir (1920) of a single layer from water surface onto solid substrates.
(2) Sequential monolayer transfer by Katharina Blodgett (1935).
(3) Synthesis and use of specifically designed molecules to achieve active molecular organizates by Hans Kuhn (1962).

By 1962 there was a renewed interest in the field of LB films, owing to the active (mainly optical) properties of especially designed molecular edifices built up by the LB technique. A considerable amount of interest arose around 1975, when LB films were suspected to give rise to specific chemical properties (Barraud *et al.*, 1975b). This was the first indication that applications could be usefully attempted. We are presently witnessing an explosion of the LB technique, owing to the numerous fields of possible applications (Roberts, 1983), such as integrated optics, nonlinear optics, new materials for electronics, microlithography, catalysis, and biology and owing to the new possibilities offered by alternate LB layers (Barraud *et al.*, 1985a).

A. The Langmuir Sequence

Langmuir films are built up from amphiphilic molecules, i.e., molecules having both a hydrophilic and a hydrophobic part. A well-known example is behenic (docosanoic) acid, in which the COOH group is the hydrophilic function and the 21-carbon-long, saturated aliphatic chain is the hydrophobic part (Fig. 1). The monomolecular film is first spread on a water surface

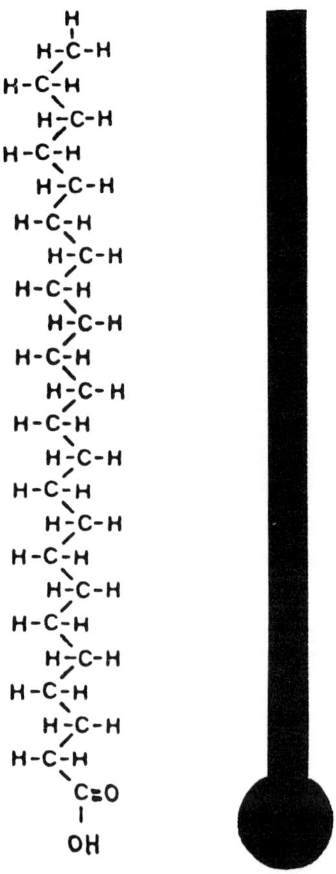

Fig. 1. An amphiphilic molecule (behenic acid) and its schematic representation.

and then transferred onto a solid support. The sequence is the following (Fig. 2):

(1) Spreading. A known amount of the amphiphilic molecule dissolved in a solvent immiscible with water is dropped onto the clean water surface of a Langmuir trough. The solvent is left to evaporate. In this state, molecules are spread on the water surface.

(2) Compression. The Langmuir trough is equipped with a mobile barrier to compress the film laterally. The film is gradually compressed to a surface pressure just below its collapse pressure (elasticity limit). Molecules cannot

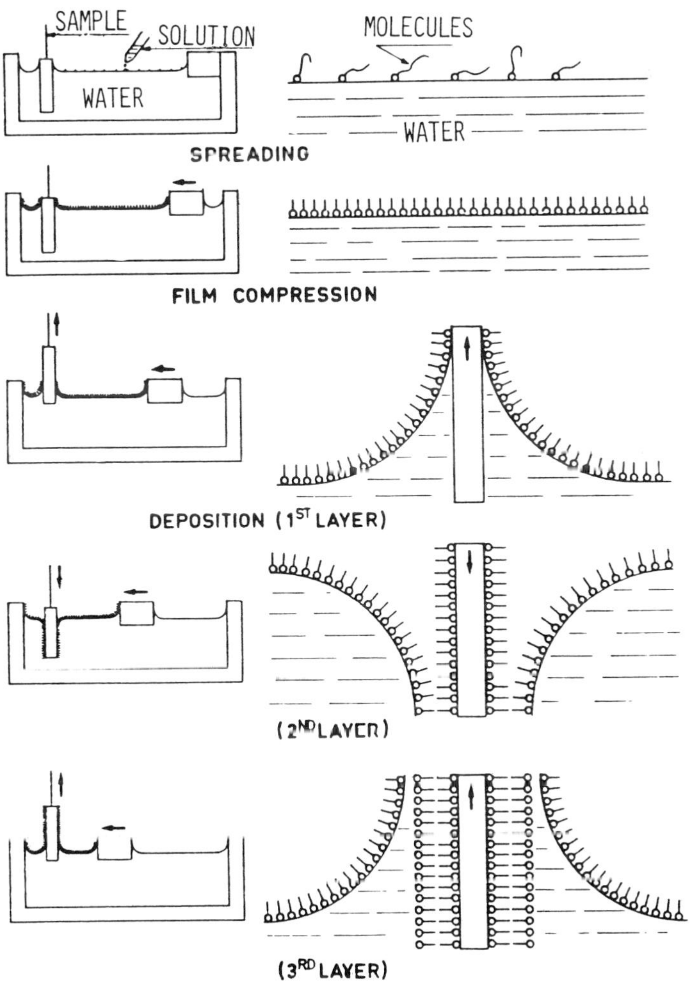

Fig. 2. The Langmuir sequence: spreading, compression, deposition.

dissolve in water because of their long aliphatic chain and cannot leave the water surface because of their hydrophilic group. Hence they form a compact, monomolecular, and in most instances stable film at the water–air interface. The lateral pressure, which can exceed 150 kg/cm^2, eliminates film pinholes, as shown by dielectric measurements (Barraud and Rosilio, 1970).

(3) Film transfer. When an hydrophilic support previously immersed is slowly raised through the water surface, the meniscus turns upward and deposition takes place. Molecules stick to the hydrophilic substrate via their hydrophilic head and squeeze out the water film that covered the support surface. During this process the surface pressure is kept constant by moving the mobile barrier. The substrate is now hydrophobic. On dipping it down, a second layer attaches to its surface. The substrate then turns hydrophilic, as it was originally. On dipping it out again, a third later deposits, and so on.

The LB technique is in essence different from the classical crystal growing methods: it offers the unique feature of layering complete planes, one after another. This allows specific molecular sequencing by simply changing the nature of the film through which the substrate is dipped.

Not only behenic acid but all members of several groups of molecules are suitable for LB deposition: dyes (Kuhn *et al.*, 1972), biological molecules (Ruaudel–Teixier *et al.*, 1983), charge-transfer complexes (Barraud *et al.*, 1986b), amphipathic polymers (Tredgold and Winter, 1983), etc., provided one conforms to certain rules in the design of the molecules.

Subphases other than water have also been successfully used to make Langmuir films: mercury (Ellison *et al.*, 1956; Ellison, 1962) and glycerol (Barraud *et al.*, 1985c) can be used as subphases with molecules that react with water.

B. Langmuir Troughs

All the laboratory Langmuir troughs have several common elements:

(1) A water tank, which can be made from glass, Perspex, or polyethylene, but preferably from polytetrafluoroethylene (PTFE) to avoid contamination by released ions or molecules from the trough walls.

(2) A mobile wall to separate the film compartment from the free water one. Two types of mobile walls are actually used: the classical sliding PTFE barrier, and the constant-perimeter PTFE stripe system. The sliding barrier is easier to clean but may leak if dirty. The constant-perimeter system cannot leak but involves pulleys rotating through the film. Figure 3 shows a schematic diagram of a commercial trough (Optilas) using a sliding barrier, and Fig. 4 shows a top view of the constant-perimeter system (Joyce Loebl).

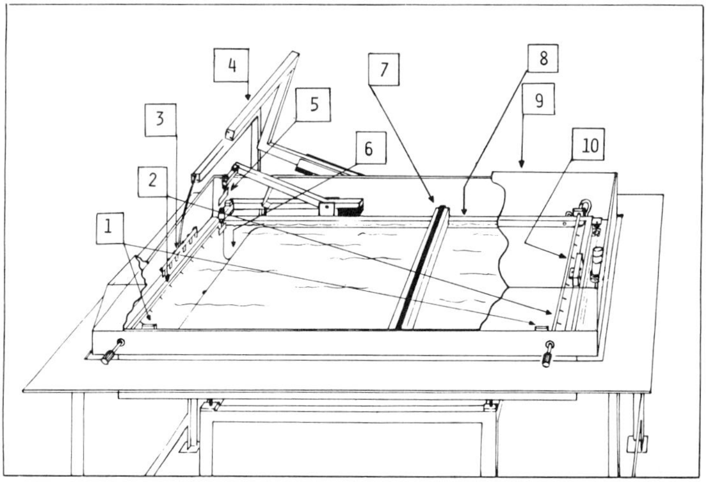

Fig. 3. Schematic view of a trough equipped with a conventional sliding barrier: 1, barrier stops; 2, sucking capillaries to clean trough surface; 3, multiple sample holder; 4, programmable dipping arm; 5, Wilhelmy balance to measure surface pressure; 6, dipping well; 7, PTFE mobile barrier; 8, PTFE trough; 9, removable cabinet flushed with nitrogen; 10, water-level control.

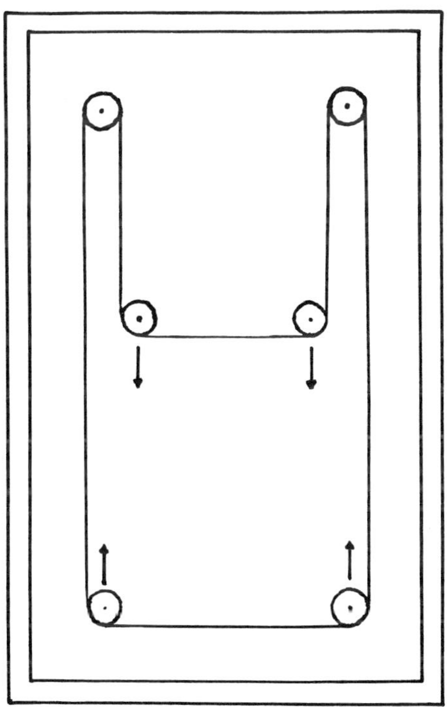

Fig. 4. Schematic top scheme of the constant-perimeter barrier.

(3) A device to measure the surface pressure applied to the film and an electronic attachment to treat the signal and monitor the mobile wall for constant pressure. Two techniques are used to measure surface pressure: the Wilhelmy balance, and the Langmuir balance (Gaines, 1966). The probe in the Wilhelmy balance is very simply made of a small sheet of filter paper partly immersed through the water surface. The vertical force due to surface pressure is either compensated by a current in a coil or converted into a displacement by a spring and then measured by a displacement sensor (balanced differential transformer, for instance). This simple technique affords a typical precision of ± 1 mN/m. The sensitive element of the Langmuir balance is a suspended barrier, which feels the horizontal thrust from the film. The horizontal force is compensated either electrically or mechanically, and the compensation serves as a measure of the surface pressure. The precision can be as high as 10^{-3} mN/m.

(4) A dipping arm, the movement of which can be programmed.

C. Alternate Layers

The common LB method gives rise to a multilayered structure (Fig. 5) in which the polar heads of two adjacent layers are paired to form hydrophilic double planes. The structure is basically centrosymmetrical and hence well suited to optical third-harmonic generation (see Chapter III-2).

For second-harmonic generation, a noncentrosymmetrical structure is needed. Such a structure can be found in alternate LB layers. This kind of layers consists of two adjacent layers of different nature AB arranged in the sequence A—B--A—B--A—B, molecules A and B being in contact by their polar groups. Alternate layers can be obtained manually on a standard Langmuir trough by changing the film at every stroke. They can also be

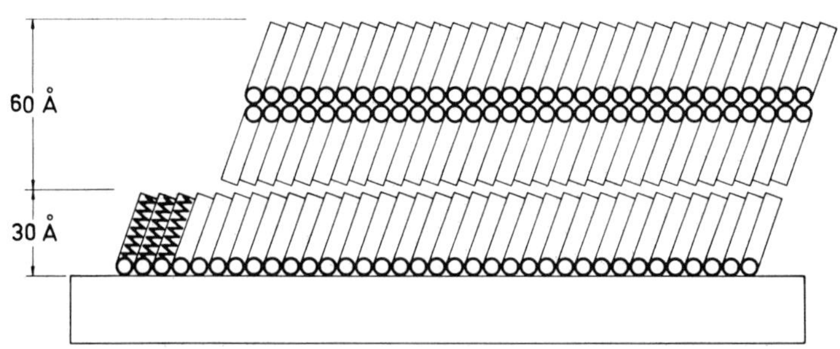

Fig. 5. Structure of classical Y (head-to-head, tail-to-tail) LB multilayers: 60 Å is the approximate spacing for molecules 20–24 carbons long with moderately tilted chains.

obtained automatically on a double trough by passing the substrate alternately through each of the films (Barraud *et al.*, 1986a). According to the path programmed for the sample, this technique allows fabrication of either centrosymmetrical layers or noncentrosymmetrical or paired alternate layers (Fig. 6).

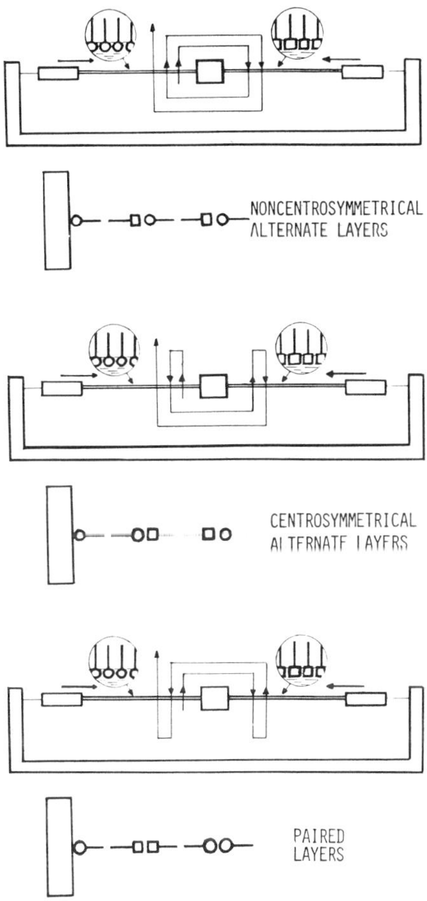

Fig. 6. Double Langmuir trough designed to make alternate LB layers. An auxiliary arm transfers the substrate from one compartment into the other. Transferring the substrate both above and below the water surface gives rise to noncentrosymmetrical alternate layers (top); transferring it only below the water surface gives rise to centrosymmetrical alternate layers (middle); and transferring it only above the water surface gives rise to paired alternate layers (bottom).

D. The Roller Trough

All the troughs based on the Langmuir sequence suffer from the basically discontinuous character of the film fabrication. This is a serious handicap for industrial applications. For industrial work, an automatic, fast, and above all continuous method of film production is needed. This is achieved with a trough that no longer works according to the Langmuir sequence but that feeds a compartment with a continuous flow of film compressed to the desired pressure (Barraud and Vandevyver, 1983). A horizontal rotating roller partially immersed in water transfers amphiphilic molecules from one side of the roller to the other and compresses them (Fig. 7). The trough is therefore divided into compartments separated by rollers. The Langmuir sequence, which is distributed in time, is replaced by a sequence distributed in space, all the elementary operations taking place simultaneously in different compartments. The throughput of this type of trough is 10-fold as high as that of a Langmuir trough of comparable size.

II. LB FILM CHARACTERIZATION

Basic information on the film can be obtained from its behavior on the subphase. Recording the compression isotherms (pressure versus area curves) (Gaines, 1966) is the first step in determining the mechanical properties of the film: collapse pressure, area per molecule, two-dimensional compressibility

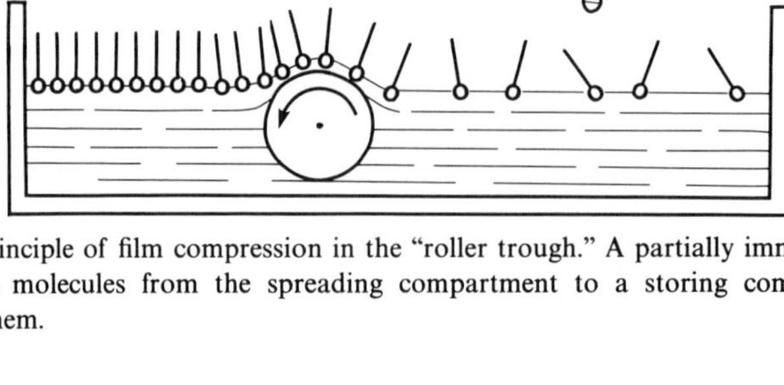

Fig. 7. Principle of film compression in the "roller trough." A partially immersed rotating roller carries molecules from the spreading compartment to a storing compartment and compresses them.

modulus, and film state (expanded, condensed, solid). It also gives information on the long-term stability of the film. In the case of mixtures of molecules, compression isotherms are particularly useful in deciding whether the mixture is a true solid solution or whatever phase segregation takes place (Gaines, 1966; Vandevyver *et al.*, 1982). A very clever example of compression isotherm exploitation is given by Polymeropoulos *et al.* (1978). In this work, compression isotherms provide evidence for molecular interlocking above a critical pressure in a mixture of complementary molecules.

Such information can be completed by optical investigations on the trough itself (Löshe and Mohwald, 1984a,b). Since these methods make use of the same techniques as for transferred films, they will be reviewed with them.

Once films have been transferred in the solid state onto a substrate, their structure is frozen and, contrary to the previous case, only static methods can be used for structure determination. On the other hand, methods such as electron microscopy, which cannot be performed in the trough, can be used. In addition, the increase in sensitivity arising from multilayer superposition makes workable several classical methods that cannot be performed in the trough because of their lack of sensitivity.

The high degree of order in these films make most methods of investigation meaningful: X-ray and electron diffraction provide lattice parameters as in crystals; optical spectroscopy gives information on the close molecular environment (aggregation, dimer formation); linear dichroism, anisotropic electron spin resonance (ESR) and Raman scattering in polarized light give access to molecular conformation; and the different kinds of microscopy can be used to study defects.

A major exception is nuclear magnetic resonance, the sensitivity of which is more than 1000 times too low to make this technique workable, even on hundreds of LB layers.

A. Direct Observation

1. Thickness Determination

Two methods are used to obtain film thickness: ellipsometry (Knoll *et al.*, 1983) and Nomarsky microscopy (Barraud *et al.*, 1980). Ellipsometry gives both refractive index and thickness. Precision on thickness can be better than 10 Å. The technique requires homogeneous samples at least 2 × 2 mm in size.

Wide-angle Nomarsky microscopy can be performed on micrometer-sized areas, but it requires a step in film thickness and it measures the thickness difference between the two sides of the step. The technique works best with metal-coated samples. Its typical precision lies around ±30 Å.

2. Optical Microscopy

Direct observation of the films with a usual optical microscope often gives no information at all, because defects in LB films are phase objects. For this reason, polarized or interference microscopy must be used. The classical crossed Nicols setting is convenient. The use of a small-angle Nomarsky accessory on a standard microscope is also of great help in revealing otherwise invisible phase defects (Fig. 8).

3. Scanning Electron Microscopy

Among the large amount of information provided by scanning electron microscopy (SEM), the most outstanding result is the possible observation of a single layer (Vandevyver and Le Gressus, 1972; Barraud *et al*., 1975a). Coating with one layer appears to enhance electron emission (Fig. 9). The enhancement still persists on going to three layers. The image grows darker when subsequent layers are superimposed (Fig. 10). It should be kept in mind that LB films are made of organic molecules and hence may be destroyed by the electron beam. Also, the thinness of the film implies a judicious choice of the accelerating voltage: the underlying substrate can be easily observed

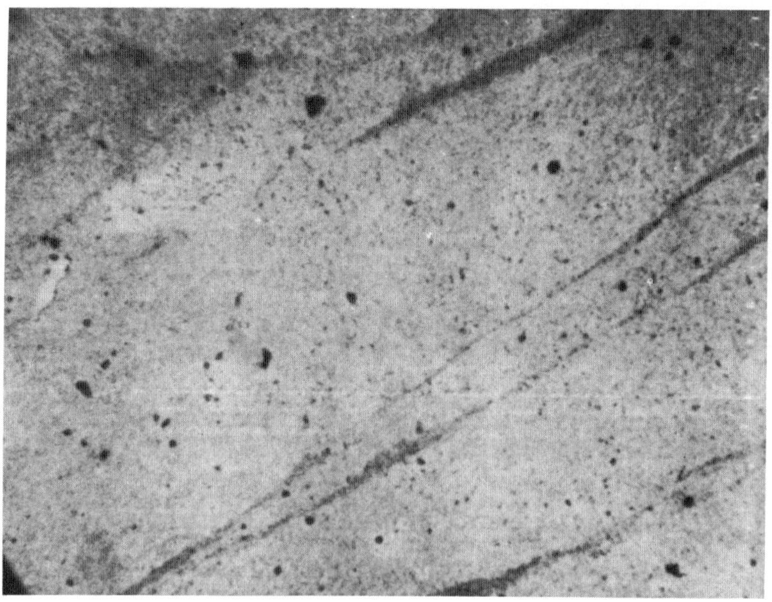

Fig. 8. Monolayer of calcium behenate collapsed by overcompression and transferred onto a silica substrate. Micrograph taken in small-angle Nomarsky interference contrast.

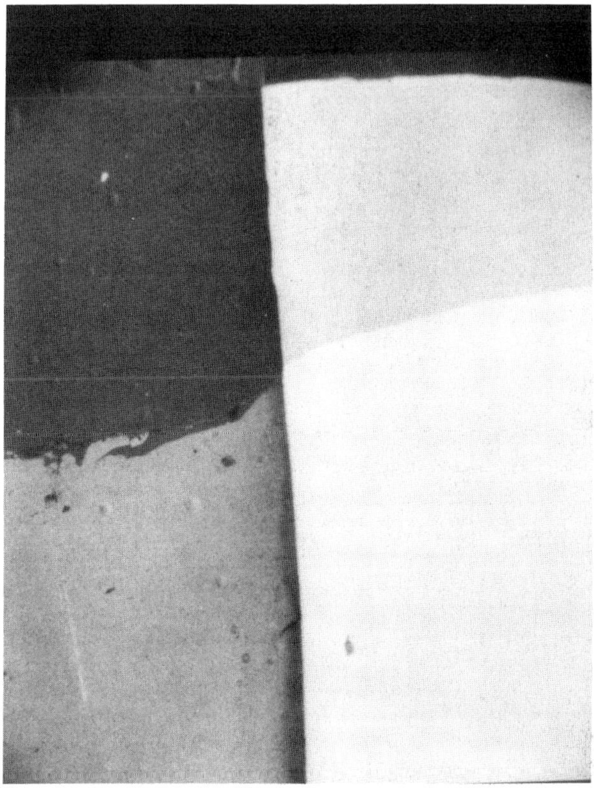

Fig. 9. Scanning electron micrograph showing the electron emission enhancement caused by a Langmuir–Blodgett monolayer. Left half of the picture: aluminum substrate. Right half: gold substrate. Top half: bare substrate. Bottom half: substrate covered by a single LB layer of calcium stearate. Accelerating voltage: 5 kV. The width of the picture is 1.5 mm.

through the film at high voltage, while the film itself is observed at low voltage. It should be noted that the film is generally thin enough for the electrostatic charge to be of no concern and so requires no conductive coating.

In addition, advantage can be taken of the highly uniform thickness of each layer in order to use Auger microscopy to monitor chemical composition of individual layers: the thickness of the film above a given layer can be adjusted according to the escape depth of the Auger electron so as to get electrons mainly from the desired layer (Barraud *et al.*, 1975a,b).

4. Transmission Electron Microscopy

LB films can be layered on virtually any solid substrate, including thin membranes such as collodion. Hence all the classical transmission electron

Fig. 10. Scanning electron micrograph showing the evolution of electron emission on increasing the number of LB layers. From top to bottom: glass, bare aluminum, and one, three, five, and seven layers of calcium stearate. Accelerating voltage: 5 kV (at 20 kV the three-layer area would be brighter than the one-layer one). The total width of the picture is 0.9 mm.

microscopy (TEM) techniques are workable on LB films: shadowing (Ries and Kimball, 1957; Ries, 1979), replication, visualization (Barraud *et al.*, 1986d), diffraction (Leloup *et al.*, 1985), or Scanning TEM (STEM). Advantage can be taken of the versatility offered by chemistry in LB films to increase electron absorption in selected parts of a given molecule. This selective decoration increases contrast and gives rise to high-resolution images, even on a single layer. As an example, Fig. 11 shows precollapse patterns generated on the trough and transferred in their native form onto a collodion membrane (Barraud *et al.*, 1985d). In this case, behenic acid was transformed into silver behenate *in situ* to selectively visualize polar planes via heavy Ag^+ ions.

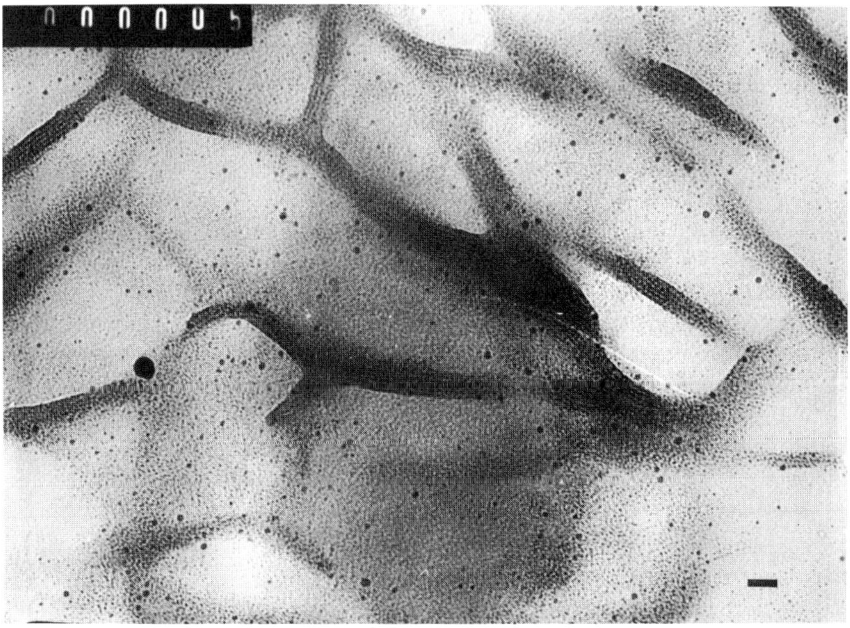

Fig. 11. Transmission electron micrograph of a film of pure behenic acid, showing the early step of "slow collapse" at low surface pressure (35 mN/m) on the trough (slow collapse of fatty acid monolayers takes place on prolonged compression at low pressure). The decoration technique was the following: after transfer onto a thin collodion membrane, the film of behenic acid was converted into silver behenate by diffusion of Ag^+. The silver heavy ions increase electron absorption selectively in the polar planes. The defects in the film appear to be randomly oriented ridges (in contrast to overpressure collapses, which are highly anisotropic as shown in Fig. 8). Every ridge is made up of several upright double layers 50 Å apart. Accelerating voltage: 50 kV. The bar represents 200 Å.

B. Optical Spectroscopy

Due to the high uniformity of film thickness, the large areas available, and the availability of very thin films in the case of highly absorbing molecules, classical transmission spectroscopy is the method of choice for studying LB films. As an example, Tetracyanoquinodimethane (TCNQ) conducting charge-transfer complexes can be investigated only by diffuse reflectivity measurements when in bulk crystals, while usual transmission spectroscopy gives the same information readily in the case of LB films (Barraud *et al.*, 1985a; Ruaudel–Teixier *et al.*, 1985). The only limiting factor is the substrate, which may absorb. In this respect, large-size, very thin alumina membranes have been shown to be nonabsorbing throughout the optical range

(A. Garin, unpublished results, 1983). Experiments have shown that this type of membrane can be coated with LB films without major problems and then examined even *in vacuo*.

1. UV–Visible Spectroscopy

As in solution, ultraviolet (UV)–visible spectroscopy can be applied to highly conjugated molecules: dye molecules such as cyanines (Kuhn *et al.*, 1972), biological molecules such as porphyrins (Ruaudel–Teixier *et al.*, 1983), charge-transfer complexes such as TCNQ salts (Ruaudel–Teixier *et al.*, 1985), and π complexes involving large molecules such as anthracene (Vincett *et al.*, 1982). In the condensed solid state, these molecules exhibit spectra generally different from those in solution due to short-range intermolecular coupling. While such interactions are difficult to study spectroscopically in crystals, measurements are straightforward on LB films. As an example, phthalocyanines in solution exhibit a sharp structured monomer peak near 680 nm

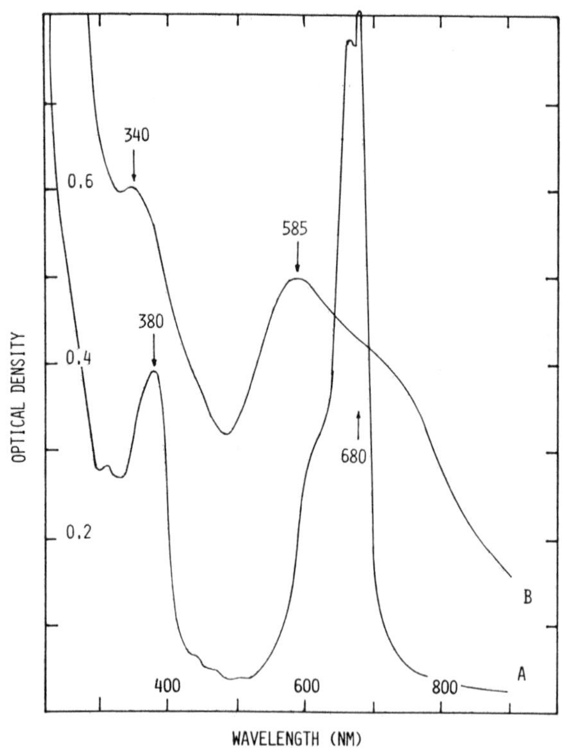

Fig. 12. UV–visible absorption spectrum of a phthalocyanin (A) in solution in dilute sulfuric acid and (B) in LB layers of the pure molecule (75 layers).

(Fig. 12). The extinction coefficient ($\sim 50{,}000\ cm^{-1}$) is too high for light to be transmitted through a crystal. LB films, on the other hand, give quantitative access to the solid-state spectrum: the monomer peak is strongly reduced, while the dimer absorption band near 585 nm is clearly visible (Barraud *et al.*, 1986a).

This method is very well suited to monitor the evolution of the solid matrix, either during thermal annealing (Barraud *et al.*, 1986a) or during a chemical or photochemical reaction (Vandevyver *et al.*, 1985).

UV–visible absorption spectroscopy can often be extremely sensitive: Mobius (1983) showed that visible spectroscopy, even performed on the trough itself, could be sensitive enough to detect less than one-hundredth of a monolayer. So far, few experiments have been performed by the Raman scattering technique. The reason for this is the very small amount of material involved (typically $0.4\ \mu g/cm^2$ per layer). Only colored molecules can be investigated by this method, owing to resonance amplification: Vandevyer *et al.* (1983) has shown that resonance Raman scattering works on poryphyrins with a sensitivity of a few layers (Fig. 13).

2. Infrared Absorption Spectroscopy

Infrared absorption spectroscopy is a sensitive (a few layers) and informative method. It is fruitfully used to monitor chemical reactions, especially polymerizations (Barraud *et al.*, 1977), as well as phase transitions

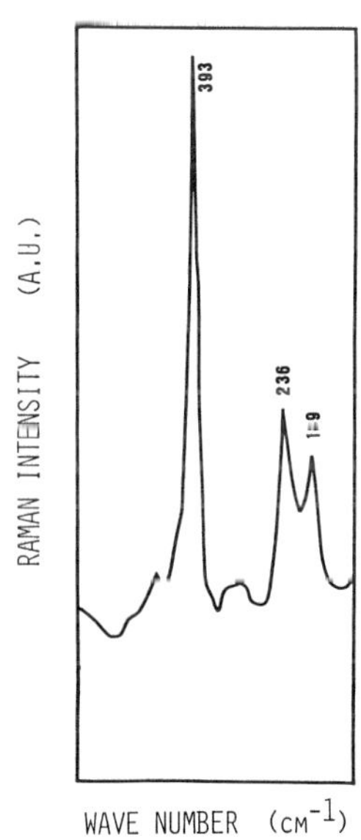

Fig. 13. Low-energy polarized resonance Raman spectra of Zn tetra-*N*-pyridyl porphyrin $(C_{20})_4$ (7 LB layers). Laser wavelength: 442 nm. Power: 1 mW. Resolution: $8\ cm^{-1}$. Sample temperature: 20 K.

(Leloup *et al.*, 1985) in LB films. As an example, the conversion of behenic acid into silver behenate by diffusion of silver nitrate into the polar planes is accompanied by two modifications of the infrared absorption spectrum (Fig. 14) (Leloup *et al.*, 1985): the 1705-cm^{-1} C=O absorption band of the acid group vanishes and is replaced by a narrow band at 1520 cm^{-1} characteristic of COO^- associated with a silver ion; simultaneously, the doublet at 1462 and 1470 cm^{-1} arising from the bending mode of CH_2 in a lattice having two molecules per unit cell is replaced by a single sharp band at 1470 cm^{-1} indicative of a change in the lattice symmetry group (one molecule per unit cell).

Infrared spectra can also provide information on the electronic properties of the films. This is exemplified by the deep modifications in the infrared spectrum of a docosylpyridinium–TCNQ charge-transfer salt upon partial oxidation of the acceptor by diffusion of iodine (Ruaudel–Teixier *et al.*, 1985). The pristine compound shows sharp absorption peaks on a rather flat baseline. In contrast, the final product (Fig. 15) shows a very broad absorption band around 3000 cm^{-1} indicative of band electron transitions and several other wide absorption bands (1550, 1320, and 1150 cm^{-1}) characteristic of an electron-conducting material (Torrance *et al.*, 1979). Among this set of wide bands several sharp peaks remain, characteristic of film regions unaffected and

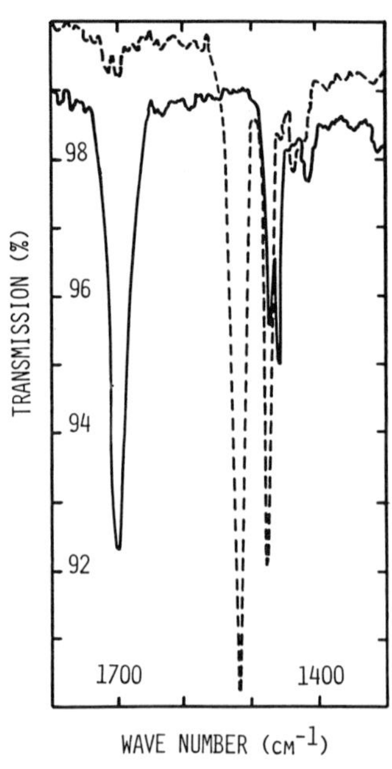

Fig. 14. Infrared absorption spectra of behenic acid (——) and silver behenate (– – –).

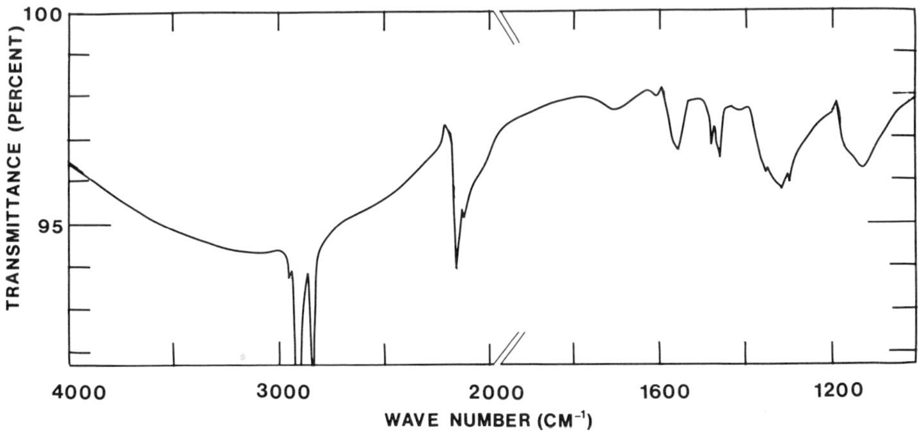

Fig. 15. Infrared absorption spectrum of a conducting LB film (docosylpyridinium–TCNQ–iodine, 2:2:3 ternary compound): conductivity (localized in the TCNQ planes, in the middle of the polar planes) gives rise to broad bands arising from electron transitions and TCNQ coupling, while molecules located out of the conducting sheet exhibit unaffected sharp absorption lines.

hence not involved in the electron path (Barraud *et al.*, 1985a): the CH_2 bands at 1470, 2860, and 2920 cm^{-1} and a pyridinium band at 1480 cm^{-1}.

3. Emission Spectroscopy

Light emission, generally fluorescence, can be used to obtain information on both the morphology and the organization of the film.

Fluorescence microscopy on the trough itself allows direct visualization of the nucleation process in crystallite formation during compression (Löshe and Mohwald, 1984a,b). Through the bottom of a shallow trough, an immersed microscope inspects the underpart of the surface under UV illumination. A light amplifier renders the tiny optical signal usable.

The first fluorescence experiments were performed by Bücher *et al.* (1967) to investigate energy-transfer mechanisms by the use of LB films as thickness gauges on the molecular scale. Since that time, LB films have appeared as a tool for fundamental research.

Sophisticated structures such as LB films of dye mixtures can also be investigated by fluorescence spectroscopy. For example, mixtures of dye A and dye B of Fig. 16 in a ratio of 300:1 exhibit the fluorescence spectrum of B when excited in the absorption spectrum of A (Mobius, 1978; Ruaudel–Teixier and Vandevyver, 1980). This arises from the high degree of organization of the film, which allows resonant migration of the excitation by a Förster mechanism: the dye molecules lying parallel to one another are optically coupled, and the excited-state explores a large number of them successively

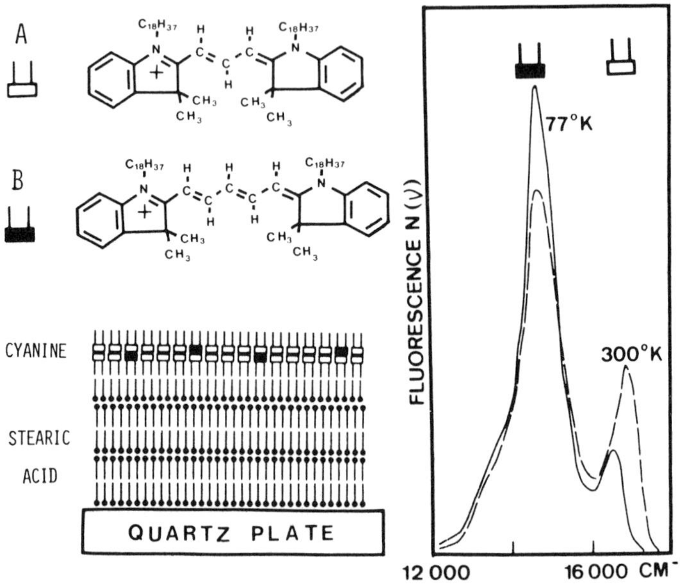

Fig. 16. The photon funnel consists of a double layer of dyes A and B mixed in a ratio 300 A : 1 B. This double layer is isolated from the silica substrate by several layers of behenic acid. When selectively excited in the absorption spectrum of A, the system exhibits almost exclusively the fluorescence of dye B (14,400 cm^{-1} instead of 16,500 cm^{-1}).

until it meets a molecule of dye B. Dye B has an excited state slightly lower in energy and hence works as a trap for the excited-state that can no longer migrate. This "photon funnel" effect was one of the first examples of highly organized molecular structures designed for a specific purpose.

C. Determination of Lattice Distances

LB films are expected to show order at least along the normal to the substrate (vertical order). This is indeed confirmed by small-angle and X-ray diffraction (Lesslauer, 1974). In some cases, order is so high that more than 20 diffraction orders can be observed, allowing precise determination of the vertical electron-density profile (Fig. 17) (Belbeoch *et al.*, 1985). This profile is a great help in determining structures.

However, even in the case of well-defined diffraction patterns, some ambiguity can remain about the centrosymmetric or noncentrosymmetric character of the layers, since the Fourier transform does not take amplitude signs into account (Gay, 1961). In this respect some help can be obtained from

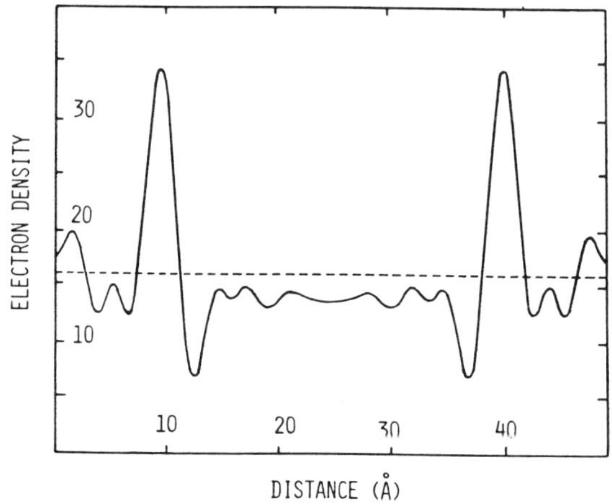

Fig. 17. Electron-density profile (electron density in electrons per angstrom) along the normal to the substrate in the conducting LB films of Fig. 15. This profile was obtained by Fourier transform of the intensities of the first 20 00*l* lamellar reflections recorded on a 64-layer sample. Distances are measured from one polar plane to the next one. This profile is quantitatively consistent with a five-step model: the two main peaks arise from iodine; the central plateau arises from interdigitated aliphatic chains; between them, the two narrow dips are characteristic of interdigitation (CH_3 chain ends); in the polar plane the two small peaks are representative of the overlap of TCNQ (electron-rich molecule); and the dips between TCNQ and iodine correspond to the pyridinium rings. [After Belbeoch *et al.* (1985).]

neutron diffraction, provided molecules have been judiciously deuterated (Highfield *et al.*, 1983).

Although it might be thought that successively deposited layers should be independent, in some instances evidence of epitaxy between layers is observed (Belbeoch *et al.*, 1985). This implies film reorganization after transfer. The three-dimensional mosaic crystals obtained preserve the original lamellar structure.

The in-plane structure can be investigated by X-ray transmission or electron diffraction. In the case of epitaxy, *hkl* lines or spots can be observed (Sarkar and Lando, 1983). Again, LB films often exhibit a high degree of in-plane order.

D. Orientation Determination

Transmission X-ray experiments are difficult to perform mostly because of substrates. Hence, only part of the structure can be obtained from X rays, and methods providing the orientation of molecules in the film are needed.

Molecular orientation can be reached by three resonance methods: linear dichroism in an absorption band (LD), resonance Raman scattering in polarized light (RRPL), anisotropic electron spin resonance (AESR). With a small number of exceptions, the normal to the substrate is also a full symmetry ($C\infty$) axis. This notably simplifies the handling of these methods. The full symmetry can be readily checked by linear dichroism at zero incidence angle (*vide infra*).

1. Linear Dichroism

The molecules exhibit absorption bands, the transition moment of which are supposed to be known in the molecular frame. Basically, linear dichroism gives access to the "order parameter," i.e., to the *average* value of $\cos^2 \theta$, where θ is the angle between a given transition moment s and the normal n to the substrate (Breton *et al.*, 1972; Chollet, 1980; Vandevyver *et al.*, 1982; Chollet and Messier, 1983; Earnshaw, 1983). In the present case of thin films on a thick substrate, multiple reflections have to be taken into account. This method is easy to perform, and its results are unambiguous when the transition is well decoupled from the others and the dipole moment is well known in the molecular frame. However, it fails in the case when $\langle \cos^2 \theta \rangle \approx \frac{1}{3}$, i.e., θ average between 50 and 60°. For example, if θ_{av} is found equal to 54° (magic angle), it is impossible to decide whether $\theta = \theta_{av}$ for all the molecules or whether θ is randomly distributed. In this case, more sophisticated (but heavier) methods have to be used (see below).

Fortunately, LB films are most of the time highly crystalline, so that $\theta = \theta_{av}$ for all the molecules; in this case, linear dichroism proves to be by far the most useful tool for structure determination. In particular, for molecules exhibiting a rich infrared spectrum, LD measurements on several transitions make structure cross-checking possible.

The power of the method is illustrated by the study of the tilt angle of the aliphatic chains in the first layers of behenic acid on an aluminum substrate. From vertical in the first layer, chains progressively tilt to the bulk angle (23°), which is reached at the seventh layer. The situation is different on gold: tilt angle is nonzero (12°) in the first layer, and the bulk tilt angle is reached at the fifth layer (Chollet, 1982). The sensitivity is high enough for the method to be performed on a single layer in the best case.

2. Resonance Raman Scattering in Polarized Light (RRPL)

RRPL is a heavy method but it exhibits maximum sensitivity precisely around the magic angle where LD fails (Lesieur, 1986). In addition to the classical order parameter (Legendre second-order polynomial) P_2, the method also gives inherently the Legendre fourth-order polynomial P_4, which gives access to the average value of $\cos^4 \theta$ (Brehamet *et al.*, 1983). Together with LD, this new method is used to refine angular distribution: the two independent accesses to quantities governed by θ remove doubt about angular distribution width (wide or narrow).

Out-of-resonance Raman scattering is far below the detection threshold, and only resonance provides enough sensitivity in the case of LB films (Fig. 13).

Besides the tremendous increase in sensitivity, resonance provides the additional advantage of selecting one species in a mixture by tuning the laser wavelength to the absorption band of the explored species (Vandevyver *et al.*, 1983). This is worthwhile for alternate layers of different dye molecules that are already being studied for nonlinear optical devices.

3. Anisotropic Electron Spin Resonance

In optical methods, all the molecules resonate at the same frequency, whatever their orientation. It is basically for that reason that only an average value of angles can be obtained. In ESR, on the other hand, the resonance frequency f_0 depends on the orientation of the magnetic field in the molecular frame. Hence for a set of molecules, only those with a correct orientation will resonate at a selected frequency. The result is that ESR, which selects molecules according to orientation, is expected to give full access to angular distribution (Vandevyver *et al.*, 1982).

The major drawbacks of the method are (1) the need for paramagnetic species and (2) the risk of spin–spin interactions in the condensed state leading to an unacceptable line broadening (Messier and Marc, 1971). However, when operative, ESR is the royal access to angular distribution. Moreover, its sensitivity is high enough to allow exploitation on a few layers if the line is a few gauss wide and on 100 layers if the signal is 1000 gauss wide, such as in the case of copper porphyrins.

The method has been successfully used on amphiphilic charge-transfer complexes not only to point out their high degree of order but also to prove a dimeric structure that could hardly be made obvious by any other method (Barraud *et al.*, 1986b).

III. CONCLUSION: WHAT ARE LB FILMS?

It has long been known that LB films are pinhole-free. In addition, all the many methods of structure determination coherently point out a very high degree of order in LB films (Kuhn, 1983; Barraud, 1985b). Their history has indeed been shaped by this order. We should keep in mind that contrary to other organized systems (crystals, membranes, micelles, microemulsions), the order is not spontaneous but is imposed from outside via the fabrication method itself, which proceeds complete layer after complete layer.

Moreover, it appears possible to bring together quite a variety of molecules in a manner that is controlled in direction and distance. The result may be a metastable structure, out of thermodynamical equilibrium. The question arises then of a possible long- or short-term rearrangement.

In the case of homolayers (layers of identical molecules) of nondiluted molecules (single phase), the large number of results now available reveals no reorganization, although reorganization has not been systematically looked for.

In the case of mixtures (two different molecules in the same layer), no reorganization is reported. However, it can take place in the case of molecules undergoing strong interaction. For example, in the system $CuTPP(Cl_2)_4$ + behenic acid examined by Vandevyver *et al.* (1982), a strong reorganization accompanied by an increase in order and arising from macrocycle to macrocycle interaction was observed over a period of several months (Fig. 18). Obviously this was a back process to thermodynamic equilibrium. Nevertheless, in most cases homolayers should be considered as stable organized solids; this behavior should be kept in mind for applications.

In the case of alternate layers, among which noncentrosymmetrical systems are of prime interest for optical second-harmonic generation, the question of stability is the first that comes to mind, since we force different molecules to cohabit. Here again the available results lead to the conclusion that alternate layers are stable solids, even in the noncentrosymmetrical form (Barraud *et al.*, 1986a). For example, in the case of porphyrin (optically active molecule) facing behenic acid (inert molecule), the evidence of alternation is given by the infrared spectrum of the acid (Barraud *et al.*, 1986a), which points out side dimerization, while facing dimerization is observed in homolayers (Fig. 19). Similarly, molecular edifices specifically built for photochemical measurements show photoreactivity in alternate layers but appear inert in paired layers, as expected (Vandevyver *et al.*, 1985). Only in the case of highly reactive molecules can a rearrangement take place. This has been observed in alternate layers of phthalocyanine and omegatricosenoic acid, even though the mechanism of the rearrangement is not elucidated yet. Even in this case, there is no evidence for the initial noncentrosymmetrical character of the structure to be lost, since no dimer peak is observed in the absorption spectrum (Fig. 20) (Barraud *et al.*, 1986a) (dimer absorption is expected to take place at 585 nm, as shown in Fig. 12).

Optical applications often need micrometer-thick films. LB films, although intrinsically ultrathin, can be layered to "thick" optical films: 1000 layers give rise to a film about 3 μm thick. Although films of fatty acids contain phase defects that make them cloudy, several other molecules provide remarkably

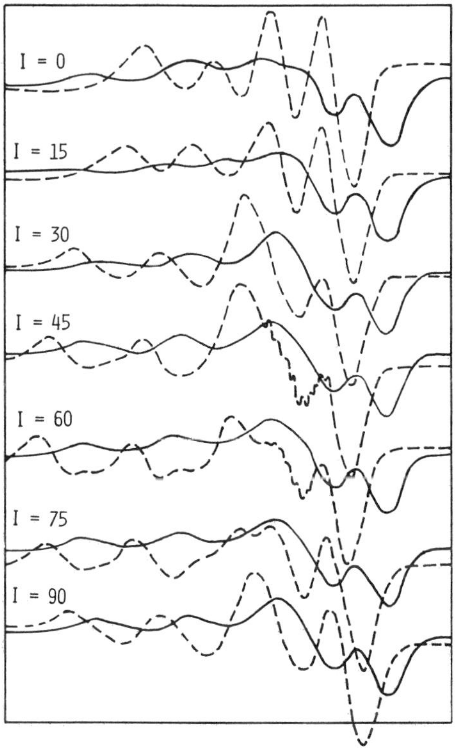

Fig. 18. Anisotropic ESR spectra of a porphyrin–behenic acid mixture (Vandevyver *et al.*, 1987) just after layering ( ) and after 6 months ( ), where I is the angle between the magnetic field and the normal to the substrate. Dotted spectra are more structured and show superhyperfine structure around $I = 45°$, indicative of film reorganization.

transparent films, even when thick. This homogeneity, together with their highly uniform thickness, makes LB films well suited to light guide applications and nonlinear optics (Aktipetrov *et al.*, 1983, Girling and Milverton, 1984).

More generally speaking, the LB method is a remarkably powerful tool for research. It is a unique tool to make specifically designed molecular edifices, since it makes use of the wide capabilities of organic synthesis and combines them with a wide choice of geometrical arrangements. One can then understand the present spectacular expansion of this method.

Nonlinear optics requires highly polarizable molecules. In addition, second-harmonic generation requires noncentrosymmetrical molecules arranged in an also noncentrosymmetrical lattice. Most highly polarizable molecules crystallize in a centrosymmetrical array. To avoid this, the permanent dipole must be carefully balanced off. A few molecules so designed have been made,

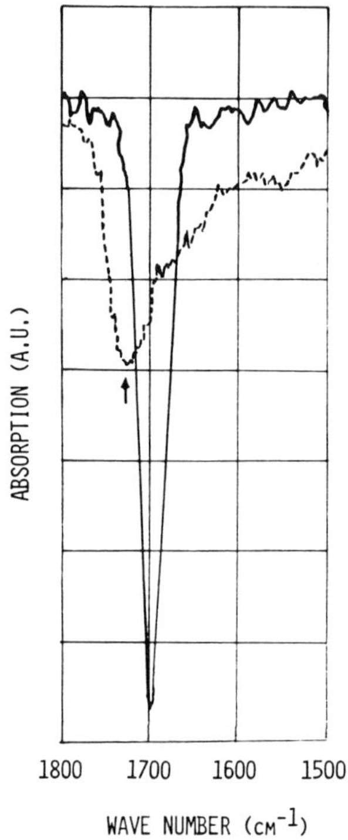

Fig. 19. Infrared absorption spectrum of benehic acid in homolayers (——) and in alternate layers (– – –).

but this stringent zero-dipole condition forbids the use of most of the promising candidates. On the contrary, in LB films, the high interchain van der Waals cohesion can force molecules into a desired molecular arrangement, whatever the polar head. This opens the way to molecules that otherwise would crystallize in a centrosymmetrical form.

New high-performance, hence markedly thinner, nonlinear optical material can then reasonably be expected from that versatile method. Thin material does not require as sharp phase tuning as bulk crystals, and handling is hence easier. In addition, the large film areas available from the LB technique open new fields in screen applications.

As a conclusion, the recent developments of LB films point them out as an entirely new material, made of active molecules deliberately arranged for specific applications. Nonlinear optics should find here a material of choice, especially for second-harmonic generation. Although methods for a fine determination of the structure and the properties of this material are now available, the field is too new for extensive exploration to have yet taken place. However, the few results already available in the field of nonlinear optics

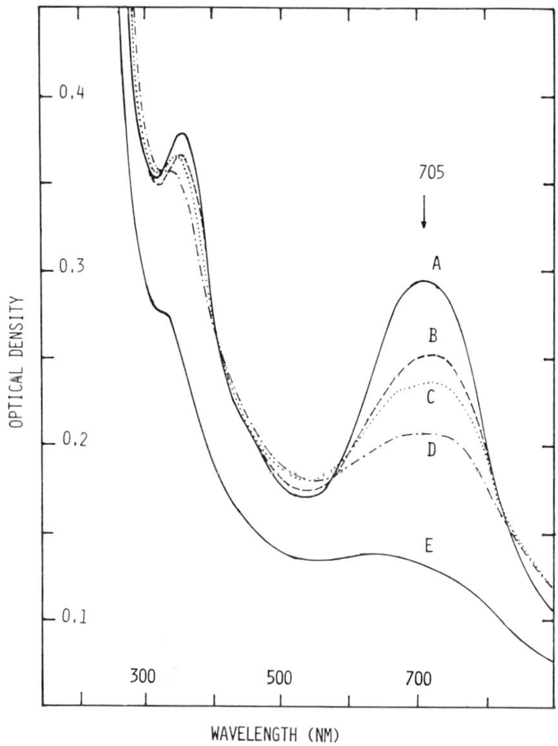

Fig. 20. UV–visible absorption spectrum of 75 alternate layers of phthalocyanine and omegatricosenoic acid. Curves A, B, C, and D are recorded at 20 min, 2 hr, 48 hr, and 5 days after layering, and curve E after a final moderate heating. The monomer peak at 705 nm progressively vanishes, but no dimer peak (expected at 585 nm) is observed.

(Aktipetrov *et al.*, 1983; Girling and Milverton, 1984) are quite encouraging and show that this method is very promising.

REFERENCES

Aktipetrov, O. A., Akhmediev, N. N., Mishina, E. D., and Novak, V. R. (1983). *JETP Lett. (Engl. Transl.)* **37B,** 207.

Barraud, A. (1985b). *J. Chim. Phys.* **82** (6), 683.

Barraud, A., and Rosilio, A. (1970). *Proc. Colloq. Int. Microelectron. Avancee, 1970,* Vol. II, p. 845.

Barraud, A., and Vandevyver, M. (1983). *Thin Solid Films* **99,** 221.

Barraud, A., Le Gressus, C., Okusumi, H., and Mogami, A. (1975a). *Proc. Int. Congr. Electron Microsc., 8th, 1974,* Part I, p. 682.

Barraud, A., Rosilio, C., and Ruaudel–Teixier, A. (1975b). *Semin. Chim. Etat. Solide* **9,** 21.

Barraud, A., Rosilio, C., and Ruaudel–Teixier, A. (1977). *J. Colloid Interface Sci.* **62,** 509.
Barraud, A., Rosilio, C., and Ruaudel–Teixier, A. (1980). *Thin Solid Films* **68,** 91.
Barraud, A., Leloup, J., Gouzerh, A., and Palacin, S. (1986a). *Thin Solid Films* **133,** 117.
Barraud, A., Lesieur, P., Richard, J., Ruaudel–Teixier, A., and Vandevyver, M. (1986b). *Thin Solid Films* **133,** 125.
Barraud, A., Leloup, J., and Lesieur, P. (1986c). *Thin Solid Films* **133,** 113.
Barraud, A., Leloup, J., Maire, P., and Ruaudel–Teixier, A. (1986d). *Thin Solid Films* **133,** 133.
Barraud, A., Ruaudel–Teixier, A., and Vandevyver, M. Lesieur, P. (1985a). *Nouv. J. Chim.* **9,** 365.
Belbeoch, B., Roulliay, M., and Tournarie, M. (1985). *C. R. Acad. Sci. Ser. 2 (Paris)* **12,** 871.
Blodgett, K. B. (1935). *J. Am. Chem. Soc.* **57,** 1007.
Brehamet, L., and Vandevyver, M. (1983). *Mol. Phys.* **48,** 315.
Breton, J., Michel-Villaz, M., Paillotin, G., and Vandevyver, M. (1972). *Thin Solid Films* **13,** 351.
Bücher, H., Drexhage, K. H., Fieck, M., Kuhn, H., Mobius, D., Schafer, F. P. Sonderman, J., Sperling, W., Tillmann, P., and Wiegand, J. (1967). *Mol. Cryst.* **2,** 199.
Chollet, P. A. (1980). *Thin Solid Films* **68,** 13.
Chollet, P. A. (1982), Thése Doctorat d'Etat, Universite de Paris-Sud (Orsay).
Chollet, P. A., and Messier, J. (1983). *Thin Solid Films* **99,** 197.
Earnshaw, J. C. (1983). *Thin Solid Films* **99,** 189.
Ellison, A. H. (1962). *J. Phys. Chem.* **66,** 1867.
Ellison, A. H., and Zisman, W. A. (1956). *J. Phys. Chem.* **60,** 416.
Gaines, G., Jr. (1966). "Insoluble Monolayers at Liquid–gas Interface." Wiley (Interscience), New York.
Gay, R. (1961). "Cours de radiocristallographie," Vol. III, p. 43. Gauthier Villars, Paris.
Girling, I. R., and Milverton, D. R. J. (1984). *Thin Solid Films* **115,** 85.
Highfield, R. R., Thomas, R. K., Cummins, P. G., Gregory, D. P., Mingins, J., Hayter, J. B., and Scharpe, O. (1983). *Thin Solid Films* **99,** 165.
Knoll, W., Rabe, J., Philpott, M. R., and Swalen, J. D. (1983). *Thin Solid Films* **99,** 173.
Kuhn, H. (1962). *Z. Naturforsch. A* **17A,** 411.
Kuhn, H. (1983). *Thin Solid Films* **99,** 16.
Kuhn, H., Mobius, D., and Bucher, H. (1972). *In* "Physical Methods of Chemistry" (A. Weissberger and B. W. Rossiter, eds.), Vol. I, Part 3B, p. 577. Wiley, New York.
Langmuir, I. (1920). *Trans. Faraday Soc.* **15,** 62.
Leloup, J., Ruaudel–Teixier, A., Barraud, A., and Maire, P. (1985). *J. Chim. Phys.* **82**(6), 695.
Lesieur, P. (1986). Thése Doctorat d'Etat, Universite de Paris VI.
Lesslauer, W. (1974). *Acta Crystallogr., Sect. B* **B30,** 1927.
Löshe, H., and Mohwald, H. (1984a). *Eur. Biophys. J.* **11,** 35.
Löshe, H., and Mohwald, H. (1984b). *Colloids Surf.* **10,** 217.
Messier, J., and Marc, G. (1971). *J. Phys. (Paris)* **32,** 799.
Mobius, D. (1978). *Ber. Bunsenges. Phys. Chem.* **82,** 848.
Mobius, D. (1983). *J. Phys. Colloq. (Orsay, Fr.)* **C10,** 441.
Polymeropoulos, E. E., Mobius, D., and Kuhn, H. (1978). *J. Chem. Phys.* **68,** 3918.
Ries, H. E., Jr. (1979). *Nature (London)* **281,** 287.
Ries, H. E., Jr., and Kimball, W. A. (1957). *Proc. Int. Congr. Surf. Act. 2nd, 1957,* Vol. 1, p. 75.
Roberts, G. G. (1983). *Sens. Actuators* **4,** 131.
Ruaudel–Teixier, A., and Vandevyver, M. (1980). *Thin Solid Films* **68**(1), 129.
Ruaudel–Teixier, A., Barraud, A., Belbeoch, B., and Roulliay, M. (1983). *Thin Solid Films* **99,** 33.
Ruaudel–Teixier, A., Barraud, A., Vandevyver, M., Belbeoch, B., and Roulliay, M. (1985). *J. Chim. Phys.* **82**(6), 711.
Sarkar, M., and Lando, J. (1983). *Thin Solid Films* **99,** 126.
Torrance, J. B., Scott, B. A., Welber, B., Kaufman, F. B., and Seiden, P. E. (1979). *Phys. Rev. B* **19**(2), 733.

Tredgold, R. H., and Winter, C. S. (1983). *Thin Solid Films* **99,** 81.
Vandevyver, M., and Le Gressus, C. (1972). *Proc. Int. Conf. X-Ray Opt. Microanal., 6th, 1971,* p. 467.
Vandevyver, M., Barraud, A., Ruaudel–Teixier, A., Maillard, P., and Giannoti, C. (1982). *J. Colloid Interface Sci.* **85,** 571.
Vandevyver, M., Ruaudel–Teixier, A., Brehamet, L., and Lutz, M. (1983). *Thin Solid Films* **99,** 41.
Vandevyver, M., Ruaudel–Teixier, A., and Barraud, A. (1985). *J. Chim. Phys.* **88**(6) 707.
Vincent, P. S., Barlow, W. A., Hann, R. A., and Roberts, G. G. (1982). *Thin Solid Films* **94,** 171.

Chapter II-6

Properties and Applications of Urea

J.-M. HALBOUT

IBM Watson Research Center
Yorktown Heights, New York 10598

C. L. TANG

Cornell University
Ithaca, New York 14850

I. INTRODUCTION

The usefulness of a nonlinear optical crystal often cannot be judged on the basis of the magnitude of its nonlinear optical coefficients alone. If a large nonlinear coefficient were the sole criterion, there would have been no reason to consider crystalline urea as a potentially useful nonlinear optical material in the first place. Urea turned out to be a useful nonlinear optical crystal [1] because it is transparent down to 200 nm and has a fairly large birefringence, so that phase matching for second-harmonic and frequency-mixing processes can be achieved well into the ultraviolet (UV) where there are few good nonlinear optical crystals. In addition, crystalline urea has a high optical damage threshold and its refractive index is relatively independent of temperature. Its main disadvantages are that it is very difficult to grow, it is mechanically relatively soft, and it is hygroscopic. On the other hand, these disadvantages are not much more serious than those associated with such widely used nonlinear optical crystals for applications in the visible, as, for example, the ADP ($NH_4H_2PO_4$) isomorphs.

Crystalline urea is one of the few organic crystals that have found practical applications in nonlinear optics so far. It can be phase-matched down to 238 nm and is superior to the ADP isomorphs for second-harmonic generation in the near UV. It is also the first organic crystal to be successfully used in an optical parametric oscillator (OPO) [2]. It has been demonstrated that an optical parametric oscillator using a single urea crystal pumped at the third-harmonic (355 nm) output of a Nd:YAG laser can be tuned continuously from 500 nm to 1.24 μm [3] with an efficiency not much lower than a pulsed dye laser. Pumped at the fourth harmonic (266 nm) of the Nd:YAG laser, it can potentially be tuned continuously from 330 nm to 1.4 μm [4]. The urea OPO is, therefore, potentially an important solid-state tunable coherent light source.

A major problem limiting the use of urea in nonlinear optics at the present time is that of crystal growth. Although relatively large single crystals of urea (over 5 cm in the [110] directions) have been grown, the growth process is very slow and difficult to control.

In this chapter, the properties and applications of molecular and crystalline urea are reviewed.

II. NONLINEAR OPTICAL PROPERTIES OF THE UREA MOLECULE

Numerous efforts have been devoted to the study of the hyperpolarizabilities of the urea molecule, both theoretical [5, 6] and experimental [7, 8]. In addition to the technological interest in this material, researchers have found in it a challenging test case for "molecular-engineering" approaches to designing new nonlinear optical materials. In urea, strong hydrogen bonding [5] in the crystalline state polarizes the electronic cloud and enhances the intramolecular nonlinear processes [6]. This makes a simple additivity model inappropriate for predicting the crystal nonlinearities from the molecular hyperpolarizabilities. Similar intermolecular interactions render the interpretation of solution measurements more involved and difficult [8].

The molecule is interesting at the theoretical level. Its relative simplicities make it tempting to apply *ab initio* calculation methods to it, even though these are usually applied to smaller molecules. The urea molecule is schematically represented in Fig. 1. The dipole moment of 4.6 D is directed

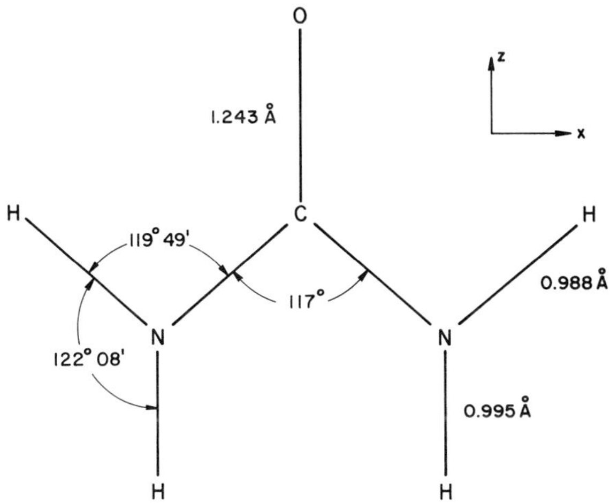

Fig. 1. Urea molecular structure and coordinate axes.

along the symmetry axis of this *mm*2 molecule, along the carbonyl group, whose dipole moment accounts for more than 80% of the overall dipole. This feature indicates a large amount of charge transfer in the ground state [6]. An interesting feature pointed out by these calculations is that in urea the nonlinearity arises from the contributions of numerous electronic states, and convergence of the summation in the perturbation theory expression is very slow [5]! This singles out this molecule from "charge-transfer" molecules where the nonlinearity is dominated by one state, which usually corresponds to a transition in the delocalized Π-electron system. In urea most of the polarizability is found in the plane of the molecule, and the contributions from the amine groups make it relatively isotropic. In all cases, the calculations show a decrease of the dipole moment in the excited states and, as the second-order hyperpolarizability β_{zzz} is proportional to the change in dipole moment, a negative sign for β_{zzz}. The magnitude of $\beta_{\parallel} = \beta_{zzz} + \beta_{zxx} + \beta_{zyy}$; which is the quantity accessible by the experiment, is between 0.5 and 1.0×10^{-30} esu [5, 6]. This value is some three orders of magnitude smaller than *para*-nitroaniline, in which a strong charge-transfer state in the near UV enhances considerably the nonlinearity at the expense of a significant narrowing of the transparency window. This trade-off between transparency and magnitude of the nonlinear response is one of the challenges of molecular engineering.

The problem of calculating the molecular hyperpolarizabilities in the crystal has been thoroughly addressed by Zyss and Berthier [6]. The authors show a strong redistribution of the electron cloud due to hydrogen bonding and a strong increase of β_{zzz}. However, the two molecules lie with their axes opposed to each other in the unit cell; the z components of the molecular hyperpolarizability do not contribute to the crystal nonlinearity, and its second harmonic coefficient is $d_{XYZ} = N(\beta_{yyz} - \beta_{xxz})$, with the conventions of Morrell *et al.* [5]. Compared to an experimental value for the crystal nonlinearity of $(3.4 \pm 1) \times 10^{-9}$ esu, a simple additivity model, not including local field corrections, tends to underestimate it $(1.57 \times 10^{-9}$ esu) [5], while taking into account the molecular environment, as in reference [6], overestimates it $(6.9 \times 10^{-9}$ esu).

Experimentally, the molecule can only be studied in solution because it decomposes at a temperature just above its fusion point, and therefore much lower than that required for vaporisation. The value of β_z can be measured by dc field-induced second-harmonic generation [6, 7] and serves mostly as a test to the theories mentioned above. However, the strong associative nature of urea makes these results hard to analyze. The measurement by Ledoux and Zyss [8] took great care to sort out the intermolecular effects arising from associations with the surrounding solvent molecules or with other solute molecules. Their value of $\beta = (0.45 \pm 0.12) \times 10^{-30}$ esu, measured in water, lower than that of reference [7], should therefore be considered more accurate and compares very favorably with the calculated values.

We thus see that urea is a molecule with a rather modest nonlinear response. A few attempts have been made at enhancing this response by substituting the amine groups by more electron-rich molecular groups [8, 9]. Although these studies helped unveil the origin of this nonlinear response, they did not identify a material with a much larger nonlinearity. The simple urea molecule remains the most promising among its family for practical applications. One direction of research that has not yet been investigated would be to lower the dipole moment of the molecule by substituting the amine groups and thereby modifying the respective alignment of the molecules in the crystal and avoiding the cancellations introduced by the present symmetry. However, one should try to preserve the transparency window that makes this molecule so attractive for practical applications.

III. LINEAR AND NONLINEAR OPTICAL PROPERTIES OF CRYSTALLINE UREA

Urea crystallizes in the tetragonal class $\bar{4}2m$ and is a positive uniaxial crystal. The unit cell contains two urea molecules that have opposing dipole moments and that have their plane rotated by 90° from each other (Fig. 2). Large single crystals of very good optical quality can be obtained. They handle in cutting and polishing very much like the KDP isomorphs. The (110) plane

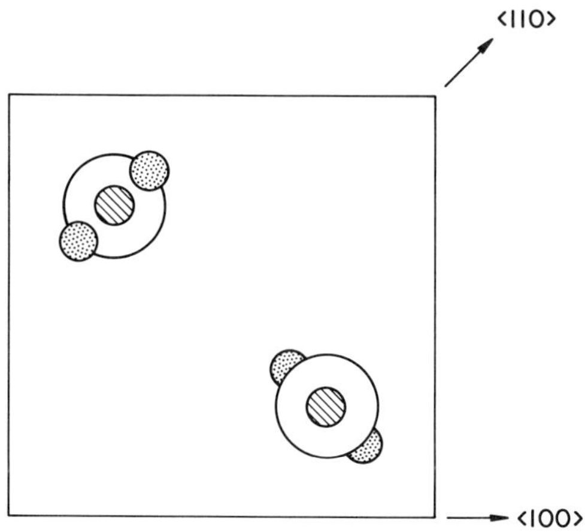

Fig. 2. The orientation of urea molecules within the crystalline unit cell. (From Halbout *et al.* [1]. Copyright © 1979 IEEE.)

of urea is a cleavage plane that represents an advantage when very thin crystals are needed, for example in doubling ultrashort pulses.

The transparency spectral window extends from 200 to 1400 nm, as shown in Fig. 3. This feature makes this material particularly suitable to the efficient up-conversion into the UV of high-power near-infrared YAG laser radiation. The absorption band centered at 1450 nm arises from harmonic absorption of the N—H stretch, while the cut-off at 200 nm marks the onset of electronic absorption with resonances centered around 150 nm.

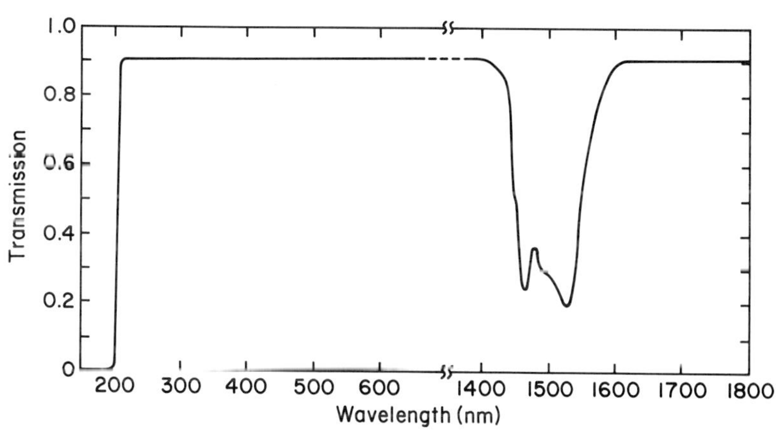

Fig. 3. Transmission of a 2-mm single crystal of urea taken with a Cary 14 spectrophotometer.

Single crystals of urea have been shown to sustain very high pulsed optical power densities [7], which makes this material particularly attractive for high-power applications. This feature is shared by many molecular crystals. The damage threshold has been measured at several wavelengths with a Q-switch Nd:YAG laser and its harmonics. The duration of the optical pulse from this laser is 10 nsec at the fundamental wavelength (1.06 μm). Pulsed power densities up to 5 GW/cm^2 at 1.064 μm, 3 GW/cm^2 at 532 nm, 1.4 GW/cm^2 at 355 nm, and 0.5 GW/cm^2 at 266 nm did not damage the polished surface or the bulk of the crystal. At higher power densities than those shown for the respective wavelengths, pitting would occur at the surface but still no bulk damage could be observed. For these experiments, the crystal was in air and the light would propagate normal to the optic axes. The decrease in the tolerable intensity with shorter wavelengths can be explained from increasing multiphoton absorption. However, it still remains unexplained why molecular crystals in general are more resistant to optical damage than ionic-type crystals [10]. For comparison, in ADP the damage threshold is about 0.5 GW/cm^2 at 1.06 μm under the same conditions.

The indices of refraction have been measured at a few selected wavelengths in the visible [11, 12], and a complete set of Sellmeier equations for the dispersion of these indices has been derived from the measured second-harmonic phase-matching angle curves [1]. The type I phase-matching angles are given theoretically by the expression

$$\sin^2 \theta_m = \frac{n_0(\lambda/2)^{-2} - n_0(\lambda)^{-2}}{n_e(\lambda)^{-2} - n_0(\lambda)^{-2}} \tag{1}$$

where λ is the fundamental wavelength. The initial values for the nonlinear fit were taken from the measured data. A modified Sellmeier equation, with three constants and a wavelength resonance denominator to reflect the UV absorption gives the best fit. As a result, the indices of refraction of urea are found to be given by

$$n_0^2 = 2.1678 + \frac{0.0139}{\lambda^2 - 0.0207} \tag{2}$$

$$n_e^2 = 2.4917 + \frac{0.0141}{\lambda^2 - 0.0240} \tag{3}$$

with λ in micrometers. Although other dispersion equations have been proposed, they do not give an overall fit as good as the expressions shown above for the second-harmonic and sum-frequency-mixing phase-matching angles. These equations predict noncritical phase-matching angles to an accuracy of 0.2–0.5 nm and give the phase-matching angle for generation of the YAG fifth harmonic to within 10°, which most other expressions fail to even allow. It should be emphasized, however, that the dispersion of the indices of refraction can vary depending upon the purity of the crystal. To help with the growth process, sometimes the crystals are slightly doped with bromine. This may lead to significantly different Sellmeier equations [2, 3] in strongly dispersive regions. Also, better fitting can obviously be achieved with more elaborate Sellmeier equations. For example, more recent measurements based on parametric luminescence tuning curves give a more accurate and also more complicated set of Sellmeier equations [4] than those given by Eqs. (2) and (3) for pure urea crystals.

The crystal symmetry imposes two independent components for the second order susceptibility tensor, $d_{14} = d_{25}$ and d_{36} [13]. In the visible, far away from resonance, Kleinman symmetry can be invoked so that the three coefficients become equal. Powder measurements [14, 15] had implied rather high values for the effective second-harmonic coefficients, ranging from 1.7 pm/V to 7 pm/V at a fundamental wavelength of 1.064 μm. The crystal nonlinear coefficient was measured by comparing the doubling efficiency in urea with that of ADA and ADP [1]. The measurements were carried out at a 90° phase-matching angle, where neither walk-off or double refraction are involved with the nonlinear generation. The comparison was made at the same wavelength and with the same unfocused beam in order to eliminate wavelength and power density calibration. This gives a fairly good estimate of the nonlinear coefficient, provided that the bandwidth of the fundamental is narrowed down [16]. The second-harmonic efficiency in urea at 90° type II phase-matching for plane waves is given by

$$\frac{P^{2\omega}}{(P^{\omega})^2} = \frac{8\pi^2 d_{14}^2 l^2}{\varepsilon_0 C (n^{\omega})^2 n^{2\omega} \lambda^2 A} \tag{4}$$

where d_{14} is the nonlinear coefficient [17], l the crystal length, λ the fundamental wavelength, ω the fundamental frequency, and A the beam area. The comparison of a 2-mm urea crystal with a 25-mm heated ammonium dihydrogen arsenate (ADA) crystal at 597 nm indicated that the d_{14} nonlinear coefficient of urea is 2.5 + 0.2 times the d_{36} of ADA, or about 1.4 pm/V [1]. It was checked that the efficiencies of ADP and ADA are identical. This value agrees very well with the value of 1.4 pm/V of the direct efficiency measurement of Bauerle *et al.* [11], but is much smaller than the values reported by Kurtz and Perry [14] and Koreneva *et al.* [18] using powder measurements. Therefore, urea should be six times more efficient than the KDP isomorphs for nonlinear interactions. However, when angle tuning is also considered, the merit of urea might be reduced, since its birefringence is greater than that of the KDP isomorphs.

We see therefore from the linear and nonlinear properties of the crystal that it represents an interesting trade-off between efficiency and transparency. Together with its high damage threshold, these properties make the crystal very attractive for nonlinear frequency generation.

IV. SECOND-HARMONIC GENERATION AND SUM-FREQUENCY MIXING IN CRYSTALLINE UREA

In this section, the performance of urea is presented for nonlinear frequency generation into the near UV by second-harmonic generation and sum-frequency mixing.

The measured phase-matching (PM) angles as a function of the fundamental wavelength for type I and II second-harmonic generation are given in Fig. 4. Noncritical (90°) phase-matching occurs at room temperature for the two types at 476 nm and 597 nm, respectively. The exact type I (90°) point is difficult to determine since the efficiency at that angle in a positive uniaxial crystal such as urea drops to zero because of the vanishing effective nonlinear coefficient [19]. Although care was taken to minimize the double refraction of the two interacting polarizations, the accuracy of the type II measurements should be considered inferior to that of the type I measurements. Our experimental results for type I are in good agreement with those of Bauerle *et al.*

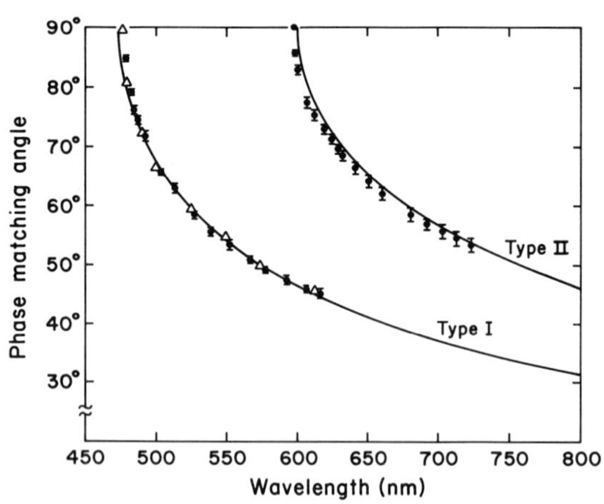

Fig. 4. Measured and calculated phase-matching angles as a function of wavelength for type I and type II SHG in urea. Data sources: ●, our measurements; Δ, Betzler *et al.* [12]; solid lines, calculated.

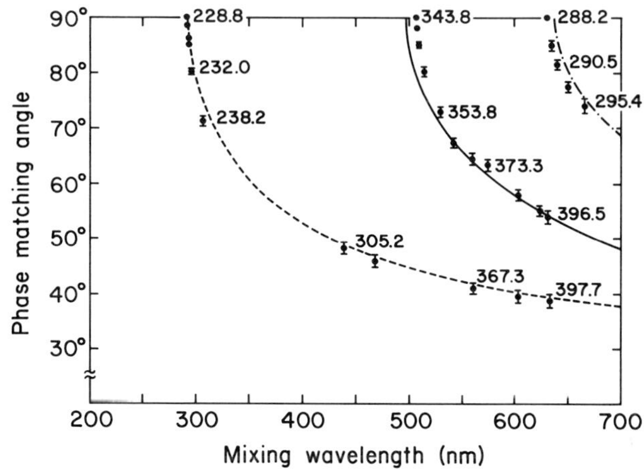

Fig. 5. Measured and calculated phase-matching angles for three type II mixing schemes: --- mixing with 1.06-μm o; -·-· mixing with 532-nm e; solid line, experimental points. The numbers next to some of the experimental points indicate the output wavelength in nanometers.

[11] (triangles in Fig. 4). Experimental values for the PM angles for three type II mixing schemes are given in Fig. 5. The 1.06-μm radiation of the Nd : YAG laser is mixed either as an ordinary (o) ray or as an extraordinary (e) ray. This yields 90° points at 291.6 nm and 507.5 nm for the o and e cases respectively. Also, the 532-nm second harmonic of the Nd : YAG was mixed, as an e ray, to yield a 90° point at 630 nm. It is seen from Fig. 5 that the 1.06 μm o-mixing scheme yielded the shortest output wavelength at 228.8 nm, when the (IR) frequency is mixed with the second harmonic of a rho damine 6G (RH6G) laser. This last scheme seems to be a very convenient one to obtain tunable coherent radiation above 229 nm. Using an Nd:YAG laser system, this scheme involves three nonlinear processes (either doubling or mixing) and one dye laser. The alternative for obtaining tunable radiation in this range is doubling a dye laser in the last stage, a process that also involves three nonlinear processes and a dye laser. However, using the proposed mixing scheme no unstable coumarin dyes have to be used. It should be mentioned, however, that such a mixing scheme is also feasible with KD*P, but with less efficiency than urea.

From the dispersion equations presented in the last section, one can construct a 90° PM mixing diagram for type II and a 65° PM mixing diagram for type I (Fig. 6). This plot shows that although type I second-harmonic generation phase-matches below type II, type II 90° mixing yields still shorter output wavelengths than type I; the infrared components for the type I mixing

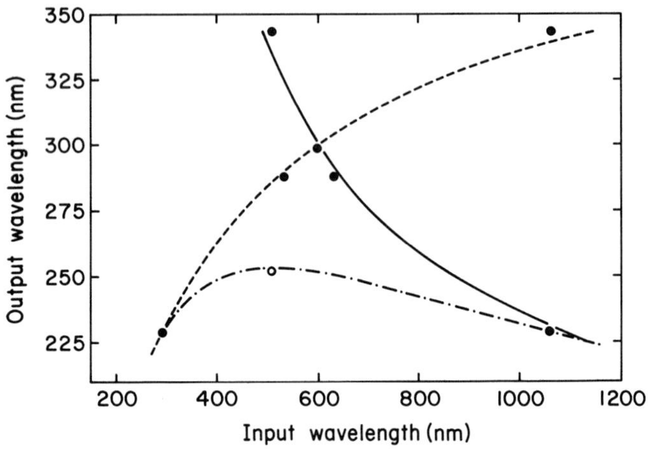

Fig. 6. Wavelength combinations for mixing: –·–· type I at 65°; ––– type II at 90° o ray. Experimental points: solid circles, type II; open circle, type I.

are either not available with present systems or fall in the absorption band of urea (see Fig. 3). Figure 6 also suggests two more type II mixing schemes utilizing the second and third harmonics of the Nd:YAG laser. One scheme uses the 532 nm as an o ray to be mixed with an e ray at 875 nm (and longer) to yield output at 330 nm (and up). The second scheme mixes 355 nm e with 870 nm o (and longer) to give output at 252 nm (and longer). These output wavelengths can be conveniently obtained with conventional second-harmonic generation and sum-frequency mixing; thus, these schemes have not been tried experimentally. Nevertheless, the reverse-difference frequency mixing of these schemes could be utilized to obtain tunable coherent radiation from 870 nm up to around 1200 nm by means of an OPO or by direct difference mixing.

Two other particularly interesting frequency-mixing schemes can be phase-matched with high conversion efficiency: the generation of the fifth harmonic of the Nd:YAG laser at 212.8 nm by mixing of the fundamental with the fourth harmonic, and the generation of the third harmonic of the ruby laser at 231.4 nm by mixing the fundamental and second harmonic. The phase-matching angle for each of these processes can be calculated from the Sellmeier equations given in Section III. It should be noted that, in this very dispersive wavelength region, the presence of any impurity can change drastically the shape of the dispersion curve for the refractive indices.

Temperature tuning has been tried at the noncritical PM point of type II and at the 65° PM point for type I [1]. The data for type II is presented in Fig. 7. Type I yielded similar results. It can be seen that urea has a very small

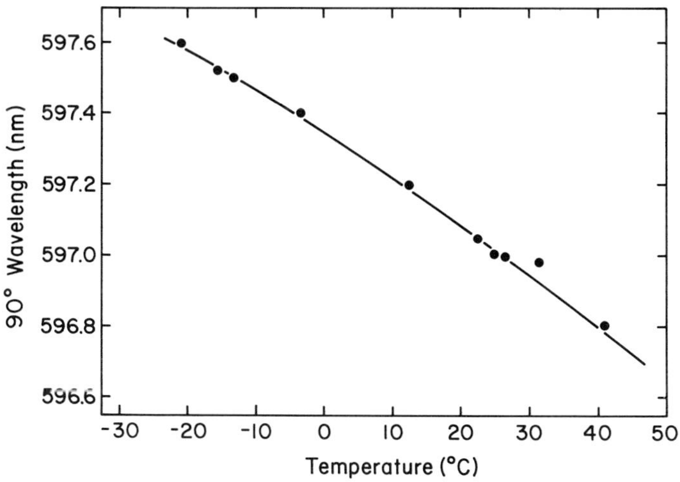

Fig. 7. Measured temperature tuning curve for 90° type II phase-matched SHG in urea.

temperature tuning capability with a negative slope of -0.013 nm/°C, as compared to 0.05 nm/°C and 0.3 nm/°C for KDP and ADP, respectively. This latter property makes urea most attractive for use in high average power systems. It is well known that residual absorption, particularly toward shorter wavelengths, will cause local heating along the beam inside the crystal [20]. This will result in continuous detuning from the optimum phase matching. It is not always possible to compensate for such detunings by external thermal regulation, particularly in the KDP isomorphs, in which the phase-matching conditions are very sensitive to temperature. This requires that such crystals be of extremely high purity to minimize unavoidable residual absorption. In urea, on the other hand, this effect is much less pronounced; indeed, it was found that the generation of very short wavelengths is highly stable, even for very long times.

The overall doubling efficiency, which is particularly important with low-continuous-wave power or (CW) sources, was assessed by comparing urea with ADP in angle-tuned second-harmonic generation [1]. Type I and type II efficiency tuning curves for urea were calculated together with type I curves for ADP, using the Boyd and Kleinman [21] theory for parametric interactions of focused gaussian beams. The type I efficiency was calculated for optimum focusing in a 2-mm crystal (i.e., a focusing lens of 35 mm focal length for a 1.5-mm-diameter laser beam). The type II calculation is an approximation in which the optimum focusing is assumed the same as in type I and in which the double refraction angle represents the walk-off angle in the Boyd and Kleinman expressions [21]. The results are presented in Fig. 8. The figure

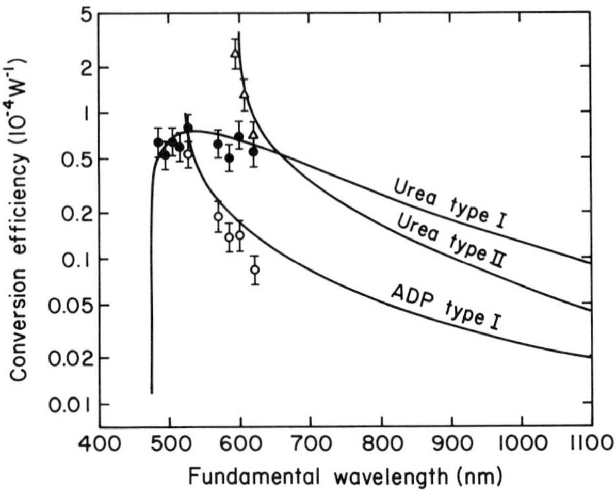

Fig. 8. Calculated and measured efficiencies at room temperature for type I and II SHG in urea and type I SHG in ADP. The crystal length in each case is 2 mm with an optimally focused beam (see text). Experimental points: ●; urea type I; Δ, urea type II; ○, ADP type I.

shows that urea exhibits maximum conversion efficiency at 90° for type II PM. However, it drops very rapidly at longer wavelengths. On the other hand, type I conversion peaks at about 530 nm at a value of about one-fifth that of the type II peak. The wavelength dependence of the type I efficiency is rather weak; over the spectral range from 480 nm to 850 nm, the doubling conversion drops by less than a factor three. It can be seen from the figure that, except for the range from 597 to 650 nm, type I second-harmonic generation is more efficient than type II. This, together with the inconvenience of alignment of type II because of double refraction, makes the type I process more convenient for general use in CW tunable second-harmonic generation than the type II one.

The comparison with ADP shows that, except very close to the noncritical PM point of ADP, urea is more efficient. Also, urea's tuning range at room temperature is larger than that of ADP. The comparison between the crystals was further pursued experimentally by doubling several argon-ion laser lines and a CW Rh6G dye laser. The radiation was optimally focused on crystals of identical length (2 mm). The calculation, including the value for the nonlinear coefficient of urea, was confirmed by the experimental results. Urea type I second-harmonic generation yielded five times more UV power than ADP when doubling Rh6G. Moreover, while at room temperature ADP cannot be phase-matched for most of the argon lines, with urea very efficient doubling was achieved down to the 488-nm line of the argon laser. The use of ADP at this wavelength requires the refrigeration of ADP very close to its Curie temperature ($-125°C$); with very good temperature stability. Avoiding all this trouble, one can obtain, at room temperature, in a 2 mm urea crystal, CW power at 244 nm of about 1 mW.

Let us finally mention three applications of the phase-matching properties of urea in the area of short optical pulse measurements. The first one takes advantage of the type I phase-matching for doubling the 514.5-nm line of the argon laser. The autocorrelation of short pulses from such a mode-locked laser can easily be obtained at a 63° phase-matching angle [22]. The two other applications involve using type II phase matching, which, in urea only, is possible in the visible. In order to time-resolve luminescence decays from photoexcited samples, the luminescence can be up-converted in a nonlinear crystal by part of the exciting pulse itself. This light gate is described in detail by Mahr and Hirsh [23] and is one of the most versatile techniques for time-resolved studies of ultrafast processes. For all the crystals presently available, in the spectral region covered by rhodamine 6G, the up-conversion must be of type I. This implies that both the strong up-converting pump pulse and the weak luminescence from the sample are in the same polarization. In the case of very weak signals, the problem then arises that the non–phase-matched second harmonic of the strong pump can attain intensity levels comparable to or even higher than the useful signal. This problem is usually solved by using high-rejection monochromators to separate out only the parasitic UV generated by the pump pulse. These instruments (such as double monochromators) are very expensive and their throughput is low. With urea, the problem can be much more easily solved by using a type II light gate. Such a configuration has been demonstrated in the near infrared [24] with a mode-locked neodymium glass laser. Its extension to the visible for quasi-CW mode-locked dye lasers is straightforward. Its main advantage lies in the fact that if the strong pump is propagating in the urea crystal as an o ray, the non–phase-matched second harmonic that it generates is polarized normal to that of the useful up-converted signal. The phase-matching angle for up-converting a signal at 750 nm with a pump pulse at 575 nm is calculated to be 74°. In addition, type II up-conversion has a higher acceptance angle and allows higher conversion efficiencies. Even with a phase-matching angle of 75°, the acceptance bandwidth is higher (~ 1 nm for a 2-cm crystal with optimum focusing). This scheme should be very useful in time-resolved spectroscopy. Finally, the higher angular acceptance bandwidth of type II phase-matching has been utilized to measure the autocorrelation of ultrashort laser pulses at 620 nm [25].

The numerous applications presented in this section emphasize the versatility of urea crystals for nonlinear frequency generation into the UV from existing powerful visible laser sources.

V. SPONTANEOUS AND STIMULATED OPTICAL PARAMETRIC PROCESSES IN CRYSTALLINE UREA

Ever since the invention of the laser, one active area of research in nonlinear optics has been the development of continuously tunable coherent radiation. Another very important and unique application of the urea crystal that has been demonstrated experimentally is in the generation of continuously tunable coherent radiation via the optical parametric process [2–4]. The parametric process [26] is the inverse of the frequency-mixing process: each pump photon breaks down into two photons of lower frequencies, known sometimes as the signal and idler waves, under phase-matched conditions. This can occur either spontaneously or through stimulated emission leading to amplification of the generated waves. If a suitable external feedback in the form of, for example, a Fabry–Perot cavity is provided, oscillation can build up from the spontaneously emitted photons. Such an optical parametric oscillator has the unique property that its output can be continuously tuned through the phase-matching condition by rotating the crystal or changing its temperature. In this section, experimental results on spontaneous and stimulated parametric processes in crystalline urea are discussed.

Optical parametric oscillators (OPO) using various nonlinear optical crystals have been studied extensively over the years [27]. However, a variety of practical problems with existing materials, especially low optical damage thresholds [28] and inadequate spectral properties, as well as the development of competing tunable-laser sources, particularly the pulse dye lasers, have limited interest in the OPO. Further developments depended on crystal data for use in OPO in the visible.

The urea crystal has a high birefringence and is positive uniaxial. This characteristic, which is in contrast to $LiNbO_3$ and the ADP isomorphs and many other nonlinear crystals, is of great importance for the design of an angle-tuned OPO. By utilizing type II ($o \rightarrow o + e$) phase matching and resonating the ordinary wave, the degree of Poynting vector walk-off of the signal from the pump due to double refraction is greatly reduced. Furthermore, for a positive uniaxial crystal, the effective *d* coefficient reaches a maxiumum at 90°, which allows for noncritical phase matching. Urea also has a high damage threshold and relatively large birefringence, allowing phase matching well into the near UV. Optical parametric oscillation and spontaneous emission in crystalline urea pumped by the third (355 nm) and fourth (266 nm) harmonics of Nd:YAG laser output have been observed [3, 4].

In the case of the 355-nm pump, Fig. 9 shows the measured and calculated tuning curves as a function of the internal crystal angle based on the Sellmeier equation given by Rosker and Tang [3]. The two generated waves cover the

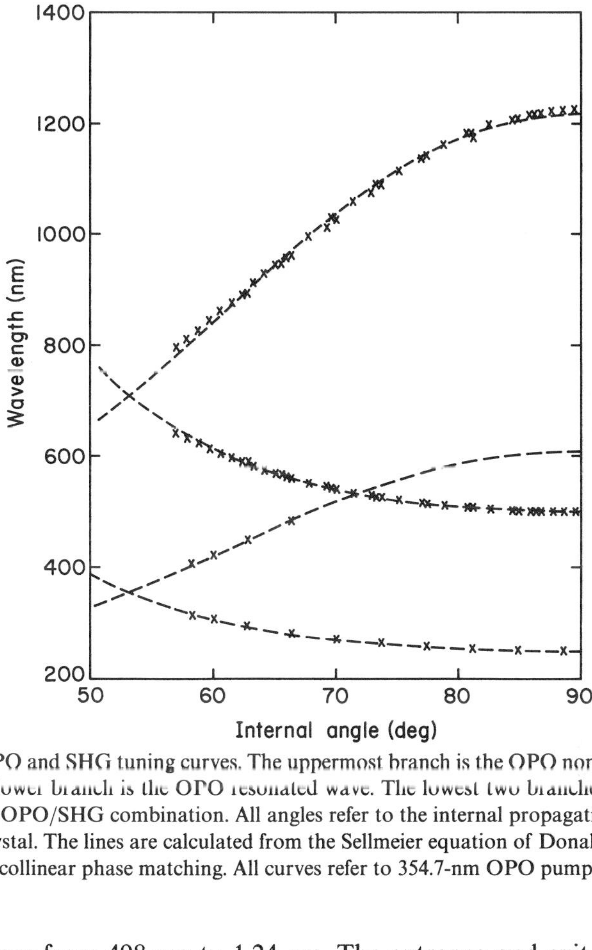

Fig. 9. OPO and SHG tuning curves. The uppermost branch is the OPO nonresonant wave, and the next lower branch is the OPO resonated wave. The lowest two branches represent the output of the OPO/SHG combination. All angles refer to the internal propagation angle in the OPO urea crystal. The lines are calculated from the Sellmeier equation of Donaldson and Tang [2] assuming collinear phase matching. All curves refer to 354.7-nm OPO pump wavelength.

spectral range from 498 nm to 1.24 μm. The entrance and exit faces of the crystal are the as-grown (110) faces. The spectral range observed corresponds to a rotation of 90° from the [001] axis to about 57°. The small gap around 700 nm is due to angular limitations imposed entirely by the dimensions of the particular crystal used. There is no limit in filling the gap in the tuning curve with a slightly larger crystal or one cut slightly differently, for example, with the faces at 15° with respect to the [110] axis.

With a 500-kW/cm^2 and 1-mm-diameter beam of 7-nsec pulses at 354.7 nm, the efficiency for both generated waves is 23% for a singly resonant (visible wave) oscillator. The pump beam intensity is far below the pulse damage threshold of 1.5 GW/cm^2 at 354.7 nm. Therefore, potentially the urea OPO can be a powerful and broadly tunable source. Much engineering work

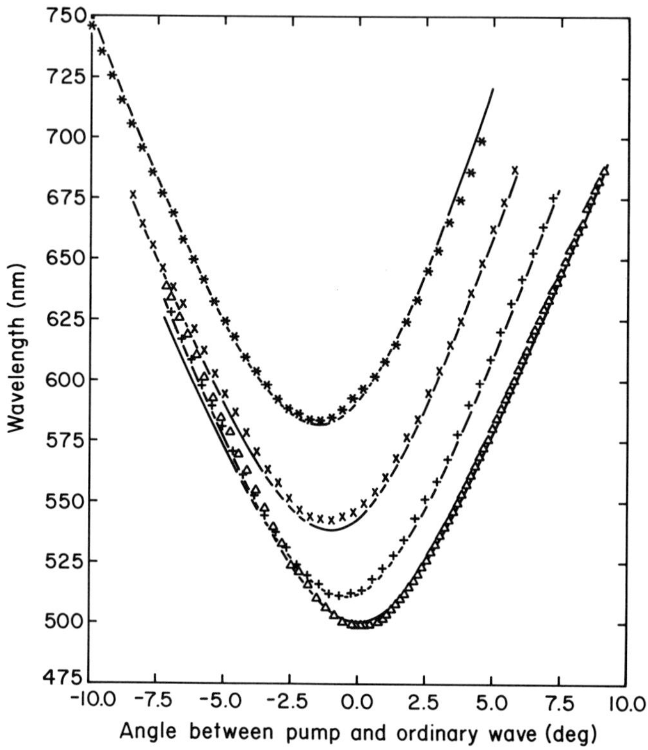

Fig. 10. Tuning curves for spontaneous parametric fluorescence in urea at several crystal orientations. From top to bottom, $\theta = 62.28°$ (∗), 69.28° (X), 78.28° (+), and 89.72° (Δ).

still remains to be done before the full potential of such an OPO is realized.

With a urea doubling crystal, the continuous-tuning range of the OPO can be further extended. Figure 9 shows also the experimentally achieved extended tuning range down to 249 nm. The conversion efficiency in this particular experiment was 8.3%. The entire spectral range from 249 nm to 1.24 μm can thus be covered with the urea OPO and second-harmonic generator.

In addition to parametric oscillation, spontaneous parametric scattering (or parametric fluorescence) of the 354.7-nm pump light in urea has been observed. Figure 10 shows the tuning characteristics of the parametrically scattered light [3].

By pumping the crystal at the fourth harmonic (266 nm) of an Nd:YAG laser output, the urea OPO can in principle be tuned continuously from 330 nm to 1.4 μm using type I phase matching ($o \rightarrow e + e$). Figures 11 and 12 show the corresponding tuning curves for type I and type II phase matching. Parametric fluorescence with a 266-nm pump in urea has already been

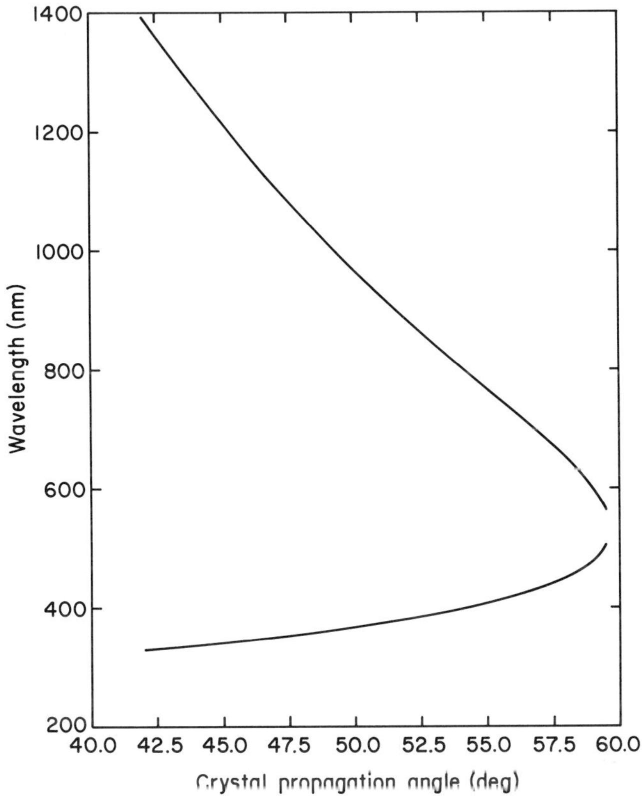

Fig. 11. Calculated OPO tuning curve for a pump wavelength of 266 nm for Type I phase-matching (o → e + e) in crystalline urea.

observed. The tuning characteristics are shown in Fig. 13. Also shown in Figure 13 are comparisons between the calculated tuning curves and experimental data points for noncolinear interaction based upon the Sellmeier equations given by Rosker *et al.* [4]. The agreement for noncollinear interaction is relatively poor, indicating the Sellmeier equations given by Rosker *et al.* [4] are not adequate to account for the measured noncollinear parametric luminescence tuning curves, which depend very sensitively on the slope of the crystal dispersion curve. For the collinear interaction case, the measured and calculated curves are quite good. To our knowledge, parametric oscillation in urea pumped at 266 nm has not yet been achieved. The only possible difficulty is mirror-coating damage at 266 nm at the pump pulse intensity required for oscillation. Damage in the crystal itself is not expected to present any difficulty.

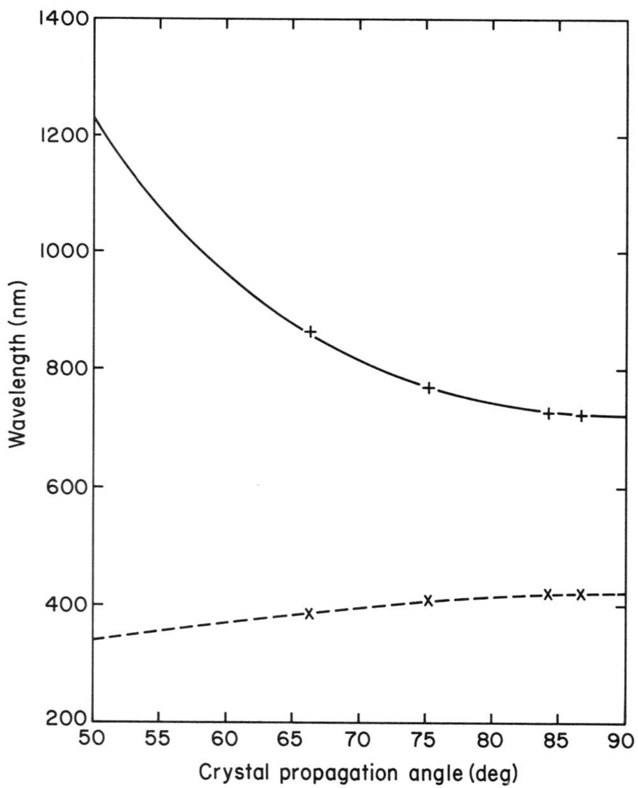

Fig. 12. Calculated (—— and ---) OPO tuning curves and measured data points (+ and x) for a pump wavelength of 266 nm for type II phase matching (o → o + e) in crystalline urea.

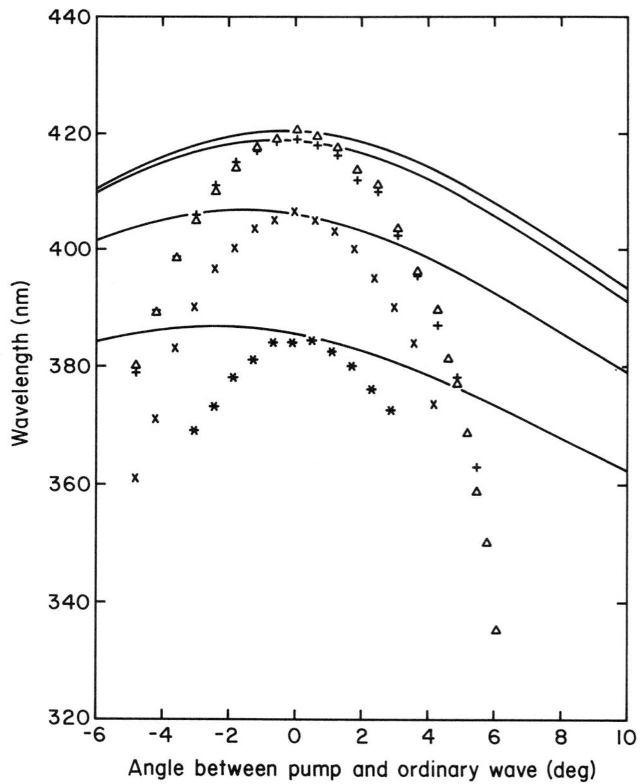

Fig. 13. Tuning curve for spontaneous parametric scattering in urea at several crystal orientations, for $\theta = 86.7°$ (Δ), 84.2° (+), 75.2° (x), and 66.3° (*), where θ is the angle between the pump direction and the crystal z axis. The solid lines are calculated using the Sellmeier equations given by Rosker *et al.* [4] for each appropriate value of θ. The pump wavelength is 266 nm.

In conclusion, it is clear that organic crystals have important applications in nonlinear optics. The urea crystal is a good example.

ACKNOWLEDGMENTS

We thank K. Cheng and M. Rosker for their contributions of this chapter. This work was supported by the National Science Foundation.

REFERENCES

1. J.-M. Halbout, S. Blit, W. Donaldson, and C. L. Tang, *IEEE J. Quantum Electron.* **QE-15,** 1176 (1979).
2. W. R. Donaldson and C. L. Tang, *Appl. Phys. Lett.* **44,** 25 (1984).
3. M. J. Rosker and C. L. Tang, *J. Opt. Soc. Am. B: Opt. Phys.* **2,** 691 (1985).
4. M. J. Rosker, K. Cheng, and C. L. Tang, *IEEE J. Quantum Electron.* **21** 10 (1985).
5. J. A. Morrell, A. C. Albrecht, K. Levin, and C. L. Tang, *J. Chem. Phys.* **71,** 5063 (1979). According to C. Willand of A. Albrecht's group, the original CNDO/s program used by Morrell and Albrecht had the wrong sign for the transition moment (both ground to excited states and excited to excited states). This was a widely available program, written originally to calculate state dipoles and oscillator strengths. Unfortunately, they had the charge of the electron as positive instead of negative. To correct for this, the sign of the state dipoles were changed in the original program. While this corrects the problem for linear dipoles, it is incorrect for induced nonlinear polarizations. Correcting this problem, Willand obtains for β_z a value of -0.054 instead of $|\beta_z| = 0.46 \times 10^{-30}$ esu obtained by Morrel (his published value was divided by a factor of 2 to allow direct comparison between the dc field-induced SHG results and the calculated β_z). Willand's corrected calculated results for urea molecule no longer agree with the reported experimental results [6–8]. There is also a significant discrepancy between the measured results for dc field-induced SHG in DMF reported in reference [6] and references [7] and [8]. The reason for the difference is not clear; it probably cannot be explained by aggregation effects alone, as similar concentration ranges were used. Recent theoretical CNDO extended-basis-set calculations by Erik Svendsen and C. Willand (submitted to *J. Chem. Phys.*) agree very well with the measured *d* coefficients of urea crystal reported in reference [1], however.
6. J. Zyss and G. Berthier, *J. Chem. Phys.* **77,** 3635 (1982).
7. C. Cassidy, J.-M. Halbout, W. Donaldson, and C. L. Tang, *Opt. Commun.* **29,** 243 (1979).
8. I. Ledoux and J. Zyss, *Chem. Phys.* **73,** 203 (1982).
9. J.-M. Halbout, A. Sarhangi, and C. L. Tang, *Appl. Phys. Lett.* **87,** 864 (1981).
10. See, for example, J. L. Oudar and R. Hierle, *J. Appl. Phys.* **48,** 2699 (1977); J. Zyss, D. S. Chemla, and J. F. Nicoud, *J. Chem. Phys.* **74,** 4800 (1981).
11. D. Bauerle, K. Betzler, H. Hesse, S. Kappan, and P. Loose, *Phys. Status Solidi A* **42,** K119 (1977).
12. K Betzler, H. Hesse, and P. Loose, *J Mol. Struct.* **47,** 393 (1978).
13. See, for exaample, C. Flytzanis, *in* "Quantum Electronics: A Treatise" (H. Rabin and C. L. Tang, eds.), Vol. 1A, p. 26. Academic Press, New York, 1975.
14. S. K. Kurtz and T. T. Perry, *J. Appl. Phys.* **39,** 3798 (1968).
15. A. Graja, *Acta Phys. Pol. (Engl. Transl.)* **A37,** 539 (1970).

16. S. Blit, E. G. Weaver, and F. K. Tittel, *Appl. Opt.* **18,** 733 (1979).
17. This expression is given in MKS units. Note that d here equals d (esu) $\times 4\pi/3 \times 10^{-4}$.
18. L. G. Koreneva, V. E. Zolin, and B. L. Davydov, "Molecular Crystals in Nonlinear Optics." Nauka, Moscow, 1975.
19. See, for example, S. K. Kurtz, *in* "Quantum Electronics: A Treatise" (H. Rabin and C. L. Tang, eds.), Vol. 1A. Academic Press, New York, 1975.
20. N. A. Barnes, R. C. Eckhardt, L. B. Edgett, and D. J. Gettemy, *LIS Inf. Exch. Meet.*, Albuquerque, New Mexico (1977).
21. G. D. Boyd and D. A. Kleinman, *J. Appl. Phys.* **39,** 3597 (1968).
22. J.-M. Halbout, N. A. Olsson, and C. L. Tang, *Appl. Phys. Lett.* **39,** 463 (1981).
23. H. Mahr and M. D. Hirsh, *Opt. Commun.* **13,** 96 (1975).
24. L. A. Hallidy and M. R. Topp, *Chem. Phys. Lett.* **46,** 8 (1977).
25. J-C. Diels, C. Y. Wang, and J. J. Fontaine, *SPIE Proc.* **322,** 166 (1982); J. Dietel, J. J. Fontaine, and J.-C. Diels, *Opt. Lett.* **8,** 4 (1983).
26. See, for example, C. L. Tang, *in* "Quantum Electronics: A Treatise" (H. Rabin and C. L. Tang, eds.) Vol. 1A, p. 419–446. Academic Press, New York, 1975.
27. See, for example, R. L. Byer, *in* "Quantum Electronics: A Treatise" (H. Rabin and C. L. Tang, eds.), Vol. 1B, p. 677–690. Academic Press, New York, 1975.
28. Y. X. Fan, R. C. Eckardt, R. L. Byer, R. K. Route, and R. S. Feigelson, *Appl. Phys. Lett.* **45,** 313 (1984).

Chapter II-7

Nonlinear Optical Properties of Guest–Host Polymer Structures

DAVID J. WILLIAMS

Research Laboratories
Eastman Kodak Company
Rochester, New York 14650

I. INTRODUCTION

The tailoring of electronic and optical properties of organic and polymeric materials along with ancillary properties such as film-forming ability, mechanical integrity, resistance to environmental stresses, and specialized requirements for specific applications has come to be known as molecular engineering. In several important examples, this approach has produced state-of-the art technology. Scientifically and commercially, organic photoconductors, in which highly specialized molecular species are doped into polymers, imparting the ability to transport photogenerated charge, are one of the best known examples of this approach (Dulmage *et al.*, 1978; Mort and Pfister, 1982). Detailed knowledge of charge transport in amorphous solids, as well as fundamental considerations of the electronic structure and dynamics of molecules and molecular solids, preceded the establishment of commercial technology, and research in these fields continues to lead to technological improvements (Bassler, 1981).

It has been recognized that many organic molecules exhibit large hyperpolarizabilities due to the delocalization of their π-electronic clouds (Franken *et al.*, 1961; Kurtz and Perry, 1968). Owing to the third-rank tensorial nature of the second-order nonlinear coefficient, the bulk medium containing these molecules must also have polar symmetry for effects arising from the molecular contribution to be useful; this places new constraints on the molecular-engineering approach. Reviews (Badan *et al.*, 1983; Williams, 1984) provide background on the relationships between crystal structure and second-order nonlinear properties and the electronic origins of these processes.

Organic crystals have not yet found widespread utility in integrated optics and waveguide optics or as bulk components for parametric processes, mainly because of the difficulty of growing large, high-quality single crystals with the required processability, mechanical integrity, and ambient stability. Recent progress in materials and crystal growth techniques suggests that this situation may be changing (Zyss *et al.*, 1984; Nayar, 1983). Nevertheless, new approaches are needed for designing materials with large second-order nonlinearities based on polymeric films that might be coated and perhaps patterned photolithographically. Since nonlinear optical responses are largely due to the properties of aromatic rings, an approach whereby electronic properties are optimized somewhat independently from orientation and other physical and chemical requirements might be appropriate. This chapter deals with electric-field-induced second-order nonlinear effects in molecularly doped polymers and polymeric chains containing nonlinear chromophores.

Nonlinear optical responses originate in the polarization induced in a medium by an electromagnetic field. The polarization response of a molecule is given by

$$P_i = \alpha_{ij}E_j + \beta_{ijk}E_jE_k + \gamma_{ijkl}E_jE_kE_l + \cdots \tag{1}$$

where α, β, γ, etc. are tensor quantities relating the induced polarization vector components P_i to the appropriate field components, and where E_j, E_k, etc. and i, j, k, and l refer to the molecular coordinate system. The odd-order ranked tensors such as β require that the molecule providing the response has no center of inversion symmetry; otherwise its contribution to the induced polarization is zero in the dipole approximation (Flytzanis, 1975). Molecules with low-lying charge-transfer states of the appropriate symmetry can give very large β. For a number of classes of chromophores, these requirements have been verified experimentally (Levine, 1976; Oudar, 1977; Oudar *et al.*, 1977; Dulcic and Sauteret, 1978; Dulcic, 1979) and theoretically (Morrell and Albrecht, 1979; Lalama and Garito, 1979; Docherty and Pugh, 1985). In

general, molecules exhibiting large β have the general structure 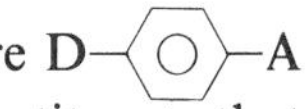
where D and A are electron-donating or electron-accepting substituents that provide a substantial charge-transfer resonance interaction through the aromatic ring. Similarly, the nonlinear polarization induced in bulk media (liquids and solids) can be expressed by the expansion

$$P_I = \chi_{IJ}^{(1)} E_J + \chi_{IKJ}^{(2)} E_J E_K + \chi_{IKJL}^{(3)} E_J E_K E_L \tag{2}$$

where the coefficients $\chi^{(n)}$ have meanings similar to their molecular counterparts but are scaled to the properties of the medium. The odd-order coefficients (expressed as scalar quantities) are related to the molecular coefficients of Eq. (1) by Eqs. (3) and (4) (Flytzanis, 1975),

$$\chi^{(1)} = N\alpha F(\omega) \tag{3}$$

$$\chi^{(3)} = N\gamma F(\omega_1)F(\omega_2)F(\omega_3)F(\omega_4) \tag{4}$$

where N is the number density and $F(\omega_i)$ is a local field factor at frequency ω_i that modifies the value of the external electric field at the molecule due to contributions from neighboring molecules. The even-order coefficients require a detailed knowledge of the projection of the molecular hyperpolarizability tensor components onto tensor components for the bulk medium. For the simple case of a rigid oriented gas with all molecules pointing in the same direction and fixed in space,

$$\chi^{(2)} = N\beta F(\omega_1)F(\omega_2)F(\omega_3) \tag{5}$$

For nonassociating liquids, Onsager (1936) determined the local field factors to be

$$F(\omega_i) = (\varepsilon_\infty^i + 2)\varepsilon_\omega/(\varepsilon_\infty^i + 2\varepsilon_\omega) \tag{6}$$

In molecular solids the structure is more complex than in a rigid oriented gas. Zyss and Oudar (1982) and Oudar and Zyss (1982) analyzed the relation between β and $\chi^{(2)}$ for molecules in the known space groups and derived detailed relationships between molecular orientation and $\chi^{(2)}$. The problem of treating polymer films with large second-order nonlinear optical coefficients is primarily that of imparting the needed polar symmetry to the medium in the absence of the stabilization energy associated with a particular crystal form. Several approaches have been explored for removing the orientational averaging that leads to cancellation of contributions from individual molecular constituents. Garito and associates (1980, 1983) used topochemical polymerization of noncentrosymmetric diacetylene crystals to prepare thin films that show ferroelectric behavior and second-order nonlinear effects. This approach has had some success, but it is fairly inflexible in terms of independent optimization of molecular orientation or other requisite properties for a range of applications.

Another promising approach for the incorporation of a variety of chromophores and independent optimization of other properties is molecular doping of the appropriate nonlinear chromophore into a polymeric host followed by electric-field alignment of the host medium and permanent fixation of the induced alignment. In this case, the expression for $\chi^{(2)}$ in Eq. (5) must be modified by an orientational factor $O(\theta, \phi)$, where θ and ϕ are polar angles, as follows:

$$\chi^{(2)} = N\beta F(\omega_1)F(\omega_2)F(\omega_3)O(\theta, \phi) \tag{7}$$

Here $O(\theta, \phi)$ provides the appropriate statistical alignment factors and relationships associated with the externally applied optical and dc fields. This process, which was explored by Meredith and associates (1982, 1983) and Singer and associates (1985), is described in greater detail here.

Havinga and van Pelt (1979) explored electric-field-induced electrochromic effects in molecularly doped thermoplastic films where the competition between field-induced alignment and thermal averaging of the molecules determines the observed spectral changes. Similar factors determine the field-induced second-order nonlinear optical phenomena described here.

In the current approach, additional cooperative effects between the nonlinear chromophore and the mesogenic chromophores in the nematic host medium are used to enhance the alignment over that obtainable in a homogeneous glassy polymeric medium. Since liquid-crystalline polymeric hosts similar to the ones used here show thermoplastic behavior (Finklemann *et al.*, 1979a,b; Finklemann and Day, 1979), it was believed that thermal quenching of nematic mesophase could trap the field-induced alignment of the host–guest medium, leading to permanent fixation of the field-induced second-order nonlinearity. Thus the primary function of the nematic environment is to provide a highly skewed potential along the axis of the alignment, which significantly enhances the maximum theoretically achievable field-induced value of $\chi^{(2)}$ over that obtainable from an isotropic host medium. In addition, the large birefringence achievable in these media can aid phase matching.

This behavior is distinctly different from that leading to other nonlinear effects commonly utilized in liquid crystals. For instance, the electro-optic effects commonly used in liquid-crystal displays and light valves arise through $\chi^{(3)}$ processes involving the low-frequency permittivity anisotropy. The enormous susceptibility, as manifested in the Kerr coefficient, comes about from cooperative interactions between molecules that cause their long axes to align in a parallel arrangement.

Although electric-field-induced second-harmonic generation has been observed in neat nematic crystals (Saha and Wong, 1979), the primary action of the electric field is to impart axial rather than polar symmetry. We therefore would not expect the nematic environment itself to contribute substantially to the second-order nonlinearity of molecularly doped nematic polymers but merely to provide a medium for enhanced alignment of the dopant molecule, as described below.

Saha and Wong (1979) substantiated this conclusion for the nematic liquid crystal *p*-(*n*-pentyl)-*p*′-cyanobiphenyl, for which the electric-field-induced value of $\chi^{(2)}$ is modestly increased over that in the isotropic phase but is considerably lower than observed for organic liquids composed of molecules with large β. On the other hand, it may be possible to design polar liquid-crystalline phases with large second-order nonlinear coefficients by virtue of their polar symmetry. Ferroelectric smectic phases prepared from a variety of chiral molecules, which are being considered for use in bistable fast-switching electro-optic light valves (Goodby and Leslie, 1984) could exhibit large second-order nonlinearities if the chromophores could be modified to increase β while retaining their ability to form the smectic c^* phase.

II. PHYSICAL DESCRIPTION OF GUEST–HOST INTERACTIONS

A. Systems To Be Considered

As stated earlier, the design of new polymeric materials for practical utilization of the large molecular hyperpolarizabilities that can be obtained in aromatic systems requires that the host–guest medium have polar symmetry. Electric-field-induced polar orientation with permanent retention is a feasible approach for imparting symmetry reduction. As will be shown, the nonlinear optical effects of interest are a strong function of the magnitude of the orientational polarization, and the competition with thermal randomization effects must be overcome.

In this section, host–guest alignment in three situations is described. The first case to be considered is a molecule in an isotropic host medium. A hypothetical example is *N*,*N*-dimethylaminonitrostilbene (DANS) doped at low loading ($\sim 1\%$) in a thermoplastic host such as poly(methyl methacrylate) (PMMA). To achieve polar orientation, the following sequence might be applied to a film in an appropriately electroded structure.

$$\begin{matrix}\text{DANS,}\\ \text{PMMA}\end{matrix} \xrightarrow{\text{heat}} T_g \xrightarrow[E\text{field}]{\text{apply}} T_g,\ E \xrightarrow[E\text{field}]{\text{cool}} \begin{matrix}\text{poled}\\ \text{sample}\end{matrix} \tag{8}$$

In this sequence a film is heated above its glass-transition temperature, a poling field is applied, and the sample is cooled to fix the induced orientation in place. The magnitude of the retained alignment will depend on several factors, including the dipole moment μ if the guest molecule, that portion of the internal field E_i that influences the direction of dipoles E_d, and the temperature T, assuming that the sample is allowed to reach equilibrium. The induced orientation can be understood quantitatively by computing the energy of the dipole in the field E_d according to the Maxwell–Boltzmann distribution law.

In the second case, a guest molecule is a low-concentration dopant in a nematic liquid-crystalline thermoplastic. A specific example is DANS in a liquid-crystalline polymer host (Fig. 1). Such polymers show nematic behavior above their T_g values (Finklemann *et al.*, 1979a), as well as homogeneous and homotropic alignment in response to externally applied electric fields (Meredith *et al.*, 1982; Finklemann *et al.*, 1979b). The ability to form a nematic mesophase in such polymers has been attributed to the decoupling of the motion of the backbone skeletal structure from that of the mesogens through the use of "spacer" units, i.e., the —$(CH_2)_6$— linkage. The sequence can be represented as

$$\begin{matrix}\text{DANS,}\\ \text{l.c. polymer}\end{matrix} \xrightarrow{\text{heat}} T_g \longrightarrow \begin{matrix}\text{nematic}\\ \text{range}\end{matrix} \xrightarrow[\text{field}]{\text{apply}} \begin{matrix}\text{cool}\\ E\text{ field}\end{matrix} \longrightarrow \begin{matrix}\text{poled}\\ \text{sample}\end{matrix} \quad (9)$$

The sample is heated to the nematic region, where an electric field is applied producing axial alignment of the nematic medium. The dopant molecule is

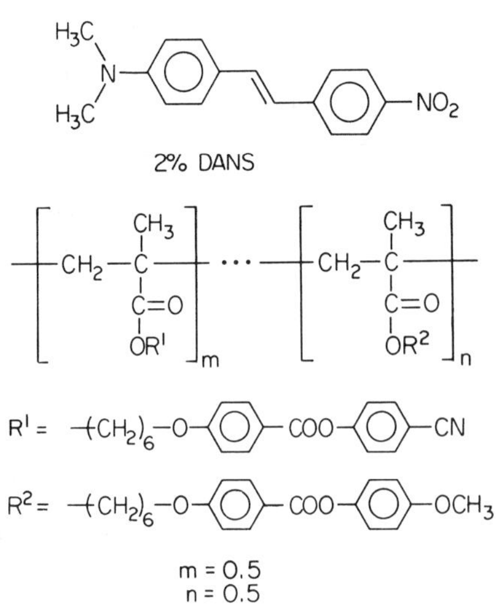

Fig. 1. Molecular structures of components of a molecularly doped polymeric liquid-crystalline thermoplastic for which second-order nonlinear optical properties have been investigated.

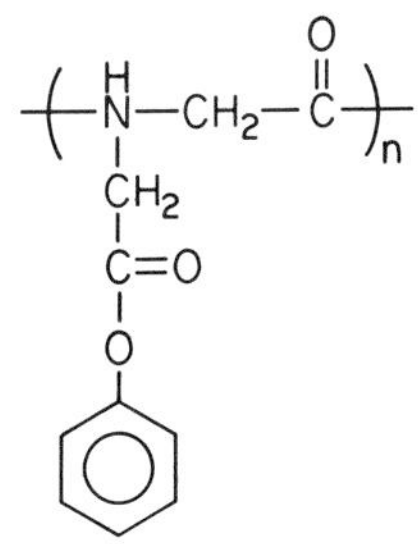

Fig. 2. Molecular structure of poly-γ-benzyl-L-glutamate, which forms an α helix in ethylene dichloride.

now in an anisotropic environment (defined by an appropriate potential), and its degree of alignment will therefore be determined by the superposition of alignment field E_d on the anisotropic potential of the medium.

A third situation is that of a "molecule" in a polar polymer chain in an otherwise isotropic medium. An example is poly-γ-benzyl-L-glutamate (PBG) (Fig. 2) in an α-helix-forming solvent. This polymer forms a rigid extended α helix in appropriate solvents and has a large dipole moment, owing to the asymmetric placement of the repeat units in the rigid polymer chain. Although the benzylglutamate repeat unit does not possess a particularly large molecular hyperpolarizability, this system (Levine and Bethea, 1976) illustrates a potentially useful concept. A hypothetical example employing this concept might be

$$\begin{matrix}\text{viscous} \\ \text{solution}\end{matrix} \xrightarrow{\text{heat}} \begin{matrix}\text{nonviscous} \\ \text{state}\end{matrix} \xrightarrow{E\,\text{field}} \begin{matrix}\text{cool} \\ E\text{ field}\end{matrix} \longrightarrow \begin{matrix}\text{poled} \\ \text{sample}\end{matrix} \qquad (10)$$

In this situation the cumulative chain dipole is sufficiently large that its energy in an alignment field E_d could be substantially larger than kT, thus permitting a much higher degree of alignment than in the molecularly doped case. Although ethylene chloride is obviously not the proper choice for stabilizing the induced alignment and PBG is not the optimum polymer, large second-order nonlinearities might be obtained by this approach.

B. Electric-Field-Induced Alignment

In this section electric-field alignment of a guest dopant molecule in a low-symmetry nematic environment is contrasted with a similar regimen in an isotropic phase. An important assumption is that an elongated rodlike molecule such as DANS will associate with the mesogenic units of the nematic environment so that the orientational potential energy of the guest molecule will be similar to that of the mesogenic units. This implies that the guest molecules do not significantly alter the stability of the nematic phase.

To make the distinction between the isotropic and nematic phases more quantitative, an order parameter, which is nonzero in the nematic environment and zero in the isotropic environment is used. The mesogenic groups and dopant molecules are assumed to behave as rigid rods, and the axis of a "molecular" rod is labeled by a unit vector **A**. The rods are assumed to have cylindrical symmetry about **A**. The direction of the nematic axis, which is the average direction of the molecular axes, will be taken as the "director" of the nematic phase **N** and Z axis of the laboratory frame. **A** can be defined by its polar angles θ and ϕ, where

$$a_x = \sin\theta\cos\phi, \qquad a_y = \sin\theta\sin\phi, \qquad a_z = \cos\theta \tag{11}$$

The state of alignment of the "rods" can be described by a distribution function $f(\theta, \phi)\,d\Omega$, which gives the probability of finding the rods in a small solid angle $\sin\theta\,d\theta\,d\phi$ around the direction (θ, ϕ). In conventional nematics it is assumed (perhaps incorrectly for thin films) that $f(\theta, \phi)$ is independent of ϕ (i.e., cylindrical symmetry about the director) so that $f(\theta)$ is all that needs to be specified (DeGennes, 1974).

For simplification we will therefore assume that the probability of finding rods (either dopant molecules or mesogens) at an angle θ with respect to the Z or the director axis is $f(\theta)\sin\theta\,d\theta$. Before illustrating the behavior of the distribution function for the nematic case, we will consider it for the isotropic medium shown in Fig. 3. In this case all directions are equally probable, and

$$f(\theta)\sin\theta\,d\theta = \frac{2\pi r\sin\theta r\,d\theta}{4\pi r^2} = \tfrac{1}{2}\sin\theta\,d\theta \tag{12}$$

The plot of $f(\theta)$ versus θ for the isotropic case (Fig. 4) illustrates the point. In a nematic medium there is preferred alignment about the directions **N** and $-$**N**, and $f(\theta)$ has the general appearance illustrated in Fig. 4. In this case $f(\theta) = f(\theta - \pi)$ (DeGennes, 1974).

Rather than attempting to specify $f(\theta)$, we wish to characterize the alignment through a related numerical quantity, the order parameter. The simplest quantity that might be specified is $\langle\cos\theta\rangle$, the average orientation of a dipole with respect to **N**,

$$\langle\cos\theta\rangle = \langle\mathbf{A}\cdot\mathbf{N}\rangle = \int_0^\pi f(\theta)\cos\theta\sin\theta\,d\theta \bigg/ \int_0^\pi f(\theta)\sin\theta\,d\theta \tag{13}$$

but this quantity vanishes for the symmetry reason stated in the previous paragraph. The first higher-order multipole giving a nontrivial answer is the

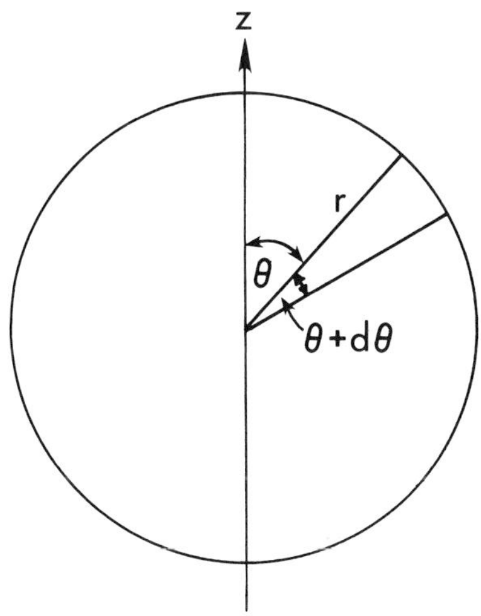

Fig. 3. The circle represents the probability of finding a molecular dipole oriented in an angular segment $d\theta$ with respect to an arbitrary axis z.

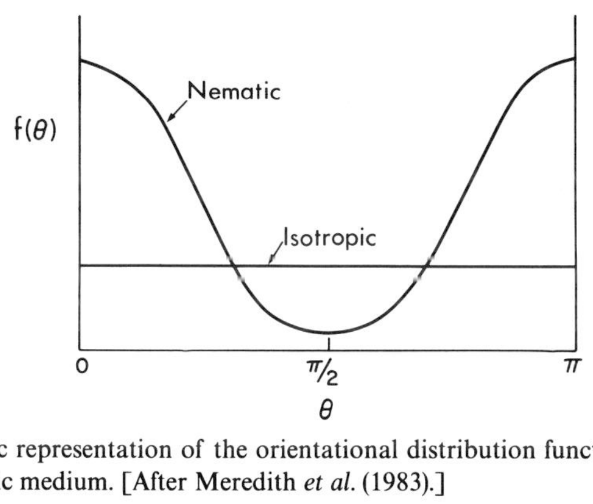

Fig. 4. Schematic representation of the orientational distribution function of a rod-shaped molecule in a nematic medium. [After Meredith *et al.* (1983).]

quadrupole defined as

$$S = \tfrac{1}{2}\langle 3\cos^2\theta - 1\rangle = \int_0^\pi f(\theta)[\tfrac{1}{2}(3\cos^2\theta - 1)]\sin\theta\,d\theta \Big/ \int_0^\pi f(\theta)\sin\theta\,d\theta \tag{14}$$

If $f(\theta)$ is strongly peaked around $\theta = 0, \pi$ (parallel alignment), then $\cos\theta \to \pm 1$ and $S \to 1$. In the physically unlikely case where $f(\theta)$ is peaked around $\pi/2$, $S = -\tfrac{1}{2}$. Finally, if the orientation is entirely random (the isotropic case), $\langle\cos^2\theta\rangle = \tfrac{1}{3}$ and $S = 0$.

In the preceding discussion it is assumed that the dopant molecule shows no preferential *polar* alignment in the $\theta = 0$ or $\theta = \pi$ directions in the nematic medium or in the isotropic environment (see Fig. 5). When an external electric field is applied to such media, polarization occurs, owing to deformation or translation of electrons relative to positive charges and atoms or groups of atoms relative to one another. Reorientational effects can also occur, with the electric field directing permanent dipoles. Rotational effects are strongly counteracted by thermal movements of molecules, whereas translational effects are only slightly temperature-dependent, being dependent on intramolecular phenomena. In a highly anisotropic medium like a nematic liquid crystal, the dielectric anisotropy ε_a ($\varepsilon_{||}$ and $\varepsilon_{\perp}$ are given as

$$\varepsilon_a = \varepsilon_{||} - \varepsilon_{\perp} \tag{15}$$

and are the components of the static dielectric constant parallel and perpendicular to the nematic director **N**) leads, in principle, to a method of macroscopic alignment by electric fields. The electric field contribution $U(\theta)_n$ to the thermodynamic potential is

$$U(\theta)_n = -\frac{\varepsilon_{\perp}}{8\pi}E^2 - \frac{\varepsilon_a}{8\pi}(\mathbf{N}\cdot\mathbf{E})^2 \tag{16}$$

The first term is independent of orientation, but the second favors parallel alignment if $\varepsilon_a > 0$ and perpendicular alignment if $\varepsilon_a < 0$. In the copolymer illustrated in Fig. 2, the comonomer with the CN group contributes to the positive dielectric anisotropy to such an extent that the electric-field-induced alignment of the nematic medium is observed (Meredith *et al.*, 1982, 1983). In this case the director **N** would be parallel to the external field, and the distribution functions in Fig. 3 would be a measure of the distribution of molecular directions with respect to the applied field. In the isotropic medium, ε is small compared with kT, and field-induced alignment of polarizable

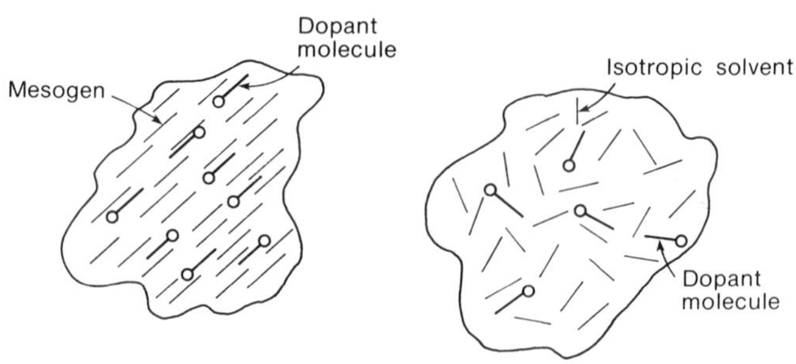

Fig. 5. Schematic representation of a thermally randomized nematic medium with dopant molecules versus an isotropic medium.

species makes no significant contribution to the potential energy of the medium.

The tendency for molecules to align with their individual dipole moments pointed in the field direction can be determined from the potential energy of an individual molecule in the applied field. In the isotropic case, the field will be $\mathbf{E}_d$ with local field corrections. In the nematic case, $\mathbf{E}_d$ will be superimposed on the nematic potential, and a difference between the total energy of the dipole in response to $\mathbf{E}_d$ might be anticipated between the two cases. The situation is illustrated qualitatively in Fig. 6, where the distribution function $f(\theta)$ [which is related to $U(\theta)$ by $f(\theta) = e^{-U(\theta)/kT}$] for an individual dipolar species such as DANS is perturbed by $\mathbf{E}_d$. In this case the potential energy of the dipole in the nematic environment becomes

$$U(\theta) = U(\theta)_n - \boldsymbol{\mu} \cdot \mathbf{E}_d \tag{17}$$

According to Eq. (13), the extent to which dipoles can be directed is given by

$$\langle \cos\theta \rangle = \int_0^\pi \cos\theta e^{-U(\theta)/kT} \sin\theta \, d\theta \bigg/ \int_0^\pi e^{-U(\theta)/kT} \sin\theta \, d\theta \tag{18}$$

Since $U(\theta)$ is different for nematic and isotropic media, it should be apparent that field-induced alignment could be enhanced by the appropriate choice of an anisotropic environment. Substituting for $U(\theta)$ gives

$$\langle \cos\theta \rangle = \frac{\int_0^\pi \cos\theta e^{\mu E_d \cos\theta/kT} f_n(\theta) \sin\theta \, d\theta}{\int_0^\pi e^{\mu E_d \cos\theta/kT} f_n(\theta) \sin\theta \, d\theta} \tag{19}$$

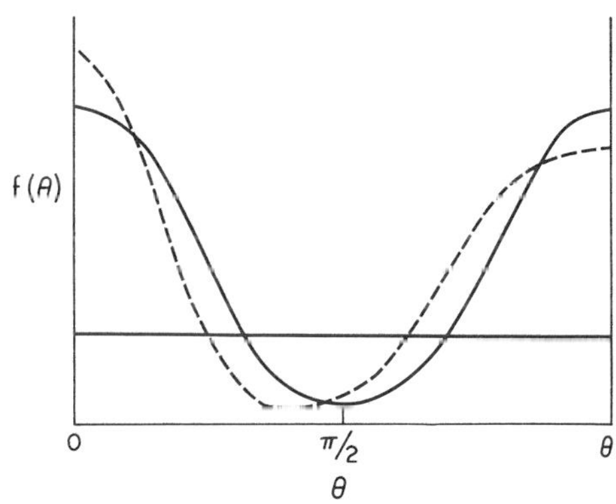

Fig. 6. Schematic representation of orientational distribution function of a rod-shaped molecule in the presence (——) and absence (– – – –) of an external electric field. [After Meredith *et al.* (1983).]

where $f_n(\theta) = e^{-U(\theta)_n/kT}$. Expansion of the exponential functions gives

$$\langle \cos\theta \rangle = \frac{\int_0^\pi [\cos^2\theta(\mu E_d/kT) + \cos^4\theta(\mu E_d/kT)^2/6 + \cdots] f_n(\theta)\, d\theta}{\int_0^\pi [1 + \cos^2\theta(\mu E_d/kT)/2 + \cos^4\theta(\mu E_d/kT)^4/24 + \cdots] f_n(\theta)\, d\theta} \tag{20}$$

This expression is identical to that for the isotropic case except for the factor $f_n(\theta)$, and the alignment for the nematic case can therefore be enhanced relative to the isotropic case. For illustration we consider the limiting case $\theta = 0, \pi$ (Ising model) corresponding to sharply peaked distributions and therefore a high-order parameter. Retaining the leading terms in Eq. (20), which is justified for $\mu E_d/kT \ll 1$, we obtain

$$\langle \cos\theta \rangle_{\text{Ising}} = \mu E/kT \tag{21a}$$

$$\langle \cos\theta \rangle_{\text{isotropic}} = \mu E/3kT \tag{21b}$$

For case three described earlier, where we have a rigid extended polymer chain with a dipole moment consisting of contributions from the individual repeat units, we may have the situation where $\mu E_d/kT \to 1$, and the neglect of higher-order terms in Eq. (20) is no longer valid. Assuming a dilute solution so there is no interaction between the rodlike chains, Eq. (18) can be written in the form of the expansion (Böttcher, 1973)

$$\begin{aligned} \langle \cos\theta \rangle &= \coth a - 1/a = L(a) \\ &= 1/3a - 1/45a^3 + 2/945a^5 - 2/9450a^7 + \cdots \end{aligned} \tag{22}$$

where $a = \mu E_d/kT$. $L(a)$ is the Langevin function, originally derived by Langevin and applied to magnetic dipoles in a directing field (Langevin, 1905a,b). This expression predicts saturation of the field-induced polar alignment near unity for sufficiently large values of the parameter a, corresponding to either large permanent dipoles or high electric fields. Under high-field conditions and high concentrations we might expect contributions from induced dipoles and, if these are anisotropic, different elements of the linear electric field polarizability tensor. Kielich (1969) wrote a more generalized Langevin function that takes these factors into account,

$$L_n(a, \pm b) = \frac{\int_0^\pi (\cos^n\theta) e^{(a\cos\theta \pm b\cos^2\theta)} \sin\theta\, d\theta}{\int_0^\pi e^{(a\cos\theta \pm b\cos^2\theta)} \sin\theta\, d\theta} \tag{23}$$

where a was defined above and b for axial systems is $E_d^2(\varepsilon_\parallel - \varepsilon_\perp)/2kT$. Assuming very low anisotropy, $L_1\ (a, 0)$ is given by Eq. (22). The next appropriate higher-order odd moment gives

$$L_3(a, 0) = (a, 0) = (1 + 6/a^2)L_1(a, 0) - 2/a + \cdots \tag{24}$$

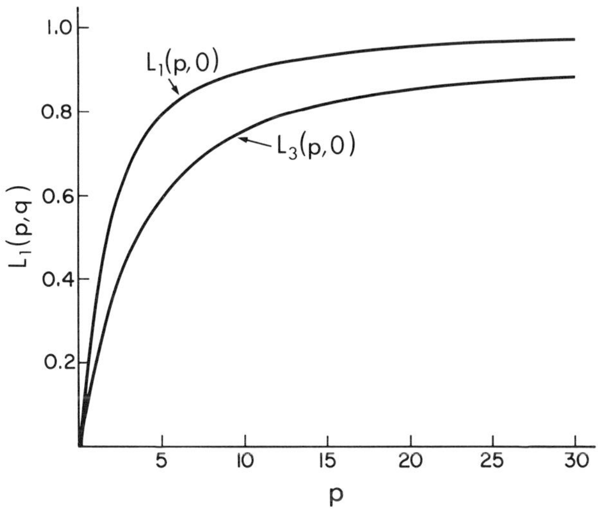

Fig. 7. First and third Legendre polynomials for an isotropic medium ($b = 0$) versus orientation parameter (a).

Figure 7 shows a plot of these two functions versus a. The effect of positive and negative dielectric anisotropy ε_a on polar alignment can be seen from the following expressions, which are valid at moderate electric fields:

$$L_1(a, \pm b) = a/3 - a^3/45 \pm 4ab/45 + \cdots \quad (25)$$

$$L_3(a, \pm b) = a/5 - a^3/105 \pm 8ab/105 + \cdots \quad (26)$$

These equations once again predict an enhancement in the polar alignment for a system of polar rodlike chains when they are allowed to interact as would be the case in the nondilute regime. Figure 8 illustrates the orientational distribution function for a system near electric-field-induced saturation.

Levine and Bethea (1976) have shown that solutions of polybenzylglutamate in moderately high concentration show electric-field-induced second-harmonic generation with an electric field dependence that scales with $[L_3(a)]^2$ (see Fig. 9). From their analysis of the data, they extracted a chain dipole moment $\mu = 6000$ D, which is somewhat less than the 8000 D expected obtained for material of similar (500,000) molecular weight in more dilute solution. It was suggested that helix–helix interactions were the possible cause of this discrepancy, but this seems inconsistent with the predictions of Eqs. (25) and (26).

Table I summarizes expected electric-field-induced orientation behavior for the various systems discussed.

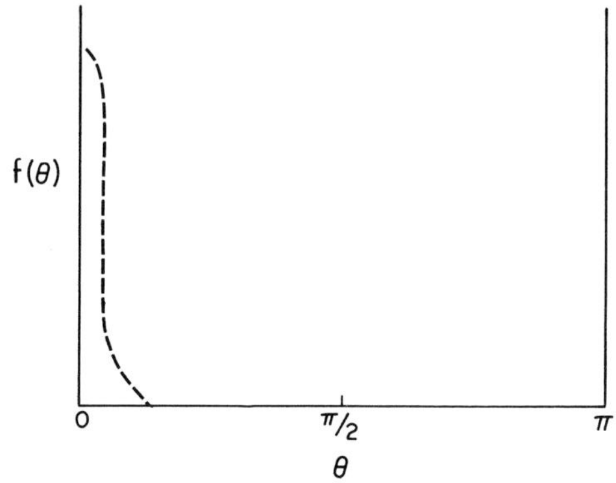

Fig. 8. Schematic representation of the orientational distribution function for a large orientational parameter ($a \to 1$).

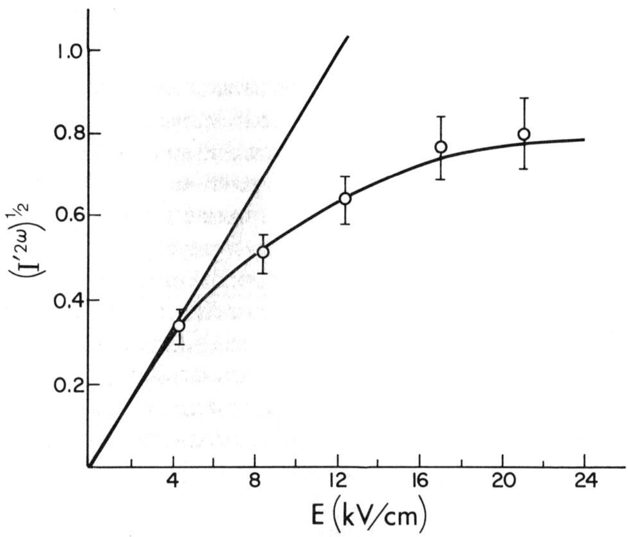

Fig. 9. Plot of square root of the measured harmonic intensity, $(I'^{2\omega})^{1/2}$, normalized to $a/5$ at low fields, versus electric field E. The solid straight line is the $a/5$, and the curved line through the data points is $L_3(a, 0)$. [After Levine and Bethea (1976).]

C. Relation of Field-Induced Alignment of $\chi^{(2)}$

It was mentioned earlier that alignment by electric-field-induced reorientation imparts polar or $C_{\infty v}$ symmetry. As with linear optical media, the second-order nonlinear susceptibility must display the symmetry properties of

TABLE I

Polar Orientation Parameter $\langle \cos\theta \rangle$ for Various Potentials $U(\theta)$

$U(\theta)$	$\langle \cos\theta \rangle$	
Isotropic low field	$1/3a$	
Ising low field	a	
Anisotropic low field	$a\langle \cos^2\theta \rangle$ or $a \cdot S$	
Isotropic, high field, large dipole	$(1/3a) - (1/45a^3) + (2/945a^5) - (2/9450a^7) + \cdots$	
Linear response		
Nonlinear response	$(1 + 1/6a^3)L_1(a, 0) - 1/2a + \cdots$	
Anisotropic, low field, large dipole	$1/3a - 1/45a^3 \pm 4/45ab + \cdots$	(for $\pm\varepsilon_a$)
Linear response		
Nonlinear response	$1/5a - 1/105a^3 \pm 8/105ab + \cdots$	(for $\pm\varepsilon_a$)

the medium. The second-order nonlinear polarization according to Eq. (2) will be a function of the tensor elements $\chi^{(2)}_{IJK}$; for $C_{\infty v}$ symmetry, the nonzero elements of this term are given in Table II. The tensor element is sometimes expressed as $\chi^{(2)}_{IJK}(\omega_3; \omega_2, \omega_1)$, where ω_3 is the resultant field component and ω_2 and ω_1 are the fields driving the interaction. For second-harmonic generation this is usually written $\chi^{(2)}_{IJK}(-2\omega; \omega, \omega)$, and for the linear electro-optic effect, $\chi^{(2)}_{IJK}(-\omega; 0, \omega)$. In addition to "crystal" symmetry restrictions, $\chi^{(2)}_{IJK}$ satisfies an additional symmetry relation in nondispersive media, which was based on a conjecture by Kleinman (1962) that permutation of the frequencies of the three interacting electric-field components along the direction I, J, K is symmetric under the permutation of the indices. With regard to Table II, this implies $\chi^{(2)}_{ZXX} = \chi^{(2)}_{XXZ} = \chi^{(2)}_{ZXZ} = \chi^{(2)}_{p\{ZXX\}}$ (p indicating all permutations of the indices).

With these simplifications, it is intuitively obvious that a cylindrical polar medium should have two independent tensor elements $\chi^{(2)}_{ZZZ}$ and $\chi^{(2)}_{p\{ZXX\}}$. For nonresonant, purely electronic processes, Kleinman symmetry is expected to be valid, so that $\chi^{(2)}$ for second-harmonic generation should be close to that for wave-mixing experiments and for the linear electro-optic effect. In nonrigid

TABLE II

Nonzero Values of the Second-Order Susceptibility Tensor (I, J, K) for which $\chi^{(2)}_{IJK}$ is Nonzero for Symmetry Classes C_{4v} (4 mm), C_{6v} (6mm), and $C_{\infty v}$ (∞mm)

$$\begin{pmatrix} 0 & 0 & 0 & 0 & XXZ & 0 \\ 0 & 0 & 0 & XXZ & 0 & 0 \\ ZXX & ZXX & ZZZ & 0 & 0 & 0 \end{pmatrix}$$

polymeric media, additional lower-frequency dielectric relaxations would be expected, with broad spectral densities of relaxation times. The net effect of these relaxation processes could be an additional contribution to the linear electro-optic coefficient beyond that associated with purely electronic processes. The magnitude of these contributions must be carefully considered for any potential application.

D. Molecular Source of $\chi^{(2)}$ in Poled Samples

The introduction stated that knowledge of orientational factors was required to relate the electronic components of the tensor β to those of $\chi^{(2)}$ [Eq. (7)]. If we assume that $\beta = \beta_{zzz}$, i.e., one predominant component of the molecular tensor, and that the z axis of the molecule is parallel to the molecular dipole moment, which interacts with the orienting field (Z direction), relatively straightforward expressions for relating β to the $\chi^{(2)}$ components can be written,

$$\chi^{(2)}_{ZZZ} = N\beta F\langle(\mathbf{k}\cdot\mathbf{K})(\mathbf{k}\cdot\mathbf{K})(\mathbf{k}\cdot\mathbf{K})\rangle \tag{27}$$

$$\chi^{(2)}_{p\{ZXX\}} = N\beta F\langle(\mathbf{k}\cdot\mathbf{K})(\mathbf{k}\cdot\mathbf{I})(\mathbf{k}\cdot\mathbf{I})\rangle \tag{28}$$

where F is the product of all local field factors, $\mathbf{k}$ is a unit vector along the molecular z direction, and $\mathbf{I}$ and $\mathbf{K}$ are vectors perpendicular and parallel to the poling (Z) direction. The nonlinear coefficients induced along the poling axis and describing the polarization induced by fields parallel ($\mathbf{K}$) or perpendicular ($\mathbf{I}$) to the poling directions are given by the orientational averages in Eqs. (27) and (28). In analogy with our previous definitions in Eq. (13),

$$\begin{aligned}\langle(\mathbf{k}\cdot\mathbf{K})(\mathbf{k}\cdot\mathbf{K})(\mathbf{k}\cdot\mathbf{K})\rangle &= \langle\cos^3\theta\rangle \\ &= \int_0^\pi f(\theta)\cos^3\theta\sin\theta\,d\theta \\ &= \int_0^\pi f(\theta)\sin\theta\,d\theta\end{aligned} \tag{29}$$

and

$$\langle(\mathbf{k}\cdot\mathbf{K})(\mathbf{k}\cdot\mathbf{I})(\mathbf{k}\cdot\mathbf{I})\rangle = \langle(\cos\theta)(\sin^2\theta)(\cos^2\phi)\rangle \tag{30}$$

Continuing with our earlier assumption of the independence of a nematic system from the polar coordinate ϕ,

$$\begin{aligned}\langle(\cos\theta)(\sin^2\theta)(\cos^2\phi)\rangle &= \frac{\int_0^\pi f(\theta)\cos\theta\sin^3\theta\,d\theta\int_0^\pi(\cos^2\phi)\,d\phi}{\int_0^\pi f(\theta)\sin\theta\,d\theta} \\ &= \frac{\frac{1}{2}\int f(\theta)(\cos\theta - \cos^3\theta)\,d\theta}{\int_0^\pi f(\theta)\sin\theta\,d\theta} \\ &= (\langle\cos\theta\rangle - \langle\cos^3\theta\rangle)/2\end{aligned} \tag{31}$$

TABLE III

Predicted Values of Tensor Components $\chi^{(2)}_{IJK}$ for Symmetry Class $C_{\infty v}$ with Kleinmann Symmetry for Different Polymeric Host–Guest Situations

Component	Poled isotropic	Poled Ising	Saturation alignment
$\chi^{(2)}_{ZZZ}$	$\beta_{zzz}\mu E/5kT$	$\beta_{zzz}\mu E/kT$	$L_n(a, \pm b)$
$\chi^{(2)}_{p\{ZXX\}}$	$\beta_{zzz}\mu E/15kT$	0	0

Following the general approach described earlier for the isotropic and Ising models, we obtain expressions for $\chi^{(2)}_{ZXX}$ and $\chi^{(2)}_{p\{ZXX\}}$ in the low-field and high-field situations (see Table III). For real samples, where the order parameter S is less than unity, we would expect values of $\chi^{(2)}$ between the isotropic and Ising limits.

III. Second-Order Nonlinear Effects in Polymer Films

To evaluate the potential utility of electric-field-induced alignment of nonlinear molecular species in polymer films, we will consider the phase-matching requirements for bulk second-harmonic generation, where phase matching occurs through the birefringence of the medium, and waveguide second-harmonic generation, where phase matching can be achieved through modal dispersion. Also, since many potentially important applications of these materials could involve the linear electro-optic effect, we will examine its characteristics in these media.

A. Bulk Phased-Matched Harmonic Generation

To illustrate the relationship between the symmetry properties and induced nonlinearity in the film and the harmonic generation occurring when incident fields propagate through the film, we consider the case of a plane wave in the low-conversion limit for which solutions for the conversion efficiency have been derived (Byer, 1977):

$$\frac{I(2\omega)}{I(\omega)} = \Gamma^2 l^2 \operatorname{sinc}^2\left(\frac{\Delta k l}{2}\right) \tag{32}$$

where

$$\Gamma^2 l^2 = \frac{2\omega^2 |d_{\text{eff}}|^2 l^2 I(\omega)}{\eta^3 c^3}$$

and by convention

$$2d_{IJK}(-2\omega;\omega,\omega) = \chi^{(2)}_{IJK}(-2\omega;\omega,\omega) \tag{33}$$

In Eq. (32), Δk is the phase mismatch between wave vectors at the harmonic and fundamental frequencies

$$\Delta k = |k_{2\omega} - k_{\omega}| \tag{34}$$

where $|k| = 2\pi\eta/\lambda$ and l is the length over which the interaction occurs. The example shows the importance of phase synchronization of the fundamental and harmonic polarization propagating through the medium, since the factor $\mathrm{sinc}^2(\Delta kl/2)$ is unity at $\Delta kl = 0$ and otherwise periodic with a coherence length l_c defined by the wave-vector mismatch ($l_c = \pi/\Delta k$). The effective nonlinear coefficient d_{eff} is determined by the propagation characteristics in the medium for various input waves, as will be discussed below. From Eq. (32) it is obvious that the interaction length and effective nonlinear coefficient should be maximized.

For wave propagation in birefringent media, an incident wave of arbitrary polarization is resolved into two polarization components in the medium. The polarizations are orthogonal and resolved into two components: η_o, which is independent of the propagation direction, and η_e, which depends on the angle of inclination (δ) of the incident beam with respect to the optic axis. This is illustrated in Fig. 10, where the refractive-index ellipsoid for a positive unaxial medium is shown schematically along with incident light of wave-vector S at the fundamental frequency ω. The allowed orthogonal polarization components are parallel to ($\mathbf{I}\cos\delta + \mathbf{K}\sin\delta$) and $\mathbf{J}$ ($\mathbf{I}$, $\mathbf{J}$, $\mathbf{K}$ were defined previously). The usual positive dispersion in refractive index is shown schematically in

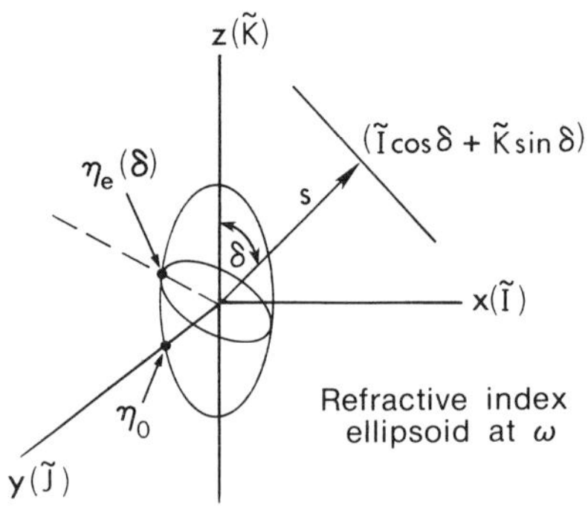

Fig. 10. Schematic representation of the refractive index of a uniaxial medium: **s** indicates the direction of an incident beam of light, and the semimajor and semiminor axes of the ellipse formed by the plane formed by the plane perpendicular to **s** and the ellipsoid define the allowed polarization directions for the propagating polarization. The refractive indices corresponding to those directions are the η_o (ordinary) and $\eta_e(\delta)$ (extraordinary). [After Williams (1984).]

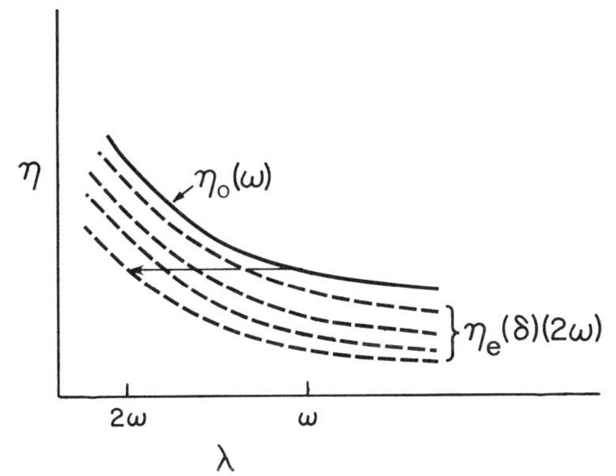

Fig. 11. Illustration of the use of birefringent phase matching of the extraordinary component of the refractive index at the harmonic wave $\eta_e(\delta)(2\omega)$ to the ordinary component $\eta_o(\omega)$ of the fundamental wave. [After Williams (1984).]

Fig. 11 for an ordinary wave at ω and an extraordinary wave at 2ω. If the dispersion is sufficient, an angle δ may be found where $\eta_o(\omega) = \eta_e(\delta)(2\omega)$ and phase matching is achieved. Owing to the anisotropic nature of the medium, the electric field vector associated with the incident beam and dielectric displacement are not necessarily parallel, and dispersion will cause this angle to vary at the two wavelengths. As a result, power flow, determined by the Pointing vector $\mathbf{E} \times \mathbf{H}$, is responsible for divergence of the fundamental and incident waves in the medium and limits the interaction length in Eq. (32) over which conversion from fundamental to harmonic can occur.

From this discussion it should be apparent that harmonic generation will involve interactions of o and e waves and that effective nonlinear coefficients $\chi^{(2)}_{p\{oee\}}$ and $\chi^{(2)}_{p\{eoo\}}$ need to be defined in terms of $\chi^{(2)}_{ZZZ}$ and $\chi^{(2)}_{p\{ZXX\}}$. This relatively straightforward procedure is described in various texts (Byer, 1977; Zernike and Midwinter, 1973). It basically involves multiplication of the $\chi^{(2)}_{IJK}$ tensor elements by the projections of the o and e waves on the IJK axes. The results of the two possibilities are

$$\begin{aligned}\chi^{(2)}_{p\{oee\}} &= P\{o \cdot \chi^{(2)} \cdots ee\} \\ &= P[J \cdot \chi^{(2)} \cdots (\mathbf{I}\cos\delta + \mathbf{k}\sin\delta)(\mathbf{I}\cos\delta + \mathbf{K}\sin\delta)] \\ &= 0 \end{aligned} \tag{35}$$

and

$$\begin{aligned}\chi^{(2)}_{p\{eoo\}} &= P\{e \cdot \chi^{(2)} \cdot oo\} \\ &= P[(\mathbf{I}\cos\delta + \mathbf{K}\sin\delta) \cdot \chi^{(2)} \cdots \mathbf{JJ}] \\ &= \chi^{(2)}_{p\{ZXX\}} \sin\delta \end{aligned} \tag{36}$$

As a consequence, only processes involving two o waves and one e wave can be expected to lead to useful bulk phase-matched harmonic generation in these films. An immediate consequence of this analysis is that the largest tensor element $\chi^{(2)}_{ZZZ}$ (see Table III) is of no use for bulk phase-matched SHG in media with this symmetry. In poled isotropic media, the smaller $\chi^{(2)}_{ZXX}$ tensor component is somewhat useful, but since the conversion efficiency is proportional to $[\chi^{(2)}]^2$, the maximum is reduced by an order of magnitude over the level that could be obtained if $\chi^{(2)}_{ZZZ}$ was phase-matchable. In highly anisotropic media, $\chi^{(2)}_{p\{ZXX\}}$ approaches zero, and the films are expected to be of little utility. One further consequence of Eq. (35) is that SHG will be a maximum under the fortuitous condition that phase matching occurred at $\delta = 90°$. (This is termed noncritical phase matching.) If the phase-matching angle is near 90°, temperature adjustment or doping the host medium with another dye may be enough to adjust it to 90°. A further advantage of 90° phase matching is that the walk-off angle discussed earlier is zero.

B. Waveguide Second-Harmonic Generation

Although we have not yet experimentally explored the possibility of waveguide second-harmonic generation in films of these materials, the approach can be explored conceptually and leads to the conclusion that if waveguides of appropriate quality can be fabricated, highly efficient SHG can be expected.

To explore this concept further, consider the slab dielectric waveguide shown in Fig. 12. In a waveguide of this type, the gradient $\partial/\partial Y = 0$ with respect to all electric and magnetic field components. The basic features of such a waveguide can be deduced from the Maxwell equations with appropriate boundary conditions applied, which leads to a wave equation having discrete solutions for allowed electric and magnetic field components. The properties of such waveguides have been analyzed

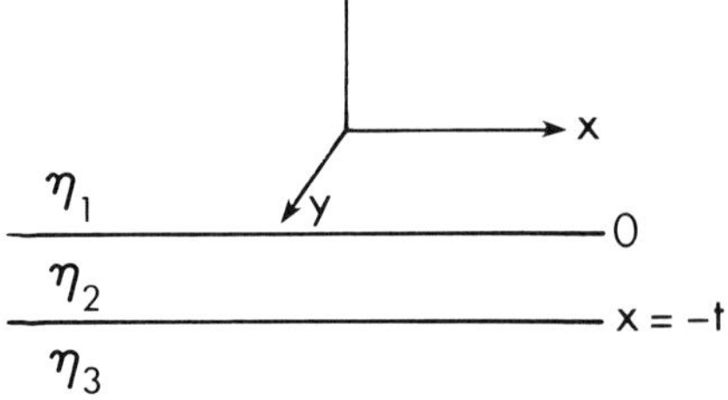

Fig. 12. Illustration of a confining waveguide medium of thickness t with $\eta_2 < \eta_1$, η_3, and the coordinate system.

extensively (Taylor and Yariv, 1974; Hunsberger, 1982), and we will illustrate some general features pertinent to our problem.

First, if $\eta_2 > \eta_3, \eta_1$ propagating light can be confined to the middle layer, its general characteristics being dependent on the thickness of the layer, the exact values of the refractive indices, the frequency of light, and a propagation constant. With the assumption that $\partial/\partial_Y = 0$, this waveguide can support a finite number of confined TE (transverse electric) modes with E_Y, H_X, and H_Z components and TM (transverse magnetic) modes with H_Y, E_X, and E_Z components. Referring to Fig. 12, it would be relatively straightforward to fabricate a slab waveguide from a polymer with electric-field-induced alignment in the Z direction of the waveguide. As indicated above, the TM modes have E_Z components suggesting the possibility of nonlinear interaction through $\chi^{(2)}_{ZZZ}$. The field components, for TM modes are given explicitly by Yariv (1975b)

$$H_y(X, Z, t) = H_y(Z)e^{i(\omega t - \beta X)} \tag{37}$$

$$E_z(X, Z, t) = \frac{\beta}{\omega\varepsilon} H_Y(Z)e^{i(\omega t - \beta X)} \tag{38}$$

$$E_x(X, Z, t) = \left(\frac{-i}{\omega\varepsilon}\right)\frac{\partial H_Y}{\partial Z} \tag{39}$$

where ε is the dielectric constant of the medium (note that the Z and X directions have been interchanged from the usual convention for ease of discussion, since we have referred to the field-induced nonlinearity as the Z direction up to this point). The solution of these equations, subject to the appropriate boundary conditions, leads to guided modes with the general characteristics shown schematically in Fig. 13. Note the thickness (t) dependence of the effective refractive index in the various modes as well as the modal dispersion between ω and 2ω. In the schematic, phase matching between the TM_0^ω and $TM_2^{2\omega}$ modes would occur at thickness t_{pm}, and because of the modal dispersion, the phase-matching thickness should be wavelength dependent. Figure 14 shows a plot of the dependence of phase-matching thickness on the fundamental wavelength of the $TM_0^\omega \rightarrow TM_2^{2\omega}$ process for a polycrystalline evaporated film of *p*-chlorophenylurea with air and glass as the upper and lower layers (Jain and Hewig, 1981).

The theory of phase-matched SHG in planar waveguides has been developed, and the second-harmonic power is given by Conwell (1973),

$$P_{2\omega} = \frac{128\pi^3\omega^2 d_{ZZZ}^2 P_\omega^2 L^2}{c^3 \eta_{\mathrm{eff},\omega}^2 \eta_{\mathrm{eff},2\omega}^2} \cdot \frac{I^2}{W_0 t_\omega^2 t_{2\omega}} \left[\frac{\sin(\Delta\beta L)}{\Delta\beta L}\right]^2 \tag{40}$$

where P_ω is the incident power, W_0 is the width of the beam in the film plane,

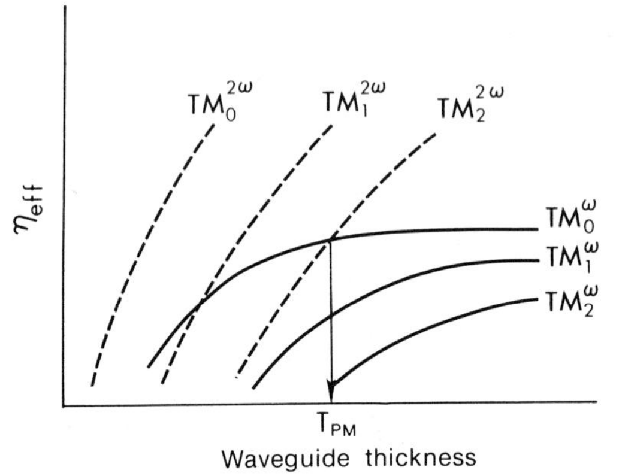

Fig. 13. Schematic representation of dispersion of waveguide modes at ω and 2ω, showing the variation of the effective index with waveguide thickness. The crossing of the $TM_0^{2\omega}$ and $TM_2^{2\omega}$ modes determines the phase-matching thickness T_{PM}. [After Jain and Hewig (1981).]

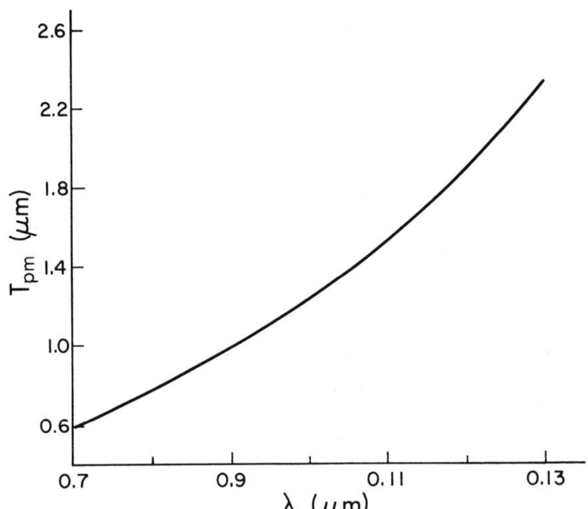

Fig. 14. Schematic representation of phase-matching thickness T_{PM} on the fundamental wavelength for the example shown in Fig. 13. [After Jain and Hewig (1981).]

$\eta_{eff,\omega}$ and $\eta_{eff,2\omega}$ are the waveguide indices at ω and 2ω, respectively, d_{ZZZ} is the appropriate nonlinear coefficient as defined previously, and t_ω and $t_{2\omega}$ are the effective film thickness at the two wavelengths defined by

$$t_{\omega,2\omega} = 2\eta_{\omega,2\omega}^2 \int_{-\infty}^{\infty} [E_{\omega,2\omega}^2(Z)/\eta_{\omega,2\omega}^2]\, dZ \tag{41}$$

and I is the transverse spatial coupling given by

$$I = \sqrt{8} \int E_{\omega}^{2}(Z) E_{2\omega}(Z)\, dZ \tag{42}$$

with $E_{\omega}(Z)$ and $E_{2\omega}(Z)$ being the normalized field amplitudes. The phase mismatch is given by

$$\Delta\beta = 2\beta_{\omega} - \beta_{2\omega} = 4\pi/\lambda(\eta_{\mathrm{eff},\omega} - \eta_{\mathrm{eff},2\omega}) \tag{43}$$

Equation (42) points out the importance of overlap of the electric field components of the harmonic with those of the fundamental field in determining the efficiency of the transfer process.

Based on the considerations developed in this section, the slab waveguide with field-induced alignment in the Z direction would appear to be a favorable structure for SHG, since the full dc field-induced nonlinearity can be used and phase matching is in principle readily achievable through thickness control. The severe practical problems associated with constructing such a waveguide with high uniformity and low scattering losses and with precise thickness control are the subject of ongoing research.

C. Waveguide Electro-Optic Modulation/Switching

The application of a controlling electric field to a waveguided structure formed in a nonlinear medium can be the basis for electro-optic modulation and switching functions. The exact function of the device would depend on control field geometries and strengths and the arrangement of input and output probes. The basic distinction between these two functions is that a device is considered to be a modulator if its primary function is to impart information on a light wave by temporarily varying one of its properties, and a switch would either change the spatial position of light or turn it on or off. The same fundamental materials properties and interactions are common to both types of devices.

The general principle on which these devices operate is the selective retardation of the phase of propagating light in a waveguided mode. Phase retardation can be used to phase-modulate light in the direction of propagation. Relatively low-powered and medium-bandwidth phase modulators have been made that use Ti-diffused $LiNbO_3$ waveguides (Kaminow *et al.*, 1973). Polarization modulators have also been demonstrated that use linear polarized incident light with propagating components subject to different degrees of phase retardation as an electric field is applied to the medium. This results in rotation of the incident polarization, which can then be detected by polarization-sensitive optics at the output.

Other useful modulator/switch structures can be formed based on the concept of a directional coupler (Hunsberger, 1982). Here, two closely spaced waveguide channels are constructed where the evanescent tails between the two channels overlap significantly over some interaction length L. Optical tunneling allows energy to be synchronously transferred from one channel to the next, provided the two channels are precisely phase-matched, i.e., identical thicknesses, geometries, propagation constants, etc. Under these conditions, energy will transfer periodically between the two waveguides over the interaction length as long as the coupling conditions are maintained. A bend in one of the channels after some length will terminate the coupling with some distribution of energy between the two channels. Application of an electric field to retard the phases is equivalent to changing the interaction length and therefore the distribution of energy in the two channels. With appropriate device geometries, both amplitude and phase-modulation effects can be observed.

In the following discussion we will consider the planar slab waveguide shown in Fig. 15 but with electrodes applied above and below the structure. The nonlinear medium with refractive index η_2 is sandwiched between layers with refractive indices lower than η_3 and η_1. Assuming the Z direction is the poling direction, the film will have $C_{\infty V}$ (∞mm) symmetry with the uniaxis in the Z direction, as discussed previously. The equation for the refractive index ellipsoid in the presence of an electric field is (Yariv, 1975a)

$$\left(\frac{1}{\eta^2}\right)_1 X^2 + \left(\frac{1}{\eta^2}\right)_2 Y^2 + \left(\frac{1}{\eta^2}\right)_3 Z^2 + 2\left(\frac{1}{\eta^2}\right)_4 YZ + \left(\frac{1}{\eta^2}\right)_5 XZ + \left(\frac{1}{\eta^2}\right)_6 XY = 1 \tag{44}$$

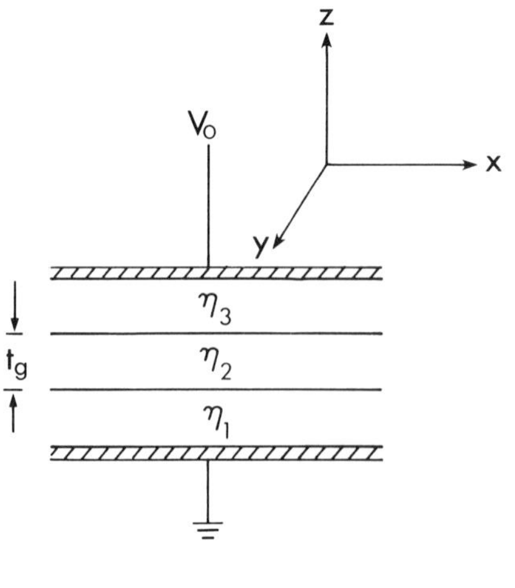

Fig. 15. Illustration of a planar slab waveguide with a confining medium of thickness t_g and electrodes applied above and below.

If we choose the coordinate system so that X, Y, and Z (corresponding to directions 1, 2, and 3 according to convention) are parallel to the principal dielectric axes of the medium, then in the absence of an applied field Eq. (44) becomes

$$\frac{1}{\eta_o^2}(X^2 + Y^2) + \frac{1}{\eta_e^2}Z^2 = 1 \tag{45}$$

with η_o and η_e indicating the ordinary and extraordinary components of the refractive index, respectively. Upon application of an electric field to a uniaxial medium, the changes in the coefficients become

$$\Delta\left(\frac{1}{\eta^2}\right)_I = \sum_{I=1}^{J} r_{IJ}E_J = \begin{vmatrix} 0 & 0 & r_{13} \\ 0 & 0 & r_{13} \\ 0 & 0 & r_{33} \\ 0 & r_{13} & 0 \\ r_{13} & 0 & 0 \\ 0 & 0 & 0 \end{vmatrix} \begin{vmatrix} E_1 \\ E_2 \\ E_3 \end{vmatrix} \tag{46}$$

where r_{IJ} is the electro-optic tensor element. For a dc field applied in the Z direction

$$\Delta\left(\frac{1}{\eta^2}\right)_1 = \Delta\left(\frac{1}{\eta^2}\right)_2 = \Delta\left(\frac{1}{\eta_o^2}\right) = r_{13}E_Z \tag{47}$$

and

$$\Delta\left(\frac{1}{\eta^2}\right)_3 = \Delta\left(\frac{1}{\eta_e^2}\right) = r_{33}E_Z \tag{48}$$

Under these circumstances, Eq. (48) becomes

$$\left(\frac{1}{\eta_o^2} + r_{13}E_Z\right)(X^2 + Y^2) + \left(\frac{1}{\eta_e^2} + r_{33}E_Z\right)Z^2 = 1 \tag{49}$$

As indicated earlier, a TM mode has E_Z, E_X, and H_Y components, so that we would expect an electric-field-induced addition to η_e. The electro-optic coefficients r_{13} and r_{33} derive from the second-order nonlinear tensor elements $\chi^{(2)}_{p\{ZXX\}}$ and $\chi^{(2)}_{ZZZ}$, respectively (Lipscomb *et al.*, 1981). We would therefore expect the phase retardation to be maximum for this field orientation for TM modes. By the same arguments developed earlier, r_{13} is considerably smaller than r_{33}, and for highly anisotropic media according to Table III, r_{13} approaches 0.

It has been proposed and verified experimentally that the tensor element r_{IJ} and $\chi^{(2)}_{IJK}(-2\omega, \omega, \omega)$ should be close in value for rigid organic media where the primary contribution to the nonlinear polarization is the π-electronic system (Lipscomb *et al.*, 1981).

The magnitude of the electro-optic (EO) effect on a Z-polarized TM mode should therefore be

$$\Delta\eta_{\mathrm{EO}} = \eta_{\mathrm{E}}^{3} r_{33} \frac{V}{2T_{\mathrm{g}}} \tag{50}$$

and since

$$\Delta\eta = \Delta\beta k = \Delta\beta\lambda/2\pi$$

the phase retardation $\Delta\phi = \Delta\beta L$ becomes

$$\Delta\phi = \frac{\pi}{\lambda}\eta_{\mathrm{e}}^{3} r_{33} \frac{V}{2T_{\mathrm{g}}} \tag{51}$$

It therefore appears that full advantage can be taken of the large induced Z nonlinearity in polymer films for a variety of electro-optic modulation and switching functions if appropriately waveguided structures can be fabricated and large enough optical nonlinearities induced. An additional benefit of rigid polymeric systems is the electronic nature of the nonlinear response, so that the bandwidth of devices based on these materials should be limited only by the electronic response times of driver circuitry.

IV. OUTLOOK FOR POLYMERIC GUEST–HOST SYSTEMS

In this section the nonlinear optical properties and associated characteristics of a guest–host system that has been characterized in detail are reviewed briefly, and research opportunities for extending the initial concepts are then discussed. Other approaches for achieving large-scale alignment in polymeric systems are also mentioned briefly.

A. Polymer Liquid-Crystalline Guest–Host Systems

Figure 1 shows the molecularly doped polymeric system that was examined in detail. The choice of the specific polymer–guest molecule combination was based on studies of dye polymer association as well as the ability to induce homogeneous alignment in the doped polymer film.

Guest–host compatibility was investigated by determination of the order parameter S through linear dichroism measurements. If the optical transition moment of the dopant molecule is parallel to the long molecular axis, which it is for DANS (dimethylaminonitrostilbene), then the linear dichroism should constitute a probe of axial alignment, owing to the influence of the nematic potential. In the experiment, a thin film of the polymer with 2% DANS cast onto a glass slide with interdigital electrodes was examined by polarization microscopy and optical absorption. As the film was heated through its glass-transition temperature (27°C), the familiar nematic scattering texture was observed through crossed polarizers. Heating above the nematic-to-isotropic transition caused total extinction of light. Application of an electric field resulted in a nearly homogeneous birefringent film between the electrodes. Measurement of the optical absorption spectrum of DANS parallel and perpendicular to the field, produced the spectrum shown in Fig. 16 (Meredith *et al.*, 1982). From this spectrum the order parameter S,

$$S = (A_{\parallel} - A_{\perp})/(2A_{\perp} + A_{\parallel}) \tag{52}$$

was found to be 0.3. The ability to induce homogeneous alignment was a necessary prerequisite for SHG measurements by transmission.

The value of $\chi^{(2)}_{ZZZ}$ was measured relative to quartz by a procedure described previously (Meredith *et al.*, 1982). A maximum value of $\chi^{(2)}_{ZZZ}$ of 3×10^{-9} esu was obtained on a film that was poled for several hours near T_g. This is ~50% of the theoretical limit from the Ising model calculation. From the order parameter determination on these films, $\langle \cos^2 \theta \rangle \approx 0.5$, so that we would

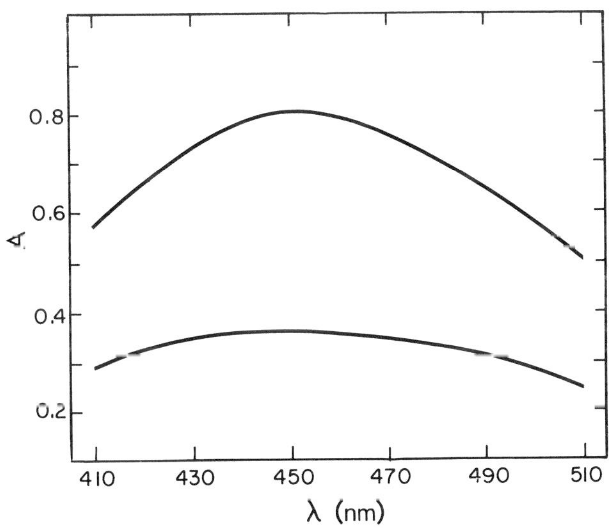

Fig. 16. Polarized optical absorption spectrum of the polymer in Fig. 1 doped with 2% DANS. The upper trace corresponds to alignment of the polarizer. [After Meredith *et al.* (1982).]

expect a reduction in the observed value from that predicted for the Ising-model limit.

Measurement of the harmonic signal intensity versus temperature under equilibrium conditions—i.e., low laser power to avoid heating of the film and long poling times—produced the unexpected and puzzling result shown in Fig. 17. The observed temperature dependence is markedly different from the $1/T$ dependence predicted from Table III and Eq. (32). The behavior is fully reversible and is observed on both heating and cooling cycles. The reduction of about two orders of magnitude in harmonic intensity from the expected value in the nematic region above ~40°C suggests the presence of competing equilibrium processes in the films. One possible explanation is an equilibrium process between DANS molecules and a weakly associating centrosymmetric dimer. As the viscosity of the medium is lowered, a dynamic equilibrium between monomer and dimer, which might be perturbed by the poling field at lower temperatures and higher viscosities, might account for the observed behavior. There is no direct proof of this explanation, but the utility of films with this behavior is suspect because of long-term stability issues.

Several approaches to solving this problem have been considered. The first

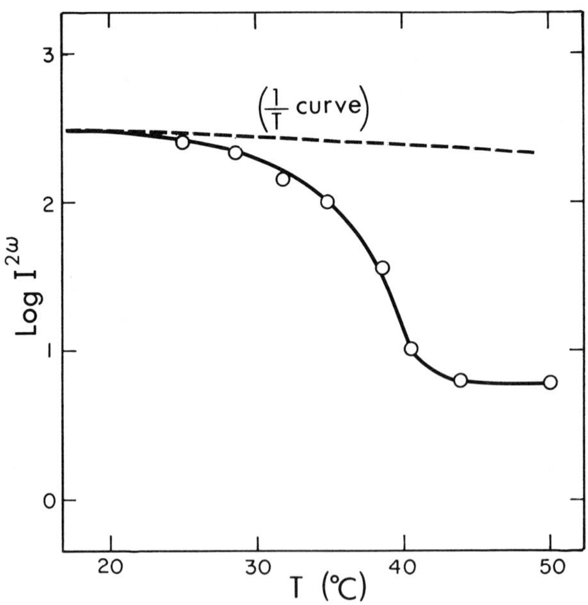

Fig. 17. Plot of the log of the harmonic intensity (arbitrary scale versus temperature) for the polymer in Fig. 1 doped with 2% DANS. The dashed line would be expected for a $1/T$ dependence. [After Meredith *et al.* (1982).]

involves raising the T_g of the host medium through chemical modification of the sidechain length. It is known, for instance (Finkleman *et al.*, 1979a; Meredith *et al.*, 1983), that shortening the spacer length between the mesogenic unit and the main chain to C_3 increases T_g to 53°C, creating a more rigid medium where field-induced alignment is less likely to undergo thermal relaxation. Detailed experiments of the higher-T_g systems would be needed to determine if homogeneous alignment and host–guest association could be maintained.

Another possible approach would be to incorporate the DANS molecule or some other appropriate nonlinear chromophore into the polymeric structure. At low concentrations, dimer formation, if it indeed occurs, would be expected to be considerably impeded relative to the molecularly doped case.

B. Other Approaches for Large-Scale Alignment

The most direct approach for obtaining nonlinear coefficients through field-induced alignment in excess of that obtainable by the molecularly doped liquid-crystalline polymer approach is to create extended polymeric chains with the nonlinear chromophore as part of the repeat unit. This approach was alluded to in Section II, where individual poly-γ-benzyl-L-glutamate (PBG) chains in α-helix-forming solvents showed virtually complete field-induced polar orientation at modest electric field strengths (Levine and Bethea, 1976). Although β for the monomeric repeat unit was only 2×10^{-31} esu, the individual chains gave $\beta \approx 500 \times 10^{-30}$ esu, owing to the large number of individual peptide units in the polymer chain. This important result suggests a molecular design approach, whereby a repeat unit of a polar polymer chain would contain a chromophore with a large β, and permanent field-induced polar alignment of the polymer chains would be imparted by chemical methods. With degrees of alignment similar to that demonstrated for PBG, one could imagine achieving $\chi^{(2)}$ of 10^{-6} esu or greater. A thin-film technology based on this approach could have considerable impact on waveguide and integrated optics technologies.

In summary, there is a considerable understanding of the sources of electric-field-induced second-order nonlinearity in molecularly doped polymers as well as in polymeric chains with intrinsic nonlinearities. The relationship between field-induced $\chi^{(2)}$ tensor elements and their utilization for nonlinear effects both in bulk and thin-film formats are also well understood. Flexible approaches exist for the design of new materials with extremely large $\chi^{(2)}$, but much remains to be done in synthetic, organic, and polymer chemistry to achieve new materials suitable for waveguide nonlinear optics and their utilization in integrated optics.

REFERENCES

Badan, J., Hierle, R., Perigaud, A., and Zyss, J. (1983). *ACS Symp. Ser.* **233,** 81–108.
Bassler, H. (1981). *Phys. Status Solidi* **107,** 9.
Böttcher, C. J. F. (1973). "Theory of Electric Polarization," Vol. 1, p. 161. Elsevier, Amsterdam.
Byer, R. L. (1977). *In* "Non-Linear Optics" (P. G. Harper and B. Wherrett, eds.), p. 61. Academic Press, London.
Conwell, E. M. (1973). *IEEE J. Quantum Electron.* **QE-9,** 867.
DeGennes, P. G. (1974). "The Physics of Liquid Crystals," p. 23. Oxford Univ. Press (Clarendon), London and New York.
Docherty, V. J., and Pugh, D. (1985). *J. Chem. Soc., Faraday Trans.* **81,** 1179.
Dulcic, A. (1979). *Chem. Phys.* **37,** 57.
Dulcic, A., and Sauteret, C. (1978). *J. Chem. Phys.* **69,** 3453.
Dulmage, W. J., Light, W. A., Marino, S. J., Salzberg, C. D., Smith, D. L., and Staudenmayer, W. J. (1978). *J. Appl. Phys.* **49,** 5543.
Finklemann, H., and Day, D. (1979). *Makromol. Chem.* **180,** 2269.
Finklemann, H., Ringsdorf, H. and Wendorf, J. H. (1979a). *Makromol. Chem.* **179,** 273.
Finklemann, H., Naegele, D., and Ringsdorf, H. (1979b). *Makromol. Chem.* **180,** 203.
Flytzanis, C. (1975). *In* "Quantum Electronics: A Treatise" (H. Rabin and C. L. Tang, eds.), Vol. 1A, pp. 9–199. Academic Press, New York.
Franken, P., Hill, A., Peters, C., and Weinreich, G. (1961). *Phys. Rev. Lett.* **7,** 118.
Garito, A. F., Singer, K. D., Hayes, K., Lipscomb, G. F., Lalama, S. J., and Desai, K. N. (1980). *J. Opt. Soc. Am.* **70,** 1399.
Garito, A. F., Singer, K. D., and Teng, C. C. (1983). *ACS Symp. Ser.* **233,** 1–26.
Goodby, J. W., and Leslie, T. M. (1984). *Mol. Cryst. Liq. Cryst.* **170,** 175.
Havinga, E. E., and Van Pelt, P. (1979). *Ber. Bunsenges. Phys. Chem.* **83,** 816.
Hunsberger, R. G. (1982). *In* "Integrated Optics Theory and Technology" (T. Tamir, ed.) p. 120. Springer–Verlag, Berlin and New York.
Jain, K., and Hewig, G. H. (1981). *Opt. Commun.* **36,** 483.
Kaminow, I. P., Carruthers, J. R., Turner, E. H., and Stultz, L. W. (1973). *Appl. Phys. Lett.* **22,** 540.
Kielich, S. (1969). *IEEE J. Quantum Electron.* **QE-5,** 562.
Kleinman, D. A. (1962). *Phys. Rev.* **126,** 1977.
Kurtz, S. L., and Perry, T. T. (1968). *J. Appl. Phys.* **39,** 3798.
Lalama, S. J., and Garito, A. F. (1979). *Phys. Rev. B* **20,** 1179.
Langevin, P. (1905a). *J. Phys.* **4,** 678.
Langevin, P. (1905b). *Ann. Chim. Phys.* **5,** 70.
Levine, B. F. (1976). *Chem. Phys. Lett.* **37,** 516.
Levine, B. F., and Bethea, C. G. (1976). *J. Chem. Phys.* **65,** 1989.
Lipscomb, G. F., Garito, A., and Narang, R. S. (1981). *J. Chem. Phys.* **75,** 1509.
Meredith, G. R., Van Dusen, J. G., and Williams, D. J. (1982). *Macromolecules* **15,** 1385.
Meredith, G. R., Van Dusen, J. G., and Williams, D. J. (1983). *ACS Symp. Ser.* **233,** 109–134.
Morrell, J. A., and Albrecht, A. C. (1979). *Chem. Phys. Lett.* **64,** 46.
Mort, J., and Pfister, G. (1982). "Electronic Properties of Polymers." Wiley, New York.
Nayar, B. K. (1983). *ACS Symp. Ser.* **233,** 153–166.
Onsager, L. (1936). *J. Am Chem. Soc.* **58,** 1486.
Oudar, J. L. (1977). *J. Chem. Phys.* **68,** 446.
Oudar, J. L., and Zyss, J. (1982). *Phys. Rev. A* **26,** 2016.
Oudar, J. L., Chemla, D. S., and Batifol, E. (1977). *J. Chem. Phys.* **67,** 1626.
Saha, S. K., and Wong, G. K. (1979). *Appl. Phys. Lett.* **34,** 423.

Singer, S. D., John, J. E., and Lalama, S. J. (1985). *In* "Extended Abstracts: Nonlinear Optical Materials" (D. A. B. Miller, ed.). Materials Research Society, Pittsburgh.
Taylor, H. F., and Yariv, A. (1974). *Proc. IEEE* **62,** 1044.
Williams, D. J. (1984). *Angew. Chem., Int. Ed. Engl.* **23,** 690.
Yariv, A. (1975a). "Quantum Electronics," 2nd ed., p. 334. Wiley, New York.
Yariv, A. (1975b). "Quantum Electronics," 2nd ed., p. 514. Wiley, New York.
Zernike, F., and Midwinter, J. E. (1973). "Applied Nonlinear Optics." Wiley, New York.
Zyss, J., and Oudar, J. L. (1982). *Phys. Rev. A* **26,** 2028.
Zyss, J., Nicoud, J. F., and Coquillay, M. (1984). *J. Chem. Phys.* **81,** 4160.

Chapter II-8

Electro-Optic Organic Materials

K. D. SINGER, S. L. LALAMA, J. E. SOHN, and R. D. SMALL

AT&T Engineering Research Center
Princeton, New Jersey 08540

I. INTRODUCTION

The first nonlinear optical phenomena observed were electro-optic effects (Kerr, 1875; Pockels, 1906). Devices based on these effects found early technological introduction as modulators, for example, Q-switches in pulsed lasers, which employ inorganic crystals, such as potassium dideuterium phosphate, KD*P. Liquid-crystal displays and light valves represent the widest application of electro-optic effects. The electro-optic processes in liquid crystals involve large-scale motion of the constituent molecules, leading to response times slower than the microsecond time scale. The discussion here focuses only on effects linear in the applied modulating electric field, arising from excitations in the phonon and electronic structure which are faster than the nanosecond time scale ($\tau < 10^{-9}$ sec). High-speed applications include spatial switches and modulators for data switching and processing, where bandwidth reduction cannot be tolerated.

Initial interest in organic crystals arose from their exceptionally large intrinsic optical nonlinearities. As shown in Fig. 1, Miller's delta (Miller, 1964),

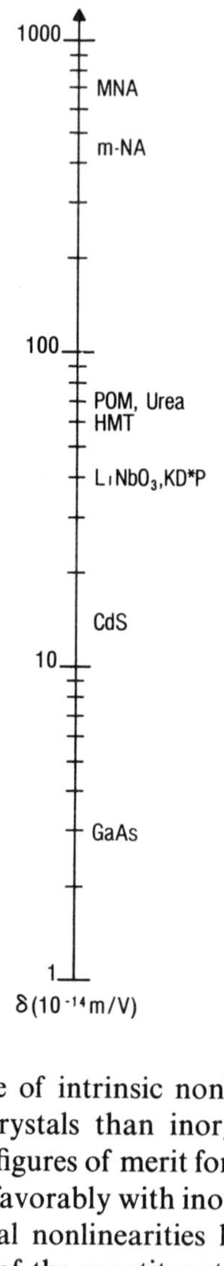

Fig. 1. Miller's delta for electro-optic materials.

a measure of intrinsic nonlinearity, can be orders of magnitude larger for organic crystals than inorganic dielectric and semiconductor crystals. In addition, figures of merit for organic materials relating to device performance compare favorably with inorganic materials. Research into the origin of these exceptional nonlinearities has revealed that they arise from the electronic structure of the constituent molecules and their arrangement in a crystalline lattice. The molecular origin in conjugated π-electron organic molecules has been established experimentally as arising from large charge reorganization in the lowest electronic excited state (Davydov *et al.*, 1970; Levine, and Bethea, 1977; Oudar and Chemla, 1977; Lalama and Garito, 1979). This understanding has led to the identification of molecules with large optical nonlinearities. The dominant role of the electronic states in the nonlinear susceptibilities of organic materials also implies a low dielectric constant, thus promising a high modulation bandwidth, a property of significant importance for high-speed applications.

In order to take advantage of the large microscopic susceptibilities of organic molecules in macroscopic devices, specific symmetry requirements must be met in the arrangement of the molecules in the bulk material. For a material to possess a nonzero second-order optical susceptibility, it must lack a center of symmetry. Further, in order to take full advantage of the molecular properties, the molecules must be arranged so that their effects add properly for the phenomenon of interest (Chemla *et al.*, 1975; Zyss and Oudar, 1982). These requirements have resulted in a great deal of molecular engineering aimed at altering the molecules in such a way that the energetics of crystallization result in the optimum symmetry (Zyss, 1982).

In Section II, the electro-optic phenomena of interest are introduced. This includes the relationship between electro-optic coefficients and nonlinear optical susceptibilities. Then the possible physical contributions to the linear electro-optic effect are described including the relationship of the linear electro-optic effect to piezoelectric and Raman phenomena. We also describe the techniques used to measure electro-optic coefficients. Section III reviews the properties of electro-optic organic crystals, with particular attention to studies into the origin of the electro-optic effect in each crystal. These studies reveal that the contribution of the molecular electronic structure dominates the electro-optic coefficient, which points the way for the engineering of new molecules and materials for electro-optics. Finally, a new class of non-crystalline organic polymer materials whose intrinsic optical quality and processability may qualify them for guided-wave, integrated electro-optic applications is presented.

II. LINEAR ELECTRO-OPTIC EFFECT

A. Definitions

The electro-optic effect was the earliest nonlinear optical phenomenon discovered. It is defined through the change in the electric impermeability induced by an applied electric field (Kaminow, 1974),

$$B_{ij}(E) - B_{ij}(0) = \left[\frac{1}{\varepsilon(E)}\right]_{ij} - \left[\frac{1}{\varepsilon(0)}\right]_{ij} \equiv r_{ij,k}E_k + s_{ij,kl}E_kE_l + \cdots \quad (1)$$

where B_{ij} is the impermeability tensor, ε the dielectric tensor, and E_m the applied electric field; $r_{ij,k}$ is the linear electro-optic or Pockels coefficient (Pockels, 1906), and $s_{ij,kl}$ is the quadratic electro-optic or Kerr coefficient

(Kerr, 1875). The early discovery of these two nonlinear optical effects is consistent with the fact that laser light is not required, and that very small changes in the optical indicatrix can be readily observed through phase retardation between the two polarizations of light propagating in a birefringent medium. For instance, applied voltages of several thousand volts may produce a refractive index change of one part in 10^7, yet a measurable phase shift results (Kaminow, 1974). The definitions in Eq. (1) are in a form that easily comports to manipulations of the optical indicatrix. Note that $r_{ij,k}$ is symmetric with respect to the first two indices, and thus $r_{ij,k}$ can be written in reduced tensor notation, r_{uk} (Nye, 1957). Kleinman (1962) symmetry does not apply for the linear electro-optic effect.

With the advent of lasers, a host of other nonlinear optical phenomena was discovered (Bloembergen, 1965). These phenomena are defined through the electric polarization in the medium, and thus are described by nonlinear optical susceptibilities. In this manner, the linear electro-optic effect is defined through the change in the optical susceptibility (Kaminow, 1974),

$$\Delta\chi_{ij} = \Delta\varepsilon_{ij} = -\varepsilon_{ii}\Delta B_{ij}\varepsilon_{jj} \tag{2}$$

for $\Delta\varepsilon_{ij} \ll \varepsilon_{ii}$ and ε_{jj}, and where χ_{ij} is the linear susceptibility and ε_{ij} the dielectric tensor. The induced polarization is then given by

$$P_i(t) = \varepsilon_0 \Delta\chi_{ij} E_j = -\varepsilon_0 \varepsilon_{ii}\varepsilon_{jj} r_{ij,k} E_j(t) E_k(t) \tag{3}$$

If the electric fields are defined as

$$\begin{aligned} E_j(t) &= E_j \cos(\omega_u t + \phi_u) \\ &= \tfrac{1}{2}[e_j(\omega_u)\exp(-i\omega_u t) + e_j^*(-\omega_u)\exp(i\omega_u t)] \end{aligned}$$

and

$$\begin{aligned} E_k(t) &= E_k \cos(\omega_w t + \phi_w) \\ &= \tfrac{1}{2}[e_k(\omega_w)\exp(-i\omega_w t) + e_k^*(-\omega_w)\exp(i\omega_w t)] \end{aligned} \tag{4}$$

with $e_l(\omega)$ the Fourier amplitudes of the electric fields including the overall phase. Taking $\omega_u = \omega$, the optical frequency, and $\omega_w = 0$, then the polarization at the optical frequency is

$$P_i(\omega) = -\varepsilon_0 \varepsilon_{ii}(\omega)\varepsilon_{jj}(\omega) r_{ij,k} e_j^{\omega} e_k^{0} \tag{5}$$

If a similar analysis is carried out in the notation of a second-order nonlinear optical susceptibility defined through

$$P_i(t) = \varepsilon_0 \chi_{ijk} E'_j(t) E'_k(t) \tag{6}$$

with the electric fields given by

$$E'_j(t) = \tfrac{1}{2}[e_j(\omega_u)\exp(-i\omega_u t) + e_j(\omega_w)\exp(-i\omega_w t) + \text{c.c.}] \tag{7}$$

where c.c. denotes the complex conjugate, then the following relationship between the nonlinear susceptibility and the electro-optic coefficient is obtained:

$$\chi_{ijk}(-\omega;\omega,0) = -\tfrac{1}{2}\varepsilon_{ii}(\omega)\varepsilon_{jj}(\omega)r_{ij,k}(-\omega;\omega,0) \tag{8}$$

where $\varepsilon_{kk}(\omega) = n_k^2$, the principal indices of refraction.

Several figures of merit associated with the linear electro-optic effect have been defined (Kaminow and Turner, 1971). The first, Miller's delta, gives an intrinsic physical measure of the material nonlinearity (Miller, 1964). This figure of merit arises from the fact that large linear susceptibilities can lead to large nonlinear susceptibilities, and it corresponds to the nonlinear susceptibility normalized to the linear susceptibilities given by

$$\delta_{ij,k} = \frac{-\varepsilon_{ii}\varepsilon_{jj}r_{ij,k}}{4(\varepsilon_{ii}-1)(\varepsilon_{jj}-1)(\varepsilon_{kk}-1)} \tag{9}$$

The variation of $\delta_{ij,k}$ with frequency and temperature is less than that of $r_{ij,k}$. Furthermore, as can be seen from Fig. 1 and Table I, the variation in $\delta_{ij,k}$ among inorganic crystals is less than that in $r_{ij,k}$. However, the organic crystals listed have exceptionally large values of $\delta_{ij,k}$, and it is this fact that originally stimulated interest in these crystals.

Two other figures of merit associated with the behavior of the electro-optic material are relevant to the performance of devices fabricated from the material. The first measures the phase-shift efficiency, which is relevant, for example, to the length–voltage product necessary to switch a directional coupler. This is the half-wave voltage, which, for a uniaxial crystal in which the optic axis is the 3 axis and the propagation direction is normal to it, is given by

$$V_{\lambda/2} = (E_3 l)_{\lambda/2} = \frac{\lambda}{n_3^3 r_{33} - n_1^3 r_{13}} \tag{10}$$

where r_{uk} is the standard contracted notation for the electro-optic tensor where u replaces ij, and λ is the optical wavelength (Kaminow and Turner, 1971). Thus the length–voltage product is minimized for large $n^3 r$ products.

The second figure of merit, F, measures the standing-wave modulation bandwidth per modulating power attainable in a crystal,

$$F = \frac{n_u^7 r_{uk}^2}{\varepsilon_k} = \left(\frac{2\varepsilon_0}{\pi^2}\right)\left(\frac{\lambda^3 \eta_u^2 \Delta f}{P}\right) \tag{11}$$

where Δf is the modulating bandwidth, P the modulating power, and η_u the phase modulation index, which is defined by the phase modulation in the output wave,

$$E_{\text{out}} = A\cos[\omega t + \eta \sin \omega_{\text{m}} t] \tag{12}$$

TABLE I

Properties of Electro-Optic Materials

Material, symmetry	$r(10^{-12}$ m/V)[g]	n	$r^{el}(10^{-12}$ m/V)[h]	$\varepsilon/\varepsilon_0$
		Organic materials		
MNA[a], m	$r_{11}^{T} = 67 \pm 25$	$n_1 = 2.0$	$r_{11}^{el} = 63$	
m-NA[b], $mm2$	$r_{33}^{T} = 16.7 \pm 0.2$	$n_1 = 1.805$	$r_{33}^{el} = 9.6$	$\varepsilon_1 = 3.2$
	$r_{23}^{T} = 0.1 \pm 0.6$	$n_2 = 1.715$	$r_{23}^{el} = 0.2$	$\varepsilon_2 = 3.3$
	$r_{13}^{T} = 7.4 \pm 0.7$	$n_3 = 1.675$	$r_{13}^{el} = 7.1$	$\varepsilon_3 = 3.2$
POM[c], 222	$r_{41}^{T} = 3.6 \pm 0.6$	$n_1 = 1.712$	$r_{41}^{el} = 3.7$	$\varepsilon_1 = 3.77$
	$r_{52}^{T} = 5.1 \pm 0.4$	$n_2 = 1.919$	$r_{52}^{el} = 6.6$	$\varepsilon_2 = 5.41$
	$r_{63}^{T} = 2.6 \pm 0.3$	$n_3 = 1.638$	$r_{63}^{el} \approx 4.5$	$\varepsilon_3 = 3.77$
HMT[d], $\bar{4}3m$	$r_{41}^{T} = 0.8$	$n = 1.5911$		$\varepsilon = 3.2$
Urea[e], $\bar{4}2m$	$r_{63}^{T} = 0.83$	$n_1 = n_2 = 1.48$	$r_{63}^{el} = 0.83$	
	$r_{41}^{T} = 1.9$	$n_3 = 1.59$	$r_{41}^{el} = 1.9$	
Azo dye/PMMA, ∞mm		$n_1 = n_2 \approx 1.52$	$r_{33}^{el} = 2.3 \pm 0.3$	$\varepsilon_3 \approx 3.6$
		$n_3 \approx 1.52$	$r_{13}^{el} = 0.8 \pm 0.3$ ($\lambda = 0.8\ \mu$m)	
		Inorganic materials		
$LiNbO_3$[f], $3m$	$r_{33}^{S} = 30.8$	$n_1 = n_2 = 2.2716$ ($\lambda = 0.7\ \mu$m)		$\varepsilon_1 = \varepsilon_2 = 43$
	$r_{13}^{S} = 8.6$	$n_3 = 2.1874$ ($\lambda = 0.7\ \mu$m)		$\varepsilon_3 = 28$
KD*P[f], $\bar{4}2m$	$r_{63}^{T} = 26.4$	$n_1 = n_2 = 1.51$		$\varepsilon_3 = 50$
	$r_{41}^{T} = 8.8$	$n_3 = 1.47$		
GaAs[f], $\bar{4}3m$	$r_{41}^{S} = 1.2$ ($\lambda = 0.9$–$1.08\ \mu$m)	$n = 3.60$		$\varepsilon = 13.2$

[a] 2-Methyl-4-nitroaniline (Lipscomb *et al.*, 1981).
[b] *m*-Nitroaniline (Stevenson, 1973).
[c] 3-Methyl-4-nitropyridine 1-oxide (Sigelle and Hierle, 1981).
[d] Various λ, hexamethylenetetramine (Lee, 1969, and references therein).
[e] Morrell *et al.*, (1979).
[f] Kaminow and Turner (1971).
[g] S denotes constant strain (clamped), T denotes constant stress (unclamped); all measurements at $\lambda \approx 0.63\ \mu$m except where noted.
[h] Electronic contribution to r.

where ω_m is the modulating frequency (Kaminow, 1968). The usefulness of any particular material in an application will depend on the figure of merit for the application, as well as other favorable properties. Thus, comparisons should be made with the demands of the application in mind.

B. Dispersion

The second-order nonlinear optical susceptibility, $\chi_{ijk}(-\omega_3; \omega_1, \omega_2)$, depends on the three frequencies involved (Ward and Franken, 1964; Wemple and DiDomenico, 1972). In fact, if any of the frequencies nears a characteristic material frequency, resonance behavior is observed. Generally, contributions to the measured susceptibility will arise from physical mechanisms appropriate to all of the frequencies involved. For optical parametric processes, such as second-harmonic generation, all of the frequencies are above fundamental acoustic and optical phonon modes and thus the parametric process is unaffected by these modes. The associated susceptibilities will arise only from the electronic structure. At frequencies above the optical phonon modes, but below the electronic excitations, the electrons contribute via virtual excitations.

For the linear electro-optic effect, the modulating frequency is below the acoustic and optical phonon modes, so that both contribute to the electro-optic susceptibility. In the Born–Oppenheimer approximation, the responses of phonon and electronic modes are independent and can be expressed as

$$r \propto \frac{d\chi(\omega_0)}{dE(\omega_m)} = \left[\frac{\partial\chi(\omega_0)}{\partial E(\omega_m)}\right]_{Q=0} + \left[\frac{\partial\chi(\omega_0)}{\partial Q}\right]_{E(\omega_m)=0} \times \frac{\partial Q}{\partial E(\omega_m)} \tag{13}$$

In the above expression, Q is a lattice displacement, $E(\omega_m)$ is the modulating field, and $\chi(\omega_0)$ is the linear response at optical frequency ω_0 (Kaminow, 1968). The first term is the primary electro-optic effect, which is of purely electronic origin. The second term is the secondary effect, which arises from the action of the modulating field on the lattice, which is related to the piezoelectric effect, and from the change in the optical susceptibility due to the lattice deformation, which is related to the photoelastic effect. The magnitudes of these contributions depend on the detailed physical mechanisms in the particular material.

The frequency dependence of the electro-optic coefficient is shown schematically in Fig. 2. At low frequencies, acoustic modes appear in the material with characteristic frequencies determined by the size and shape of the particular sample. At these frequencies, the material will deform with the applied field, so that the unclamped or stress-free coefficient r^T is appropriate. The coefficient r^T contains contributions from low-frequency acoustic modes, r^B; from optical modes, r^R; and from electronic modes, r^{el}. Above the acoustic modes and up to the optical phonon modes, the material strains are small, yielding the clamped or strain-free coefficient r^S, with contributions from r^R and r^{el}. Above the optical phonon restrahl, only r^{el} is measured, and this is then considered an optical parametric process. It should be noted that the magnitude and sign of the various contributions are material-dependent.

In general, then, the electro-optic effect can depend on three distinct processes depending on the modulating frequency, so that an understanding of the origin of the electro-optic effect in organic materials must take the possible contributions into account. The contribution of the electronic states

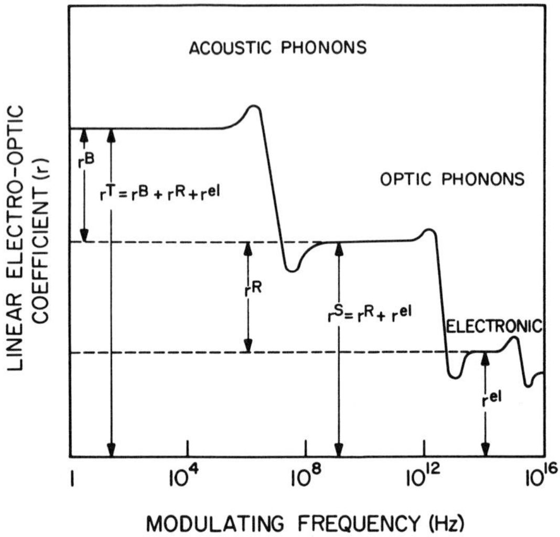

Fig. 2. General dispersion for the linear electro-optic coefficient including contributions from the electronic, r^{el}, the optical phonon modes, r^{R}, and the acoustic phonon modes, r^{B}. The term r^{T} is the unclamped and r^{S} the clamped electro-optic coefficient. [From Lipscomb (1980), with permission.]

to the electro-optic coefficient can be determined by measuring an optical parametric susceptibility, such as the second-harmonic coefficient (Ward and Franken, 1964; Wemple and DiDomenico, 1972). The residual electro-optic susceptibility is then due to the phonon modes. The contribution of the optical modes can be determined independently by Raman scattering, since it measures fluctuations in the refractive indices due to thermally excited optical phonons. Similarly, the acoustic phonon contribution is relatable to information deduced from Brillouin scattering, since it probes thermally excited acoustic phonon modes. The relative contributions of the acoustic and optical phonons can be probed by measuring the clamped and unclamped electro-optic coefficients. Thus, a variety of measurements elucidate the origin of electro-optic properties in organic materials.

C. Measurement Techniques

The experimental determination of linear electro-optic coefficients is controlled by

$$\Delta n = n^3 rE/2 \tag{14}$$

where tensor subscripts have been omitted for clarity and where Δn is the induced change in the refractive index, n the refractive index, r the electro-optic coefficient, and E the applied modulating field. The morphology and orientation of the sample as well as the crystal symmetry determine the tensor components that can be identified in a particular experiment. Furthermore, the change in refractive index Δn is typically small. For instance, if $n^3 = 5$, $r = 5 \times 10^{-12}$ m/V, and $E = 10^6$ V/m, then $\Delta n = 1.25 \times 10^{-5}$.

These small changes in the refractive index are measured by detecting the phase shift of light traversing the material, where the phase shift is defined by

$$\Phi = (2\pi/\lambda)\,\Delta n l \tag{15}$$

where l is the optical path length. In order to detect the phase shift due to the electro-optic effect, the phase must be fixed or referenced. Two methods to do this have been employed. The first is to fix the phase through birefringent propagation in the crystal; that is, the relative phase of the ordinary and extraordinary rays is fixed by the crystal boundary. The second method is to refer the phase to an external interfering beam. This method, although requiring source coherence over the interfering length, has the advantage that the reference is external to the material being measured. This means that only one ray needs to be propagating in the electro-optic material, so that polarizations along the principal dielectric directions can be chosen. Thus, individual tensor components can be measured independently. When the phase is fixed by birefringence, it is possible that the phases of both rays are changing with voltage, which leads to a "moving" reference. Thus, the measurements often mix tensor components, requiring multiple measurements to deduce individual components. This is often inconvenient due to crystal geometry and quality. The advantages of birefringent phase referencing is that a coherent source is not required, and it is identical to the device geometry of bulk modulators. Both techniques can use static or ac modulating fields.

The birefringent phase-retardation method, which rotates the polarization of a propagating beam, relies on a measurement of the material birefringence as a function of applied modulating voltage. The relationship between the induced birefringence and the electro-optic coefficients is given in the description of the refractive index ellipsoid (Wemple and DiDomenico, 1972):

$$\sum_{ij}\left[\left(\frac{1}{n^2}\right)_{ij} + \Delta\left(\frac{1}{n^2}\right)_{ij}\right]x_i x_j = 1 \tag{16}$$

where the change in dielectric impermeability is related to the electro-optic coefficient through the first term in Eq. (1),

$$\Delta\left(\frac{1}{n^2}\right)_{ij} = r_{ij,k}E_k \tag{17}$$

resulting in

$$\sum_{ij}\left[\left(\frac{1}{n^2}\right)_{ij} + \sum_k r_{ij,k}E_k\right]x_i x_j = 1 \tag{18}$$

Absorbing the electro-optically induced change in refractive index into the refractive index as $1/(n)^2 \to 1/(n + \Delta n)^2$, taking Δn to be small, and expanding $1/(n + \Delta n)^2$ leads to Eq. (14). The crystal structure and morphology restrict which tensor components of the electro-optic coefficient can be measured. One configuration that can be used consists of an incident light beam striking an electro-optic crystal with the polarization of the beam at 45° to a principal dielectric axis of the crystal. In this way, the incident beam breaks up into two equal amplitude components polarized along the two principal dielectric axes. Upon exiting the crystal, the beam is elliptically polarized as a result of the different refractive indices along the different principal dielectric axes. In addition to the zero-field birefringence that the crystal may possess, an induced birefringence through the linear electro-optic effect is imposed. This birefringence is measured by the phase retardation Γ and is defined as the difference between the phase velocities along each principal dielectric axis. This is a function of the relevant electro-optic coefficients, the modulating electric field applied to the crystal, the sample dimensions, and wavelength of the optical field. One way that the phase retardation can be measured employs a Sénarmont compensator (Lee, 1969) in which an incident polarizer, the crystal, a quarter-wave plate, and a polarization analyzer are arranged for extinction of the transmitted intensity. The term Γ is related to the angle between the incident and analyzing polarizers. In another method (Lipscomb *et al.*, 1981), the phase modulation is converted to an amplitude modulation through the use of fixed orthogonal polarizers at the input and output.

The determination of the electro-optic coefficients of a crystal in a birefringence phase-retardation experiment can be improved if the modulating field is applied at a nonzero frequency. In such a configuration, the optical propagation is the same as the static case; however, the output optical intensity through the polarization analyzer oscillates at the same frequency as the modulating field. The properties of this optical modulator can then be analyzed to deduce the electro-optic coefficient from relations such as the figure of merit defined in Eq. (11) above. Among the advantages of an ac technique are increased measurement sensitivity through the use of optical detectors (Carpenter, 1950) and heterodyning techniques (Kaminow, 1965), as well as the ability to vary the modulating frequency into both clamped and unclamped regions of the material's frequency response, thereby probing the contributions of phonon structure to the electro-optic coefficient.

In the interferometric phase modulation technique, the sample is placed into one arm of an interferometer, and then, as a modulating field is applied, its refractive index changes, thereby changing the optical path length of that arm, which produces interferometric fringe shifts at the output of the interferometer. As mentioned previously, by choosing the orientation and polarization properly, the individual tensor components can be measured independently in this configuration. As in the other method, an interferometer can be set up for either static (Stevenson, 1973) or ac (Sugie and Tada, 1976; Sigelle and Hierle, 1981) operation.

The experimental layout of the Sigelle and Hierle (1981) experiment is shown in Fig. 3. Two electro-optic crystals are employed, and a null modulation condition is established using a calibration crystal in one arm and the sample in the other. The output intensity I of the interferometer is related to the input intensity I_0 by

$$I = I_0(A + B\cos\phi)/2 \tag{19}$$

where A and B, both between 0 and 1, are parameters related to losses in the interferometer and ϕ is the phase difference between the two arms. The phase difference consists of an initial phase difference ϕ_0 and a time-dependent part given by

$$\phi = \phi_0 + (\Delta\phi_2 - \Delta\phi_1)\cos\Omega t \tag{20}$$

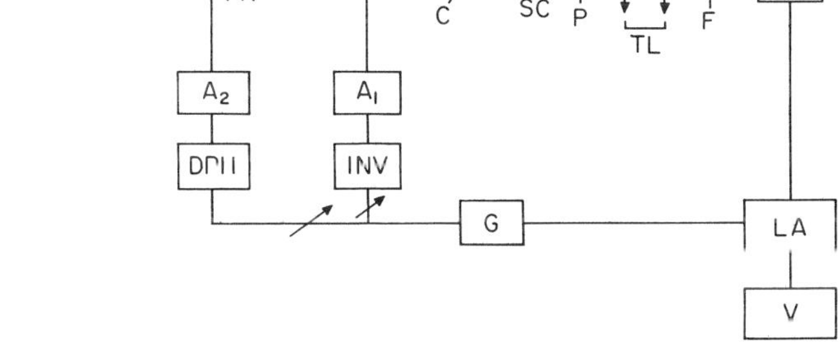

Fig. 3. Schematic of Sigelle and Hierle's (1981) ac interferometric measurement. He–Ne, helium–neon laser; P, polarizer; L_1, converging lens (focal length $f_1 = 0.5$ m); L_2, diverging lens (focal length $f_2 = -0.33$ m); $\Delta = 26.5$ cm; SC, splitting cube; PR, prism; C_1 and C_2, reference and sample crystals; C, compensator (rotating microscope slide); TL, telescope lunette; F, filters; PM, photomultiplier; A_1 and A_2, amplifiers; DPH, dephasor; INV, inverter; G, low-frequency generator ($\Omega = 1$ kHz); LA, lock-in amplifier; V, dc voltmeter. [From Sigelle and Hierle (1981), with permission.]

where Ω is the modulating frequency provided by the generator G in Fig. 3 and $\Delta\phi_i$ is the phase shift introduced by the ith crystal and is related to the crystal's electro-optic coefficient and the applied voltages provided by the amplifiers. If the applied voltages do not produce the same phase shifts, then there will be a time-dependent contribution at frequency Ω to the output intensity. The voltages are adjusted until the ac component of the output intensity vanishes, at which point the sample's electro-optic coefficient can be determined from the applied voltage ratio and the known value of the reference crystal's electro-optic coefficient. This ac interferometer combines the advantages of sensitivity and isolation of tensor coefficients. Clamped and unclamped coefficients can be determined with this experiment by proper choice of the modulating frequency.

In addition to these direct methods for measuring the electro-optic coefficients, an indirect method has been reported (Tokura *et al.*, 1985). This method relates electroreflectance spectra ($\Delta R/R$ versus photon energy) to the electro-optic effect by

$$(\Delta R/R)_{ij} = 4n\,\Delta n_{ij}/(n_i^2 - n^2) \tag{21}$$

where R is the reflectance, n the refractive index of the surrounding medium, n_i the refractive index of the sample, and the electro-optic coefficient is related to Δn_{ij} by Eq. (14). The results obtained in this work were in reasonable agreement with direct methods of measuring the electro-optic coefficients and showed relations of the coefficients to absorption-spectrum features through a Kramers–Kronig analysis of the ordinary reflectance spectra. An advantage of this method is that crystal surface quality, not bulk quality, is required.

An important consideration in all measurements is the application of the modulating field (Lee, 1969; Lipscomb, 1980). The applied field must be uniform in the volume containing the laser beam. Fringe fields in a finite capacitor are appreciable within one electrode spacing of electrode edge. Thus the laser-beam cross section should be much smaller than the cross section of the input and output faces of the crystal, both for longitudinal and transverse measurements. Electrodes in intimate contact with the sample are preferred, but not necessary.

III. ORGANIC CRYSTALS

A. General Properties

The linear electro-optic effect has been studied in several organic crystals. The results of measurements of the linear electro-optic coefficients are listed in Table I. An important feature that distinguishes the electro-optic effect from optical parametric processes is the contribution of the acoustic and optical phonon modes to the electro-optic coefficient. Thus, any model for the origin of the linear electro-optic coefficient must take these contributions into account.

In order to assess the relative contribution of the electrons and phonons to the electro-optic effect, the electro-optic and the optical parametric susceptibilities can be compared (Ward and Franken, 1964). Equation (8) expresses the relationship between the electro-optic coefficient and its corresponding susceptibility. By considering the optical parametric polarization and taking into account the permutations of the electric field components given in Eq. (7), the following relationship between the optical second-harmonic coefficient and susceptibility is obtained (Kaminow, 1974):

$$\chi_{ijk}(-2\omega;\omega,\omega) = 2d_{ijk}(-2\omega;\omega,\omega). \tag{22}$$

The electro-optic and optical parametric susceptibilities are related by the dispersion, as discussed in Section II.B. In making the comparison, the contribution of the electronic dispersion to the nonlinear susceptibility as well as the contribution of the local fields must be considered. Sigelle and Hierle (1981) have analyzed this in detail. The electronic contribution to the second-order susceptibility is given by the van der Waals sum (Zyss and Oudar, 1982)

$$\begin{aligned}\chi_{ijk}^{\mathrm{el}}&(-\omega_3;\omega_1,\omega_2)\\ &= Nf_i^{\omega_3}f_j^{\omega_1}f_k^{\omega_2}\sum_{I,J,K}\sum_{s=1}^{n}\cos[i,I(s)]\cos[j,J(s)]\cos[k,K(s)]\beta_{IJK}(s)\end{aligned} \tag{23}$$

where s is summed over molecules in the unit cell, N the number of unit cells per unit volume, the terms f are local field factors, the cosines transform the molecular to crystalline axes, and β_{IJK} is the electronic molecular nonlinear optical susceptibility. A van der Waals crystal is one where the intermolecular interactions are much weaker than intramolecular interactions, and the crystal susceptibility can be expressed as a sum over molecules multiplied by dielectric local field factors. When the molecular susceptibility β_{ijk} is dominated by the change in dipole moment between the ground and the first excited electronic states along the z axis, then it is given by

$$\beta_{zzz}(-\omega_3;\omega_1,\omega_2) = \frac{e^3|z_{01}|^2(z_{11}-z_{00})}{\hbar^2} \times \frac{\omega_0^2(3\omega_0^2+\omega_1\omega_2-\omega_3^2)}{(\omega_0^2-\omega_1^2)(\omega_0^2-\omega_2^2)(\omega_0^2-\omega_3^2)} \tag{24}$$

where the z_{01} is a transition moment, z_{00} and z_{11} the ground- and excited-state dipole moments, respectively, and ω_0 gives the energy of the first excited state $\hbar\omega_0$. Since only the electrons contribute to second-harmonic generation, the electronic contribution to the electro-optic coefficient is related to the second harmonic coefficient by

$$r^{\mathrm{el}}_{ij,k}(-\omega;\omega,0) = -\frac{4d_{kij}}{n_i^2(\omega)n_j^2(\omega)} \times \frac{f^{\omega}_{ii}f^{\omega}_{jj}f^{0}_{kk}}{f^{2\omega'}_{kk}f^{\omega'}_{ii}f^{\omega'}_{jj}} \times \frac{(3\omega_0^2-\omega^2)(\omega_0^2-\omega'^2)(\omega_0^2-4\omega'^2)}{3\omega_0^2(\omega_0^2-\omega^2)^2} \tag{25}$$

where ω is the frequency of the electro-optic experiment and ω' the fundamental frequency in the second-harmonic generation measurement. The relationship depends, then, on the choice of local field model. These relationships were applied in varying degrees of rigor by the original authors to the organic crystals with the results listed in Table I, each of which will now be discussed.

B. Hexamethylenetetramine (HMT)

The first organic crystal whose linear electro-optic properties were studied in detail was hexamethylenetetramine (HMT). Crystals were grown from saturated solution, and from vapor. The resulting van der Waals crystal is transparent from 300 to 2200 nm. It belongs to the cubic crystal class $\bar{4}3m$, which has one independent electro-optic coefficient, namely, $r_{41} = r_{52} = r_{63}$. Several workers have studied the electro-optic properties of this crystal (McQuaid and Watkins, 1963; Sliker, 1964; Heilmeier, 1964; Buhrer and Ho, 1964; Pisarevskii *et al.*, 1965; Gottlieb, 1965; Lee, 1969), and Lee (1969) has tabulated the results of various workers. The results vary over an order of magnitude from 0.7×10^{-12} m/V (Lee, 1969) to 7.3×10^{-12} m/V (McQuaid, 1963). Since early workers ascribed the irreproducibility to strain in the crystals, Gottlieb (1965) employed thermal annealing to relieve the strain. He found that annealing removed visible inhomogeneous strain, which did generally result in a larger measured coefficient. Lee (1969) evaluated earlier measurements and found the electro-optic coefficient to be relatively insensitive to uniform strain, but he did observe effects attributable to nonuniform strain. Other measurements of improved quality crystals reproduced Lee's results (Forno and Jones, 1973; Rodgers, 1969). All workers report a positive correlation between the observed crystal quality and the magnitude of the electro-optic coefficient. Lee (1969) suggested that the large variation in the early measurements may also be ascribable to inhomogeneities in the applied field due to electrode geometry and deposition, and to flaws and cracks. The origin of the large variation in measured electro-optic coefficients remains an open question.

Accurate measurements of the second-harmonic coefficients of HMT have not been reported, and frequency-dependent electro-optic measurements have been inconclusive (McQuaid and Watkins, 1963; Rodgers, 1969), so no direct comparison of the relative contribution of the electrons and phonons can be made. Lebon (1970) studied the origin of the electro-optic effect in HMT. He assumed that the origin of the effect resides in the molecular electronic structure and that the crystal susceptibility was a sum over molecules taking local field effects into account. HMT, shown in Fig. 4, though noncentrosymmetric itself, does possess a high degree of symmetry. Lebon obtained the molecular susceptibility by calculating the moments of individual molecular bonds using linear combinations of Slater atomic orbitals. The tensor sum of the individual bonds was then taken. This method has been shown to be successful for other covalent, non-π-electron materials (Levine, 1973). The calculated coefficient is $r_{41}^{\mathrm{el}} = 1.4 \times 10^{-12}$ m/V, which is in agreement with the later measurements of Gottlieb (1965), Lee (1969), and Forno and Jones (1973). These results along with a low dielectric constant suggest that the origin of the electro-optic effect is mainly electronic.

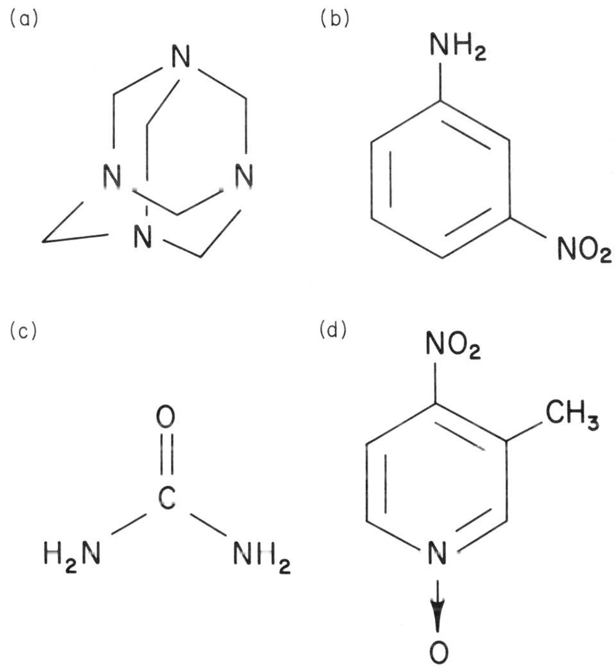

Fig. 4. Molecular units of electro-optic organic crystals: (a) hexamethylenetetramine (HMT), (b) *m*-nitroaniline (*m*-NA), (c) urea, (d) 3-methyl-4-nitropyridine 1-oxide (POM).

C. *m*-Nitroaniline (*m*-NA)

Larger susceptibilities are found in organic molecular systems possessing the charge-correlated properties of polarizable π-electrons. The first π-electron conjugated electro-optic molecular crystal studied was *m*-nitroaniline (*m*-NA), as shown in Fig. 4. The detailed crystal structure was studied by Stevenson and Skapski (1972; Skapski and Stevenson, 1973). The crystal is orthorhombic, belonging to the space group $Pbc2_1$ and the point group $mm2$. There are then five independent components of the electro-optic tensor: r_{33}, r_{13}, r_{23}, r_{42}, and r_{51}.

The linear electro-optic coefficients were measured by several authors (Kalymnios, 1972; Ayers *et al.*, 1972; Stevenson, 1973). Variations of a factor of two in the measured electro-optic coefficients were ascribed to crystal quality. Irreproducibility was also observed in measurements along differing light paths in the same crystal. These variations were correlated with the appearance of in-grown inhomogeneous strains. Stevenson, Ayers, and co-workers indicated that the crystals grown from vapor were of higher quality than those grown from the melt, and exhibited somewhat larger linear electro-optic coefficients (Ayers *et al.*, 1972; Stevenson, 1973).

Stevenson (1973) reported that the linear electro-optic susceptibility is essentially equal to the second-harmonic susceptibility as reported by Southgate and Hall (1971); thus the linear electro-optic effect in *m*-NA is electronic in origin. Stevenson and co-workers (1973) modeled the electro-optic effect by relating it to the electrochromic effect. They assumed that, as a van der Waals crystal, the origin of the linear electro-optic effect can be traced to the individual molecules. The molecular susceptibility was estimated from electrochromic measurements and an oscillator model. Based on the crystal structure, the molecular contributions are then added as a molecular gas, ignoring local fields. The electrochromic behavior indicates that the linear electro-optic effect arises from a large change in dipole moment between a low-lying excited state and the ground state. This view was supported in subsequent studies of second harmonic generation (Bergman and Crane, 1977; Carenco *et al.*, 1977).

D. Urea

Even though urea does not possess an extended π-electron system, its nonlinear optical properties are of interest due to its transparency into the ultraviolet (Halbout *et al.*, 1979). Morrell and co-workers (1979) have studied its electro-optic properties. Urea forms a uniaxial crystal with $\bar{4}2m$ tetragonal symmetry; therefore, it has two independent electro-optic tensor components,

r_{63} and $r_{41} = r_{52}$. The crystal is composed of columns of parallel molecules with hydrogen bonding between and within the stacks. The results of their electro-optic measurements are shown in Table I.

Morrell *et al.* (1979) have applied the electronic dispersion model (Section III.A) to urea in order to compare the second harmonic and electro-optic susceptibilities. It was found that the electro-optic coefficient and the electronic contribution calculated by the model are equal, leading to the conclusion, again, that the electro-optic effect is of electronic origin. They also studied the electronic properties of the molecule, whose structure is shown in Fig. 4. Using molecular-orbital calculations to analyze the molecular electro-optic susceptibility, they calculated the crystal susceptibility as the van der Waals sum over the molecules, however, the agreement was not completely satisfactory. It was determined later by Zyss and Berthier (1982) that the extensive hydrogen bonding within the crystal must be considered; thus urea is not strictly a van der Waals crystal. Urea is an example of an organic crystal whose nonlinear optical properties are not due solely to individual molecular properties.

E. 3-Methyl-4-nitropyridine 1-Oxide (POM)

The molecular crystal of 3-methyl-4-nitropyridine 1-oxide (POM), whose molecular structure is shown in Fig. 4, was molecularly engineered to simultaneously optimize molecular properties and favorable crystal growth (Zyss *et al.*, 1981). The local dipole moment of the *N*-oxide tends to cancel that of the nitroaromatic portion, leading to a vanishingly small ground-state dipole moment. The small dipole moment reduces the tendency of dipoles to crystallize in opposing pairs and reduces energies of association in crystal growth solutions. POM possesses a finite excited-state dipole moment, even though its ground-state dipole moment is small. Thus, a moderately large change in dipole moment between the ground and excited state imparts POM with a substantial molecular nonlinear susceptibility.

Large POM crystals have been grown from solution. The space group is $P2_12_12_1$, and the point group is 222. As such, there are three independent electro-optic coefficients, r_{41}, r_{52}, and r_{63}. The coefficients were measured by Sigelle and Hierle (1981; Sigelle *et al.*, 1982) and are listed in Table I. They carried out the analysis described in Section III.A and found that only r_{41} is totally of electronic origin within experimental error. They found that the electronic states contribute 85–90% for r_{52}, and around 60% for r_{63}. They then calculated the mean contribution of the electronic states by computing the tensor norms of the second-harmonic and electro-optic tensors and found that the mean electronic contribution is 35% larger than the measured electro-optic susceptibility, implying that the phonons make a negative contribution to the electro-optic tensor. This large phonon contribution along with the low values of electro-optic coefficients indicate that the large contributions of the π-electrons are substantially diminished in the reduction of the ground-state dipole moment.

F. 2-Methyl-4-nitroaniline (MNA)

The organic crystal with the largest measured electro-optic coefficient is 2-methyl-4-nitroaniline (MNA), whose molecular structure is shown in Fig. 5 (Lipscomb *et al.*, 1981; Tokura *et al.*, 1985). The figure of merit for phase retardation defined by Eq. (10) is 2.7 times that of lithium niobate ($LiNbO_3$) (Lipscomb, 1980). In addition, assuming a low dielectric constant typical of organic molecular crystals, the modulator bandwidth will be substantially higher than $LiNbO_3$. A detailed study of electro-optic properties and the crystal structure was carried out by Lipscomb *et al.* (1981). The space group is *Cc* with point group *m*, and as such, there are 10 independent electro-optic tensor components, $r_{11}, r_{21}, r_{31}, r_{13}, r_{23}, r_{33}, r_{42}, r_{51}, r_{53}$, and r_{62}. As shown in the stereoscopic view of the unit cell in Fig. 6, the molecules are nearly aligned about a polar axis. This highly polar arrangement is nearly ideal for

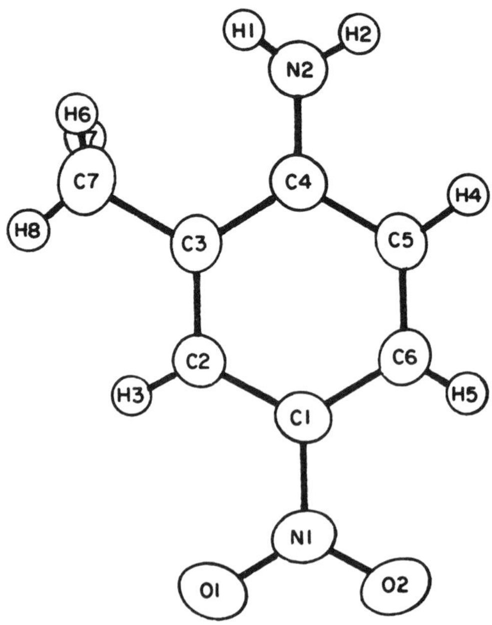

Fig. 5. Molecular structure of 2-methyl-4-nitroaniline (MNA).

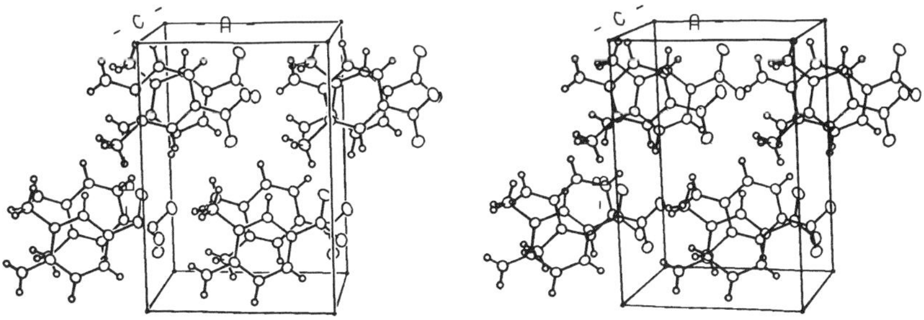

Fig. 6. MNA unit cell. [From Lipscomb *et al.* (1981), with permission.]

maximizing the electro-optic coefficient coincident with the polar axis. As shown in Fig. 7, the molecules orient so that the charge-transfer axis is 20° about the polar axis, and the polar axis is 8° from the x dielectric axis. The ideal case consists of all of the charge-transfer axes aligned with the crystal polar axis, and MNA comes closer to this than any other organic material studied.

High-quality crystals of MNA were grown from solution, and the r_{11}

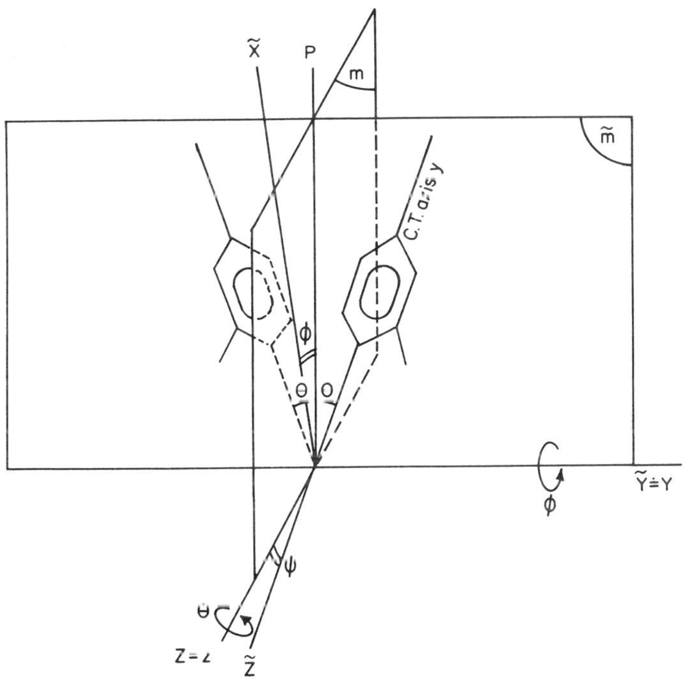

Fig. 7. MNA crystal coordinate system. P is the polar axis, m the crystallographic mirror, $\tilde{m}$ the pseudomirror, $\tilde{X}$, $\tilde{Y}$, and $\tilde{Z}$ the principal dielectric axes, $\theta = 20°$, and $\phi = 8°$. [From Zyss and Oudar (1982), with permission.]

coefficient was measured to be $r_{11} = 67 \pm 25 \times 10^{-12}$ m/V, as shown in Table I (Lipscomb *et al.*, 1981). Lipscomb (1980) reported that crystal quality severely affected the measured values, observing order-of-magnitude variations. In addition, after aging and use, the crystal quality deteriorated with the appearance of surface defects, which may have been caused by water vapor. The appearance of massive bulk strains was also noted. Subsequent measured values of the electro-optic coefficients by Tokura and co-workers (1985) were lower than that reported by Lipscomb and co-workers. Again, crystal quality was cited as a possible cause of irreproducibility.

The electronic contribution to the electro-optic coefficient was estimated by Lipscomb and co-workers (1981) from the second-harmonic coefficients measured by Levine and co-workers (1979). They measured a second-harmonic coefficient d_{11} corresponding to a nonlinear susceptibility of $\chi_{111}(-2\omega; \omega, \omega) = 500 \pm 125 \times 10^{-12}$ m/V. The electro-optic susceptibility was measured to be $\chi_{111}(-\omega; \omega, 0) = 540 \pm 100 \times 10^{-12}$ m/V. Once again, the electro-optic effect is dominated by the electronic contribution.

The origin of this electronic susceptibility arises from the molecular electronic properties, which were studied by Garito and co-workers using molecular-orbital calculations and molecular measurements (Garito *et al.*, 1984; Teng and Garito, 1983; Singer and Garito, 1981; Lalama and Garito, 1979). The molecular susceptibility was obtained using quantum-mechanical perturbation theory in terms of transition dipole moments and excitation energies, and can be evaluated once a quantum description of the molecule is available. This description was obtained by semiempirical Hartree–Fock molecular orbital calculations (Lalama and Garito, 1979). These calculations include all valence electrons, which are used to obtain the molecular ground and excited states from excited-configuration interactions of antisymmetrized products of the one electron states. The parameterization is done with respect to gas-phase photoemission and ultraviolet absorption data and is effective in describing the molecule's low-lying excited states. The results of these calculations indicate that the molecular susceptibility is dominated by a single excited state, which is characterized by a large change in dipole moment. This is the term depicted in Eq. (24), and this feature is shown graphically in Fig. 8, where electron wavefunction contour maps for molecular orbitals dominating the molecular ground and first excited state in MNA are shown (Garito *et al.*, 1984). Since the square of the wavefunction gives the electronic density, one sees the reorganization of charge along the molecular axis from the amino group to the nitro group upon excitation. The change in dipole moment along this axis confirms the nearly one-dimensional nature of this molecule.

The results of these calculations are in excellent agreement with molecular susceptibility measurements over a wide wavelength range as shown in Fig. 9 (Teng and Garito, 1983). The measured dispersion is consistent with Eq. (24). The agreement between calculations and experiment is due to the dominance

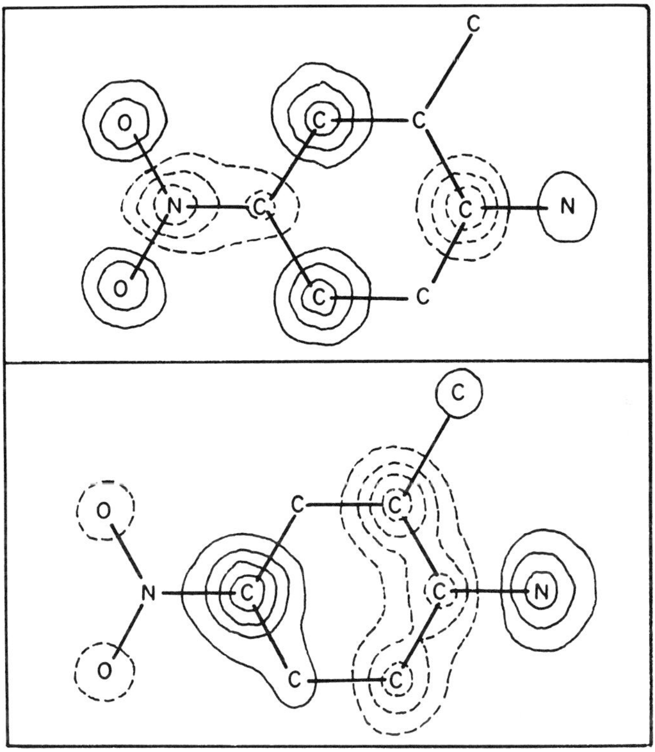

Fig. 8. Contour diagrams at $z = 0.6$ Å above the molecular plane for the one-electron molecular wavefunctions. The bottom diagram is a one electron wavefunction belonging to the ground state. The top belongs to a principal excited state. [From Garito *et al.* (1984), with permission.]

of these charge-correlated states, since, in the subtraction of the ground-state moments from the excited-state moments, the contribution of the core states cancels. Then, since the transition moment is also only dependent on valence electrons, the susceptibility arises mainly from these electrons. The states near the Fermi level are precisely the ones these Hartree–Fock calculations are best at determining. In addition to elucidating the origin of the nonlinear susceptibility, these calculations of β_{ijk}, along with accurate measurements, can make β_{ijk} a sensitive probe of valence-level states in π-electron conjugated molecules (Singer and Garito, 1981).

The validity of the van der Waals sum has been evaluated by calculating the molecular susceptibility component β_{zzz} from the crystal values by inverting Eq. (23). In determining β_{zzz}, spherical-cavity local fields were used, as were the relative contributions of the various molecular tensor components as determined from the molecular calculations. The result was found to be $\beta_{zzz} =$

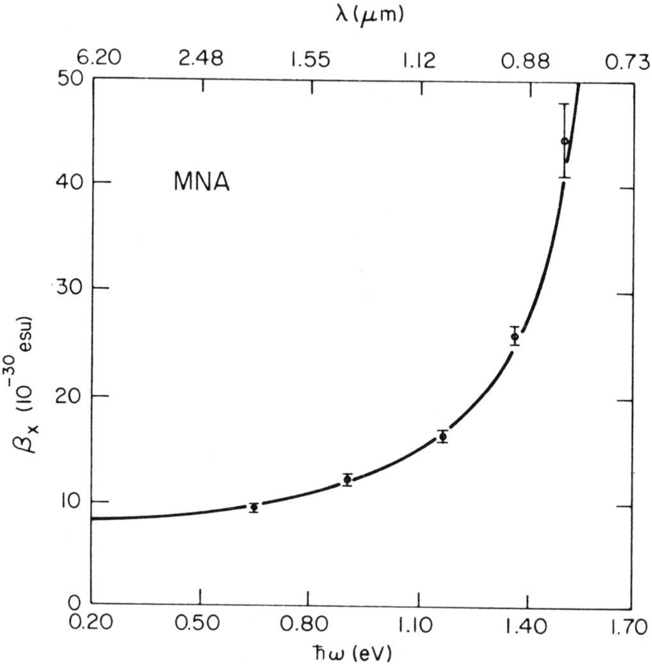

Fig. 9. Dispersion of the vector part of β_{ijk} for MNA. The data points were measured using dc induced second-harmonic generation, and the solid line was calculated using molecular Hartree–Fock theory. [From Teng and Garito (1983), with permission.]

$20 \pm 6 \times 10^{-30}$ cm^5/esu at $\lambda = 1.06$ μm (K. D. Singer and A. F. Garito, unpublished, 1982). This is in agreement with the value obtained from dc-induced second-harmonic generation measurements of solutions of MNA ($\beta_{zzz} = 16 \pm 3 \times 10^{-30}$ cm^5/esu). The value of β_{zzz} was calculated from the solution-measured vector part of β_{ijk} using the results of the quantum-mechanical calculations. Thus, a complete picture of the electro-optic coefficient arising from specific molecular electronic properties is firmly established.

IV. NEW TRENDS

A. Molecular Crystals

Exploration of new organic crystals continues (Garito *et al.*, 1983; Meredith, 1983; Twieg and Jain, 1983; Badan *et al.*, 1983). Much of this research focuses on optimizing the electro-optic coefficient. The studies described in the previous section and summarized in Table I establish that the electro-optic coefficient is dominated by the electronic contribution, and this will be more pronounced in crystals composed of molecules with larger molecular susceptibilities. An important implication of the electronic origin is that the same considerations apply to the development of second harmonic and optical parametric materials as to electro-optic materials with two important relaxed constraints. One is that the transparency window required in electro-optic materials is reduced to only one wavelength. The second is that optimization of the crystal structure with respect to parametric wave phase matching is not necessary. Thus, the optimum crystal structure for a one-dimensional molecule, where a single diagonal molecular-susceptibility tensor component dominates, is one in which the directions of the difference between the excited- and ground-state dipole moments of all of the molecules are oriented as close as possible to the polar axis of the crystal. In higher-dimensional molecules, the sum of the projections of β_{ijk} components onto a polar crystal axis should be maximized.

Research into crystal growth is of considerable importance to the implementation of organic crystals into commercial electro-optic devices, since the reproducibility of electro-optic measurements was lacking in all the crystals where it was investigated. The irreproducibility was ascribed mainly to crystal quality. Some efforts are directed at producing crystals bound by forces stronger than dipolar interactions, such as the growth of electro-optic single crystal polymers (Garito and Singer, 1982; Garito *et al.*, 1980, 1983). Such crystals would be more robust than molecular crystals, and the crystallization may be controlled by the covalent polymer rather than dipolar interactions. Additionally, ionic crystals were investigated by Meredith (1983, and references therein) and hydrogen-bonded crystals by Zyss and coworkers (Zyss, 1985).

Crystals now find application in bulk modulators, such as laser *Q*-switches. However, future applications for electro-optic devices are in switching and modulation in integrated optical systems. These require that a suitable optical waveguide technology be developed. Along these lines, several workers have reported organic molecular crystal growth in hollow-core optical fibers (Nayar, 1983, and references therein). These generally involve modified Bridgeman growth techniques. Stevenson (1977) was able to grow *m*-NA crystals in a hollow fiber, but the morphology did not allow coupling to the largest susceptibility component. Nayar (1983) was able to observe second-harmonic generation in a fiber containing benzyl. No one has indicated that structures fabricated to date are suitable for device purposes.

Other attempts at fabricating structures in organic materials suitable for integrated optical applications include nonlinear waveguide cladding, and the growth of organic molecular crystals in channels (Tomaru *et al.*, 1984;

Sasaki *et al.*, 1984). As described elsewhere in this volume, monomolecular deposition (Langmuir–Blodgett) techniques promise fine control over planar structures. Williams (1984) has given an excellent assessment of the various approaches.

B. Noncrystalline Materials

Heretofore, electro-optic properties have only been discussed in crystalline materials. The requirement of a lack of inversion center for second-order processes does not restrict possible materials to crystals. Williams (1984) has suggested the possibility of fast electro-optic processes in orientationally ordered glasses, and this has been demonstrated by Singer and co-workers (1985, 1986). In these materials, the inversion center is removed by aligning dipolar electro-optic molecules using an external electric field. As shown in Fig. 10, the temperature is raised until the material is in a state where the application of an external electric field will force the nonlinear optical species to rotate in a thermodynamic sense in the host, which, for a glassy polymer, is heating above its glass–rubber transition temperature. During application of the electric field, the material is returned to the lower temperature state where

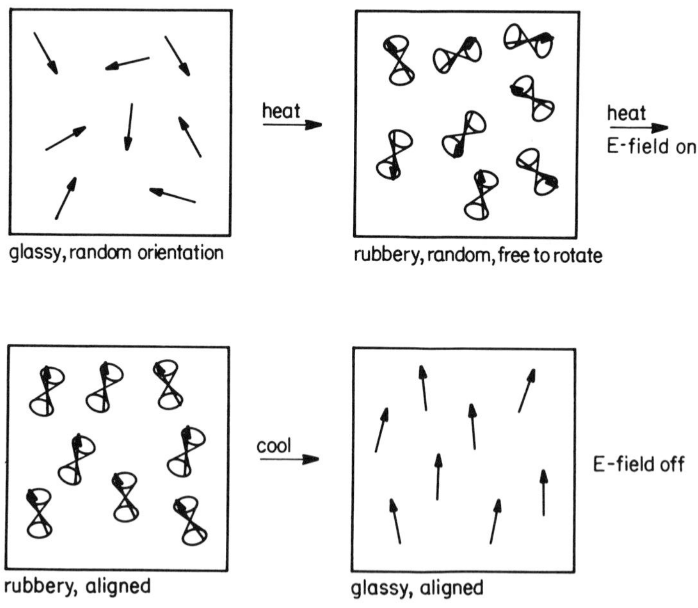

Fig. 10. Poling concept for glassy polymers.

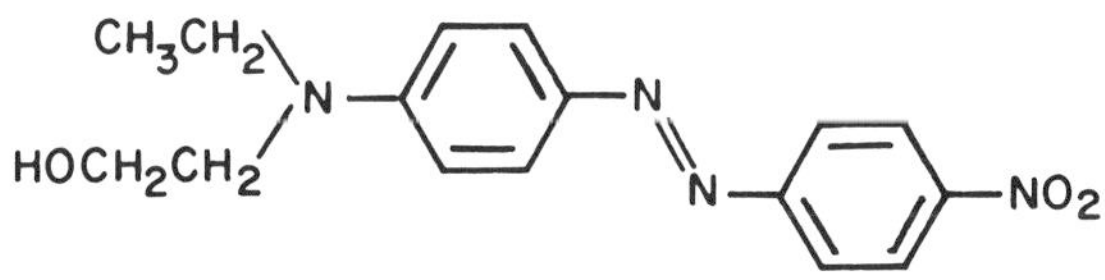

Fig. 11. Disperse red 1, an azo dye.

the molecular motion is severely restricted. The resulting material is electro-optic, of high optical quality, and suitable for guided-wave electro-optic devices (Lalama *et al.*, 1985). The demonstration by Singer and co-workers utilized an azo dye, a derivative of 4-amino-4′-nitroazobenzene (Fig. 11), which was dissolved in poly(methyl methacrylate) (PMMA). The material was then treated according to the process depicted in Fig. 12. The nonlinear optical properties were measured by second-harmonic generation in transmission through the thin film. The electronic contribution to the electro-optic coefficient was then evaluated using Eq. (24) and found to be $r_{33}^{\mathrm{el}} - 2.3 \pm 0.3 \times 10^{-12}$ m/V at $\lambda = 0.8$ μm, as indicated in Table I.

The oriented molecular dipoles produce a polarization, $P = N\mu\langle\cos\theta\rangle$, where N is the number density of the molecules, μ the ground state dipole moment, and θ the angle the dipoles subtend with respect to the direction of

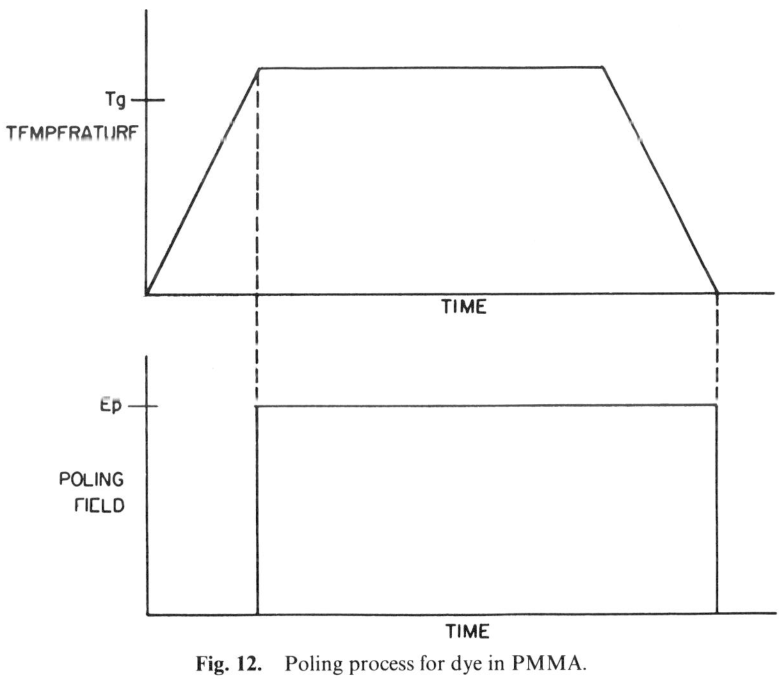

Fig. 12. Poling process for dye in PMMA.

the poling field. The brackets denote the canonical average of the direction cosine. If the material is prepared as a uniform film, and the poling field is applied by parallel plate electrodes, then the resulting material will possess a unique axis in the direction parallel to the field direction. This direction defines an infinite-fold rotation axis, and an infinity of mirror planes. The material then belongs to the point group ∞mm. Applying the symmetry operations to the third-rank tensor describing the lowest-order nonlinear susceptibility, with the 3 axis the unique axis, one finds that there are five nonzero tensor components, three of which are independent. For the electro-optic tensor r_{uk}, these components and their relationships are $r_{13} = r_{12}, r_{33}$, and $r_{42} = r_{51}$.

The magnitude of the electro-optic coefficient can be estimated by considering the thermodynamics, including the effect of local fields. In order to calculate the largest electro-optic coefficient, r_{33}, an expression for the bulk polarization at optical frequency ω in the presence of incident electromagnetic fields at frequencies ω and 0 must be derived. This is done by considering the molecular polarization at these frequencies,

$$p_i(\omega) = \alpha_{ij}(\omega)E_j^{(\mathrm{local})}(\omega) + \beta_{ijk}(-\omega;\omega,0)E_j^{(\mathrm{local})}(\omega)E_k^{(\mathrm{local})}(0) + \gamma_{ijkl}(-\omega;\omega,0,0)E_j^{(\mathrm{local})}(\omega)E_k^{(\mathrm{local})}(0)E_l^{(\mathrm{local})}(0) + \cdots \quad (26)$$

where α, β, and γ are molecular susceptibility tensors of successive order. Assuming that intermolecular interactions are much weaker than intramolecular interactions allows the bulk susceptibility to be calculated from the sum of molecular polarizabilities,

$$P_i(\omega) = \sum_{s=1}^{n} p_i^s(\omega) \quad (27)$$

In order to evaluate the sum in a amorphous body, the thermodynamic average over the molecular ensemble is taken, which leads to (Kielich, 1969)

$$P_i(\omega) = Nf^{\omega}f^{\omega}f^{0}\langle b\rangle_{ijk}E_j(\omega)E_k(0) \quad (28)$$

where N is the number density of the nonlinear optical molecules, the f terms are local field factors at the indicated frequencies, and $\langle b\rangle_{ijk}$ is the thermodynamic average of the nonlinear polarizability assuming the third-order susceptibility, γ_{ijkl} is negligible, which is given by

$$\langle b\rangle_{ijk} = \int (a_{iu}a_{jv}a_{kw}\beta_{uvw})G(\Omega,\mathbf{F}_{\mathrm{p}})\,d\Omega \quad (29)$$

where the a_{iu} are the components of the coordinate transformation from the molecular to the laboratory reference frames, Ω the angle between the induced molecular dipole $\mathbf{m}$ and the local poling field $\mathbf{F}_{\mathrm{p}}$, and $G(\Omega,\mathbf{F}_{\mathrm{p}})$ the distribution function. In Eq. (28), the following definition of the local field has been assumed:

$$f^{\omega} = \frac{E_i^{(\text{local})}(\omega)}{E_i(\omega)} = \frac{N\langle\alpha\rangle}{1 + \eta N\langle\alpha\rangle} \tag{30}$$

where α is the linear polarizability and η the depolarization factor. Equation (29) can be evaluated by considering the function

$$G(\Omega, \mathbf{F}_p) = \frac{\exp\{-[(\mathbf{m}\cdot\mathbf{F}_p + U_1)/kT]\}}{\int \exp\{-[(\mathbf{m}\cdot\mathbf{F}_p + U_1)/kT]\}\, d\Omega} \tag{31}$$

where U_1 is the potential due to the interaction of the molecule with its environment, kT the Boltzmann factor, and where

$$\mathbf{m} = \mu + \alpha\mathbf{F}_p \tag{32}$$

Assuming the molecule is axial and planar, $\beta_{zzz}(-\omega;\omega,0)$ is dominant, and intermolecular interactions are negligible (U_1 is small), taking E_p, the external poling field, in the 3 direction, and carrying out the integrations, the following expression for the susceptibility is obtained (Singer *et al.*, 1985, 1986):

$$\chi_{333}(-\omega;\omega,0) = Nf^{\omega}f^{\omega}f^{0}\beta_{zzz}(-\omega;\omega,0)L_3(p) \tag{33}$$

where L_3 is the third-order Langevin function, which, assuming that molecular anisotropy is small, is given by

$$L_3(p) = \frac{p}{5} - \frac{p^3}{105} + \cdots \tag{34}$$

where

$$p = \left[\frac{\varepsilon(n^2+2)}{n^2+2\varepsilon}\right]\frac{\mu E_p}{kT} \tag{35}$$

where ε is the static dielectric constant and n the index of refraction.

Appropriate local field factors for the optical field and the static fields (including E_p) are (Singer and Garito, 1981)

$$f^{\omega} = \frac{n^2+2}{3} \quad \text{and} \quad f^{0} = \frac{\varepsilon(n^2+2)}{n^2+2\varepsilon} \tag{36}$$

The molecular susceptibility $\beta_{zzz}(-\omega;\omega,0)$ is related to the molecular second harmonic coefficient through

$$\beta_{zzz}(-\omega;\omega,0) = 2\beta_{zzz}(-2\omega';\omega',\omega') \times \frac{(3\omega_0^2-\omega^2)(\omega_0^2-\omega'^2)(\omega_0^2-4\omega'^2)}{3\omega_0^2(\omega_0^2-\omega^2)^2} \tag{37}$$

where ω_0 is the frequency of the first molecular excited state, ω' the fundamental frequency for second harmonic generation, and ω the electro-optic frequency.

The molecular second-harmonic coefficient of the azo dye was measured at $\lambda = 1.9\ \mu$m using dc induced second-harmonic generation in solution, yielding a value of $\beta \cdot \mu = 525 \pm 100 \times 10^{-30}$ cm^5 D/esu (Singer *et al.*, 1986). This value, along with Eqs. (33)–(37), Eq. (8), and Eq. (24), was used to calculate the electronic contribution to the electro-optic coefficient for the azo dye/PMMA film. The values of the other pertinent material parameters are given in Table II as well as a comparison between the calculated and measured r_{33}^{el}. There is good agreement between the thermodynamic model and the film measurements. This indicates that the system behaves like a simple solution where interactions between molecules and between the molecules and the polymer at the poling temperature are not substantial. Thus, the thermodynamic model, along with the microscopic understanding of the origin of second-order nonlinearities in organic molecules, can be employed to develop polymer glass systems with larger electro-optic coefficients.

Since glassy polymers are not in thermodynamic equilibrium, relaxation processes occur. These processes include the randomization of the incorporated molecules, which depend on detailed polymer dynamics as well as polymer–molecule interactions. These relaxations will diminish the electro-optic coefficient. Our results indicate that the polarization does not substantially decay over 1 year, but detailed studies are required.

This azo dye/PMMA system is viewed as a model for a new class of electro-optic materials, namely, oriented glasses. Polymer glasses are materials of demonstrated high optical quality (Lalama *et al.*, 1985, and references therein). Furthermore, polymers can be fabricated into integrated optical structures (Lalama *et al.*, 1985, and references therein) with a wide range of planarization and processing techniques.

Another possible approach to noncrystalline electro-optic materials is the incorporation of electro-optic moieties into "frozen" liquid crystals and liquid crystal polymers, as discussed by Meredith and co-workers (1982; 1983). The interaction between the host and the electro-optic molecule may enhance the molecular alignment. Meredith and co-workers (1982; 1983) reported results on the guest molecule, 4-*N*,*N*-dimethylamino-4′-nitrostilbene (DANS) in a

TABLE II

Properties of Azo Dye/PMMA Electro-Optic Glass

$N(10^{20}/\mathrm{cm}^3)$	$E_p(10^6$ V/cm)	$\varepsilon/\varepsilon_0$	n	r_{33}^{el}(expt)(10^{-12} m/V)	r_{33}^{el}(calc)(10^{-12} m/V)
2.74	0.62	3.6	1.52	2.4 ± 0.6	3.4 ± 0.7

liquid-crystalline copolymer host. The material was poled in order to measure second-harmonic generation, but the stored polarization was released on application of the laser light.

V. CONCLUSIONS

Opto-electronic integrated circuits require the electronic control of light, which will incorporate guided wave modulators and switches. This implies that appropriate linear electro-optic material technologies will be required as circuitry evolves toward all-optical processing. The intrinsically large electro-optic response, the low dielectric constant, and the nearly limitless number of chemical and material combinations should ensure that organic materials will play an important role in the development of lightwave circuit technologies.

The basic phenomena associated with the linear electro-optic effect, in-including properties required for the emerging electro-optic technology have been reviewed. The general contributions to the linear electro-optic effect in material systems and methods of measurement of the linear electro-optic effect have been described. The electro-optic properties of several materials have been studied, with the molecular cystal of 2-methyl-4-nitroaniline having the largest effect measured to date. Studies into the origin of the linear electro-optic effect in organic materials have traced it to the electronic structure in all of the organic cyrstals measured. In the class of molecules possessing highly conjugated π-electron systems, such as *m*-NA, POM, and MNA, the microscopic susceptibilites arise from charge reorganization from the ground state to a low-lying excited state of the molecule. Further, tensor additivity of the molecules in determining the bulk electro-optic response has been verified for van der Waals crystals. This additivity implies that a highly polar arrangement of nonlinear optical molecules leads to the maximum bulk electro-optic coefficient. The understanding gained from the studies described above and from studies of second order optical parametric processes has led to the elucidation of general principles to guide the development of new organic electro-optic materials.

This has led to the development of new classes of materials. Specifically, a new class of electro-optic materials, poled polymer glasses, has been demonstrated. A nonlinear polarization has been imparted in a polymer glass with a dissolved nonlinear optical molecule. This demonstration and an appreciation of the thermodynamics can guide research into materials with technological capabilities. The high intrinsic optical quality and easy planarization and processing of polymer glasses may qualify them for integrated optical applications.

In general, research directed at applications of electro-optic materials must focus on all properties important to the technology. High on this list are optical quality and reproducibility. Molecular crystals studied to date have fallen short in these respects. In crystals where multiple measurements have been reported, variations have been found from a factor of two to an order of magnitude from crystal to crystal, and even in different optical paths in the same crystal. In addition, several of the workers reported in-grown inhomogeneous strain and other defects that reduce the optical quality of the materials. Steps along the way to improvement of crystal quality include the studies of urea and POM. Both urea and POM exhibit features where the dipolar interactions between molecules are reduced, which has led to improvements in crystal growth and quality. In the case of urea, hydrogen bonding reduces the dominance of the molecular dipole energy. In POM, a very small ground-state dipole moment reduces the dipolar energy. These crystals, however, have smaller electro-optic coefficients than the more highly polar materials, thereby limiting their applicability for electro-optic technology. Moreover, Langmuir–Blodgett films as an alternative to crystals are also lacking in device optical quality, so that research into processing is required (Carter *et al.*, 1983).

Another important property for integrated optical applications is integrability. Polymers and liquid crystalline materials have the potential to be integrated with semiconductor sources and detectors, and polymer waveguides have already been demonstrated (Lalama *et al.*, 1985, and references therein). Bulk crystals suffer from difficulty in integration with sources and detectors. However, as mentioned above, work is proceeding in growing crystals in grooves in semiconductor substrates, but again, crystal quality has not been demonstrated.

Resistance to optical damage is another important requirement for electro-optic materials. Irreversible changes in refractive index related to photorefraction are observed in several inorganic crystals. These changes result in unacceptable operating voltage drifts in electro-optic devices. Photorefraction and optical damage studies have not been reported for organic materials.

One of the most attractive features of organic materials for electro-optic applications is their switching times. Switching times in electro-optic materials are limited by their dielectric properties. The electronic contribution dominates the electro-optic coefficient in many materials, both organic and inorganic. However, the electronic structure also dominates the dielectric properties in organic materials, in contrast to many inorganics. It is this, the low dielectric constant, that promises short switching times in organics, since switching speeds will be limited by capacitive time constants and microwave group velocities.

The research described has elucidated the origin of the electro-optic coefficient in organic materials. This knowledge has been employed to develop new material approaches. Fundamental research can be directed at finding new molecular electronic phenomena and at novel material systems, such as liquid crystals and polymer glasses. An organic material with an adequate electro-optic coefficient, low dielectric constant, high optical quality, and processability could make a substantial impact on emerging electro-optic technologies.

REFERENCES

Ayers, S., Faktor, M. M., Marr, D., and Stevenson, J. L. (1972). *J. Mater. Sci.* **7,** 31.

Badan, J., Hierle, R., Perigaud, A., and Zyss, J. (1983). *ACS Symp. Ser.* **233,** 81–107.

Bergman, J., and Crane, G. R. (1977). *J. Chem. Phys.* **66,** 3803.

Bloembergen, N. (1965). "Nonlinear Optics." Benjamin, Reading, Massachusetts.

Buhrer, C. F., and Ho, L. (1964). *Appl. Opt.* **3,** 1500.

Carenco, A., Jerphagnon, J., and Perigaud, A. (1977). *J. Chem. Phys.* **66,** 3806.

Carpenter, R. O'B. (1950). *J. Opt. Soc. Am.* **40,** 225.

Carter, G. M., Chen, Y. J., and Tripathy, S. K. (1983). *ACS Symp. Ser.* **233,** 213–228.

Chemla, D. S., Oudar, J. L., and Jerphagnon, J. (1975). *Phys. Rev. B* **12,** 4534.

Davydov, B. L., Derkacheva, L. D., Dunina, V. V., Zhabotinskii, M. E., Zolin, V. F., Koreneva, L. G., and Sarmokhina, M. A. (1970). *JETP Lett.* (*Engl. Transl.*) **12,** 16.

Forno, C., and Jones, O. C. (1973). *J. Phys. E* **7,** 101.

Garito, A. F., and Singer, K. D. (1982). *Laser Focus* **80,** 59.

Garito, A. F., Singer, K. D., Hayes K., Lipscomb, G. F., Lalama, S. J., and Desai, K. N. (1980). *J. Opt. Soc. Am.* **70,** 1399.

Garito, A. F., Singer, K. D., and Teng, C. C. (1983). *ACS Symp. Ser.* **233,** 1–26.

Garito, A. F., Teng, C. C., Wong, K. Y., and Zammani'Kharmiri, O. (1984). *Mol. Cryst. Liq. Cryst.* **106,** 219.

Gottlieb, G. E. (1965). *J. Electrochem. Soc.* **112,** 903.

Halbout, J. M., Blit, S., Donaldson, W., and Tang, C. L. (1979). *IEEE J. Quantum Electron.* **QE-15,** 1176.

Heilmeier, G. H. (1964). *Appl. Opt.* **3,** 1281.

Kalymnios, D. (1972). *J. Phys. D* **5,** 667.

Kaminow, I. P. (1965). *Appl. Phys. Lett.* **7,** 123.

Kaminow, I. P. (1968). *IEEE J. Quantum Electron.* **QE-4,** 23.

Kaminow, I. P. (1974). "An Introduction to Electrooptic Devices." Academic Press, New York.

Kaminow, I. P., and Turner, E. H. (1971). *In* "Handbook of Lasers" (R. J. Pressley, ed.), pp. 447–459. Chem. Rubber Publ. Co., Cleveland, Ohio.

Kerr, J. (1875). *Philos. Mag.* [4] **50,** 337.

Kielich, S. (1969). *IEEE J. Quantum Electron.* **QE-5,** 562.

Kleinman, D. A. (1962). *Phys. Rev.* **126,** 1977.

Lalama, S. J., and Garito, A. F. (1979). *Phys. Rev. A* **20,** 1179.

Lalama, S. J., Sohn, J. E., and Singer, K. D. (1985). *SPIE Proc.* **578,** 168.

Lebon, F. (1970). *Phys. Status Solidi* **41,** 297.

Lee, R. W. (1969). *J. Opt. Soc. Am.* **59,** 1574.
Levine, B. F. (1973). *Phys. Rev. B* **7,** 7.
Levine, B. F., and Bethea, C. G. (1977). *J. Chem. Phys.* **66,** 1070.
Levine, B. F., Bethea, C. G., Thurmond, C. D., Lynch, R. T., and Bernstein, J. L. (1979). *J. Appl. Phys.* **50,** 2523.
Lipscomb, G. F. (1980). Ph.D. Dissertation, University of Pennsylvania, Philadelphia.
Lipscomb, G. F., Garito, A. F., and Narang, R. S. (1981). *J. Chem. Phys.* **75,** 1509.
McQuaid, R. W. (1963). *Appl. Opt.* **2,** 320.
McQuaid, R. W., and Watkins, M. C. (1963). *J. Opt. Soc. Am.* **53,** 1339.
Meredith, G. R. (1983). *ACS Symp. Ser.* **233,** 27–56.
Meredith, G. R., Van Dusen, J. G., and Williams, D. J. (1982). *Macromolecules* **15,** 1385.
Meredith, G. R., Van Dusen, J. G., and Williams, D. J. (1983). *ACS Symp. Ser.* **233,** 109–133.
Miller, R. C. (1964). *Appl. Phys. Lett.* **5,** 17.
Morrell, J. A., Albrecht, A. C., Levin, K. H., and Tang, C. L. (1979). *J. Chem. Phys.* **71,** 5063.
Nayar, B. K. (1983). *ACS Symp. Ser.* **233,** 153–166.
Nye, J. F. (1957). "Physical Properties of Crystals." Oxford Univ. Press, London and New York.
Oudar, J. L., and Chemla, D. S. (1977). *J. Chem. Phys.* **66,** 2664.
Pisarevskii, Yu. V., Tregubov, G. A., and Shaldin, Yu. V. (1965). *Sov. Phys.—Solid. State (Engl. Transl.)* **7,** 530.
Pockels, F. (1906). "Lehrbuch der Kristalloptik." Teubner, Leipzig.
Rodgers, K. F. (1969). *Appl. Opt.* **8,** 2369.
Sasaki, K., Kinoshita, T., and Karasawa, N. (1984). *Appl. Phys. Lett.* **45,** 333.
Sigelle, M., and Hierle, R. (1981). *J. Appl. Phys.* **52,** 4199.
Sigelle, M., Zyss, J., and Hierle, R. (1982). *J. Non-Cryst. Solids* **47,** 287.
Singer, K. D., and Garito, A. F. (1981). *J. Chem. Phys.* **75,** 3572.
Singer, K. D., and Garito, A. F. (1982). (Unpublished).
Singer, K. D., Lalama, S. J., and Sohn, J. E. (1985). *SPIE Proc.* **578,** 130.
Singer, K. D., Sohn, J. E., and Lalama, S. J. (1986). *Appl. Phys. Lett.* **49,** 248.
Skapski, A. C., and Stevenson, J. L. (1973). *J. Chem. Soc., Perkin Trans. 2,* p. 1197.
Sliker, T. R. (1964). *J. Opt. Soc. Am.* **54,** 1348.
Southgate, P. D., and Hall, D. S. (1971). *Appl. Phys. Lett.* **18,** 456.
Stevenson, J. L. (1973). *J. Phys. D* **6,** L13.
Stevenson, J. L. (1977). *J. Cryst. Growth* **37,** 116.
Stevenson, J. L., and Skapski, A. C. (1972). *J. Phys. C* **5,** L233.
Stevenson, J. L., Ayers, S., and Faktor, M. M. (1973). *J. Phys. Chem. Solids* **34,** 235.
Sugie, M., and Tada, K. (1976). *J. Appl. Phys.* **15,** 421.
Teng, C. C., and Garito, A. F. (1983). *Phys. Rev. Lett.* **50,** 350.
Tokura, Y., Kurita, A., and Koda, T. (1985). *Phys. Rev. B* **31,** 2588.
Tomaru, S., Kawachi, M., and Kobayashi, M. (1984). *Opt. Commun.* **50,** 154.
Twieg, R. J., and Jain, K. (1983). *ACS Symp. Ser.* **233,** 57–80.
Ward, J. F., and Franken, P. A. (1964). *Phys. Rev.* **133,** A183.
Wemple, S. H., and DiDomenico, M., Jr. (1972). *Appl. Solid State Sci.* **3,** 263.
Williams, D. J. (1984). *Angew. Chem., Int. Ed. Engl.* **23,** 690.
Zyss, J. (1982). *J. Non-Cryst. Solids* **47,** 211.
Zyss, J. (1985). *J. Molec. Electron.* **1,** 25.
Zyss, J., and Berthier, G. (1982). *J. Chem. Phys.* **77,** 3635.
Zyss, J., and Oudar, J. L. (1982). *Phys. Rev. A* **26,** 2028.
Zyss, J., Chemla, D. S., and Nicoud, J. F. (1981). *J. Chem. Phys.* **74,** 4800.

Index

D

E

F

G

O

P

Q

R

S

Y

Z